RECENT ADVANCES IN ELECTRON AND LIGHT OPTICAL IMAGING IN BIOLOGY AND MEDICINE

ANNALS OF THE NEW YORK ACADEMY OF SCIENCES
Volume 483

RECENT ADVANCES IN ELECTRON AND LIGHT OPTICAL IMAGING IN BIOLOGY AND MEDICINE

Edited by Andrew P. Somlyo

The New York Academy of Sciences
New York, New York
1986

Cover: The cover shows a stereo pair of intensity contour displayed in pseudocolor (see page 397).

Library of Congress Cataloging-in-Publication Data

Recent advances in electron and light optical
imaging in biology and medicine.

(Annals of the New York Academy of Sciences; v. 483)
Based on papers presented at a conference held in
New York City by the New York Academy of Sciences on
Mar. 18–21, 1986.
Includes bibliographies and index.
1. Electron microscopy—Methodology—Congresses.
2. Microscope and microscopy—Methodology—Congresses.
3. Electron probe microanalysis—Methodology—Congresses.
I. Somlyo, Andrew P. (Andrew Paul), 1930–
II. New York Academy of Sciences.   III. Series.
[DNLM: 1. Electron probe microanalysis—Methods—
Congresses.   2. Microbiological technics—Congresses.
3. Microscopy, Electron—Methods—Congresses.
W1 AN 626YL v. 483 / QH 212.E4 R295 1986]
Q11.N5 vol. 483   500 s   86-33158
[QH212.E4]   [578'.45]
ISBN 0-89766-361-6
ISBN 0-89766-362-4 (pbk.)

SP
Printed in the United States of America
ISBN 0-89766-361-6 (cloth)
ISBN 0-89766-362-4 (paper)

ANNALS OF THE NEW YORK ACADEMY OF SCIENCES

Volume 483
December 31, 1986

RECENT ADVANCES IN ELECTRON AND LIGHT OPTICAL IMAGING IN BIOLOGY AND MEDICINE[a]

Editor
ANDREW P. SOMLYO

CONTENTS

[a]The papers in this volume were presented at a conference entitled Recent Advances in Electron and Light Optical Imaging in Biology and Medicine, which was held by the New York Academy of Sciences in New York City on March 18–21, 1986.

Part V. Analytical Electron Microscopy: Electron Energy-Loss Spectroscopy and Photoelectron Microscopy

Part VI. Light Microscopy and X Ray Microscopy

Financial assistance was received from:
- ALLIED CORPORATION
- CARL ZEISS, INC.
- DIVISION OF RESEARCH RESOURCES—NATIONAL INSTITUTES OF HEALTH (NIH)
- E. I. DU PONT DE NEMOURS & COMPANY
- JEOL (U.S.A.), INC.
- NATIONAL HEART, LUNG, AND BLOOD INSTITUTE—NIH
- NATIONAL INSTITUTE OF GENERAL MEDICAL SCIENCES—NIH
- NATIONAL SCIENCE FOUNDATION

Overview

ANDREW P. SOMLYO

Pennsylvania Muscle Institute
University of Pennsylvania
Philadelphia, Pennsylvania 19104

The remarkable growth of structural biology through optical methods during the past decade brought important contributions to biomedical research. The purpose of this meeting was to introduce, to the nonspecialist, some of the novel and improved optical techniques and their biological applications, and to call the attention of the specialist to further needs in technical developments and to new biological problems amenable to structural methods.

The inclusion of light, electron, and X ray optical research in the same conference reflects the common aims of these approaches and their convergence through the evolution of the shared methodology and vocabulary of image processing. Image processing of electron micrographs has significantly improved the resolution of the structure of radiation-sensitive biological materials, and has been equally useful for the analysis of "energy-filtered" electron micrographs that represent the composition of cells and their constituents. Computer processing is also assuming an important role in the three-dimensional reconstruction of images, whether obtained by high and intermediate high voltage electron microscopy, scanning confocal, or "conventional" digital light microscopy. New signal-averaging methods that have been developed for electron microscopy of noncrystalline biological materials will also find application in light microscopy. The time limitations of the Conference, unfortunately, precluded covering a wide range of important research on cell differentiation and molecular assembly and transport achieved through microinjection of fluorescently labeled proteins and lipids. A major loss to us all was the absence, due to illness, of one of the major contributors to microscopy, Dr. R. D. Allen (see In Memoriam), who passed away shortly after the Conference ended.

The goal of modern structural biology is to understand function through the visualization of both composition and structure. Electron optical methods have accomplished this by detecting secondary particles (X ray photons) generated by incident electrons or through the detection of the characteristic energy losses of inelastically scattered electrons. These methods can resolve different elements at approximately 8.7 nm in X ray maps and to at least 3.4 nm in energy-filtered electron micrographs. Electron-probe analysis and X ray mapping, electron energy-loss spectroscopy, and energy-filtered imaging each represent the convergence of spectroscopy with imaging to obtain compositional information. A similar combination of spectroscopic and imaging methods is rapidly evolving in light microscopy, as indicated by the use of fluorescent and other types of indicators for imaging dynamic changes in pH, cytoplasmic free Ca^{2+}, proteins, and lipids. It can be anticipated that light microspectrophotometric methods will also be applied to the imaging of native structures in living cells.

The importance of rapid freezing methods in modern electron microscopy is evident and made many of the new results reported possible. Only rapid freezing and cryoultramicrotomy preserve the distribution of diffusible elements in a manner suitable for high-resolution analytical electron microscopy. Improvements in rapid freezing have led to better visualization of native structures by either "conventional" electron microscopy or through freeze-fracturing and freeze-etching, and permit the "freeze-capture" of proteins, such as cross bridges in muscle, in different dynamic states. Cryoelectron microscopy, the imaging of frozen-hydrated molecules and organelles, aims to achieve the highest possible resolution close to the living state, and can be applied to the "freeze-trapping" of different conformational states of enzymes, receptors, and other proteins.

The Conference, obviously, owed its greatest debt to the participants, but its realization would not have been possible without the financial support listed in these proceedings. The Organizer is particularly grateful to Dr. John Wooley of the National Science Foundation and Drs. Suzanne Stimler and James Dingell of the National Institutes of Health for their help and encouragement, to the staff of the New York Academy of Sciences for logistic support, and to Mr. Hans-Dieter Klink of Zeiss USA for organizing a social event for the speakers that gave them a view of the New York panorama at its best.

In Memoriam

Robert Day Allen (1927–1986)

Robert Day Allen, a modern-day pioneer in electronic-aided light microscopy, died at his Hanover, New Hampshire home on March 23, 1986, after a long struggle with pancreatic cancer. Bob invested a large percentage of his enormous research talent on light microscopic methods, since he always felt that direct measurements on living cells or functional components of cells would yield more profound insights than strictly ultrastructural or biochemical approaches. He succeeded in developing new light microscopic methods and in applying these methods to fundamental problems in cell biology.

Dr. Allen began using polarized light microscopy in the early 1950s to investigate the structure of cytoplasm in sea urchin eggs and in a variety of amoeboid cells. He demonstrated strain birefringence in the endoplasm of the giant amoeba, *Chaos carolinensis,* which indicated that cytoplasm was a non-Newtonian medium with viscoelastic properties. He improved the polarizing microscope during the course of his work on amoeboid movement by employing a servocontrolled electrooptical light modulator to measure phase retardations. This "electronic" polarizing microscope permitted Bob and his colleagues to measure phase retardations down to the level of a few angstroms, reproducibly.

In the 1960s, Bob met Dr. George Nomarski, the French optical scientist, who developed the differential interference contrast (DIC) microscope that bears his name. Bob recognized the potential value of DIC in biology and, together with Dr. George David and a borrowed set of optical components, demonstrated the applicability of DIC in biology and medicine. This study illustrated one of Dr. Allen's main strengths, an uncanny instinct in all of his research concerning the questions to ask and the methods to use.

Bob became interested in the general utility of polarized light microscopy to investigate the structure of cells in the late 1960s. He and his colleagues extended the techniques of phase modulation microscopy based on the electrooptical light modulator. This more advanced instrument was capable of dissecting complex polarization signals into components of linear birefringence, linear dichroism, optical rotatory dispersion, and circular dichroism. The new instrument, called the phase modulation microspectrophotometer (PMMSP), enabled Bob to measure optical activity in both vitally stained and in fixed materials. In addition, this same instrumentation was used to analyze cross-bridge orientations in vertebrate striated muscle.

In 1981, Dr. Allen made his most important contribution to the field of light microscopy. Together with Nina S. Allen, he discovered Allen Video-Enhanced Contrast microscopy (AVEC). This technique takes advantage of both analog and digital enhancement of video signals and the ability of image processing to improve the total contrast. Therefore, objects even below the limit of resolution of the light microscope could be detected in the presence of light-scattering objects. This fundamental technical discovery led to the demonstration that single microtubules could support particle transport and could themselves move along a substrate.

Dr. Robert D. Allen was a creative force in the fields of optical light microscopy and in cell biology. His work on light microscopy over the years played a major role in the recent revolution in light optical methods.

D. Lansing Taylor
Carnegie-Mellon University
Pittsburgh, Pennsylvania 15213

The Use of Cryoelectron Microscopy in Elucidating Molecular Design and Mechanisms

NIGEL UNWIN

Department of Cell Biology
Stanford University School of Medicine
Stanford, California 94305

Imaging and evaluating rapidly frozen specimens[1,2] is a new approach to electron structure analysis by which one can observe molecules directly in their original ionic environment unaffected by chemical fixatives or stains. It provides a unique opportunity to investigate fine molecular details in one or more defined physiological states.

This contribution outlines some important features of the approach, and describes some recent results from our own work which demonstrate its advantages and potential for exploring other systems.

PRESERVATION AND CONTRAST

Changes in ionic environment and level of hydration brought about by most methods of sample preparation alter or destroy subtle features of most proteins and molecular assemblies. Very often, too, structures partially collapse on drying down in negative stain. However, the rapid freezing of a specimen within its native, wet surroundings largely eliminates these adverse influences, at least when cooling rates are sufficient to produce amorphous ice (about $10^{5\circ}$ C/second) and the observation temperatures are low enough to prevent it from crystallizing (less than $-130°$ C). This technique, therefore, is expected to give the truest representation of the structure in solution.

Also important in terms of what the images show, are the different electron scattering strengths of each of the contributing components. Membrane proteins appear in their entirety, because protein scatters more strongly than either lipid or ice. On the other hand, lipids are not so visible, because their average scattering strength is similar to that of ice. Nucleoproteins show the internal distribution of their chemically distinct components as well as their surface outlines. In general, images of this kind are devoid of effects leading to interpretative difficulties, such as indeterminate degrees of stain penetration, and loss of information in parts from which the stain is excluded.

MOLECULAR DESIGN

These improvements in preservation and imaging characteristics have allowed us to identify a narrow, low-density region within the large ribosomal subunit in three-dimensional maps of frozen-hydrated ribosome crystals.[3] This region extends about 150 Å from the small-large subunit interface, where the polypeptide is synthesized, to

1

an area close to the endoplasmic reticulum membrane, identified by immunolabeling as the nascent protein exit site.[4] It probably represents a channel or deep groove within the large subunit along which the growing polypeptide must travel on its way to the external surface. Since only thirty to forty amino acid residues are protected from proteolysis[5] over this approximately 150-Å length, the polypeptide chain within the channel must be in an extended configuration. To maintain this configuration the bore of the channel should be about 10 Å or less. Such dimensions are too small to be seen at the resolution of our analysis (about 50 Å), so it seems likely that the channel is lined by protein, which together with the channel creates a zone of low density within a dense core of RNA.

The pseudopentagonal symmetry created by the subunits of the acetylcholine receptor is most conspicuous in maps calculated from frozen-hydrated crystals.[6] Moreover, the extended synaptic portion of the channel is almost constant in width and not funnel-shaped as some earlier studies of stained specimens suggested. These more precise features bear upon the mechanism of gating and the efficiency by which the channel permits diffusion of cations across the cell membrane.

In three-dimensional maps obtained from frozen-hydrated gap junctions[7] the cross section of the channel-lining subunits within the bilayer has a shape different from that outside of it. Thus the underlying secondary structure in the separate domains is probably distinctly different, a deduction not possible for stained specimens where details in the bilayer are not visualized.

MECHANISMS

Many biological assemblies undergo specific conformational changes in response to applied chemical stimuli. Ligand-gated ion channels are examples of such assemblies: the gap junction channel closes in response to increased calcium levels, and the acetylcholine receptor channel opens momentarily (for about one millisecond) when acetylcholine binds. Visualizing structure captured in a particular chemical environment enables one to investigate such transitions directly.

Two subunit configurations result from the gap junction oligomer frozen in Ca^{2+}-free and Ca^{2+}-containing solutions.[7] The relationship between them points to a simple mechanism by which the channel might switch between open and closed states with a concerted tangential tilting of the subunits around its axis. Similar studies of the acetylcholine receptor require experimentation on the millisecond time-scale, which can be conducted by having the grid containing the specimen encounter a fine spray of acetylcholine, appropriately diluted, during its rapid transfer to the freezing medium.

PROSPECTS

Although these are examples of cryomicroscopy at only moderate levels of resolution, there appears to be no fundamental reason why analysis of these (and other comparable) crystals cannot be extended to considerably finer levels. Radiation damage is a significant problem, but not overwhelming given the large number of unit cells in these crystals available to overcome the poor counting statistics associated with the low doses. With improved design of cold stages and correction of lattice distortions,[8] the tertiary organization of the polypeptide chain may soon become visible in structures other than bacteriorhodopsin.

REFERENCES

1. TAYLOR, K. A. & R. M. GLAESER. 1974. Electron diffraction of frozen, hydrated protein crystals. Science **186:** 1036–1037.
2. ADRIAN, M., J. DUBOCHET, J. LEPAULT & A. D. MCDOWALL. 1984. Cryoelectron microscopy of viruses. Nature **308:** 32–36.
3. MILLIGAN, R. A. & P. N. T. UNWIN. 1986. Location of exit channel for nascent protein in 80S ribosome. Nature **319:** 693–695.
4. BERNABEU, C., E. M. TOBIN, A. FOWLER, I. ZABIN & J. A. LAKE. 1983. Nascent polypeptide chains exit the ribosome in the same relative position in both eucaryotes and procaryotes. J. Cell Biol. **96:** 1471–1474.
5. BLOBEL, G. & D. D. SABATINI. 1970. Controlled proteolysis of nascent polypeptides in rat liver cell fractions. I. Location of the polypeptides within ribosomes. J. Cell Biol. **45:** 130–145.
6. BRISSON, A. & P. N. T. UNWIN. 1985. Quaternary structure of the acetylcholine receptor. Nature **315:** 474–477.
7. UNWIN, P. N. T. & P. D. ENNIS. 1984. Two configurations of a channel-forming membrane protein. Nature **307:** 609–611.
8. HENDERSON, R., J. M. BALDWIN, K. H. DOWNING, J. LEPAULT & F. ZEMLIN. 1986. Structure of purple membrane from Halobacterium halobium: Recording, measurement and evaluation of electron micrographs at 3.5 Å resolution. Ultramicroscopy, in press.

DISCUSSION OF THE PAPER

D. F. PARSONS (*New York State Department of Health, Albany, New York*): It's very encouraging to see pictures without the negative staining, but you should also consider the possibility of looking at the wet material at room temperature. There's a misconception around that there's large radiation damage when one looks at wet objects; I do not believe that this is true, judging from electron diffraction patterns of wet organic materials. They are easy to use, once they are set up, and they enable one to work with an extremely thin water layer only about 100 to 200 Å thick.

I would be interested to see what the differences in contrast mechanisms are between the two methods. I think there are going to be some differences between looking at frozen or immobilized hydrocarbon chains and fluid ones more at the physiological temperature.

UNWIN: My feeling was that there is a gain of a factor of about five in radiation damage between the low temperatures and the room temperature wet specimens.

PARSONS: I think it's much less than that, probably less than two.

K. A. TAYLOR (*Duke University Medical Center, Durham, North Carolina*): When we first measured the radiation damage in frozen-hydrated materials we couldn't compare, in the same instrument, frozen-hydrated crystals with crystals hydrated at room temperature. We did measure glucose-embedded specimens, and presumably the difference between glucose-embedded specimens at room temperature and glucose-embedded specimens at low temperature ($-120°$ C) is comparable to the difference between frozen specimens. That was at least a factor of five, and that number's been confirmed by others. I would not think that the radiation resistance of hydrated specimens at room temperature would exceed that of glucose. I would suspect it to be worse.

PARSONS: Could one argue that glucose isn't a very natural environment and that they might be extra sensitive in the glucose?

TAYLOR: No. I think that glucose probably preserves some sort of a hydrated environment. Specimens dried in glucose maintain a fair amount of bound water. It's basically a syrup.

PARSONS: Dr. Unwin, have you looked at acetylcholine channels in the presence of acetylcholine? If one were looking for conformational states of the molecule activated or not activated by acetylcholine, the physiology shows that the duration of these conformational states may only be milliseconds at room temperature. One may want to catch them at a very low temperature.

UNWIN: We're working on that right now. The changes in the acetylcholine receptor are likely to be quite small, because the channel only opens a very small amount, and we have to improve the resolution of our cold stage. The freezing method does give you the opportunity of doing very short time scale studies. With the acetylcholine receptor, for instance, as you're freezing, you can have the specimen pass through a spray of acetylcholine, and have it frozen and fixed in a certain state within about a millisecond or so of actually applying the acetylcholine.

Recent Advances in Cryoelectron Microscopy

E. ZEITLER AND F. ZEMLIN

Fritz-Haber-Institut der Max-Planck-Gesellschaft
Faradayweg 4-6
D-1000 Berlin 33, Federal Republic of Germany

Cryoelectron microscopy has become a fashionable term—widely known, yet lacking precise definition. It has been used for many purposes, the meaning of the prefix "cryo" varying depending on the instrument or specimen in question. Originally it meant that the objective lens coils were superconducting. Then the sense was broadened to refer to cooled specimen chambers or specimen holders below room temperature, and eventually to the specimen itself, prepared at temperatures lower than ambient. Superconducting lenses were expected to produce a stable lens current, which, however, directed attention to the optical performance of the microscope rather than to the specimen. It was soon realized that cooling the specimen reduced the rate of chemical reaction triggered by the impact of the electron, thus limiting radiation damage. The ultimate expectation, the fulfillment of a dream, is that the natural hydrated state of biological structures can be preserved (frozen-in) and examined. In view of this a close definition of cryoelectron microscopy may perhaps be unnecessary. For since the goal of electron microscopy is to gain information about structures by means of electron diffraction and imaging, any trick, any method, any approach is permitted as long as it remains scientific—that is, as long as it can be reproduced and communicated.

A complete investigation should state the temperature of the specimen. But since this is difficult to measure, it is seldom determined. The use of liquid helium by no means assures that the temperature of the specimen is that of the coolant. After having devoted many years to achieving low temperatures in the electron microscope, Heide[1] published a summary of his findings, which help to solve the problem. Nevertheless the cryo stage in the objective of a handmade 250-kV electron microscope (Deeko 250) permits investigations of specimens at temperatures from 6 K to 300 K. With it, therefore, we can observe the various states of frozen water and their transitions.[2] We have found four modifications—two crystalline ones (cubic and hexagonal), which are useless for embedding biological material, and two amorphous ones, known to have different specific gravity. The results of this study are given in FIGURE 1. Variations in temperature were chosen so that only unidirectional transitions occurred. These data can be directly translated into a procedure for successful embedding, that is, one whereby crystallization of the water is avoided. This method, perfected by Dubochet[3] and illustrated by many beautiful micrographs, can now be employed by any skilled investigator.

The data in FIGURE 1 also show how electron irradiation influences the ability of ice to protect the embedded specimen. It is known that free radicals produced in the embedding ice react with the specimen. Therefore, separation of the ice and the specimen limits these reactions.[4] When we speak of advance in science, we should not forget that the idea of embedding specimens in ice has been around for some thirty years and was, in 1974, proved by Taylor and Glaeser to be feasible with ice-embedded catalase crystals.[5] Progress is, after all, not sudden; it requires the effort of many scientists over many years.

5

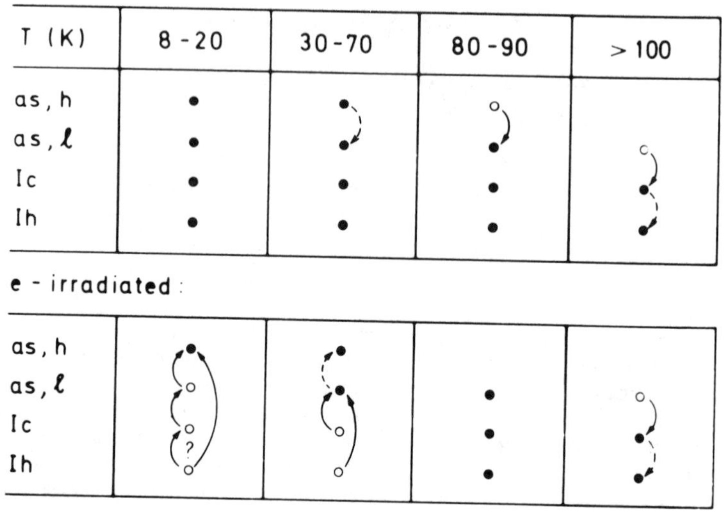

FIGURE 1. Ice in four temperature ranges. *Solid circles* indicate (meta)stable forms of ice in their temperature ranges. *Open circles* indicate unstable forms. *Arrows* show transitions between forms. Note that irradiation can cause reversed transitions.

Activation energy of chemical reactions is derived from their temperature dependence in the well-known Arrhenius plot. Using the variable-temperature cryoobjective, Giersig employed this method to determine the radiation damage to copper phthalocyanine in the temperature range 6 K to ambient.[6] The plot of the critical dose obtained from fading curves of diffraction patterns versus the reciprocal temperature is two distinct straight lines, that is, two activation energies (FIGURE 2). The energy of 10 meV between 80 K and room temperature compares well with van der Waals' binding energies, whereas the 0.3-meV energy between 6 K and 80 K is not fully understood.

The importance of instruments to advances in cryoelectron microscopy is obvious. We are fortunate to have at our disposal a Siemens prototype 120-kV superconducting cryomicroscope, which can be calibrated for high resolution. A practical question, namely, what temperature of the specimen is reached at what time, is answered in FIGURE 3. It shows that within half an hour the temperature of the specimen holder drops from that of liquid nitrogen to 4.5 K as a result of liquid helium cooling. This rapid rate of cooling permits a full day's use of the microscope! In addition, reliable temperature measurements essential for comparisons can be made with a built-in calibrated Ge diode. Better still would be to know the temperature of the specimen itself, which, in the case of biological samples, is very difficult to assess due to unpredictable and uncontrollable thermal contact and conductivity. Today this statistical uncertainty can be overcome only by numerous (stultifying) repetitions, at least until somebody can think of a better way.

A particular advantage of the superconducting lens is that the deflection system and the stigmators are also superconducting devices.[7] The absence of Joule heat brings about the desired thermal isotropy in the specimen chamber, which in turn leads to remarkable mechanical stability and a drift rate of only one atomic radius per ten seconds. This chain of cause and effect is normal in low-temperature microscopy and should be kept in mind when results from different microscopes are compared. Minimum exposure techniques can be employed with a stable microscope. A lower dose, which requires a longer time for sufficient exposure, causes only a negligible rise in temperature in the specimen and less radiation damage.

High-resolution images of paraffin taken with minimum radiation damage at less than 4.5 K look disappointingly pale and structureless. Only a laser beam producing a diffraction pattern of the image content reveals whether the "portrait" session was successful. Whereas the search for the specimen in the microscope is carried out only in the diffraction mode, there is no way of knowing whether or not the image will be good until after it has been taken. And even then high-resolution images can only be presented after contrast enhancement or "falsification." We mention all these things to reemphasize that the mere cooling of the specimen by no means constitutes electron microscopy. It is not information retrieval!

If one calculates the power spectra from adjacent image sections of a paraffin crystal sheet, one finds distinct and disturbing differences between them (FIGURE 4). The patterns range from being intact, like that of the entire image, to having only three reflexes instead of the dominant eight. The cause of differences cannot be determined, because they result from damage during both preparation and irradiation (movement, distortion, and lack of flatness). The present remedy is a post-factum computer therapy called cross-correlation averaging.[8] This idea, too, has been around for some years, though the practical problem of managing the large number of picture elements has only lately been resolved by computers and advanced software.[9]

By means of an ideal image square the correlation technique corrects the deviations of other squares from the lattice, as it were, rebuilding an ideal crystal. We have averaged over three hundred squares for reconstruction, taking into account the optical parameters that determine the transfer of contrast. The results of this "facelift" have been published[10] and are here given in FIGURE 5, in which the white elements correspond to the alkane chains of $C_{44}H_{90}$ paraffin. These pictures agree quite well with images calculated by the multislice method.

One curve of a specimen, a purple membrane, has through the dedicated work of

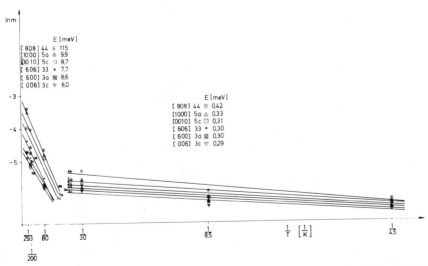

FIGURE 2. Arrhenius plot of the radiation damage to copper phthalocyanine ($C_{32}H_{16}CuN_8$) from 6 K to 293 K. Two activation energies are discernible in the two groups of straight lines.

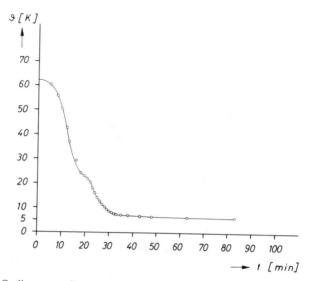

FIGURE 3. Cooling curve of a specimen stage in the superconducting cryoelectron microscope. The curve begins with a LN_2-cooled cryostat.

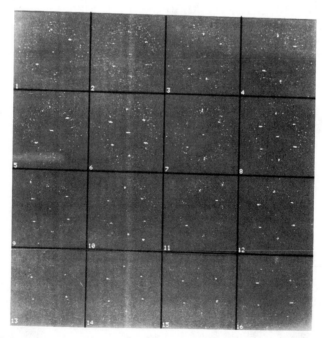

FIGURE 4. Power spectra (Fourier transforms) calculated from the 17.5-nm squares of a paraffin image and inserted into corresponding image squares.

Unwin and Henderson[11] acquired the quality of a measuring rod (FIGURE 6). The glucose-embedded specimen comes from a collaborative effort with Henderson,[12] and the measurements were performed in our superconducting microscope at 4.5 K. The film's optical density was evaluated as a function of the impinging dose, which in turn was measured with a Faraday cup. It is well known that the diffraction intensities of the various spacings fade differently. Yet all three curves confirm an exponential decay wherein the critical doses have a well defined meaning and value. It is remarkable that the critical doses are higher than 1,000 e/nm², which should produce a respectable signal-to-noise ratio in the images.

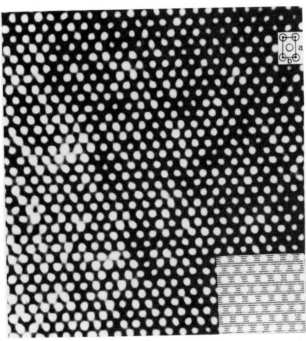

FIGURE 5. Paraffin: averaged image of 372 cross-correlated image squares. The *upper inset* shows the known unit cell; a = 0.750 nm and b = 0.495 nm. The *lower inset* is the result of multislice model calculation, which agrees well with the discovered structure. The white elements are the top view of single paraffin chains.

So far images have been taken at exposures of only 100 e/nm². The best micrographs of two-dimensional crystals show optical diffraction spots to a resolution of about 0.6 nm. After correcting the imperfections of the crystalline sheets by the method of cross-correlation already mentioned in the paraffin study, one can test for the significance of weak high-resolution spots. The recorded electron diffraction intensities (FIGURE 6) play a decisive role here,[13] because they are not influenced by the contrast transfer function or other optical aberrations such as misalignment (axial coma) and residual astigmatism. Thus retrieving and refining all the data in real and reciprocal space produces an excellent consistency of the amplitudes and phases of the purple membrane to a resolution of 0.35 nm.[12]

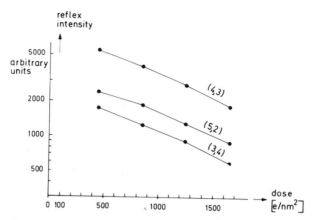

FIGURE 6. Fading curves of purple membrane for the selected reflexes (4,3), (5,2) and (3,4), all belonging to periods of about 0.9 nm. Critical dose: $1,300 \pm 400 \, e/nm^2$.

We have not discussed single-particle work, because with the usual negative staining techniques radiation damage is not a great problem. Also relatively low resolution might obscure radiation effects on fine details.

In conclusion, there is reason to hope that beam-sensitive materials can be analyzed successfully by direct imaging of two-dimensional crystals (membranes), and that particulate specimens can also be assessed. Both types of specimen require averaging of the raw data. Fourier techniques can be used for the former and statistical classification for the latter.[14] Although misalignment and some optical aberrations can be corrected, only the best preparations and the best images should be treated. Cryoelectron microscopy as a multifaceted approach is on its way to becoming true electron crystallography and electron tomography.

REFERENCES

1. HEIDE, H. G. 1982. Ultramicroscopy **10**: 125–154.
2. HEIDE, H. G. & E. ZEITLER. 1985. Ultramicroscopy **16**: 151–160.
3. DUBOCHET, J., J.-J. CHANG, R. FREEMAN, J. LEPAULT & A. W. McDOWALL. 1982. Ultramicroscopy **10**: 55–62.
4. TALMON, Y. 1984. Ultramicroscopy **14**: 305–316.
5. TAYLOR, K. A. & R. M. GLAESER. 1974. Science **186**: 1036–1037.
6. GIERSIG, M. 1985. Diplom-Thesis, Dept. of Physics, Free University of Berlin.
7. LEFRANC, G., E. KNAPEK & I. DIETRICH. 1982. Ultramicroscopy **10**: 111–124.
8. CROWTHER, R. A. & U. B. SLEYTR. 1977. J. Ultrastruct. Res. **58**: 41–49.
9. FRANK, J. 1982. Optik **63**: 67–89.
10. ZEMLIN, F., E. REUBER, E. BECKMANN, E. ZEITLER & D. DORSET. 1985. Science **229**: 461–464.
11. UNWIN, P. N. T. & R. HENDERSON. 1975. Nature **257**: 28–29.
12. HENDERSON, R., J. M. BALDWIN, K. H. DOWNING, J. LEPAULT & F. ZEMLIN. 1986. Ultramicroscopy, in press.
13. HENDERSON, R. & R. M. GLAESER. 1985. Ultramicroscopy **16**: 139–150.
14. VAN HEEL. M. & J. FRANK. 1981. Ultramicroscopy **6**: 167–194.

DISCUSSION OF THE PAPER

K. A. TAYLOR (*Duke University Medical Center, Durham, North Carolina*): Presumably your etching effect, the electron dose, is orders of magnitude greater than the dose to cause loss of crystalline diffraction. Actually, it wouldn't offer much just being sure that you still had ice above and below your crystals.

ZEITLER: It's true, but you might then say, it's embedded and I can look as long as I want to. So I would just want to caution people that this etching happens.

E. MANDELKOW: (*Max-Planck-Unit for Structural Molecular Biology, Hamburg, Federal Republic of Germany*): You distinguish the amorphous ice on the basis of its diffraction pattern. Can you distinguish between the two density forms on the basis of their diffraction patterns?

ZEITLER: We know this is done by Rice *et al.* in Chicago, who determined the densities. From them we know the X ray diffraction patterns.

MANDELKOW: If water condenses in the amorphous form, it's thought to be microporous and has a much lower density. Could one distinguish that?

ZEITLER: No, I don't think you can distinguish between them.

J. FRANK (*New York State Department of Health, Albany, New York*): For the segmented area of the paraffin micrographs you showed different diffraction patterns, from which you concluded that there may have been movement. Don't you have statistical fluctuations if you analyze such small areas? And if you have only a small number of unit cells in each of these small patches, then I would expect that there would be statistical fluctuations in the reflections.

ZEITLER: But they're still on the order of hundred unit cells.

FRANK: Okay. I didn't appreciate that.

ZEITLER: So they would be spread out. It's really the quality of the crystal. I don't fully agree with the movement theory, but certainly there are differences between adjacent patches.

FRANK: Have you done similar studies for other crystals?

ZEITLER: No, we haven't.

J. S. WALL (*Brookhaven National Laboratory, Upton, New York*): My question is common to both of you. To obtain such nicely contrasting images of molecules in ice, what focus value do you use?

ZEITLER: I couldn't say off hand. I think it was a very large defocus. This is the problem that I alluded to—that the imaging is another story now.

WALL: Yes. (I also want to direct this question to Dr. Unwin.) In channels?

ZEITLER: Yes, it's just the defocus that gives the maximum in the contrast-transfer function near the highest spatial frequency in the picture that's biologically relevant.

WALL: Did you adjust your defocus value to the spatial frequency you want to image?

ZEITLER: Yes. The ribosomes, which were lower resolution, were defocused more. This may be a completely new situation with this imaging where you are selecting spatial frequencies that are the right ones. Unfortunately, in the ribosome work we don't have such a high resolution yet, so we don't have to be that critical.

WALL: Can you measure the electrical conductivity in any direct way?

ZEITLER: No.

WALL: And would you comment on the possibility of charging a frozen specimen?

ZEITLER: We only surmise this from the movement of the ice, since it's very strange that the two forms—the low density and the high density—move completely differently. This was the sluggish one—the low density. And the other one goes like this—swish! But, no, we don't know the conductivity.

WALL: Most things become much better insulators as you cool them down, don't they? Is it possible that the paraffin is charging locally and deflecting the beam slightly?

ZEITLER: It's true, there could be such a charge pattern.

Quick-Frozen Microtubules Studied by Cryoelectron Microscopy and Image Processing

ECKHARD MANDELKOW

AND EVA-MARIA MANDELKOW

Max-Planck-Unit for Structural Molecular Biology
Ohnhorststrasse 18
D-2000 Hamburg 52, Federal Republic of Germany

INTRODUCTION

Microtubules form part of the cytoskeleton and are involved in diverse cellular processes (mitosis, axonal transport, etc.; for a review see REFERENCE 4). They are built from globular protein subunits (A and B tubulin, relative molecular mass (M_r) 50,000), which combine into longitudinal protofilaments. Thirteen or more protofilaments form hollow tubules. Their outside is decorated by microtubule-associated proteins (MAPs). The protein subunits can be reassembled *in vitro* into microtubules or polymorphic assembly forms. The structure and assembly of microtubules has been studied by several methods, for example by negative stain electron microscopy combined with image reconstruction,[2] X ray fiber diffraction,[7] solution X ray scattering,[8] and cryoelectron microscopy.[9]

METHODS

Rapid freezing and cryoelectron microscopy were done by the blotting method[3,6,11] using a Philips EM 400 microscope equipped with a PW 6591/100 cold stage. Freezing in amorphous ice is capable of preserving nearly atomic resolution,[12] but the contrast is weak, and the image represents a projection of the structure.[1,14] The preparation of microtubules, electron microscopy, and image processing were done as previously described.[9,10]

RESULTS AND DISCUSSION

Microtubule Structure

An example of a frozen-hydrated microtubule is shown in FIGURE 1a. The main features may be summarized as follows:

Image Interpretation. The contrast of an unstained particle is weak, and it is opposite to a negatively stained one (FIGURE 1d). A low contrast implies that there is more noise and thus a reduction in resolution, even when the particles are well preserved. The effective resolution is similar in the presence and absence of stain (about 2–3 nm). Since the contrast is largely determined by phase contrast rather than

13

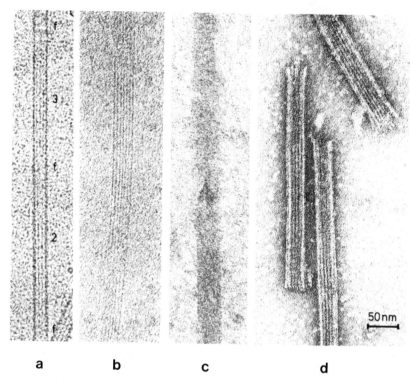

a   b   c   d

FIGURE 1. (a) Frozen-hydrated microtubule in a vitrified layer about 100 nm thick. Note the small and uniform diameter (about 23 nm), the dark edges due to superposition of several protofilaments, the light halo outside of particle due to strong underfocus (about 3.5 μm), and the longitudinal striations inside the edges due to protofilaments. One observes a regular alternating succession of two or three internal striations, separated by fuzzy regions showing no striations. This feature arises from a gentle twist of the protofilaments around the tubule axis. The clear striations arise from the superposition of protofilaments on the front and back of the particle, whereas they are out of register in the fuzzy regions. The alternation between two and three internal striations is typical of microtubules containing 14 protofilaments. (b) Partly flattened microtubule with preserved substructure. Up to seven striations are visible, as expected in a microtubule with 14 protofilaments. There are two fuzzy regions (top and bottom) where protofilaments cross over one another. (c) Microtubule after freeze-drying in the electron microscope. Note the flattening and the loss of substructure. (d) Negatively stained microtubules (1% uranyl formate). Compared to (a) the contrast is strong and reversed. Note the greater apparent diameter due to flattening and the variable number of striations. (From Mandelkow and Mandelkow.[9] Reprinted by permission from *Journal of Molecular Biology*.)

amplitude contrast (resulting in a down-weighting of low-resolution features), optimal imaging requires a strong underfocus.[13] This generates the light halo around the edges of many particles.

Shape of Microtubules. Unstained frozen-hydrated microtubules are usually long and straight and show no kinks or breaks, indicating that freezing does not destroy the structure if it is rapid enough. The mean diameter is 22–23 nm, in good agreement with

X ray diffraction data[7] but smaller than in negatively stained samples (FIGURE 1d). This shows that staining and drying leads to flattening, *i.e.* the shape is preserved only in a hydrated environment. When a solution of frozen-hydrated microtubules is dried in the electron microscope one also observes particles in various degrees of flattening (FIGURE 1b), resulting eventually in the loss of any recognizable substructure (FIGURE 1c).

Protofilament Number and Supertwist. Negatively stained and flattened micro-tubules show a variable number of longitudinal striations (up to 7), and it is impossible to determine the number of protofilaments accurately. By contrast, hydrated microtu-bules show a pattern of striations characteristic of the protofilament number. Inside the two dark edges of FIGURE 1a (comprising several protofilaments in projection) one can count two or three striations, representing pairs of protofilaments in projection (front and back). The pattern is periodic; regions with clear striations alternate with fuzzy ones where no striations can be discerned. There are two types of pattern: 3-fuzzy-3-fuzzy, etc. (not shown), and 3-fuzzy-2-fuzzy, etc. (FIGURE 1a), characteris-tic of microtubules with 13 or 14 protofilaments, respectively. These features are explained by a gentle twist of the protofilaments around the microtubule axis, which causes a periodic change in the superposition pattern from the front and back.

Polarity. The supertwist allows the polarity of a microtubule to be determined by inspection. When one glances down at a shallow angle the particle in FIGURE 1a shows an arrowhead pointing downward, which is formed by the superposition of twisting protofilaments.

Microtubule Dynamics. Since rapid freezing amounts to rapid physical fixation, one can observe time-dependent processes such as assembly or disassembly. For example, images of disassembling microtubules show that the ends fray apart, and that short protein oligomers appear in the vicinity of fraying ends. One also observes ringlike structures, which are common with microtubule protein at low temperature. This suggests that microtubules disassemble into oligomers smaller than rings, and that ring formation at low temperature is based in part on oligomers. This is consistent with time-resolved X ray scattering experiments from solutions.[8]

In summary, cryoelectron microscopy offers several advantages in the study of microtubules. They include the investigation of microtubule substructure, their supramolecular conformation, dynamics of reactions, and distributions of particle sizes, lengths, shapes, polarities, etc. All of these are based on the reliable preservation of the native state. The major disadvantage is the low contrast, which limits the resolution and makes the study of disordered or single molecules (*e.g.* MAPs) difficult.

Diffraction Pattern and Contrast Transfer

When an image is digitized and Fourier transformed one obtains amplitude distributions such as in FIGURE 2. The curve represents the diffraction along the equator and contains the information on the particle density projected down its axis. This figure is related to the zero layer line of the X ray pattern of oriented microtubules.[7] The main differences are the following:

- The computed transform yields both amplitudes and phases and allows a reconstruction (in contrast to the X ray pattern, which yields only amplitudes).

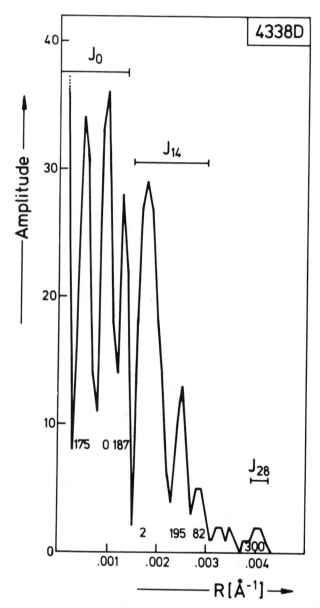

FIGURE 2. Computed amplitude distribution along the equator, obtained by Fourier transformation of an area with three internal protofilament striations (similar to the upper part of FIGURE 1a, a microtubule with 14 protofilaments). Bessel orders are given *above,* and phases *below* the peaks. Note that the first four side maxima are of comparable magnitude. The Bragg scattering vector is given in reciprocal Ångstroms.

- The computed transform is obtained from a single particle and thus is noisier than an X ray pattern. This limits the effective resolution.
- The computed transform is sensitive to the angle of view and the corresponding superposition of protofilaments on the front and back of microtubules (by contrast, X rays measure the rotational average, since the irradiated volume contains particles in all rotational orientations). The angular dependence is illustrated in FIGURE 3a, which shows projected density distributions of microtubule models of 14 protofilaments differing by about 6° of rotation. FIGURE 3b shows the corresponding equatorial amplitude distributions. The three innermost side maxima are identical because they represent the average cylinder. However, the peak around 0.02 $Å^{-1}$ is strong only when the protofilaments from the front and back reinforce each other in projection (FIGURE 3a and b, top and bottom curves).
- In the X ray pattern one observes a rapid decrease of successive maxima, whereas the peaks of the pattern computed from an electron image are of a similar order of magnitude (compare FIGURES 2 and 3b). This is explained by the behavior of the contrast transfer function and its dependence on the defocus (FIGURE 4). For negatively stained particles the optimal underfocus is a few hundred nm,[5] whereas a few thousand nm are appropriate with frozen-hydrated specimens.[1]

FIGURE 4a (top) shows the model diffraction pattern (Fourier-Bessel transform) of a single microtubule of 13 protofilaments.[10] The solid line represents the average cylinder (as in FIGURE 3b, center), and the dotted line a view where the protofilaments from front and back overlap (similar to FIGURE 3b, top or bottom). In the electron microscope this pattern is modulated by the transfer function (FIGURE 4a, bottom). In the case of unstained specimens this is approximated by the solid line (phase contrast for 1,500-nm underfocus), while the dotted line approximates the case for negative staining.

The observed image is obtained by a multiplication of the top and bottom curves of FIGURE 4a, followed by an inverse Fourier transform (FIGURE 4b). With pure phase contrast the "average cylinder" is multiplied by a function approaching zero, resulting in very low or vanishing contrast. However, the higher spacings are enhanced as the contrast transfer increases, resulting in a good representation of internal density fluctuations (FIGURE 4b, dotted line). With negative stain the situation is different, because the transfer function approaches a finite value at low scattering vectors. This means that both the low-resolution structure and the substructure are imaged reasonably faithfully within a large range of defoci, i.e. negative-stain images represent a much wider bandpass of spatial frequencies than unstained ones.

The three projections shown in FIGURE 4b correspond to pure amplitude contrast, 50% amplitude contrast, and pure phase contrast. The top curve represents the "true" object in projection with protofilaments of front and back in register. Each of the three interior peaks comes from the overlap of two protofilaments (front and back); the large peak on the left contains four edge protofilaments not resolved in the projection, the peak on the right contains three edge protofilaments, and the zero level represents the average background density. A similar picture is obtained at 50% amplitude contrast (dashed), except that the whole profile appears to be lowered. The bottom curve shows the projection with pure phase contrast. The mean density is now equal to that of the background, i.e. the "average structure" has nearly disappeared, but the contrast between protofilaments is still appreciable.

In the image of FIGURE 1a the unstained protofilaments appear darker than the background, while the gap in between is lighter. There is an edge-sharpening effect at

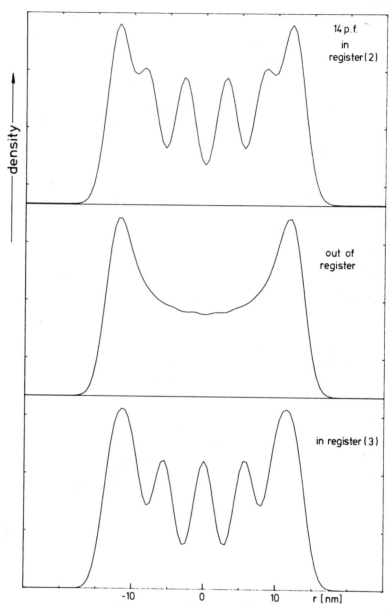

FIGURE 3a. Projected density distribution of a microtubule with 14 protofilaments in three different views, rotated by 360°/56 (equivalent to a quarter of the angle between two adjacent protofilaments). In the *top figure* one observes two edges containing five protofilaments each, and two clear internal peaks arising from two pairs of protofilaments on the front and back (compare lower part of FIGURE 1a). In the *center figure* the protofilaments from front and back are rotated out of register. The *bottom figure* shows another superposition with three internal striations (compare upper part of FIGURE 1a).

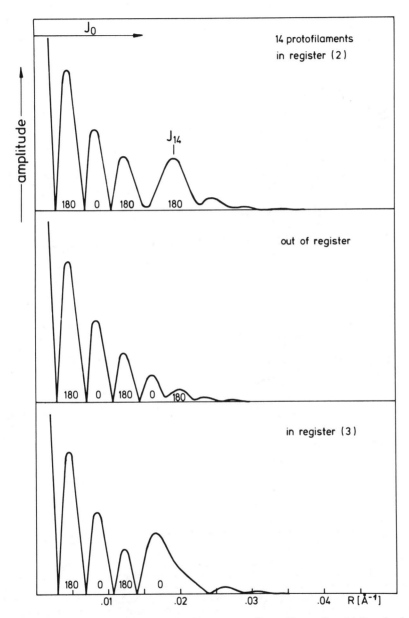

FIGURE 3b. Equatorial amplitude distributions corresponding to FIGURE 3a, with Bessel orders and phases indicated. Note that the J_{14}-term is visible only in the *top and bottom curves*.

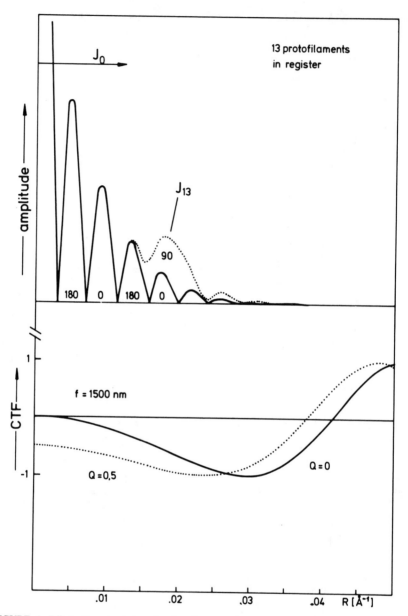

FIGURE 4. Effect of contrast transfer function on microtubule diffraction pattern. (a) *Top:* Computed equatorial diffraction of a microtubule with 13 protofilaments. The *solid line* describes the situation where the protofilaments from front and back interdigitate in projection (see the fuzzy region in FIGURE 1a and the center curve in FIGURE 3a), whereas the *dotted line* refers to the case where the protofilaments from front and back overlap in projection (clear region), generating a strong peak at a spacing of 5 nm. The phases are given in degrees. *Bottom:* Contrast transfer functions for an underfocus of 1,500 nm and pure phase contrast (Q = 0, *solid line*) or 50% amplitude contrast (Q = 0.5, *dotted line*). At low resolutions the phase contrast approaches zero, whereas the amplitude contrast remains finite. The effective amplitude responsible for image formation is a product of the top and bottom curves. Note that with pure phase contrast the average cylinder (J_0-term) is depressed, relative to the contrast between protofilaments (J_{13}-term).

the particle boundaries, visible as a light lining of the microtubules due to the strong underfocus. The reverse situation is observed with negatively stained particles (FIGURE 1d): the protofilaments are lighter than the background (= absence of stain), and the particles are surrounded by a dark outline caused by stain accumulation. Note that the opposite appearance of stained and unstained particles is caused by different effects: physical presence of metal salts versus edge enhancement due to contrast transfer.

In summary, there are clear differences between the patterns obtained by X ray diffraction or computed from microtubule models, compared to those derived from images of frozen-hydrated particles. However, this can largely be accounted for by the differences in experimental conditions, *i.e.* contrast transfer, single views versus

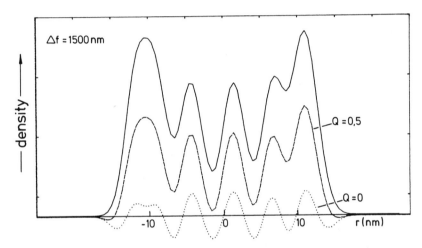

FIGURE 4b. Projected structure resulting from the product of the curves of FIGURE 4a, with protofilaments in register (J_{13} strong). The *edges* of the plot show the edges of the microtubule containing 4 (left) and 3 (right) protofilaments, the *3 interior peaks* correspond to three pairs of overlapping protofilaments. Top to bottom, 100%, 50%, and 0% amplitude contrast. Note the high contrast between protofilaments relative to the mean density, in the *bottom curve,* and the outlines with negative density. The *middle curve* has an amplitude contrast similar to that of negatively stained particles, except that the distribution of stain itself is not included. (From Mandelkow *et al.*[10] Reprinted by permission from *Journal of Microscopy.*)

average over many views, etc. The similarities of the corrected patterns supports the view that the electron images are faithful representations of the native structure of microtubules.

SUMMARY

Frozen-hydrated microtubules have been studied by cryoelectron microscopy. We have described some structural features of microtubules, their relationship with the computed or X ray diffraction patterns, and the influence of the contrast transfer function.

ACKNOWLEDGMENTS

We thank Drs. N. Unwin and R. Milligan (Stanford University) and Drs. J. Dubochet and J. Lepault (EMBL Heidelberg) for making available their facilities for cryoelectron microscopy.

REFERENCES

1. ADRIAN, M., J. DUBOCHET, J. LEPAULT & A. W. McDOWALL. 1984. Nature **308:** 32–36.
2. AMOS, L. A. & A. KLUG. 1974. J. Cell Sci. **14:** 523–549.
3. DUBOCHET, J., J. LEPAULT, R. FREEMAN, J. BERRIMAN & J.-C. HOMO. 1982. J. Microsc. **128:** 219–237.
4. DUSTIN, P. 1984. Microtubules. Springer Verlag. Heidelberg.
5. ERICKSON, H. P. & A. KLUG. 1971. Philos. Trans. R. Soc. London Ser. B **261:** 105–118.
6. LEPAULT, J., F. P. BOOY & J. DUBOCHET. 1983. J. Microsc. **129:** 89–102.
7. MANDELKOW, E., J. THOMAS & C. COHEN. 1977. Proc. Natl. Acad. Sci. USA **74:** 3370–3374.
8. MANDELKOW, E.-M., A. HARMSEN, E. MANDELKOW & J. BORDAS. 1980. Nature **287:** 595–599.
9. MANDELKOW, E.-M. & E. MANDELKOW. 1985. J. Mol. Biol. **181:** 123–135.
10. MANDELKOW, E.-M., R. RAPP & E. MANDELKOW. 1986. J. Microsc. **141:** 361–373.
11. MILLIGAN, R. A., A. BRISSON & P. N. T. UNWIN. 1984. Ultramicroscopy **13:** 1–10.
12. TAYLOR, K. A. & R. M. GLAESER. 1974. Science **186:** 1036–1037.
13. UNWIN, P. N. T. & R. HENDERSON. 1975. J. Mol. Biol. **94:** 425–440.
14. UNWIN, P. N. T. & P. D. ENNIS. 1984. Nature **307:** 609–613.

DISCUSSION OF THE PAPER

N. UNWIN (*Stanford University School of Medicine, Stanford, California*): I wonder if one will be able to do correlation averaging between your individual microtubules to improve the signal-to-noise range here to a resolution in the images to 5 or 10 Å. Do you think one will ever be able to get that out of the microtubules?

MANDELKOW: I don't believe to 5 or 10 Å. I would imagine perhaps to 15 Å. That would be a realistic correlation averaging. Since this is not in a range that is really interesting for those who want high resolution, I'm not sure it's worth the effort.

UNWIN: What's the limitation?

MANDELKOW: The limitation is the irregularity in the microtubule structure itself. We labored for a long time to understand differences between different areas and different particles. The fact is that X ray diffraction averages over all of these and gives an artificially good picture, whereas, if you look at individual pictures, there's quite a lot of genuine variation, perhaps very similar to what Dr. Zeitler mentioned before.

T. S. REESE (*Marine Biological Laboratory, Woods Hole, Massachusetts*): Is it possible that the high molecular weight MAPs are going to come into view at some point?

MANDELKOW: Of course, our hope was to see the MAPs. The disappointment was that the MAPs are just as poorly visible as they are in negative stain.

REESE: How about in the future?

MANDELKOW: Well, thinner layers of ice, perhaps, but I'm not so sure about that aspect.

J. FRANK (*New York State Department of Health, Albany, New York*): I recall that Richard Henderson and his collaborators found a big discrepancy between the fall off in power between X ray crystallography and ice-embedded electron microscopy. Is this only in the higher-resolution range? Is there more of an agreement in the range that you've been looking at?

MANDELKOW: Yes, they are talking mainly about the high-resolution range, whereas this comparison is at very low resolution.

FRANK: What is the resolution?

MANDELKOW: Up to about 40 Å.

D. F. PARSONS (*New York State Department of Health, Albany, New York*): At what point are the differences in form and in structure at small angles and the differences between electron diffraction scattering and X ray going to be important now that the comparisons have been made between real mass distributions without stain and X ray? Have you considered that the image is being contributed to by electron scattering and not by X ray scattering, and that at small angles the form factors are completely different? In fact, they're inverted in terms of the Z numbers at low angles. Have you brought this into your calculations?

MANDELKOW: First of all, we didn't do electron diffraction as such.

PARSONS: That doesn't matter. The contribution to the image is here, the Fourier transform of the electron diffraction pattern, so you have to consider the form factors of electron diffraction.

MANDELKOW: Indeed. Our practical approach to this has been to take the contrast transfer function as measured from the Thon's fringes and apply them to the measured transform of frozen-hydrated microtubules. We then reproduce the X ray intensities within 20% and at that level, given the noise one has, we stop worrying. I know that if one looked at higher detail, one would probably find discrepancies.

W. CHIU (*University of Arizona, Tucson, Arizona*): To respond to the question Dr. Unwin raised about the prospect of going to high-resolution, recently we tried to use the ice-embedded method for the TMV. One of the problems is that even when ice embedded, TMV is still bent a little bit. We applied the technique that Dr. Zeitler mentioned using unbending, and there was an improvement in signal-to-noise beyond 10 Å. On the diffraction pattern we get up to about 4.5 Å, suggesting that in the microscope we are using, these cryoembedded techniques preserve the specimen quite well.

Studies of Eukaryotic Flagella by Cryoelectron Microscopy

JOHN M. MURRAY

Department of Anatomy
University of Pennsylvania
Philadelphia, Pennsylvania 19104

Flagella and cilia of eukaryotes have presented a challenge to microscopists since the earliest observations of cells.[1] These organelles are constructed from 11 parallel tubular elements, arranged in the familiar "9 + 2" configuration. The nine peripheral fibers are doublet microtubules, whereas the central pair are singlet microtubules similar to those commonly found in cytoplasm of cells. The primary motion-generating component is found periodically arranged on the outer doublets, forming a set of inner and outer arms reaching from the A subfiber of one doublet toward the B subfiber of its neighbor (FIGURE 1). This component is a large protein that has been isolated from a number of different sources.[2] Its role is to produce a relative sliding motion of adjacent doublets, deriving energy for this from ATP hydrolysis. Normally this sliding motion is restricted, so that the organelle is forced to bend;[3] controlled activation of the dynein at different locations is presumed to be responsible for the characteristic sinusoidal bending of normal flagella motion. Various radially and azimuthally oriented cross-links stabilize the axially arranged microtubules, and probably contribute mechanical constraints that increase the effectiveness of the flagellar beat cycle.[1]

Cilia and flagella are the most highly ordered systems known for generating motion based on microtubules. As such, they have attracted continuous attention from cell biologists, utilizing a variety of electron microscopic methods. Although highly ordered, their size has hindered effective application of Fourier averaging techniques. The intact organelles are too large and complex for negative staining with subsequent image processing, yet too small to yield a useful X ray pattern. Most of our structural information has been derived from thin sections of plastic embedded whole cilia, or negatively stained images of small fragments. Much has been learned,[1] but our knowledge is still incomplete. It would be extremely useful to have a complete three-dimensional reconstruction of an intact flagellum, to integrate all of the separate facts derived from studying small pieces of the whole. For this reason, and to take advantage of the capability for imaging native hydrated structures, we have been examining flagella by cryoelectron microscopy.[4]

FIGURE 2a shows two sea urchin sperm flagella, unstained, unfixed, and embedded in amorphous ice at 100° K. The image is dominated by a series of longitudinal striations arising presumably from the peripheral doublets and central pair of microtubules. No clear axial periodicity can be seen. This rather simple appearance is in fact the result of complex superpositions that camouflage the enormous amount of structural information present in the images. The computed Fourier transform (FFT) from images such as this shows a clear pattern of layer lines indicative of an axial periodicity of approximately 1,000 Å (FIGURE 2b). All orders of this repeat out to the 24th are present, as are several additional higher orders. The amplitude and phase variation along the layer lines is extremely complicated, as expected for an object of this size. Layer lines 4, 12, and 24 are particularly prominent. Similar though weaker patterns have been observed in images of negatively stained flagella.[5]

It is somewhat surprising to obtain images of this quality from specimens that must be at least 2,500 Å thick. FFTs of images taken at different amounts of defocus show that the periodic contrast variations in the image are strongly modulated by a transfer function that varies with defocus and falls to zero for some values of spatial frequency at each defocus setting. We conclude from this behavior that the periodic contrast variations in these images arise predominantly from a phase contrast interference mechanism between elastically scattered electrons.[6] This gives us some hope that the contrast variations in the image can be interpreted in terms of electron density variations in the specimen, and thus contain information sufficient for three-dimensional reconstruction.

The FFT contains information to a resolution of approximately 30 Å. To compute a three-dimensional reconstruction at 30-Å resolution, we will have to combine approximately 250 different views of the flagellum.[7] Major problems associated with determining relative orientation and accurate alignment of this large number of images must be addressed for reconstruction to be successful.

We have also examined smaller pieces of flagella as intermediate goals on the way to our final objective of three-dimensional reconstruction of the intact structure.

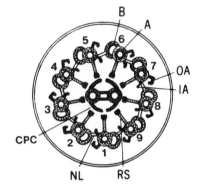

FIGURE 1. Schematic drawing of a cross section through a typical eukaryotic flagellum: **A,** subfiber A of a peripheral doublet; **B,** subfiber B of a peripheral doublet; **OA,** outer dynein arm; **IA,** inner dynein arm; **NL,** nexin link; **RS,** radial spoke; **CPC,** central pair complex.

FIGURE 3a shows an isolated central pair complex, unstained, unfixed, and embedded in amorphous ice. Both microtubules and a striking array of cross bridges and projections are clearly seen. This image is consistent with pictures obtained from sectioned[8] and negatively stained[9,10] material. The computed transform (FIGURE 3b) shows stronger periodic information than was seen in the fixed and stained images. An axial repeat of approximately 335 Å is indicated. Even order layer lines are much stronger than the odd orders. The tilted views necessary for three-dimensional reconstruction of this central pair complex have been partially collected at present.

Fragmented flagella often yield individual or small clusters of outer doublet microtubules. FIGURE 4a shows a group of three outer doublets, bound together by the dynein arms, embedded in amorphous ice. The A and B subfibers of each doublet are clearly visible, along with a set of striations from the superimposed protofilaments. Transforms from the image show a weak axial repeat of about 1,000 Å (FIGURE 4b). Although the region between the doublets, where dynein arms would be expected to lie, do not appear to be particularly well ordered, selective transformation again shows that this is due to a camouflaging effect of superposition. The FFTs from a region lying entirely between the doublets show a strong, sharp 250-Å repeat, with no significant

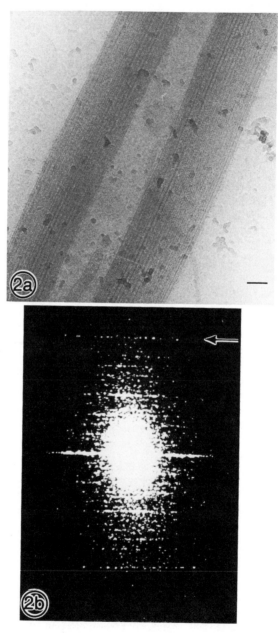

FIGURE 2. (a) Electron micrograph of two sea urchin sperm flagella, unstained, unfixed, and embedded in amorphous ice. *Bar:* 100 nm. **(b)** Computed Fourier transform of a digitized image of flagella prepared as in (a). The *arrow* indicates the $1/40$-Å^{-1} layer line.

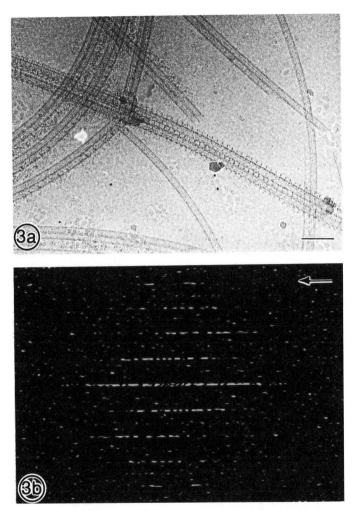

FIGURE 3. (a) Electron micrograph of fragments of flagella, unstained, unfixed, and embedded in amorphous ice. The central pair complex runs diagonally downward from the top left. *Bar:* 100 nm. (b) Computed Fourier transform from a short length of the central pair complex. The *arrow* indicates the $1/40$-Å^{-1} layer line.

contribution from longer spacings. Thus considerable order is preserved in the dynein-arm arrangement even after the flagellum is broken into fragments.

Also seen on properly oriented isolated doublets are the radial spokes, grouped axially into triplets as seen in thin sections.[8] We have not yet sufficient good images of these to reliably describe their arrangement.

The dominant axial period of the dynein-doublet microtubule arrangement (250 Å) is widely different from the repeat length of the central pair complex (335 Å). The smallest commensurable distance between these two structures would span 3 central-

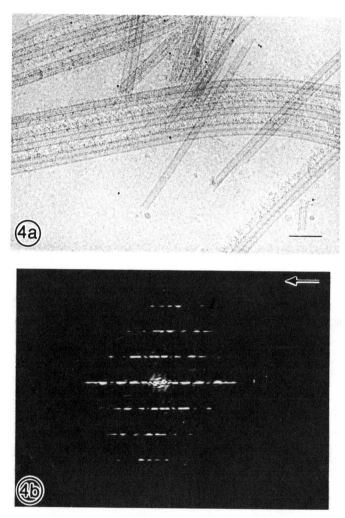

FIGURE 4. (a) Peripheral doublet microtubules and associated dynein prepared by fragmenting whole flagella, unstained, unfixed, and embedded in amorphous ice. *Bar:* 100 nm. **(b)** Computed Fourier transform from a digitized portion of the central member of the group of three doublets shown in (a). The *arrow* indicates the $1/40$-Å^{-1} layer line.

pair repeats and 4 dynein-doublet repeats, approximately 1,000 Å. This is a useful confirmation of the very long axial period observed in the intact flagella. At the moment we cannot rule out the existence of other periods that would make the true flagella unit cell a multiple of 1,000 Å.

In summary, we have shown that cryoelectron microscopy offers useful new information on the structure of eukaryotic flagella in their intact, native state, information that should be adequate to calculate a three-dimensional electron density map to moderate resolution.

REFERENCES

1. GIBBONS, I. R. 1981. J. Cell Biol. **91:** 107s–124s.
2. JOHNSON, K. 1986. Am. Rev. Biophys. Chem. **14:** 161–188.
3. SATIR, P. 1965. J. Cell Biol. **26:** 805–834.
4. TAYLOR, K. A. & R. GLAESER. 1975. Science **186:** 1036–1037.
5. AMOS, L. 1976. *In* Proc. 6th European Congress on Electron Microscopy: 14–20.
6. MISELL, D. L. 1978. Image Analysis, Enhancement and Interpretation. 23. North-Holland. New York, NY.
7. CROWTHER, R. A., D. J. DeROSSIER & A. KLUG. 1970. Proc. R. Soc. London, Ser. A. **317:** 319–340.
8. WARNER F. D. & P. SATIR. 1974. J. Cell Biol. **63:** 35–63.
9. CHASEY, D. 1972. Exp. Cell Res. **74:** 471–479.
10. OLSON, G. E. & R. W. LINCK. 1977. J. Ultrastruct. Res. **61:** 21–43.

DISCUSSION OF THE PAPER

B. McEwen (*New York State Department of Health, Albany, New York*): Thin sections generally show a repeat of 16 nm from the central pair projections. Your 33-nm repeat suggests a pairing of the projection, possibly a complexing with the spokeheads. Would you comment on that?

MURRAY: The central pair repeat seems to vary a bit from organism to organism, and in some cases it is 16 nm and in other cases it's longer. Periods have also been found in some of the negative stained other organisms to be 32 nm, which I take to be close enough to my 33.5.

U. AEBI (*Johns Hopkins University, Baltimore, Maryland*): Was I right that in the diffraction pattern of the ice embedded doublet, the central doublet appeared rather one-sided? What is the reason for that? Would you expect that in ice everything would be perfect?

MURRAY: Your expectations may be slightly higher than mine, but you must remember we're not dealing with a helical structure here. The doublet is indeed a one-sided structure.

AEBI: I think it's very impressive how clear a detail you can see even with the ice being so thick. It seems to make it very promising for people who want to look at just whole cells or the thin parts of whole cells on microscope grids. What do you think the prospects are?

MURRAY: I think that perhaps at high voltage one might be able to do that indeed.

J. B. PAWLEY (*University of Wisconsin, Madison, Wisconsin*): I'd like to follow up on that question, because I know that you went to Boulder and made some micrographs of your structure in the high voltage microscope. Could you say a few words about those?

MURRAY: I did go to Boulder and used the cold stage that Dr. Fotino had made for the Boulder microscope, which worked very nicely. The pictures were not as good as the best I've been able to do at 100 kV. But as I learned later, the microscope operating parameters were not ideal, in the sense that we were using very large condensor apertures at that time. Therefore, the beam coherence would have been considerably worse than it normally is in my microscope. Given those facts and that it was a first try, it's quite promising.

E. MANDELKOW (*Max-Planck-Unit for Structural Molecular Biology, Hamburg, Federal Republic of Germany*): One finds when one is studying supposedly native helical particles, that there is a lot of deviation from what we have anticipated in terms of helical symmetry. We have found very few particles that show, for example, symmetric layer lines. Although part of this may be due to noise, I think another part is due to the fact that even helical particles are not strictly helical. This is a function, of course, of the resolution.

The other thing that was surprising to me was that the doublet microtubules were lying mostly side by side. In ice I would have expected a random distribution. Do you have an explanation for that?

MURRAY: I think they're still attached probably by nexin links or something, to their neighbors, or perhaps even by the dynein. Even in the reconstituted microtubule-dynein complex, it's much more common for them to cluster than to find individual microtubules. So there is some interaction there.

MANDELKOW: But then wouldn't you sometimes see them in projection, one on top of the other?

MURRAY: I'm not showing you all the pictures, but one does indeed, and that projection is quite useful, because then you can see the radial spokes sticking out.

Now I'll make a quick comment on the helical symmetry. You have to remember that in ice, as opposed to negative staining, you're looking through an entire structure, so you have things at a wide range of radii contributing. If you calculate for a microtubule, the layer lines look entirely one-sided, because you get beating Bessel functions, which on one side wipe out the intensity.

J. S. WALL (*Brookhaven National Laboratory, Upton, New York*): There's a lot of denaturation of proteins on an air-water interface. Have you considered the possibility that the microtubules are stuck to one of these interfaces or to the surface of the ice? Would it result in flattening and distortion of the helix?

MURRAY: Yes, that would be true. I don't think I have a good answer to that yet.

Image Analysis of the Ca^{2+}-ATPase from Sarcoplasmic Reticulum[a]

KENNETH A. TAYLOR,[b] MING-HSIU HO,[b] AND
ANTHONY MARTONOSI[c]

[b]*Anatomy Department*
Duke University Medical Center
Durham, North Carolina 27710
and
[c]*Department of Biochemistry*
State University of New York Upstate Medical Center
Syracuse, New York, 13210

The Ca^{2+} transport ATPase of skeletal muscle sarcoplasmic reticulum is an intrinsic membrane protein with a molecular weight of 110,000[1] that catalyzes the ATP-dependent accumulation of Ca^{2+} into sarcoplasmic reticulum. The process involves cyclic phosphorylation of the enzyme associated with changes in conformation between the E$_1$ and E$_2$ states.[2]

The Ca^{2+}-ATPase is asymmetrically distributed in the phospholipid bilayer with much of its mass exposed on the cytoplasmic surface. Electron microscopy of negatively stained or freeze-dried, rotary shadowed sarcoplasmic reticulum membranes show a dense packing of 40-Å surface particles with an average density as high as 20,000–30,000 per μ^2.[3] Freeze-fracture studies of sarcoplasmic reticulum membranes show 75-Å particles that are more numerous on the cytoplasmic than on the luminal fracture face.[4] The density of surface particles is several times greater than that of the intramembranous particles, which led to the suggestion that the Ca^{2+}-ATPase may be present in the membrane as an oligomer.[5,6] Recent fluorescence, electron spin resonance, and enzyme kinetic data on detergent-solubilized and reconstituted Ca^{2+}-ATPase systems are consistent with interactions observed between ATPase molecules, but leave the nature of these interactions and their functional significance largely undefined.[7,8]

The crystallization of the Ca^{2+}-ATPase in scallop sarcoplasmic reticulum vesicles[9] and in rabbit sarcoplasmic reticulum exposed to Na$_3$VO$_4$[10–12] defined the stoichiometry of ATPase-ATPase interactions by identifying the Ca^{2+}-ATPase dimers as structural units. This report summarizes our progress in the three-dimensional image reconstruction of unstained Ca^{2+}-ATPase crystals induced by vanadate in rabbit sarcoplasmic reticulum, and compares it with previous results from negatively stained crystals.

[a]Supported by National Institutes of Health (NIH) Grants AM 26545 and GM 30598, National Science Foundation (NSF) Grant PCM 84-03679, research grants from the Muscular Dystrophy Association and R. J. Reynolds Tobacco Co., and an Established Investigatorship (to K. A. T.) from the American Heart Association. The PDS 1010M densitometer was purchased with funds provided by NSF Grant PCM 84-00167 and NIH Grant S10-RR-02283-01 and the VAX 11/750 by NSF Grant PCM 83-06638.

ANALYSIS OF NEGATIVELY STAINED Ca²⁺-ATPase CRYSTALS

Ordered assays of Ca^{2+}-ATPase molecules in rabbit sarcoplasmic reticulum vesicles can be induced by treatment with 5 mM Na_3VO_4 in the presence of 10–100 mM KCl, 10 mM imidazole, 5 mM $MgCl_2$, 0.5 mM EGTA at pH 7.4 at 0° C for one to two days.[13,14] Vanadate functions as an analog of inorganic phosphate[15]; therefore the structure of the molecule in the vanadate-induced crystals should reflect the E_2 conformation.[16]

The crystalline arrays induced with Na_3VO_4 are made up of chains of Ca^{2+}-ATPase dimers that form an oblique lattice on the surface of elongated tubules. The unit cell dimensions of the lattice are a = 65.9 Å, b = 114.4 Å and γ = 77.9°.[10] In filtered images of uranyl-acetate-stained crystals, the Ca^{2+}-ATPase molecules are arranged in chains of dimers in a P2 lattice.[9,10] The molecules in projection have a pear-shaped profile with a length of 66 Å and a width of 46 Å. The crystalline arrays so far observed in sarcoplasmic reticulum membranes of rabbit or scallop muscles have been usually tubular.

Three-dimensional image reconstruction from multiple tilted views of flattened, negatively stained, crystalline tubules provided the first information on the three-dimensional shape of the molecule.[11] The long axis of the cytoplasmic domain is oriented parallel with the plane of the phospholipid bilayer at an angle of about 15° to the a axis of the unit cell. The Ca^{2+}-ATPase molecules that form a dimer are connected by a bridge centered about 40 Å above the presumed cytoplasmic surface of the bilayer. The gap between the major cytoplasmic domains of the dimers is likely to be accessible to solutes. In all crystalline tubules examined so far, the orientation of the bridge is parallel to the long axis of the tubule; perhaps in this orientation changes in the radius of the curvature of the tubules place little or no strain on the structure of the dimer. Some density features were observed within the lipid bilayer region of the reconstruction, but because of the difficulty associated with distinguishing lipid from protein, both of which are assumed to be stain-excluding, the interpretation of these features is uncertain.

ELECTRON MICROSCOPY OF UNSTAINED, FROZEN-HYDRATED Ca²⁺-ATPase CRYSTALS

To better define the structural features of the Ca^{2+}-ATPase within the bilayer we have embarked on a study of crystalline sarcoplasmic reticulum membranes preserved frozen-hydrated in amorphous ice. The crystalline membranes that were selected for this study were produced in a low-salt medium (10 mM KCl), spread on carbon films glow-discharged in amylamine, and frozen in a liquid ethane slush essentially as described previously.[17] The grids were loaded under liquid nitrogen into a Philips EM300 cooling holder adapted for the EM400 instrument and viewed on the Philips EM400 electron microscope in Dr. Nigel Unwin's laboratory at Stanford University in the fall of 1983. Despite its simplicity, this arrangement worked well, and numerous good images were collected. Reproducible defocus values for the micrographs were obtained by focusing images at a magnification of 117,000 under minimal-dose conditions and selecting defocus settings of -2 to -5 μ. At high magnification Gaussian focus can more easily be determined than at the lower magnification at which the images were recorded. After selecting the desired defocus at 117,000X the magnification was reduced to 33,000X without defocus adjustment and the image recorded.

Tubules for three-dimensional reconstruction were selected through initial screening by optical diffraction. Of the many tubule images obtained, only five appeared to be worth processing, and we report here the results obtained from just one of these. The diameters of the best diffracting tubules ranged from 600 Å to 900 Å. Each tubule

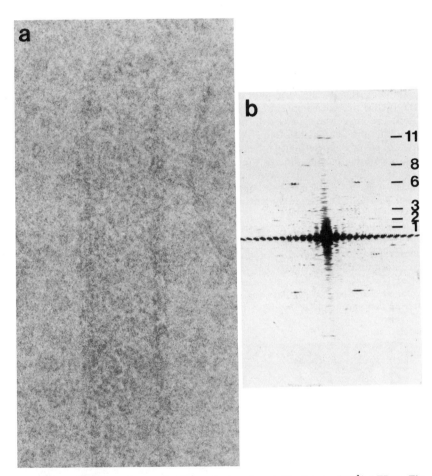

FIGURE 1. (a) Electron micrograph of frozen-hydrated crystalline tubule of Ca^{2+}-ATPase. The tubule is about 600 Å in diameter and has a repeat period of 358 Å. The statistical noise in the image makes difficult the perception of periodic features, which are nevertheless indicated by the optical diffraction pattern. 410,000X. **(b)** Optical diffraction pattern of the tubule shown in (a). The positions and numbers of the layer lines used in the reconstruction are labelled to the *right*.

examined has a different helical structure and must therefore be dealt with individually.

The image analyzed here was selected because it had the smallest diameter, and therefore fewer possible indexing schemes required testing. In addition all of the diffraction data fell within the first zero of the contrast transfer function. The relatively low defocus (−11,000 Å) yielded an image with little visible periodic structure (FIGURE 1a) due to the high statistical noise in the image and low inherent contrast in the specimen. The optical diffraction pattern (FIGURE 1b) yields evidence for good preservation of periodic structure as well as cylindrical shape. Computer-averaged images (FIGURE 2) have been calculated to separate the contributions of near and far sides.

FOURIER-BESSEL THREE-DIMENSIONAL IMAGE RECONSTRUCTIONS OF FROZEN-HYDRATED VANADATE-INDUCED CRYSTALS

Initially, the selected area of the tubule was digitized using a Perkin-Elmer PDS-1010M microdensitometer at a pixel size of 7.58 Å with respect to the original object. The image was then boxed, floated, and padded to 512 × 1,024 points before calculating the Fourier transform. The rotation angle of the particle relative to the lattice of points was determined by least squares fitting of a family of parallel lines with equal separation passing through the maxima of each layer line. With this rotation correction, the layer-line data were calculated using bilinear interpolation in Fourier space.[18] A second method was also applied, using the same image digitized at a pixel size of 3.65 Å and corrected for particle rotation and integral sampling of the 8 axial repeats using bilinear interpolation in real space.[19] This interpolated image was then subjected to Guassian edge apodization and padding to 784 × 784 points prior to Fourier transformation.[20] The layer lines then fell exactly on rows of sample points in Fourier space, and further interpolation was not required. Altogether, 6 nonequatorial layer lines were obtained extending out to the 11th order of the 358-Å axial repeat period (FIGURE 3). Density maps were calculated from layer-line data that were averaged over near and far sides and for which a twofold rotation axis was enforced after an appropriate phase origin search.

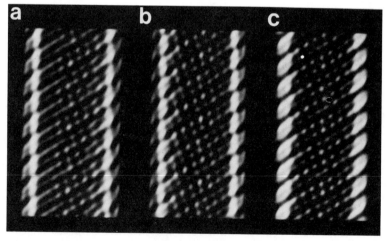

FIGURE 2. Computer-filtered images of the layer-line data used in the reconstruction. Filtering (averaging) is necessary to bring out the periodic features of the image. Far-side data has been converted to near-side for purposes of comparison. Each averaged image is 1,074 Å in axial extent. (**a**) Far side, (**b**) near side, and (**c**) the average of the two.

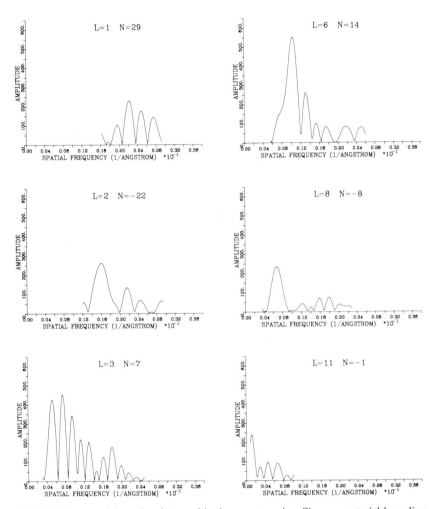

FIGURE 3. Averaged layer-line data used in the reconstruction. Six nonequatorial layer lines were extracted and used in the reconstruction. Each layer line shows multiple peaks that are indicative of cylindrical structure, although the observation that on some layer lines the innermost peak is weaker than some of the outer peaks suggests that some flattening may have occurred.

With the exception of the equatorial layer-line data, no attempt was made to correct for the contrast transfer function except to insure that no layer-line data extended beyond the first zero in it. With this particular structure, the contrast transfer function mainly affected the equator by underweighting the data near the origin. To correct for this effect, we calculated the equator of a model uniform density cylinder with inner and outer diameters similar to those of the crystalline vesicle and determined the ratio of the native equator to the model equator. A smooth, monotonically increasing spline-fitted curve gave the effective contrast transfer function near the origin, in this case from 0.0 to 0.009 Å^{-1}. The remaining equatorial data were left unaltered. The corrected equatorial layer line was then included at half weight in the reconstruction. However, such a correction appeared to have little effect on the three-dimensional density map.

Several sets of information were used to deduce the helical structure of this particular tubule. The positions of the peaks on the various layer lines in the transform can be used to at least set limits on the possible Bessel order for a particular layer line. In this case the most useful layer line was the 11th, which consisted of a pair of peaks particularly close to the meridian at radii of 0.001 and 0.0014 Å^{-1}. If this layer line was due to a first order (J_1) Bessel function, then the radius of the feature giving rise to it would be between 210 and 294 Å (TABLE 1). If the layer line was a J_2, the diffracting feature would be between 346 and 404 Å. The J_1 index for this layer line is also supported by the antisymmetric phase relationship across the meridian. Finally, this layer line occurs on the descending arm of the first helix cross and for this particular structure must therefore arise from densities that follow a left-handed helical path (FIGURE 4).

Possible indexing schemes for the 2nd and 3rd layer lines that will produce a J_{-1} on the 11th layer line include $(-19, 6)$, $(-22, 7)$, and $(-25, 8)$. Published lattice

TABLE 1. Analysis of Diffraction Radii for Three Helical Indexing Schemes

Layer line		1st Peak Position Å^{-1}	Bessel Order	Radius Å	Bessel Order	Radius Å	Bessel Order	Radius Å
1	Near	.0220	25	198	29	228	33	258
	Far	.0218		200		230		260
2	Near	.0158	-19	213	-22	244	-25	276
	Far	.0154		219		251		283
3	Near	.0052	6	230	7	263	8	295
	Far	.0052		230		263		295
6	Near	.0104	12	212	14	244	16	276
	Far	.0106		208		240		271
8	Near	.0070	-7	195	-8	219	-9	244
	Far	.0064		213		240		266
11	Near	.0010	-1	294	-1	294	-1	294
	Far	.0014		210		210		210

constants for the two-dimensional net estimated from flattened, negatively stained tubules were available[10,21-23] and were used to construct a range of surface lattices from which possible helical nets could be determined.[24,25] The best prediction of radius axial repeat and pitch was obtained for a two-dimensional net with lattice constants of a = 64.5 Å, b = 105.6 Å, and γ = 81.7° and indices of J_{-22} and J_7 for the 2nd and 3rd layer lines respectively (FIGURE 4). The selection rule obtained was l = 80m − 11n. The circumferential vector was $V = 22a − 7b$ and the axial vector $C_0 = 2a + 3b$. The structure has a left-handed genetic helix of 80 subunits in 11 turns. This indexing scheme also produces the best prediction of the mean radius of the membrane (TABLE 1).

THE THREE-DIMENSIONAL STRUCTURE OF THE UNSTAINED, FROZEN-HYDRATED MEMBRANES

The three-dimensional reconstruction of the unstained Ca^{2+}-ATPase tubule has density features contained within a cylindrical shell between 193 and 303 Å radius and

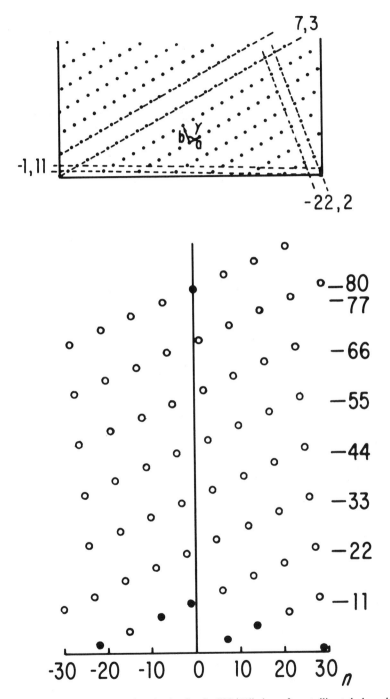

FIGURE 4. *Top panel* is the surface lattice for the "80/11" class of crystalline tubule and the *bottom panel* its corresponding (n,l) plot, which describes the diffraction from this lattice. The surface lattice is drawn as if viewed from the outside looking toward the center of the tubule. Dimer chains follow the (7, 3) helical tracks and are right-handed. The unit cell vectors are labelled **a,b** and helical tracks corresponding to $(-1, 11)$, $(-22, 2)$, and $(7, 3)$ are indicated with *broken lines*. The first meridional reflection, though not observed, would occur on the 80th layer line. Those reflections used in the reconstruction are indicated by *solid circles*.

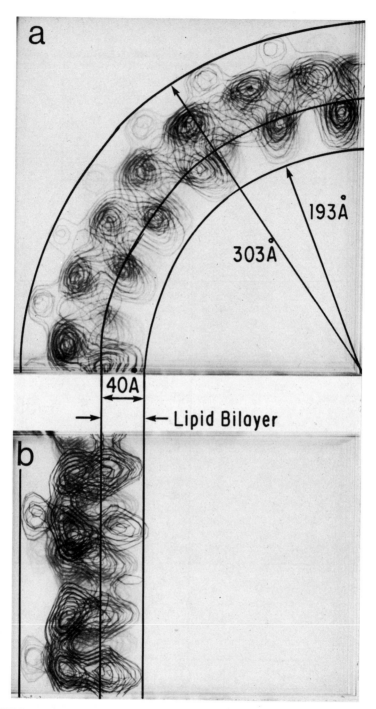

FIGURE 5. Three-dimensional contour maps corresponding to (**a**) a view down the helical axis and (**b**) a view perpendicular to the helical axis showing a vertical slice through the wall of the tubule. At the contour level chosen, 100% of the predicted volume of the Ca^{2+}-ATPase dimer is contained in the region between 193 and 303 Å radius from the tubule axis. The position of the lipid bilayer (193 to 233 Å), although not resolved, is suggested by comparison with the multiple tilt view three-dimensional reconstruction of negatively stained, flattened tubules, which primarily showed features of the molecule that are positioned in the aqueous environment (233 to 303 Å).

shows both the cytoplasmic and intramembranous regions of the protein (FIGURES 5 and 6). The overall thickness of the membrane is in reasonable agreement with values derived from other techniques.[26] The most striking feature of the cytoplasmic surface of the reconstruction is the deep helical groove between dimer chains, while on the luminal surface, a deep groove extends down the center of the dimer chains. The cytoplasmic region of each Ca^{2+}-ATPase molecule contains domains that bind

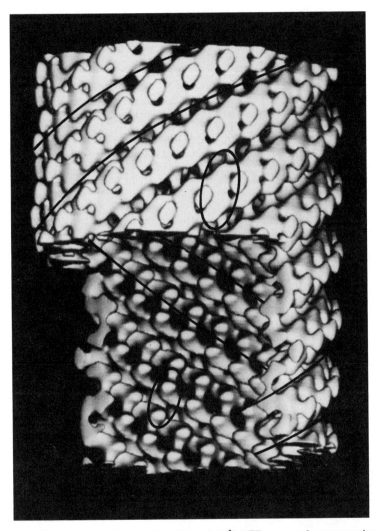

FIGURE 6. Surface view of the vanadate-induced Ca^{2+}-ATPase crystal reconstruction. The contour cutoff envelops 100% of the predicted volume of the molecule. A section of the map has been cut out to reveal the inside surface of the tubule. Views of Ca^{2+}-ATPase dimers have been outlined on both the cytoplasmic and luminal surfaces. The helical path followed by a single dimer chain has been drawn in. Deep helical grooves separate Ca^{2+}-ATPase dimer chains on the cytoplasmic surface, while on the luminal surface a deep helical groove runs down the middle of the dimer chains. Axial dimension of tubule shown is 716 Å.

molecules to form the dimer and that link dimers to form the dimer chains. The intradimer connection forms a bridge centered at a radius of 270 Å from the tubule axis. The long axis of this bridge is oriented parallel with the tubule axis, and is the only point of contact between the two ATPase molecules that make up the dimer. Dimers are connected into chains primarily through the lobe domain located at a radius of 253 Å. Finally, a weak density between dimers within each dimer chain is centered at a radius of 297 Å.

Within the lipid bilayer region of the density map the major feature is a somewhat cylindrical domain about 40 Å in diameter that extends through the bilayer at a slight angle. Thus, the intrabilayer domains of the Ca^{2+}-ATPase molecules within the dimer are farther apart on the luminal side than on the cytoplasmic side. There is no additional contact inside the bilayer between ATPase molecules within the dimer or the dimer chains. Only low density is present in the region running down the middle of the dimer chains.

A single connection at a radius of about 230 Å oriented along steep left-handed helical tracks links dimer chains together to form the surface lattice. This connection is near the cytoplasmic surface of the assumed position of the lipid bilayer. In an earlier reconstruction, in which the layer-line data were interpolated from the Fourier transform, there was an additional interdimer chain connection running approximately circumferentially. At the contour cutoff chosen for this most recent reconstruction, which includes 100% of the predicted protein volume (FIGURES 5 and 6), this circumferential connection is lost, so its significance in the earlier reconstruction is doubtful. The loss of the circumferential density and the reduction in the level of noise peaks occurring in the luminal region of the tubule constitute the major differences between the density map obtained with data interpolated in Fourier space and the more recent result, which utilized interpolation of the image in real space. All other features of the two reconstructions are essentially the same.

COMPARISON WITH RESULTS FROM NEGATIVELY STAINED Ca^{2+}-ATPase CRYSTALS

The three-dimensional structure of the unstained Ca^{2+}-ATPase crystal contains features that are similar to that seen in reconstructions of negatively stained membranes,[9,11] which are expected to yield structural information primarily of the protein regions that are accessible to solvent. All of the three-dimensional reconstructions done to date of negatively stained Ca^{2+}-ATPase crystals show a pronounced connection—the bridge—between ATPase molecules forming the dimer. Our reconstruction of the unstained Ca^{2+}-ATPase tubules shows that the bridge is the single point of interaction between molecules in the dimer. No additional connections have so far been revealed within the lipid bilayer region. Indeed, it appears instead that the transmembrane domains become farther apart with decreasing radius.

Connections between ATPase dimers that help form the dimer chains as well as the two-dimensional surface lattice are seen in the three-dimensional reconstructions of both negatively stained and frozen-hydrated membranes. The most significant connection between dimers to form the dimer chains occurs in the cytoplasm through the lobe domain. A second weaker connection of higher radius occurs in the three-dimensional reconstruction of unstained membranes. The Fourier-Bessel reconstruction of negatively stained scallop sarcoplasmic reticulum crystals, in particular the 53/11 class tubule, showed a similarly placed density that, however, did not form a complete connection between pairs of dimers.[9] The connection in the present case is produced by the enforced twofold rotation axis normal to the tubule axis. This latter class of interdimer connection was not observed in our multiple tilt view three-dimensional reconstruction of negatively stained, flattened tubules.[11]

The greatest differences, not surprisingly, occur in the lipid bilayer region. In three-dimensional reconstructions of specimens preserved in negative stain it is usually assumed that structural features arise from stain exclusion; *i.e.*, aqueous regions are filled with negative stain, and hydrophobic regions of the structure are stain-excluding. With some stains, notably uranyl acetate, positive staining may occur, complicating the interpretation. In biological membranes, the hydrophobic regions of the protein and lipid are both stain-excluding and would therefore be indistinguishable according to this model. In the frozen-hydrated membranes, structural features appear based on their relative electron density (or, more correctly, on their inner potential.[27]) Protein will usually appear denser than the aqueous phase, which in turn will be denser than the hydrocarbon phase.

The present reconstruction is remarkably well-defined within the proposed lipid bilayer region. The organization of dimer chains into the two-dimensional surface lattice is through a single connection oriented along steep left-handed helical tracks. This connection is also the most conserved interdimer connection in the scallop Ca^{2+}-ATPase reconstruction.[9] A second class of interdimer chain interaction, oriented circumferentially, was observed in our earlier multiple tilt view reconstruction of negatively stained membranes.[11] The reconstruction reported here does not show this connection, thereby supporting the conclusion that it may represent a stain-excluding region of lipid.

The three-dimensional reconstruction shown here needs to be confirmed by analysis of further tubules. We conclude, however, that the structures obtained so far from unstained and negatively stained membranes show remarkable consistency. This gives confidence in the reality of the features observed, and demonstrates the complementarity and usefulness of the two techniques.

ACKNOWLEDGMENTS

We thank Dr. Murray Stewart for useful discussions on the utility of real space interpolation and edge apodization and Dr. John Murray for his Fourier-Bessel synthesis program.

REFERENCES

1. BRANDL, C. J., N. M. GREEN, B. KORCZAK & D. H. MACLENNAN. 1986. Cell **44**: 597–607.
2. MARTONOSI, A. & T. J. BEELER. 1983. *In* Handbook of Physiology. Skeletal Muscle. L.D. Peachey & R.H. Adrian, Eds. 417–485. American Physiological Society. Bethesda, MD.
3. FRANZINI-ARMSTRONG, C. & D. G. FERGUSON. 1985. Biophys. J. **48**: 607–615.
4. DEAMER, D. W. & R. J. BASKIN. 1969. J. Cell Biol. **42**: 296–307.
5. JILKA, R. L., A. N. MARTONOSI & T. W. TILLACK. 1975. J. Biol. Chem. **250**: 7511–7524.
6. SCALES, D. & G. INESI. 1976. Biophys. J. **16**: 735–751.
7. TANFORD, C. 1984. Crit. Rev. Biochem. **17**: 123–152.
8. INESI, G. & L. DE MEIS. 1985. *In* The Enzymes of Biological Membranes. A. Martonosi, Ed. Vol. 3: 157–191. Plenum. New York, NY.
9. CASTELLANI, L., P. M. D. HARDWICKE & P. VIBERT. 1985. J. Mol. Biol. **185**: 579–594.
10. TAYLOR, K. A., L. DUX & A. MARTONOSI. 1984. J. Mol. Biol. **174**: 193–204.
11. TAYLOR, K. A., L. DUX & A. MARTONOSI. 1986. J. Mol. Biol. **187**: 417–427.

12. TAYLOR, K. A., L. DUX, S. VARGA, H. P. TING-BEALL & A. MARTONOSI. 1987. Methods Enzymol., in press.
13. DUX, L. & A. MARTONOSI. 1983. J. Biol. Chem. **258:** 2599–2603.
14. DUX, L. & A. MARTONOSI. 1983. J. Biol. Chem. **258:** 11896–11902.
15. BOYD, D. W. & K. KUSTIN. 1985. Adv. Inorg. Biochem. **6:** 311–365.
16. PICK, U. & S. J. D. KARLISH. 1982. J. Biol. Chem. **257:** 6120–6126.
17. DUBOCHET, J., J. LEPAULT, R. FREEMAN, J. A. BERRIMAN & J.-C. HOMO. 1982. J. Microsc. (Oxford) **128:** 219–237.
18. DEROSIER, D. J. & P. B. MOORE. 1970. J. Mol. Biol. **52:** 355–369.
19. AEBI, U., P. R. SMITH, J. DUBOCHET, C. HENRY & E. KELLENBERGER. 1973. J. Supramol. Struct. **1:** 498–522.
20. STEWART, M., R. W. KENSLER & R. J. C. LEVINE. 1981. J. Mol. Biol. **153:** 781–790.
21. BUHLE, E. L., B. E. KNOX, E. SERPERSU & U. AEBI. 1984. J. Ultrastruct. Res. **85:** 186–203.
22. CASTELLANI, L. & P. M. D. HARDWICKE. 1983. J. Cell Biol. **97:** 557–561.
23. FERGUSON, D. G., C. FRANZINI-ARMSTRONG, L. CASTELLANI, P. M. D. HARDWICKE & L. J. KENNEY. 1985. Biophys. J. **48:** 597–605.
24. YANAGIDA, M., E. BOY DE LA TOUR, C. ALFF-STEINBERGER & E. KELLENBERGER. 1970. J. Mol. Biol. **50:** 35–58.
25. BAKER, T. S. & D. L. D. CASPAR. 1984. Ultramicroscopy **13:** 137–152.
26. BLASIE, J. K., L. G. HERBETTE, D. PASCOLINI, V. SKITA, D. H. PIERCE & A. SCARPA. 1985. Biophys. J. **48:** 9–18.
27. STEWART, M. & G. VIGERS. 1986. Nature **319:** 631–636.

DISCUSSION OF THE PAPER

N. UNWIN (*Stanford University School of Medicine, Stanford, California*): Can you say anything at all about the relationship between the amino acid sequence and the structure?

TAYLOR: MacLennan and co-workers (Nature 316:696–700, 1985) have recently proposed, based upon analysis of the amino acid sequence, a model for the Ca^{2+}-ATPase that has 10 transmembrane alpha helices connected to 3 cytoplasmic domains with a penta helix stalk. The transmembrane domain of the reconstruction, which is 40 Å in diameter, would accommodate their 10 alpha helices if they were closely packed. It may be that with further work we can say more, but with only one reconstruction in hand, it is difficult to make any firm statements at this time.

W. CHIU (*University of Arizona, Tucson, Arizona*): Have you compared your model with the one at Brandeis. Are they different?

TAYLOR: Based upon the published reconstruction by Castellani and co-workers, I would say that our reconstruction is pretty similar to theirs. This feature at the outer surface of our reconstruction that I am disregarding as Fourier ripple, is also present in their reconstructions. They may wish to comment.

L. CASTELLANI (*Brandeis University, Waltham, Massachusetts*): Just looking at your ice reconstruction and the negatively stained reconstruction from flattened tubes, I would say that the two structures look extremely similar. A lot of the features are conserved in terms of the cytoplasmic bridge at the interdimers or interdimer ribbon connections, which we found vary slightly between tubes of different diameter. There is one type conserved all the way through, but another one seems to be more friable. Pertaining to what you call ripple, when we analyzed the structure, we did not enforce the twofold symmetry perpendicular to the plane of the membrane. Instead we analyzed tubes of different diameters and therefore different helical parameters and checked that in fact the structure would look similar in different tubes. We haven't really made a final decision what that is. It could very well be a ripple. We just don't know, and we don't feel that we can say one way or another.

TAYLOR: Yes, we have done a symmetry operation that makes this ripple feature appear as if successive dimers within the dimer chains contribute density to it. In your case, the twofold was not enforced, so this feature appears associated with only one ATPase molecule within the dimer.

CASTELLANI: No, a twofold would not apply.

TAYLOR: Also, in your case you were able to calculate averaged reconstructions for two different classes of helical particle. Frankly, we have only found one tubule within our population of images with this helical structure. So I figured we needed all the help we could get to reduce the noise level in the reconstructions, and enforcing the twofold helps in that regard.

The Density and Disposition of Ca-ATPase in *In Situ* and Isolated Sarcoplasmic Reticulum[a]

CLARA FRANZINI-ARMSTRONG,[b]
DONALD G. FERGUSON,[c]
LORIANA CASTELLANI,[d] AND LINDA KENNEY[e]

Departments of [b]Anatomy, [c]Biology, and [e]Physiology
University of Pennsylvania
Philadelphia, Pennsylvania 19104
and
[d]Rosenstiel Basic Medical Sciences Research Center
Brandeis University
Waltham, Massachusetts 02254

The sarcoplasmic reticulum (SR) membrane contains a high concentration of a Ca-ATPase that pumps calcium ions during muscle relaxation. The calcium pump protein is an intrinsic membrane protein with a molecular weight of 110,000 that is inserted asymmetrically in the membrane. Several segments of the molecule are exposed on the cytoplasmic surface of the membrane[1,10] together forming projections 5 nm tall called heads, which are accessible to trypsin digestion and are visible by negative staining and shadowing. (For a recent review see REFERENCE 11.)

Early morphological studies on the distribution of ATPase in the SR of muscles from vertebrates indicated that the ATPase may form small oligomers within the membrane. When the SR membrane is freeze-fractured the density of intramembranous particles is three or four times less than that of ATPase molecules predicted from biochemical data and also less than that of the projections on the cytoplasmic surface of the SR seen by negative staining and deep-etching. However, despite its high density, the ATPase does not form orderly arrays in either the intact or isolated SR, so that for many years the size of the intramembranous molecular domain of the ATPase and the precise arrangement of the molecule in the plane of the membrane could not be determined.

Recently, orderly aggregates of ATPase were found in vanadate-treated SR from rabbit[2] and native SR from striated scallop muscle.[5] Filtered images of electron micrographs from negatively stained vesicles[4,5,19] establish a dimeric basis for ATPase polymers and allow one to calculate the surface area of the membrane occupied by an individual polypeptide. The structural details of the cytoplasmic portion of these molecules in three-dimensional reconstructions from images of negatively stained

[a]Supported by National Institutes of Health Grants HL-15835-12 to the Pennsylvania Muscle Institute and AM-17346-13 to Carolyn Cohen, National Science Foundation (NSF) Grant DMB-85-02233 to Carolyn Cohen and Peter Vibert, Muscular Dystrophy Association Grants to the Henry M. Watts Research Center and to Carolyn Cohen, and NSF Grant PCM-8400140 for instrumentation.

[c]Present address: Department of Physiology, University of Cincinnati, Cincinnati, Ohio 45267.

vesicles are quite similar.[6,20] However, scallop SR in contrast to that of vertebrate muscles does show a regular arrangement of ATPase without exposure to either vanadate or phosphate.[5]

Four questions arise from an examination of *in situ* SR from various vertebrate muscles, native and vanadate-polymerized isolated SR vesicles from rabbit muscle, and *in situ* and isolated SR from scallop muscle. 1) What is the density of ATPase in native SR from vertebrate muscles? 2) Is there any evidence for the formation of oligomers in the native and isolated but functional SR from vertebrate muscles? 3) Is the scallop SR polymerized in its native state, or is the observed polymerization the result of the isolation? And 4) Is the scallop SR polymerized when the calcium pump is turning over?

Freeze-fracture and metal shadowing are the only methods that show the disposition of ATPase within the membrane of the intact muscle. For the isolated vesicles negative staining has the advantage of higher resolution. Nevertheless it cannot reveal the disposition of unpolymerized molecules, since no algorithm is available for separating the superimposed images from the two sides of the vesicles. Therefore, metal shadowing techniques must be used with them as well.

Initially we used polymerized vesicles from rabbit and scallop to establish the resolution of our shadowing technique.[8] We polymerized ATPase in a light microsomal fraction from rabbit muscle in the presence of vanadate,[2] and isolated scallop SR from muscles skinned with saponin and gently homogenized.[5] Then we applied a drop of vesicles suspension to the surface of freshly cleaved mica, and, after washing with 100 mM ammonium acetate, fixed the vesicles with 1–2% uranyl acetate, rinsing with either water or 30% methanol. Next, we froze a very thin layer of final solution in liquid nitrogen and then freeze-dried the vesicles and rotary shadowed them with platinum at 25°.

FIGURES 1a and 1b show the surface disposition of ATPase in the vanadate-polymerized rabbit SR and the native scallop SR. In the former polymerization has occurred starting from several centers, and thus groups of polymers have different orientations; in the latter the vesicle has a tubular shape, and its surface is covered by spirally arranged rows of ATPase. The stereomicrograph (FIGURE 1a) shows that the spherical rabbit vesicle is flattened and partly distorted when laid on the mica. The central region of the larger vesicles, however, is not distorted, and optical diffractions of images from such areas have reflections that index on a slightly skewed lattice of (7×11) nm^{-1}. Patterns from the best areas manifest reflections up to the third and fourth order of the 11-nm spacing, indicating that the replicas have sufficient resolution to identify individual ATPase molecules.[8] Indeed, the elongated morphological unit within each unit cell is composed of two distinct subunits (FIGURES 1a, 1b, 2 inset), which correspond to the two ATPase molecules forming the dimeric arrangement seen in filtered images from negatively stained vesicles. In FIGURE 1b individual dimers are more clearly resolved where the shadow is lighter at the bottom of the image.

We made a final check of the suitability of the shadowing technique for the study of ATPase disposition by freeze-drying and shadowing reconstituted vesicles with variable lipid-to-ATPase ratios provided by Dr. S. Fleischer. In these vesicles we found that at lipid-to-protein ratios comparable to those in the native SR the entire surface is randomly covered with dots. In vesicles with a higher lipid-to-protein ratio some of the surface is covered with dots while the shadow in other areas has a much finer texture and is at a lower level. We conclude that the ATPase has a tendency to aggregate at a high density and to leave protein-free patches of membrane, which can be seen in freeze-dried vesicles as well as freeze-fractured membranes (see below).

Thus freeze-drying and shadowing are appropriate methods for visualizing individ-

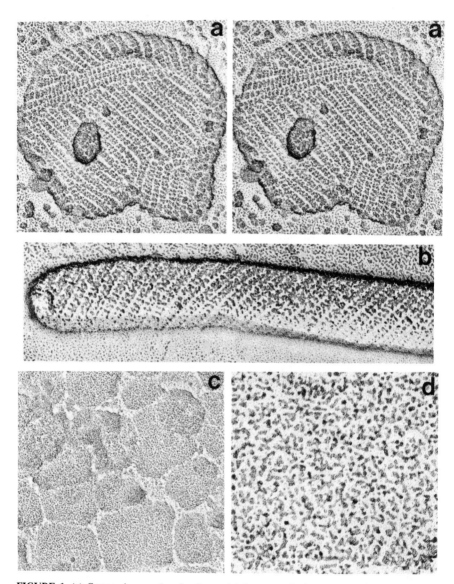

FIGURE 1. (a) Stereomicrographs of a freeze-dried, rotary shadowed vesicle from vanadate-treated rabbit SR. The long rows are formed by the polymerized ATPase, each small segment of which contains two ATPase molecules. See FIGURE 5 for details. 198,000X. (b) SR tube gently isolated from scallop SR, freeze-dried, and rotary shadowed. At the bottom of the image, where, the shadow is lighter, the individual components of the ATPase dimers are resolved. 285,000X. (c) and (d) Freeze-dried, rotary shadowed vesicles from a standard light SR fraction from rabbit muscle. Dots represent individual ATPase molecules. Shadow is approximately as light as the lower half of (b). (c) 120,000X; (d) 500,000X. Note that FIGURE 1(c) and FIGURE 3 are at similar magnifications, but the texture of the split membrane is much coarser than that of the cytoplasmic surface.

ual ATPase molecules and can be used to study their distribution in native vesicles from vertebrate SR, where there is no orderly arrangement. FIGURES 1c and 1d are low and high magnification images of vesicles from a light SR fraction from rabbit muscle. The surface is uniformly covered with dots, each presumably representing an ATPase molecule. Counting the dots gives a density of 31–34,000/μm.[2,9] This estimate is valid, for when rabbit SR vesicles are polymerized in vanadate, the entire surface of most vesicles is covered with a complete array of polymers. From the sizes of the unit cells given in the literature, assuming that each contains an ATPase dimer, one can calculate a density that is consistent with those obtained from direct counts of nonpolymerized vesicles.

The surface disposition of ATPase molecules of isolated SR is not totally random. Dimeric, tetrameric, and occasionally higher-order arrangements can be recognized

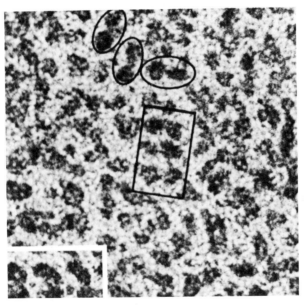

FIGURE 2. Pattern of ATPase on the surface of native and vanadate-treated (*inset*) rabbit SR. Dimers and larger aggregates are *outlined*. 1,400,000X.

when comparing the disposition of ATPase in nontreated vesicles (FIGURE 2) with that in vanadate-treated ones (FIGURE 2, inset).

Freeze-fracture has been used for many years to describe the disposition of ATPase within the SR in the intact muscle.[11] However, a precise relationship between the intramembranous particles observed on the exposed cytoplasmic leaflet of the membrane after fracture and individual ATPase molecules has been difficult to establish. Part of the problem is that the particles, although evenly distributed over the surface, are neither uniform in size nor regularly disposed. We compared the images of the true cytoplasmic surface with those of the fractured cytoplasmic leaflet using similar rotary shadowing conditions. For freeze-fracture we fixed muscles of frogs, glass fish, and rabbits in glutaraldehyde, cryoprotected them in glycerol, froze them in freon, and fractured and shadowed them at 25°. In all three muscles rotary shadowing reveals the reason for the difficulty in assigning a specific molecular identity to the intramembranous particles (FIGURE 3). The particles are of different size and at different heights above the fractured surface as indicated by the variable amounts of platinum associated with them. We also compared groups of ATPase molecules or dots seen on the free cytoplasmic surface with intramembranous particles. The largest particles cover an area equivalent to that occupied by groups of four molecules. Yet the smaller particles must either be coextensive with smaller aggregates of molecules or result from uneven fracturing of groups of them. This result disagrees with a recent calculation[14] that assigns a dimeric content to all of the observed particles. The probable reasons for this disparity are discussed in a recently published paper.[9]

In fast-twitch muscles the entire surface of the cytoplasmic leaflet of the free SR is covered by a carpet of particles. Since the ATPase in the isolated vesicles aggregates at a density of approximately $30,000/\mu m,^2$ we assume that this is also the density of ATPase in the intact SR. That ATPase-free lipid patches are present in the intact SR and are visible with rotary shadowing confirms this (FIGURE 3). We are now comparing the SR in fibers of a slow-twitch and a fast-twitch muscle—the soleus and the white bundle of the *vastus lateralis* in the guinea pig—using freeze-fracture and rotary shadowing. The number of patches varies greatly, the fast muscle having an ATPase density close to the maximum and the slow one much less.

The fast adductor in scallop is a cross-striated muscle composed of very small fibers, each containing a single myofibril.[15,17] The SR is located immediately under the plasmalemma and in thin sections, and like that in vertebrate muscles is composed of longitudinal elements and junctional cisternae. The discovery that it has an ordered disposition of ATPase was unexpected, since phosphate and vanadate, which might have induced crystallization during the isolation procedure, were not present in the

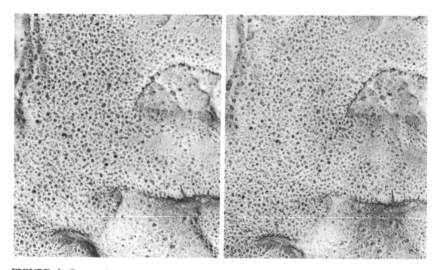

FIGURE 3. Stereomicrographs of freeze-fractured, rotary shadowed fish SR showing the cytoplasmic leaflet. The intramembranous particles vary in size and height. The SR tubule at *center right* has two featureless lipid patches devoid of ATPase. The rest of the membrane is uniformly crowded. 100,000X.

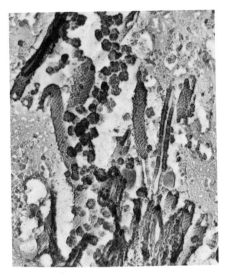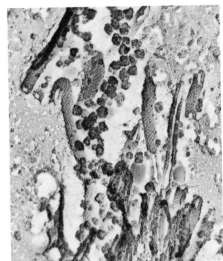

FIGURE 4. Stereomicrographs of deep-etched, rotary shadowed scallop SR fixed after saponin skinning of the muscle fibers. The surface of the long SR tubules is covered by a spiral pattern of ATPase polymers. The large granules are glycogen granules in the cytoplasm of the muscle fiber. 60,000X.

ATP or solutions.[5,6] To define the conditions under which it remains crystalline in the intact muscle, we variously fixed bundles of fibers in glutaraldehyde. We skinned one set of bundles in a solution containing 0.1% saponin, 100 mM NaCl, 8 mM $MgSO_4$, 5 mM EGTA, 5 mM ATP, and 10 mM TES buffer, pH 7.4 for one hour and then fixed them. The other set we fixed immediately after dissection. After fixation we infiltrated the bundles in 30% methanol (which acts as a cryoprotectant), froze them in freon, freeze-fractured them, deep-etched them for fifteen minutes at $-110°$ (nominal) C, and rotary-shadowed them at 25°, which allowed us to see the true cytoplasmic surface of the SR tubules (FIGURE 4). We cryoprotected other bundles in 30% glycerol and shadowed them at 45° without etching or rotation. Images from these showed the fractured interiors of the membranes (FIGURE 5). In saponin-skinned bundles the true cytoplasmic surface is covered by rows of dimeric ATPase (FIGURE 4). The cytoplasmic and luminal leaflets (FIGURE 5) have rows of helically arranged particles and pits, which are characteristic of fractures through SR membrane with polymerized ATPase.[12,16,18] However, in freshly fixed bundles the SR has a disorderly disposition of the ATPase, even though it preserves a tubular shape (not shown).

In order to understand the difference in the ATPase of the skinned and the intact muscles, we investigated the effects of saponin on the SR membrane by making several SR isolations in the solution specified above, though without ATP, both with and without saponin. However, FIGURE 6 shows that vesicles isolated without saponin (a and c) are indistinguishable from those with it (b and d) when examined by negative staining and shadowing. Therefore, the ATPase is polymerized *in situ,* and the only requirement for maintaining that disposition is gentle homogenization. Saponin's lack of effect is to be expected, for though it perforates the surface membranes of many

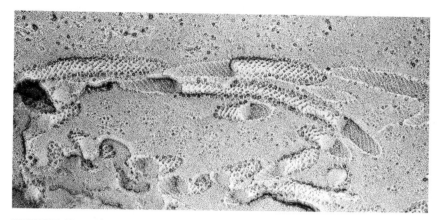

FIGURE 5. Freeze-fracture of scallop SR fixed after saponin skinning. Cytoplasmic and luminal leaflets of the long SR tubules show helical rows of particles and complementary pits characteristic of the polymerized ATPase. 93,000X.

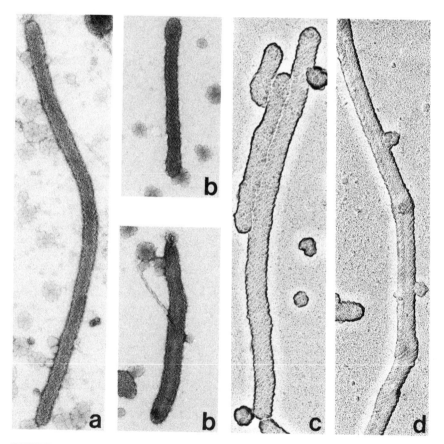

FIGURE 6. Negative staining (**a** and **b**) and rotary shadowing following air drying (**c** and **d**) of SR tubules isolated from scallop muscle without (**a** and **c**) and with (**b** and **d**) exposure to saponin. 73,000X.

cells, it fails to penetrate internal membranes, particularly the SR.[7] The requirement of saponin skinning for preserving the orderly arrangement when the muscle is fixed in glutaraldehyde may therefore be related either to the need to maintain low calcium concentrations or to better fixative penetration.

Since vanadate is not needed to obtain polymerization of the ATPase in scallop SR, we wondered whether endogenous vanadate might be present in these animals. Dr. Henry Shuman from the Pennsylvania Muscle Institute kindly analyzed some of our freshly isolated scallop vesicles with the electron microprobe and found no vanadium ions in them. If vanadium had been associated with the scallop ATPase in a one-to-one molar ratio, it would have been detected. Therefore, polymerization of the scallop ATPase is not dependent on the presence of vanadate.

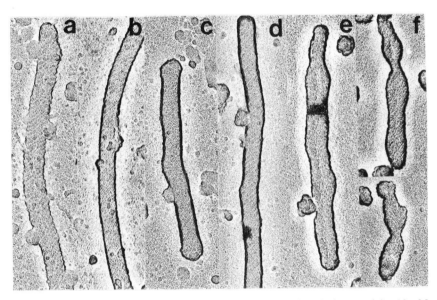

FIGURE 7. (a) Control. Vesicles were laid on mica and washed with a solution containing 10 mM TES, pH 7.4, 100 mM NaCl, 8 mM MgSO₄, 5 mM EGTA. (b,c,d) Vesicles were laid on mica, then exposed for 8 minutes to solutions containing 10 mM TES, pH 7.0, 2.25 mM ATP, 2.6 mM Mg, 0.5 mM EGTA, and 0.25, 0.13, and 0.06 mM Ca respectively. Note the helical arrangement of ATPase. (e) Vesicles were exposed for 20 seconds to a solution containing 0.25 Ca, no EGTA, and (f) were exposed as (e) but rinsed with the high EGTA, no Ca solution. 73,000X.

Vanadate-polymerized rabbit ATPase is sensitive to calcium.[3] To determine the effect of different calcium concentrations on the disposition of the ATPase in scallop SR, we exposed freshly isolated vesicles both while in suspension and after being adhered to the support for shadowing or negative staining to various buffered Ca-EGTA solutions in the presence of ATP. After a given incubation time with the calcium, we either fixed the vesicles in uranyl acetate, washed them in 10% glycerol, air dried and rotary shadowed them at 25°, or simply negatively stained them with uranyl acetate.

We exposed the air-dried and shadowed vesicles to solutions for eight minutes. The solution for the control vesicles contained 5 mM EGTA (FIGURE 7a) and that for the others ratios of Ca to EGTA corresponding to a pCa of 7.2 (FIGURE 7b), 6.8 (FIGURE 7c), and 6.2 (FIGURE 7d). (See the figure legend for solution details.) All tubules display the prominent diagonal striping characteristic of the polymerized ATPase, indicating that calcium concentrations up to a pCa of approximately 6 do not affect the disposition of the ATPase. When calcium is increased to 0.25 mM, however, the shadowed images show no apparent order of the ATPase (FIGURE 7e). We now hope to determine whether the effect of calcium at high concentrations is reversible. Partial recovery of the regular arrangement is seen in a vesicle washed with EGTA solution after exposure to high calcium (FIGURE 7f).

For vesicles incubated with a solution at a pCa of approximately 5 for increasing periods of time some indication of order is still present at exposures of up to 10 minutes

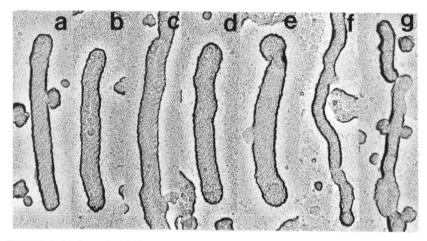

FIGURE 8. (a) Control, as in FIGURE 7. (b to f) Vesicles were exposed to a solution containing 0.5 mM Ca and 0.5 mM EGTA (pCa approximately 5) for 50 seconds, 2 minutes 5 seconds, 5 minutes 20 seconds, 10 minutes, and 20 minutes respectively. Helical rows of ATPase are visible in all but the longest exposures, (f). The vesicle in (g) was exposed to unbuffered calcium (see FIGURE 7f).

(FIGURE 8b–e) but is lost at 20 minutes (FIGURE 8f). Again, after brief exposure to unbuffered 0.5 mM Ca there is no obvious order (FIGURE 8g).

Negatively stained vesicles washed briefly on the grid with a solution containing ATP at a pCa of approximately 5 show a large proportion of stain-filled tubules in comparison to the control and a population of tubules only partly flattened. The stain-filled tubules have a well preserved surface arrangement of dimeric rows of ATPase (not shown). The stain distribution of the partly flattened tubules, however, is unusual, since the rows of units are visible in a large number of vesicles, but the dimeric arrangement of the rows is observed only occasionally (FIGURE 9b–e). That the two methods of visualization are inconsistent may be because in rotary shadowing the two classes of appearances of SR tubules observed in negative staining—stain-filled and flattened—are not clearly distinguished.[5,6] The stain-filled tubule type may be partly collapsed during air drying and therefore corresponds to the shadowed tubules that

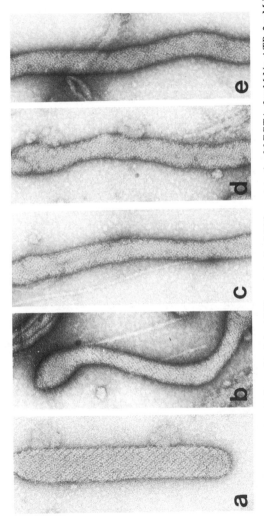

FIGURE 9. Negatively stained SR vesicles. (a) Control in 0.1 mM Na acetate, 3 mM Mg acetate, 1 mM EGTA, 2 mM Na$_2$ATP, 2 mM imidazole pH 7.0; (b to e) Vesicles washed with the same buffer at pCa 5.0 before staining with 2% uranyl acetate. 109,000X.

display dimer ribbons on their surface (FIGURE 8). The partly flattened tubules that show little penetration of stain between ATPase molecules (FIGURE 9) may correspond to vesicles that lack contrast in the shadowed images. We want eventually to determine whether a rearrangement of the surface structure has occurred in these tubules, or limited conformational changes in the ATPase molecules have produced a different stain distribution.

Thus the scallop ATPase maintains a polymerized organization in solutions at pCa of up to 6, though at approximately 5 there is some reorganization of the surface array. The calcium sensitivity of scallop SR is similar to that of vanadate-treated SR vesicles,[3] but with the major difference that scallop ATPase is functional,[5] whereas that of vanadate-treated rabbit vesicles is inhibited.[11] In frog muscle fibers complete reloading of depleted SR occurs in from one to five minutes at 5×10^{-7} M calcium concentration.[13] In our experiments vesicles such as those illustrated in FIGURE 7d should have actively pumping ATPase and be at least partly loaded with calcium, unless scallop SR is significantly different from that of frog. It seems, therefore, that pump activity in scallop SR may not require loss of the intermolecular attachments that are responsible for keeping the ATPase in the dimeric configuration.

ACKNOWLEDGMENT

We thank Dr. Fanny Itshak for collaboration in the isolation of scallop SR used in the calcium sensitivity experiments.

REFERENCES

1. ALLEN, G., B. J. TRINNAMAN, & N. M. GREEN. 1980. Biochem. J. **187:** 591–616.
2. DUX, L. & A. MARTONOSI. 1983. J. Biol. Chem. **258:** 2599–2603.
3. DUX, L., K. A. TAYLOR, H. P. TING-BEALL, & A. MARTONOSI. 1985. J. Biol. Chem. **260:** 11730–11743.
4. BUHLE, E. L., Jr., B. E. KNOX, E. SERPERSU, U. AEBI. 1983. J. Ultrastruct. Res. **85:** 186–203.
5. CASTELLANI, L. & P. M. D. HARDWICKE. 1983. J. Cell Biol. **97:** 557–561.
6. CASTELLANI, L., P. M. D. HARDWICKE & P. VIBERT. 1985. J. Mol. Biol. **185:** 579–594.
7. ENDO, M. & M. IINO. 1980. J. Muscle Res. Cell Motility **1:** 89–100.
8. FERGUSON, D. G., C. FRANZINI-ARMSTRONG, L. CASTELLANI, P. M. D. HARDWICKE & L. J. KENNEY. 1985. Biophys. J. **48:** 597–605.
9. FRANZINI-ARMSTRONG, C. & D. G. FERGUSON. 1985. Biophys. J. **48:** 607–615.
10. MacLENNAN, D. H., C. J. BRANDL, B. KORCZAK & N. M. GREEN. 1985. Nature **316:** 696–700.
11. MARTONOSI, A. N. & K. TAYLOR. 1986. *In* Electron Microscopy of Proteins. J. R. Harris & R. W. Horne, Eds. Academic Press. New York, NY. In press.
12. MAURER, A. & S. FLEISCHER. 1984. J. Bioenerg. Biomembr. **16:** 491–505.
13. NAKAJIMA, Y. & M. ENDO. 1973. Nature New Biol. **246:** 216–218.
14. NAPOLITANO, C. A., P. COOKE, K. SEGELMAN & L. HERBETTE. 1983. Biophys. J. **42:**119–125.
15. NUNZI, G. & C. FRANZINI-ARMSTRONG. 1981. J. Ultrastruct. Res. **76:** 134–148.
16. PERACCHIA, C., L. DUX, & A. MARTONOSI. 1984. J. Muscle Res. Cell Motility. **5:** 431–442.
17. SANGER, J. W. 1971. Z. Zellforsch. Mikrosk. Anat. **118:** 156–161.
18. SCALES, D. & S. HIGHSMITH. 1984. Z. NATURFORSCH. Teil C **39:** 177–179.
19. TAYLOR, K. A., L. DUX & A. MARTONOSI. 1984. J. Mol. Biol. **174:** 193–204.
20. TAYLOR, K. A., L. DUX & A. MARTONOSI. 1986. J. Mol. Biol. **187:** 417–427.

DISCUSSION OF THE PAPER

G. C. RUBEN (*Dartmouth College, Hanover, New Hampshire*): Do you think that the active ATPase is in the dimer form as two molecules of calcium ATPase?

FRANZINI-ARMSTRONG: We cannot say that for sure. All that we can say is that in the presence of a concentration of 10^{-6} moles of calcium, the ATPase remains in the dimeric configuration. On skinned frog fibers, Endo showed quite clearly that with 10^{-6} molar calcium, the SR takes it up very rapidly. In the frog, the ATPase is in the dimeric configuration and should be actively pumping. It's not been done carefully in the scallop under those conditions yet. That's one of the things we have to do.

RUBEN: Were the freeze-etched pictures you showed us rotary shadowed at about -100 degrees?

FRANZINI-ARMSTRONG: Freeze-fractured. Yes, they were rotary shadowed about -110 degrees.

RUBEN: The rotary shadowing at about that temperature builds up platinum carbon nuclei, and it does alter the shapes and sizes of the particles rather dramatically.

FRANZINI-ARMSTRONG: Yes. On the other hand, if one fractures these Ca ATPase vesicles under the same conditions when it's in a regular disposed dimerical arrangement, one really clearly sees that in the freeze-fractured images. This is totally different from the disposition that was shown in the disorderly arrangement of the vertebrate muscle.

R. N. MCBURNEY (*Medical Research Council, Newcastle-upon-Tyne, England*): Although you've isolated in EGTA, etc., can you be sure that your vesicles are empty and that they're not fairly loaded with calcium already?

FRANZINI-ARMSTRONG: The calcium uptake and ATP splitting have been checked in those vesicles, but whether they are empty at the end of isolation procedure I don't think we know.

CASTELLANI: One thing we know for sure is that membranes isolated by this method are leaky, so it would be very unlikely that they're full of calcium. We do get calcium uptake, but we get an ATPase rate that is usually higher than that found in rabbit. This suggests that in fact the membranes are leaky and therefore the pump keeps pumping, because you do not get saturation inside. When the experiments that Dr. Franzini-Armstrong showed are done in the presence of ATP, the general appearance of the tubules suggests that something is accumulating in the terminal part of the tubule.

L. MAKOWSKI (*Columbia University, New York, New York*): I want to comment on the comparison between data from electron microscopy and X ray diffraction. Particularly with frozen-hydrated specimens on the electron microscope, except for the possibility of differences in the scattering factors of atoms to electrons and to X rays, one might expect quite similar results. In the case of microtubules, where Eckhardt showed that there was a very close comparison, this is the sort of thing we might expect. At 30- or 40-Å resolution we're going to have proteins with a particular scattering density that is more or less uniform and solvent with a particular electron density that is more or less uniform. With microtubules we get quite good comparisons between X ray diffraction and electron microscopy. This is crucial, because X ray diffraction will often have a better signal-to-noise ratio and better counting statistics, whereas with electron microscopy you get the phase information.

Where you have a membrane involved as well, it may not be quite so simple, because the ratio of the scattering factors for lipid hydrocarbon solvent and protein are different for electrons and for X rays. The hydrocarbon has a higher density relative to solvent with electrons than with X rays. So we may not always get exact correspondence between the results of X ray diffraction and electron microscopy.

There's one other issue to mention with regard to Dr. Taylor's talk and other talks on membranes today. Usually when you look at a membrane in the electron microscope, there is a difference between the scattering in the lipid hydrocarbon and in lipid polar head groups. However, if you have a flat membrane on your grid, this differential may not be observed, because you have a problem with a missing cone of data. You may not have that problem with a cylindrical membrane. You may, in fact, be able to image density in ice, which may correspond to the lipid. It's important to look at the comparison of negative stain and frozen-hydrated results, as Dr. Taylor was saying.

N. UNWIN (*Stanford University School of Medicine, Stanford, California*): Yes, the head groups are rather smeared out.

Metal Shadowing and Decoration in Electron Microscopy of Biological Macromolecules

WOLFGANG BAUMEISTER, REINHARD
GUCKENBERGER, HARALD ENGELHARDT, AND
CHRISTOPHER L. F. WOODCOCK[a]

Max-Planck-Institut für Biochemie
D-8033 Martinsried, Federal Republic of Germany

This symposium coincides with the fortieth anniversary of the introduction of shadowing, the use of an evaporated heavy metal film to impart contrast in electron microscopy.[1] Since its inception, the basic method has changed little, although improvements in instrumentation and a fuller knowledge of the physics involved have led to a gradual increase in reproducibility and quality.[2-4] Slayter recently presented in these *Annals* an account of the various landmarks in the evolution of shadowing from the perspective of the fibrinogen structure.[5] Two technical advances are particularly significant—the development of the electron gun for evaporating high-melting-point metals and the availability of quartz crystal film thickness monitors. It is perhaps surprising that the use of electron beams to evaporate high-melting-point material was implemented some fifteen years ago,[6] and yet has been widely practiced only relatively recently.

The majority of the thousands of biological applications of shadowing that have been published during the last twenty-five years have been associated with two quite separate methodological developments—the surface spreading technique for nucleic acids (and other negatively charged polymers)[7] and techniques of freeze-fracturing and freeze-etching.[8,9] Electron microscopy for nucleic acid analysis is now being replaced in many instances by more convenient biochemical techniques, but for many years it has been the most reliable way of sizing DNA, detecting A-T rich regions, mapping heteroduplex molecules, and detecting spliced segments in mRNA.[10] Its requirements for shadow quality are relatively undemanding. In addition to stabilizing the molecules at an air-water interface, the envelope of cytochrome c at least doubles the effective diameter of DNA, so that the fibers are readily visualized at low shadowing angles even when the metal grains are large by high-resolution standards. Thus, although higher-resolution methods for visualizing nucleic acids in their naked state either through shadowing[11] or positive staining[12] have been developed, there has been little incentive to adopt them due to the adequacy of simpler techniques.

The introduction of freeze-fracturing and related methods, for which shadowing is the sole contrasting method, has played a decisive role in cell biology, especially in understanding membrane structure. Prior to the first freeze-fracture results most knowledge of cellular ultrastructure was based on thin sectioning, and there was always the disturbing possibility that the "alchemy" involved in fixation, dehydration, and embedding might introduce gross artifacts. Only when similar membrane systems

---

[a]Permanent address: Zoology Department, University of Massachusetts, Amherst, Massachusetts 01003.

were observed in the fast-frozen state were these worries finally quieted. Maximum resolution is often critically important in freeze-fracture studies, and attempts have been made to exceed the shadow/specimen quality obtainable with conventional instruments. These attempts have mostly employed very low specimen temperature during fracturing and shadowing, requiring ultra-high vacua and elaborate instruments.[13]

SHADOWING VERSUS NEGATIVE STAINING

Other contrasting methods are available for studying many biological components, macromolecules, and supramolecular assemblies. Negative staining is the most widely used alternative, mainly because of its technical facility. However, since shadowing is ideally a simple surface coating, it should provide a more reliable picture of the external morphology of the specimen than negative staining, with which complex and unpredictable interactions between stain and specimens are encountered. Also, as the negative stain dries, the specimen is exposed to rapid changes in ionic strength, which can affect the structure substantially. Moreover, unlike negative stains, the metal film in shadowed preparations is quite resistant to damage from the electron beam. Nevertheless, the advantages of metal shadowing do not always give superior results! Some specimens that are virtually impossible to contrast with negative stains are pleasingly imaged by shadowing, while others yield little information with shadowing in comparison with negative staining. Only rough guidelines exist for predicting the behavior of a particular specimen, though fortunately some structures fall into the middle ground, clearly observable both by shadowing and negative staining. Fibrinogen, for example, which probably has the distinction of being the most-shadowed molecule, has been observed in a bewildering variety of forms after shadowing. However, under conditions of complete dehydration, an extended tripartite structure,[5] also observed with negative staining and in unstained freeze-dried samples,[14] is the predominant form.

Fibrinogen and other fibrous proteins such as laminin[15] also illustrate the power of high-resolution shadowing when combined with biochemical information, since the end product is a fairly detailed molecular map. Regular bacterial layers are another example of the convergence of shadowing and negative stain data (FIGURE 1).[16]

Ribosomal subunits are representative of structures that have a distinct surface outline when negatively stained but are relatively featureless ellipsoids when shadowed. Globular proteins also tend to fall into this category, of which the *Arenicola marina* hemoglobin is an example (FIGURE 2). Not only is the morphology of the shadowed particles more variable, but the clear six-fold symmetry seen with negative stain is no longer so obvious. From the results with ribosomes, one might assume that other nucleic-acid protein complexes would behave similarly, but such is not the case. Shadowing has been at least as informative as negative staining for chromatin fibers, especially in studying internucleosomal relationships.[17,18] At the other extreme, shadowing is usually much more successful for fibrous proteins, since these very thin structures may not exclude enough negative stain to be visualized. Even for fibrinogen, a satisfactory negatively stained preparation is difficult to achieve. As the predominantly fibrous components of the basal lamina and extracellular matrix come under closer scrutiny, shadowing will no doubt remain the preferred contrasting method. It is also possible to use shadowing for more than merely the determination of structure; antibodies and other ligands can be used to map specific sites and to locate the positions of flexible domains.[19]

PROBLEMS IN SPECIMEN PREPARATION FOR METAL SHADOWING

Metal shadowing produces quite different results with different specimens.[20] While there is little evidence that shadowing itself, in particular metal condensation or the heat load radiated from the evaporated source, is a major hazard under proper conditions,[3] the steps preparatory to shadowing are crucial. In order to obtain a surface coat that faithfully portrays the surface morphology, the specimen must be uncontaminated with residual solutes. It is best to remove the specimen from its "normal" ionic milieu, transfer it to distilled water, dehydrate it, and place it in vacuo for shadowing.

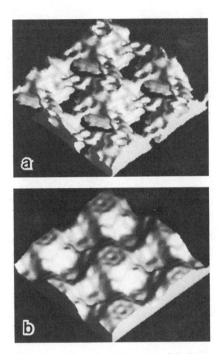

FIGURE 1. Comparison of the three-dimensional structure of the inner surface of the HP1-layer of *Deinococcus radiodurans* as obtained by tilt series reconstruction of a negatively stained preparation (**a**) and surface relief reconstruction of a unidirectionally shadowed preparation (**b**). For details see REFERENCE 17. The two structures appear congruent, though resolution is higher in (**a**) than in (**b**).

However, only robust structures will withstand this without serious changes in morphology. Prefixation with cross-linking agents or the use of volatile buffers may or may not rectify this; moreover, the conditions necessary for driving off "volatile" buffers or other contaminants may interfere with subsequent preparatory steps (see below). Dehydration by simple air drying will be deleterious to most specimens. In addition to (systematic) flattening due to the surface tension forces that act upon the specimen as the water-air interface passes it, unprotected drying produces a loss in uniformity amongst the individual molecules (FIGURE 3). This has been analyzed in

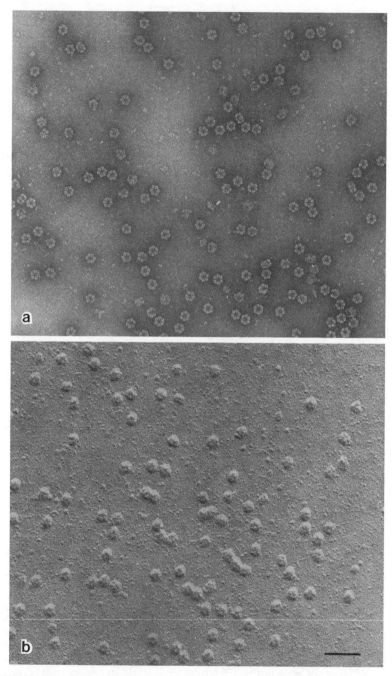

FIGURE 2. Hemoglobin molecules from *Arenicola marina* after negative staining with neutral sodium phosphotungstate (**a**) and shadowing with tantalum-tungsten (**b**). The *scale* is 100 nm. Despite every precaution having been taken to minimize specimen damage (gentle freeze-drying at 188 K and low-temperature (15 K) shadowing in ultrahigh vacuum) the shadowed particles show greater heterogeneity and a less distinct outline than after negative staining.

some detail for a two-dimensional protein crystal.[21] With single molecules the situation is even worse, since the orientation dependence of dehydration creates a disturbing amount of structural heterogeneity. Fortunately, metal shadowing, as opposed to negative staining, can be combined with freeze-drying in a straightforward manner. Slow freeze-drying[22] eliminates the worst effects of dehydration but leaves behind a rather labile specimen.[23,24] Highly hydrated biopolymers are especially prone to postdrying collapse. One could envisage, however, optimal freeze-drying conditions in which instead of complete dehydration, the specimen retained a minimum hydration level. In addition to the internal tightly bound water[25] a monolayer of 0.2—0.4 g water per g protein covering the surface is considered essential for a functional conformation.[26,27] Also, for cryomicroscopy of unstained molecules this is an alternative to the

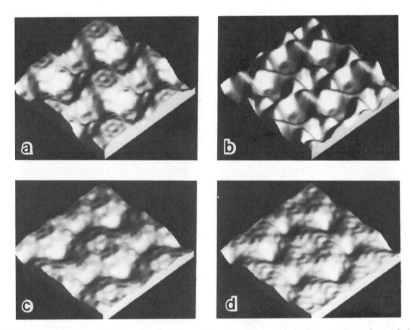

FIGURE 3. Surface relief reconstructions of the inner (**a, c**) and outer (**b, d**) surface of the HP1-layer of *Deinococcus radiodurans,* comparing the freeze-dried (**a, b**) and air-dried (**c, d**) situation. For experimental details see REFERENCE 21. While the inner surface shows only a moderate degree of flattening and appears slightly blurred upon simple air-drying, the outer surface is more prone to distortion; the characteristic pronounced relief collapses, becoming almost featureless.

examination of fully ice-embedded specimens. Though finding favorable conditions of (de)hydration is difficult, it may well be fruitful for many specimens.

A variation of the preshadowing preparation introduced by Tyler and Branton[28] has enabled many fragile particulate specimens to be shadowed successfully. The sample in its normal milieu is mixed with glycerol, and the mixture is sprayed on a substrate of mica or carbon film before being placed in vacuo for shadowing. During the drying of the microdroplets the liquid interface contracts, sweeping salts to the center and leaving clean, apparently well-preserved particles in its wake. It is not known to what extent glycerol substituting for water is retained by the molecules under normal vacuum conditions. The success of the procedure, which involves more unknown parameters than one would wish, perhaps underscores our present ignorance of these matters.

Another poorly understood aspect of specimen preparation for both shadowing and negative staining is the initial step, in which the particulate specimen is attached to a support film under conditions that can, and in many instances do, lead to surface denaturation. Since shadowing gives information primarily on the portion of the specimen away from the support film, the results with it are less distorted by surface denaturation than with negative staining. On the other hand, negative stains and other admixtures such as glucose provide some protection against surface forces that may continue to perturb the freeze-dried specimen.

METAL SHADOWING AND DECORATION

In everyday microscopy shadowing and decoration effects intermingle with each other. Since the information conferred by the two mechanisms is of a fundamentally different nature, this may cause serious problems in interpretation. Thus there is an incentive to strive for a clear experimental separation. As discussed in the last section, an *a posteriori* separation of shadowing and decoration can be achieved at least for unidirectionally shadowed periodic specimens.

Ideal shadowing is easy to define but difficult to attain in practice. Upon striking their target the atoms of the evaporated material are immobilized instantaneously, *i.e.* they adhere to the surface of the specimen where they land. Thereby a continuous film of uniform thickness (modulated only by statistical fluctuations) is deposited. If it is viewed from the direction of the source of the evaporation no contrast is observed, but if the direction of evaporation and viewing do not coincide, contrast related strictly to the surface topography appears (see the next paragraph). In reality, however, the process of thin film formation is a much more complex series of events, which includes thermal accomodation, surface diffusion, physicochemical interactions with the target surface, nucleation, crystal growth, and coalescence. For a detailed discussion of these physical processes, the reader should consult pertinent reviews.[29-31] The selective interaction of mobile material with specimen sites of specific physicochemical properties (high affinity sites) causes decoration, which occurs predominantly during the nucleation phase but continues during later stages of film growth. With decoration, therefore, final distribution of the deposited material reflects the pattern of locations with different affinities rather than relief geometry. However, decoration is not entirely governed by the intrinsic properties of the specimen and the evaporated material, but also by other factors including the presence of ions or multiatomic clusters in the vapor, the deposition rate, the temperature of the specimen, and vacuum conditions. It is the interplay of all these factors that determines the final appearance of the metal film (FIGURE 4). Their analysis and experimental control, painstaking though it is, can be expected to lead to either pure decoration or ideal shadowing.

For ideal shadowing one must arrest the impinging evaporated material by strategies such as impurity stabilization, co-deposit quenching, or low specimen temperature.[30] The latter approach has recently been taken by Gross and his collaborators[32] using equipment suitable for shadowing at a very low temperature (about 15 K) with an ultrahigh vacuum (p 10^{-9} mbar). With platinum-carbon—the preferred shadowing material in freeze-etching—improvements in resolution and interpretability of the micrographs have been made as a result of a reduced grain size. With tantalum-tungsten, which is still the best choice for high-resolution shadowing, effects of temperature prove to be much less conspicuous. Amongst the alternative

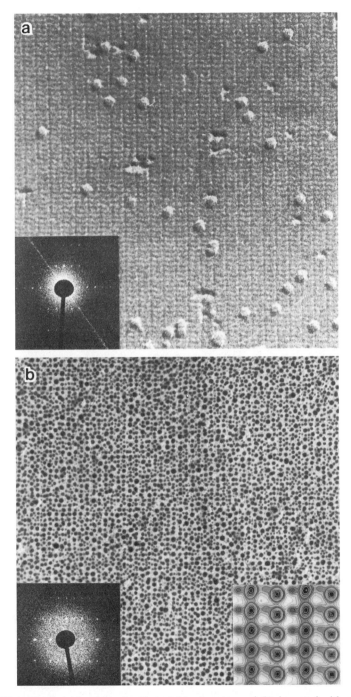

FIGURE 4. Catalase (**a**) shadowed with tantalum-tungsten and (**b**) decorated with gold. For experimental details see REFERENCE 43. While shadowing with tantalum-tungsten reveals the surface topography of the crystal, the gold decorated catalase is dominated by a coarse, grainy texture. Nevertheless the optical diffractogram and the average (*insets*) reveal a surprising amount of intrinsic order, *i.e.* the gold grains have an affinity for specific surface sites.

materials tried, pure carbon was a particularly interesting choice.[33] The apparently amorphous films deposited on periodic objects conveyed, according to the optical diffraction patterns, high-resolution information. Averaging reveals, however, that carbon deposition does not obey the rules for ideal shadowing, because of a partial reflection of the impinging carbon upon the specimen. This edge enhancement may be considered as a decoration effect, though the underlying mechanism is probably different.

Ion and atom beam sputtering have also been tested for their capacity to produce fine-grain films suitable for high-resolution work.[34-37] Sputtered atoms, since they have a higher kinetic energy than thermally evaporated ones, are assumed to penetrate the surface of the specimen, reducing their surface diffusion, and hence diminishing chances of decoration. Thin, fine-grain films can be produced with various metals by sputtering techniques, the deposition of which closely follows the shadowing geometry. Resolution, however, is disappointingly low, and it is suspected that the implantation of the sputtered atoms into the specimen surface is destructive.[38]

Metal decoration is, unwittingly or not, common practice in biomolecular microscopy, particularly with filiform molecules such as DNA, collagen, and myosin, in which cases it is usually accentuated by self-shadowing of the primordial grains. This is because decoration has the potential to render structural features visible that are beyond the resolving power of true shadowing. In fact, the first intentional use of metal decoration was for the visualization of monoatomic steps on NaCl crystals![39] Another illustrative example of the intentional use of decoration is work with the purple membrane. Differences between the two surfaces (and hence determination of sidedness) can easily be detected by decoration with silver;[40] shadowing, however, becomes quite demanding because of the shallow surface profiles.[41,33] It is often more advantageous to encourage decoration (for example by using lower melting point metal) than to struggle for the best possible shadowing. If one is just looking for simple information such as the length of filamentous structures or binary sidedness, it is often quite adequate to rely on decoration, provided one is aware of the phenomenon and its consequences for interpretation. Decoration may also work in spite of surface impurities. With a bacterial surface layer it has been shown that the mobility of silver clusters allows them to decorate the protein lattice even if it is covered by a carbohydrate coat that totally obscures it when tantalum-tungsten shadowing is employed.[42]

It is tempting to carry interpretation of decoration patterns further, relating the location of individual grains to surface features. However, this requires either more *a priori* information about the specimen than is usually available or a more detailed understanding of decoration mechanisms. Although a modicum of experience has now been accumulated,[43-46] elucidation of the mechanisms must be regarded as an arduous if not impossible task in view of the vast number of parameters involved. Moreover, to make progress, it seems mandatory to improve the quality of decoration, not so much in terms of lateral resolution as in terms of affinity resolution.[43] Increasing the positional fidelity of decoration and suppressing nonspecific binding could allow one to use decoration as a means of positive staining from the vapor phase or to rely on the grains as guiding dots for the alignment of low-dose pictures of molecules. Differential decoration using more than one material is another interesting variant worth exploration.

RECONSTRUCTION OF SURFACE RELIEFS

One advantage of metal shadowing is that three dimensional information is contained in individual micrographs and is thus retrievable at little expense. However, the extraction of this information from the micrographs usually relies on intuitive (visual) perception, which may be adequate at low resolutions and for simple structures but is unsatisfactory for complex ones. Also, rotary shadowing may seem to produce images that are more easily interpreted than those obtained by unidirectional shadowing. However, we believe that this preference is unjustified, since incorrectly interpreted micrographs of rotary shadowed specimens abound in the literature. One typical pitfall is illustrated in REFERENCE 47.

Such problems of interpretation as well as advances in shadowing technology have led to the development of schemes for reconstructing surface profiles from shadowed preparations[48-53] that promise improved resolution.[22] Standardizing and incorporating them into image processing systems should make shadowing a viable and economical alternative (or, in some cases, complement) to three-dimensional reconstruction derived from tilting negatively stained preparations (FIGURE 5).

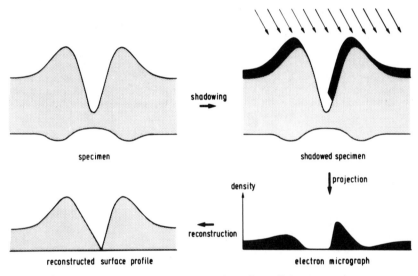

FIGURE 5. Schematic illustration of a surface relief reconstruction.

The following equation describes ideal unidirectional shadowing for thin (relative to resolution) metal films, assuming a coordinate system whose z-axis coincides with the imaging direction:

$$t(\vec{r}) = t_0 = \vec{a} \, \vec{\nabla} \{z(\vec{r})\} \tag{1}$$

where $\vec{r} = \begin{pmatrix} x \\ y \end{pmatrix}$ two component vector in the x,y plane;

$z(\vec{r})$ is the z-coordinate of the relief depending on $\vec{r}$;
$t(\vec{r})$ is the thickness of the metal film measured in the imaging direction;
t_0 is the thickness of the metal film on a plane surface parallel to the x,y plane;

$\vec{\nabla} = \begin{pmatrix} d/dx \\ d/dy \end{pmatrix}$ Nabla operator;

$\vec{a} \cdot \vec{\nabla} \{z(\vec{r})\}$ is the derivative of $z(\vec{r})$ in the direction of shadowing multiplied by $|\vec{a}|$;

$|\vec{a}| = t_0 \cdot \cot(\theta)$;

θ is the angle of shadowing measured to the x,y plane;

the direction of $\vec{a}$ is the direction of shadowing projected normal to the x,y plane.

However, shadowing never is ideal, because thin metal deposits are composed of single grains and are therefore discontinuous when observed at high resolution. Nevertheless, if decoration effects are small, averaging yields a quasi-homogenous film for which Equation 1 is applicable. Equation 1 has been verified by rigorous three-dimensional reconstructions that show that at a 2-nm resolution level Ta/W films are consistent with the simple uniform film model (FIGURE 6). The congruence of tilt series reconstructions of a regular protein layer rotary shadowed with Ta/W and surface relief reconstructions of it shadowed unidirectionally (FIGURE 7) is another indication that this approach is legitimate.[54]

It is frequently suggested that rotary shadowing might be a preferable choice for surface relief reconstruction. Unfortunately, however, this method is not amenable to a straightforward reconstruction scheme. Applying the simple model on which Equation 1 is based to rotary shadowing results in a constant thickness of the metal coat over the specimen apart from zones that in some sectors of the evaporation cone are not reached by the metal beam.[55] The correction of Equation 1 for this effect is a nonlinear procedure that makes a simple mathematical description of rotary shadowing impossible. Theoretical calculations of layer geometries for some rotary shadowed model structures are given in REFERENCE 56. Moreover decoration and genuine (i.e. total) shadows are important contrast mechanisms in rotary shadowing that also interfer with relief reconstruction.

The surface profile can be derived from Equation 1 by integration.[49-53] However, the complete solution requires a term that must be calculated from supplementary information. For a molecule on a plane support the unknown function can be determined, since $z(\vec{r})$ is constant around the molecule. If the environment of the molecule is unknown, as for a unit cell within a periodic object, the missing information must be obtained from a second view in which the object is shadowed from a different direction.

This second view can be taken from another micrograph of an identically prepared specimen, but with a different shadowing direction relative to the object, in which case two different specimens are utilized for a complete surface reconstruction. In order to get two views from the same specimen, it can be tilted or rotated in the electron microscope. If metal is deposited on the specimen at an angle of 90°, it yields no contrast ($\theta = 90°$, therefore $|\vec{a}| = 0$, see Equation 1). However, contrast appears when

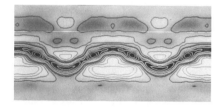

FIGURE 6. Vertical section through a rotary shadowed (tantalum-tungsten) protein layer. This section is derived from a tilt series reconstruction. For experimental details see REFERENCE 54. The metal film appears as a continuous dark band on top of the protein.

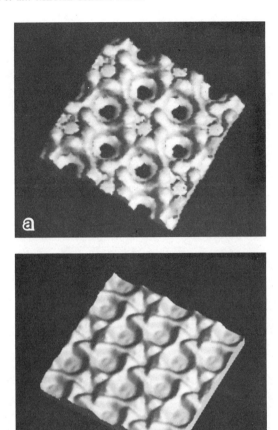

FIGURE 7. Views of the three-dimensional structure of the tantalum-tungsten film deposited on the outer surface of the HP1-layer by rotary shadowing; full tilt series reconstruction (**a**) and a surface relief reconstruction of the same specimen from unidirectional shadowing (**b**). For experimental details see REFERENCE 54. Note the congruence of the two reconstructions obtained by totally different routes.

the specimen is tilted, so that by rotating the tilted specimen in the specimen plane one gets in effect different shadowing directions relative to the objects.[57,58] The complication for its surface reconstruction caused by a tilt of the specimen is discussed below.

For objects with symmetries higher than twofold, the two views necessary for a complete surface reconstruction can be derived from a single specimen (FIGURES 8 to 10). Given n-fold symmetry, rotation through an angle of 360°/n results in an identical view of the structure, which for unidirectionally shadowed surfaces is equivalent to leaving the object at rest. Two views can be combined by solving a linear system of n equations, n denoting the number of pixels per image. This method is proposed for application in scanning electron microscopy and described by Carlson[59] for two views

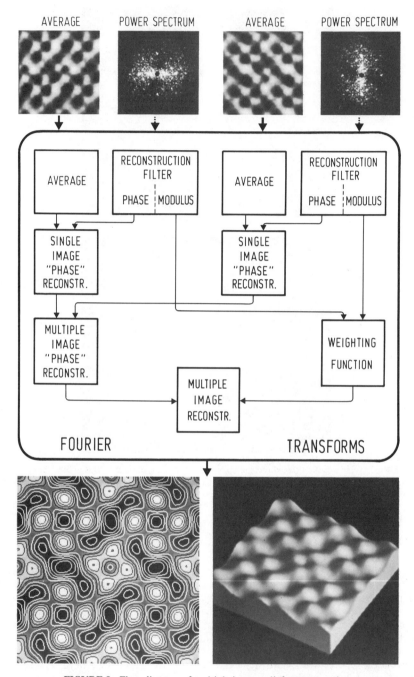

FIGURE 8. Flow-diagram of multiple image relief reconstruction.

with shadowing directions differing by 90°, but it can be adapted to other angles as well.

Such a large system of equations can be avoided by employing a Fourier domain reconstruction scheme. In this way Equation 1 can easily be Fourier transformed[48] to yield;

$$T(\vec{k}) = t_0 \delta(\vec{k}) + Z(\vec{k}) \cdot F(\vec{k}) \tag{2}$$

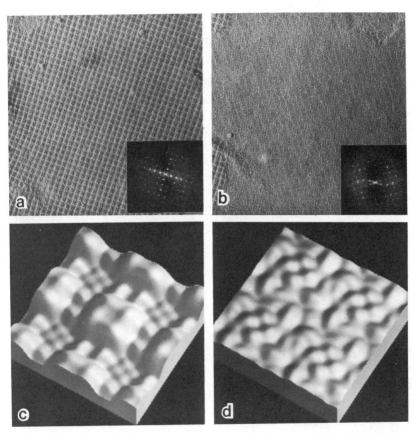

FIGURE 9. T-layer of *Bacillus sphaericus*. (a) and (b) are micrographs of the inner and outer surfaces unidirectionally shadowed with tantalum-tungsten. The micrographs were kindly provided by Dr. J. Lepault. (c) and (d) are the corresponding surface relief reconstructions.

$$F(\vec{k}) = 2\pi i \vec{k} \vec{a} \tag{3}$$

where capital letters denote Fourier transforms and $t_0 \cdot \delta(\vec{k})$ equals t_0 at the reciprocal space origin, vanishing elsewhere. The surface profile can be found simply by dividing Equation 2 by $F(\vec{k})$,[1] which gives a single image reconstruction. However, for $F(\vec{k}) = 0$ there is no solution. The filter function $F(\vec{k})$ is zero on a line perpendicular to the

FIGURE 10. Three-dimensional representation of the cytoplasmic face of the photosynthetic membrane of *Ectothiorhodospira halochloris*. Surface relief reconstruction was performed from micrographs of freeze-dried, unidirectionally shadowed (tantalum-tungsten) samples.

shadowing direction. Thus for ideal shadowing the power spectrum should show very small values in a strip along the zero-line. This missing information therefore precludes reconstructing the correct surface from one view alone. In order to obtain the missing Fourier coefficients additional information is needed, as in the real-space approach. Two views can be combined in reciprocal space for a complete reconstruction using the following simple equation:[60]

$$Z(\vec{k}) = T_1(\vec{k})F_1^*(\vec{k}) + T_2(\vec{k}) \cdot F_2^*(\vec{k})/[F_1(\vec{k})]^2 + [F_2(\vec{k})]^2 + c \qquad (4)$$

where the asterisk denotes the complex conjugate, and c is a small constant that can be optimized with respect to noise suppression. Except for c, Equation 4 is the same as the method described by Carlson[59] and can easily be extended to combine any number of views. This Fourier approach is outlined in more detail by Guckenberger.[60]

For objects with more than twofold symmetry, all symmetrical relationships are exploited for the reconstruction. Twofold symmetry allows one to detect decoration effects in the images as well as contrast caused by underlying biological material. These effects preserve the symmetry of the object, whereas the contrast in ideal shadowing changes twofold symmetry to antisymmetry. A complete reconstruction utilizing any multiple of twofold symmetry automatically suppresses decoration effects and the residual contrast of underlying biological material. This suppression, however, is greatly altered in regions that receive no metal. Such total or true shadows are a major problem, since no information about the surface profile is available for these regions, and Equation 1 is not valid. These regions can, however, be detected if they are not too small compared to resolution. In these cases one can use information derived from symmetrical properties of the object or from other views.

Another difficulty with true shadows lies in determining centers of symmetry in the image. Shadow zones cause the center of symmetry to be displaced in the direction of shadowing. Therefore, such areas should be masked off and excluded prior to the calculation of a center of symmetry. Nevertheless, shadow zones are exploited for scaling exact heights in the reconstruction of a surface. For producing a scaled reconstruction, it is not necessary to know the exact thickness of the metal coat measured in the imaging direction. An arbitrary scaling factor for the layer thickness disappears in Equation 4, because it affects $T(\vec{k})$ as well as $\vec{a}(1)$ in $F(\vec{k})$ (3).

Furthermore, knowledge of exact thickness is not required for the integration of Equation 1. In order to derive the correct relative thickness from the density values of a micrograph, one has to know the density value at areas not reached by metal, and the density value corresponding to t_0 at an area normal to the imaging direction (see Equation 1). The former can be measured in shadow regions, and the latter is furnished by the picture mean, provided that the orientation of the specimen is normal to the imaging direction during microscopy.

The Fourier method described by Guckenberger[60] and used to combine different views of an object is not restricted to periodic specimens, but can be applied to single molecules as well. To overcome the granularity of the metal film, averaging over several molecules is mandatory, as with periodic specimens. However, with single molecules it is often preferable to incorporate averaging into the reconstruction,

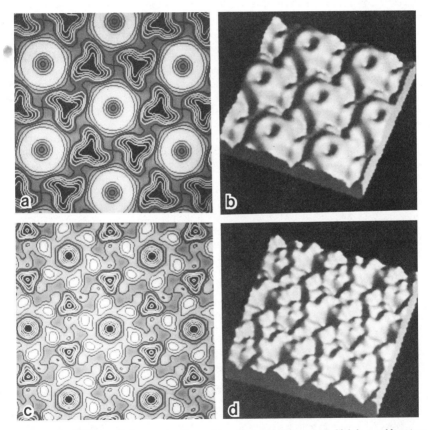

FIGURE 11. Surface relief reconstructions of two bacterial S-layers: *Sulfolobus solfataricus* (**a, b**) and *Sulfolobus acidocaldarius* (**c, d**). Reconstructions were performed with micrographs of deep-etched, unidirectionally shadowed (tantalum-tungsten) preparations; the carbon backing was deposited by rotary evaporation.

because most of the molecules to be processed are oriented differently with respect to the shadowing and imaging directions. In the case of disk-shaped molecules with two preferred orientations on the support film, the orientation in relation to the imaging direction is the same for all molecules, apart from an up-down effect. Once the relation of the azimuthal orientation to the shadowing direction is known (which can be determined from imperfect single image reconstructions), one can process the images separately, again using the extension of Equation 4.

This strategy fails, however, if the molecules assume more than two orientations upon adsorption. The scheme then has to include separate tilting of each molecule, which can only be done in a straightforward manner for complete surface reconstructions. A complete reconstruction, however, can be undertaken for the whole digitized image (comprising many molecules) simultaneously if the support film is plane (see the complete solution of Equation 1 for single molecules). After completing the surface reconstruction all the individual reoriented reconstructions are combined. This approach, however, is more affected by noise.

Freeze-etching, which leaves molecular structures undisturbed by any isolating procedure, furnishes information at moderate resolution on supramolecular topography that is not available from experiments with the isolated component (FIG. 11). Unfortunately, the experimental geometry during the production and microscopy of a replica is less well-defined than with shadowing of isolated molecular structures. In most cases the surface of a given area of a replica is not normal to the imaging direction. For periodic specimens this inclination can be determined from the apparent lattice base vectors. Equation 2 and the whole reconstruction in Fourier space must be corrected for this inclination;[60] however, for small angles of inclination it is sufficient to correct for just the geometric distortion in the projection and then to use Equation 4 as before.

A more serious problem arises from the need to deposit a carbon coat to stabilize the replicas. In addition to phenomena reminiscent of the decoration effect,[33] whose underlying mechanisms are not yet fully understood, this coat often produces significant contrast wherever the direction of carbon evaporation and imaging do not coincide.[43] In order to minimize this undesired and unpredictable contrast, the carbon backing should be made as thin as possible. Rotary deposition of the carbon enables the backing to be kept as low as 5 nm even with deeply-etched surfaces. Nevertheless, rotary deposition creates some contrast even in the ideal situation, where the imaging direction and the rotary axis of evaporation coincide. For objects with twofold, fourfold or sixfold symmetries the symmetrical part of the contrast from the carbon coat can be eliminated in the course of the surface relief reconstruction.

REFERENCES

1. WILLIAMS, R. C. & R. W. G. WYCKOFF. 1946. Applications of metallic shadow casting to microscopy. J. Appl. Phys. **17**: 23.
2. HENDERSON, W. J. & K. GRIFFITHS. 1972. Shadow casting and replication. *In* Principles and Techniques of Electron Microscopy. M. A. Hayat, Ed. Vol. 2: 151–196. Van Nostrand. New York, NY.
3. ABERMANN, R., M. M. SALPETER & L. BACHMANN. 1972. High resolution and shadowing. *Ibid.*: 197–220.
4. WILLISON, J. H. M. & A. J. ROWE. 1980. Replica, shadowing, and freeze-etching techniques. *In* Practical Methods in Electron Microscopy. A. M. Glauert, Ed. Vol. 8. N. Holland. Amsterdam.
5. SLAYTER, H. S. 1983. Electron microscopic studies of fibrinogen structure; historical perspectives and recent experiments. Ann. N.Y. Acad. Sci. **408**: 131–145.

6. BACHMANN, L. 1962. Verdampfung durch Elektronenbeschuss zur hochauflösenden Beschattung elektronenmikroskopischer Präparate. Naturwissenschaften **49**: 34.

7. KLEINSCHMIDT, A. K. & R. K. ZAHN. 1959. Über Desoxyribonucleinsäure-Molekeln in Protein-Mischfilmen. Z. Naturforsch. **14b**: 770.

8. STEERE, R. L. 1957. Electron microscopy of structural detail in frozen biological specimens. J. Biophys. Biochem. Cytol. **3**: 45.

9. MOOR, H., K. MÜHLETHALER, H. WALDNER & A. FREY-WYSSLING. 1961. A new freezing ultramicrotome. J. Biophys. Biochem. Cytol. **10**: 1.

10. CHOW, L. T. & T. R. BROKER. 1981. Mapping RNA: DNA heteroduplexes by electron microscopy. *In* Electron Microscopy in Biology. J. D. Griffin, Ed. Vol 1: 139–188. John Wiley. New York, NY.

11. ABERMANN, R. & M. M. SALPETER. 1974. Visualization of DNA molecules by protein film adsorption and tantalum-tungsten shadowing. J. Histochem. Cytochem. **22**: 845–855.

12. KOLLER, T., M. BEER, M. MÜLLER & K. MÜHLETHALER. 1973. Electron microscopy of selectively stained molecules. *In* Principles and Techniques of Electron Microscopy. M. A. Hayat, Ed. 53–111. Van Nostrand. New York, NY.

13. GROSS, H., E. BAS & H. MOOR. 1978. Freeze-fracturing in ultrahigh vacuum at −196° C. J. Cell Biol. **76**: 712–728.

14. WALL, J., J. HAINFELD, R. H. HASCHEMEYER & M. W. MOESESSON. 1983. Analysis of human fibrinogen by STEM. Ann. N.Y. Acad. Sci. **408**: 164–179.

15. TIMPL, R., J. ENGEL & G. R. MARTIN. 1983. Laminin—a multifunctional protein of basement membranes. Trends in Bioch. Sci. **8**: 207–209.

16. BAUMEISTER, W., M. BARTH, R. HEGERL, R. GUCKENBERGER, M. HAHN & W. O. SAXTON. 1986. Three-dimensional structure of the regular surface layer (HP1 layer) of *Deinococcus radiodurans*. J. Mol. Biol. **187**: 241–253.

17. THOMA, F., T. KOLLER & A. KLUG. 1979. Involvement of histone H_1 in the organization of the nucleus and the salt-dependent structures of chromatin. J. Cell Biol. **83**: 403–427.

18. WOODCOCK, C. L. F., L.-L. Y. FRADO & J. B. RATTNER. 1984. The higher order structure of chromatin: evidence for a helical ribbon arrangement. J. Cell Biol. **99**: 42–52.

19. ENGEL, J., E. ODERMATT, A. ENGEL, J. A. MADRI, H. FURTHMAYR, H. ROHDE & R. TIMPL. 1981. Shapes, domain organizations and flexibility of laminin and fibronectin, two multifunctional proteins of the extracellular matrix. J. Mol. Biol. **150**: 97–120.

20. BUHLE, E. L., U. AEBI & P. R. SMITH. 1985. Correlation of surface topography of metal-shadowed specimens with their negatively stained reconstructions. Ultramicroscopy **16**: 436–450.

21. WILDHABER, I., H. GROSS, A. ENGEL & W. BAUMEISTER. 1985. The effects of air-drying and freeze-drying on the structure of a regular protein layer. Ultramicroscopy **16**: 411–422.

22. WILDHABER, I., H. GROSS & H. MOOR. 1982. The control of freeze-drying with deuterium oxide. J. Ultrastruct. Res. **80**: 367–373.

23. KELLENBERGER, E. & J. KISTLER. 1979. The physics of specimen preparation. *In* Adv. in Structure Research, Unconventional Electron Microscopy for Molecular Structure Determination. W. Hoppe & R. Mason, Eds. Vol. 3: 49–79. Vieweg Verlag. Braunschweig.

24. RACHEL, R., U. JAKUBOWSKI & W. BAUMEISTER. 1986. Electron microscopy of unstained, freeze-dried macromolecular assemblies. J. Microsc. **141**: 179–191.

25. RAO, P. B. & W. P. BRYAN. 1975. Measurement of strongly held water of lysozyme. J. Mol. Biol. **97**: 119–122.

26. BONE, S. & R. PETHIG. 1985. Dielectric studies of protein hydration-induced flexibility. J. Mol. Biol. **181**: 323–326.

27. FINNEY, J. L. 1977. The organization and function of water in protein crystals. Philos. Trans. R. Soc. London, Ser. B. **278**: 3–32.

28. TYLER, J. M. & D. BRANTON. 1980. Rotary shadowing of extended molecules dried from glycerol. J. Ultrastruct. Res. **71**: 95–102.

29. MAISSEL, L. I. & R. GLANG, Eds. 1970. Handbook of Thin Film Technology. McGraw-Hill. New York, NY.

30. CHOPRA, K. L. 1969. Thin Film Phenomena. McGraw-Hill. New York, NY.

31. VENABLES, J. A., G. D. T. SPILLER & M. HANBÜCKEN. 1984. Nucleation and growth of thin films. Rep. Prog. Phys. 47: 399–459.
32. GROSS, H., T. MÜLLER, I. WILDHABER & H. WINKLER. 1985. High resolution metal replication, quantified by image processing of periodic test specimens. Ultramicroscopy 16: 287–304.
33. MÜLLER, T., H. GROSS, H. WINKLER & H. MOOR. 1985. High resolution shadowing with pure carbon. Ultramicroscopy 16: 340–348.
34. ADACHI, K., K. WOJOU, M. KATON & K. KANAYA. 1976. High resolution shadowing for electron microscopy by sputter deposition. Ultramicroscopy 2: 17–29.
35. FRANKS, J., C. S. CLAY & G. W. PEACE. 1980. Ion beam thin film deposition. Scanning Electron Microsc. 87: 155–162.
36. PETERS, K.-R. 1980. Penning sputtering of ultrathin metal films for high resolution electron microscopy. Scanning Electon Microsc. 87: 143–153.
37. COLQUHOUN, W. R. & L. U. CASSIMERIS. 1985. Sputter shadowing improved by a tungsten target. J. Ultrastruct. Res. 91: 138–146.
38. WILDHABER, I., H. GROSS & H. MOOR. 1985. Comparative studies of very thin shadowing films produced by atom beam sputtering and electron beam evaporation. Ultramicroscopy 16: 321–330.
39. BASSET, G. A. 1958. A new technique for decoration of cleavage and slip steps on ionic crystal surfaces. Philos. Mag., Ser. 8: 1042.
40. NEUGEBAUER, D.-C., & H. P. ZINGSHEIM. 1978. The two faces of the purple membrane. Structural differences revealed by metal decoration. J. Mol. Biol. 123: 235–246.
41. STUDER, D., H. MOOR & H. GROSS. 1981. Single bacteriorhodopsin molecules revealed on both surfaces of freeze-dried and heavy metal-decorated purple membranes. J. Cell Biol. 90: 153–159.
42. BAUMEISTER, W., O. KÜBLER & H. P. ZINGSHEIM. 1981. The structure of the cell envelope of Micrococcus radiodurans as revealed by metal shadowing and replication. J. Ultrastruct. Res. 75: 60–71.
43. BACHMANN, L., R. BECKER, G. LEUPOLD, M. BARTH, R. GUCKENBERGER & W. BAUMEISTER. 1985. Decoration and shadowing of freeze-etched catalase crystals. Ultramicroscopy 16: 305–320.
44. WINKLER, H., I. WILDHABER & H. GROSS. 1985. Decoration effects on the surface of a regular layer protein. Ultramicroscopy 16: 331–339.
45. WALZTHÖNY, D., H. M. EPPENBERGER & T. WALLIMANN. 1984. Shadowing of elongated helical molecules (myosin, tropomyosin, collagen, and DNA) yields regular molecule-dependent heavy metal grain patterns. Eur. J. Cell Biol. 35: 216–225.
46. GROSS, H., O. KUEBLER, E. BAS & H. MOOR. 1978. Decoration of specific sites on freeze-fractured membranes. J. Cell Biol. 79: 646–656.
47. NEUGEBAUER, D.-C. & H. P. ZINGSHEIM. 1979. Apparent holes in rotary shadowed proteins: dependence on angle of shadowing and replica thickness. J. Microsc. 117: 313–315.
48. SMITH, P. R. & J. KISTLER. 1977. Surface reliefs computed from electron micrographs of heavy metal shadowed specimens. J. Ultrastruct. Res. 61: 124–133.
49. KRBECEK, R., C. GEBHARDT, H. GRULER & E. SACKMANN. 1979. Three dimensional microscopic profiles of membranes reconstructed from freeze etching electron micrographs. Biochim. Biophys. Acta 554: 1–22.
50. SMITH, P. R. & I. E. IVANOV. 1980. Surface reliefs computed from micrographs of isolated heavy metal shadowed particles. J. Ultrastruct. Res. 71: 25–36.
51. CHALCROFT, J. P. 1980. Three dimensional information from microdensiometric analysis of preshadowed replicas. Mikroskopie 37 (Suppl.): 198–203.
52. GRULER, H. 1981. Quantitative picture analysis of freeze-fracture electron micrographs. Acta Histochem. Suppl. 23: 55–74.
53. RASIGNI, M., G. RASIGNI, J.-P. PALMARI & A. LLEBARIA. 1981. Study of surface roughness using a microdensitometer analysis of electron micrographs of surface replicas; I, Surface profiles. J. Opt. Soc. Am. 71: 1124–1133.
54. WILDHABER, I., R. HEGERL, M. BARTH, H. GROSS & W. BAUMEISTER. 1986. Three dimensional reconstruction of a freeze-dried and metal shadowed bacterial surface layer. Ultramicroscopy. 19: 57–68.

55. SMITH, P. R. 1980. Surface reconstruction from micrographs of metal shadowed specimens. *In* Proc. 7th European Congress on Electron Microscopy. Vol. 2: 698–699.
56. LANDEMANN, L. & J. ROTH. 1985. Theoretical calculation of layer geometry in rotary shadowed model of membrane-associated particles. J. Microsc. **139:** 221–238.
57. CHALCROFT, J. P. 1984. Towards rational terminology and improved techniques in the quantitative analysis of surface replicas. *In* Proc. 8th European Congress on Electron Microscopy. Vol. 2: 1297–1298.
58. GUCKENBERGER, R. 1985. Reconstruction of surface profiles from shadowing data. *In* Proc. 8th European Congress on Electron Microscopy. Vol. 2: 1281–1282.
59. CARLSON, I. C. 1985. Reconstruction of true surface topographies in scanning electron microscopes using backscattered electrons. Scanning **7:** 169–177.
60. GUCKENBERGER, R. 1985. Surface reliefs derived from heavy-metal-shadowed specimens—Fourier space techniques applied to periodic objects. Ultramicroscopy **16:** 357–370.

DISCUSSION OF THE PAPER

G. C. RUBEN (*Dartmouth College, Hanover, New Hampshire*): What conditions did you use for your unidirectional shadowing and rotary shadowing for image reconstruction? And what were the conditions used for decoration?

BAUMEISTER. Most of the unidirectional shadowing was done at very low temperature, around 15 K, and under UHV conditions. The rotary shadowing was not done under the same conditions, because there's no device available yet for rotary shadowing at these temperatures. This was done at liquid nitrogen temperature. That's as far as temperature is concerned. Otherwise preparations were identical; the nominal thickness was five Å.

RUBEN: What about the shadow angles?

BAUMEISTER: It was all 45 degrees.

RUBEN: What temperature did you use for decorating the specimen?

BAUMEISTER: It's temperature dependent. We used a variety of temperatures to see what effect it had; most of those were done at slightly above liquid nitrogen temperature. All the decoration I showed was done with *gold,* and those were done at liquid nitrogen temperature.

RUBEN: Have you done any with platinum or platinum carbon?

BAUMEISTER: There was one example with platinum plus silver. As far as I remember, that was done at room temperature and the two metals were evaporated simultaneously.

A. C. STEVEN (*National Institutes of Health, Bethesda, Maryland*): Is there any information available about what amount of damage might be done to the exposed protein's surface by the incident evaporated metal?

BAUMEISTER: That's not the critical step. I'm talking about thermal evaporation, not sputtering. Sputtering can give very nice films but, because of the higher kinetic energy, you get some penetration into the specimen. So despite the fine, grainy-looking films, resolution is disappointingly low.

With thermal evaporation the main concern is with the thermal load. We are pretty sure that the temperature rise is not above 50° C as measured with a thermocouple in place of the specimen or with triglycerides, which have well defined melting points. You will see immediately upon shadowing whether you have sharp crystal edges or whether you have some indication of melting. From these experiments we would conclude that the temperature rise is fairly small, which, given the fact that the specimen is normally kept at a very low temperature, is quite reassuring.

STEVEN: Have you done any calculations about how much kinetic energy has to be dissipated from the incident metal as opposed to the amount of energy that would be required in kilocalories per mole to affect some significant conformational change at the protein structure?

BAUMEISTER: No, we haven't done that.

A New Method for Three-Dimensional Reconstruction of Single Macromolecules Using Low-Dose Electron Micrographs[a]

JOACHIM FRANK, MICHAEL RADERMACHER,
TERENCE WAGENKNECHT, AND
ADRIANA VERSCHOOR

Wadsworth Center for Laboratories and Research
New York State Department of Health
Albany, New York 12201

INTRODUCTION

Three-dimensional reconstruction[1] continues to be reinvented. Despite recent claims[2] the first "tomographic" three-dimensional reconstructions of *single* macromolecules were done ten years ago.[3] There is, however, a good reason why these methods have not found widespread use: when applied to a single molecule, the reconstruction yields a three-dimensional "image" of the particular stain distribution surrounding the one molecule selected, resulting in approximate information on this particular molecule's shape, but with no indication of how representative it is (see REFERENCE 4). The noise, in other words, is part of the physical structure being imaged. Hence, it cannot be eliminated unless the reconstruction is repeated for a sizeable number of different molecules, and appropriate three-dimensional averages are formed.[5]

The second shortcoming of this approach is that a large radiation dose accumulates during the tilting experiment. The higher the resolution, the larger the number of projections, and hence the larger the electron dose "seen" by a molecule in the course of the entire experiment. Accumulated doses of more than 1,000 e/A^2 have been used in such experiments.[3,5,6] However, even though the use of stain mitigates the need for very low dose (0.5 e/A^2) techniques, the dose should be kept below 20 e/A^2 to maximize the information yield: it has been shown[7-9] that extensive stain migration and recrystallization take place when the accumulated dose reaches this limit, and that many high-resolution features disappear.

In the most recent advance, electron microscopy of ice-embedded (or *frozen-hydrated*) specimens[10,11] has made it possible to visualize protein in its hydrated and undenatured state, making the need for three-dimensional reconstruction methods that employ low-dose techniques even more obvious.

What is needed, then, is a general method of reconstruction that combines projections with well-defined angles of a large number (hundreds or thousands) of identical macromolecules in a way that each molecule receives no more than a small fraction of the total dose (a certain minimum total dose is required for statistical reasons). Attempts have been made previously to collect three-dimensional information from many different, randomly oriented particles;[12] however, the limited accuracy

[a]Supported by National Institutes of Health Grant 1R01 GM 29169 and National Science Foundation Grant PCM 8313045.

of angle estimation limits the resolution and the general applicability of this approach. In the following, we describe a new method of data collection and three-dimensional reconstruction that overcomes these limitations and is applicable to a large number of macromolecules that could not previously be studied by quantitative methods of electron microscopy.

CONICAL-TILT DATA COLLECTION IN A SINGLE-TILT ELECTRON MICROGRAPH

Theoretically, the collection of projections on a full angular cone leads to a more satisfactory reconstruction than the collection of projections generated by tilting around a single axis,[13] because the former technique minimizes the size of the gap in Fourier space due to missing information. Several years ago, Frank and coworkers[14,15] proposed a method of reconstruction that takes advantage of the fact that many biologically interesting macromolecules exhibit strongly preferred orientations on the carbon grid: since they normally lie in random (in-plane) azimuthal orientations, the image of a tilted specimen grid effectively contains a conical tilt series (FIGURE 1). According to the proposal, the data actually used as input to the reconstruction come exclusively from one low-dose micrograph of the tilted grid, while a second micrograph of the same field at zero tilt furnishes the azimuthal angles with the aid of rotational correlation functions.[14,16]

It has taken several years for this method to be successfully implemented.[17] (A detailed description of the technique will be published elsewhere.)[18] Among the reasons for the delay were the inadequacy of computing facilities available at the time, and the lack of algorithms that could deal with the particular angular sampling employed in this method. Since the solution to the latter problem[17] has been crucial for the success of this approach, we will describe it in greater detail.

Since the angular coverage relies on the random occurrence of in-plane orientations of the molecules, the projection angles as a rule are nonequispaced on the angular cone. Previously, there existed no efficient algorithm for three-dimensional reconstruction from nonequispaced projections. Algebraic reconstruction techniques (ART[19]) work with any geometry but require several iterations for convergence. Because the number of projections to be combined is in the range of 500 to 1,000, the time for computation taken by ART or similar iterative real-space techniques would be prohibitive. The classical weighted back-projection method,[20] on the other hand, requires regular tilt intervals for both single axis[20] and conical collection geometry[13] and makes use of an analytical expression for the weighting function in Fourier space. The new algorithm, by contrast, uses a weighting function that is calculated *numerically* by superposition of *sinc* functions in Fourier space. These functions are defined by the actual geometrical locations of the measured projections.

APPLICATION OF THE NEW METHOD TO THE 50S RIBOSOMAL SUBUNIT

The 50S ribosomal subunit of *Escherichia coli* possesses a stable orientation, in which it exhibits the characteristic crown view.[21] In this view (FIGURE 2), three features of the structure are clearly discernible: the L7/L12 stalk, the head protrusion, and the L1 shoulder. The stability of this orientation was deduced by A. Verschoor *et al.*[23] from a study of the crown view with single particle averaging methods: while the

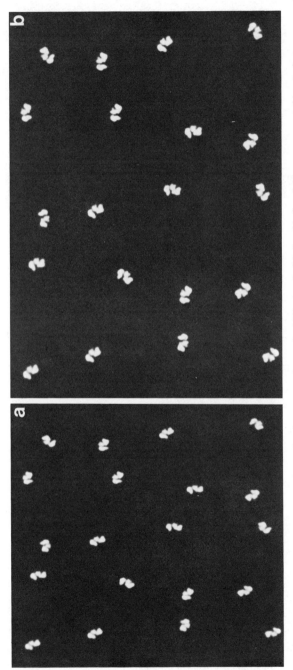

FIGURE 1. Data collection used in the three-dimensional reconstruction method illustrated with the aid of model images generated by placing an averaged lateral view of the 40S ribosomal subunit from HeLa[16] into random orientations. (a) Image field with untilted specimen. (b) The same field with the specimen tilted 50°.

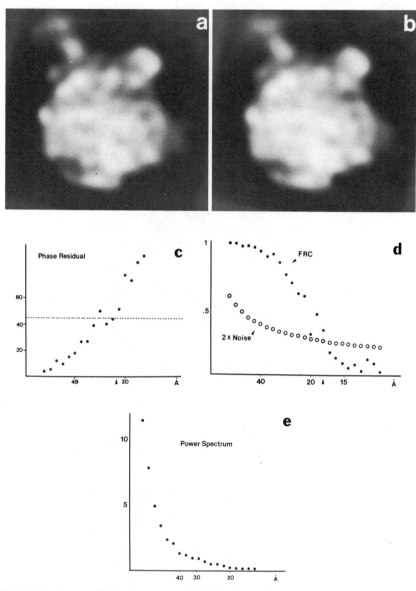

FIGURE 2. Reproducibility of averages of the 50S ribosomal subunit from *E. coli* in its crown view. The two sets of averaged images originated from two different experimentş. Specimen preparation, electron microscopy, and computer processing were done independently. (a) Average of the first set obtained with 245 images (T. Wagenknecht, F. Plastini, J. Frank, work in progress). (b) Average of the second set obtained with 114 images. (c) Differential phase residual comparison of averages (a) and (b) in Fourier space.[16] The 45° limit is reached at a resolution of 23 Å. (d) Fourier ring correlation (FRC) of (a) and (b) in Fourier space. Comparison with the theoretical noise curve, allowing for a factor of 2,[22] gives a resolution of 18 Å. (e) Radial profile of the rotationally averaged power spectrum of the average in (a).

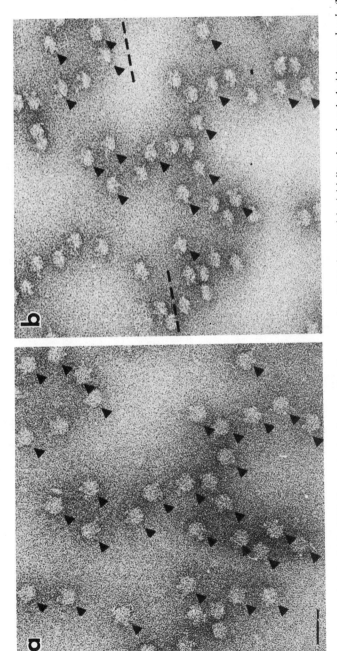

FIGURE 3. A portion of a field of 50S ribosomal subunits presenting the crown view. (**a**) 0° view: particles initially selected marked with *arrow heads*. (**b**) With the specimen grid tilted 50°; particles finally used are marked with *arrow heads*. *Broken lines* indicate the direction of the tilt-axis. A sandwich technique[21] was used, and the electron microscope was operated at 49,000X magnification and 100 kV. The particles in (b) received a total radiation dose of 10 e/A² . *Scale bar* (in (a)) represents 500 Å.

definition of the two side protrusions and the level of peripheral staining vary considerably among the images, the appearance of the main body of the particle remains remarkably stable. Therefore, the 50S ribosomal subunit is ideal for testing the new method of reconstruction.

The reproducibility of the crown view in two independent averages (FIGURE 2a-d) and the radial profile of the power spectrum (FIGURE 2e) suggests that significant detail can be recovered to a resolution of 20 Å, provided that the number of particles is sufficient to fill the 360° range of the cone statistically. The exact resolution value depends on the choice of resolution measure; in our experience, the phase residual criterion[16] gives a figure that is rather conservative, whereas the Fourier ring

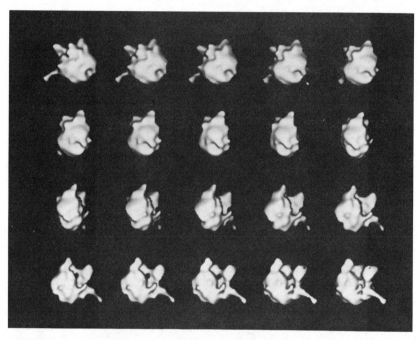

FIGURE 4. The 50S ribosomal subunit at 40-Å resolution reconstructed from 490 projections. The model was obtained by applying a surface representation algorithm[24] to the reconstruction after low-pass filtration, and was rotated 10° between images.

correlation (FRC) as a rule is overly optimistic. At 18-Å resolution obtained by the FRC criterion (FIGURE 2d), the phase residual (FIGURE 2e) reaches 90°, which is the phase residual found for random data.

We are presenting preliminary results of a reconstruction from 490 crown views. Portions of the micrographs of untilted and tilted ($\theta = 50°$) specimen are shown in FIGURE 3. Reconstructions that were done from two subsets of the projections proved to be reproducible to a resolution of 40 Å as determined by the phase residual criterion; the reconstruction from the entire set of projections (FIGURE 4) was therefore limited to this resolution by low-pass filtration.

Obviously, the resolution tests based on two subsets underestimate the resolution present in the complete reconstruction. Other factors that affected the resolution adversely in the reconstruction shown are residual alignment errors, which can be overcome by refinement, and variations of the particle orientation and particle shape, which can be minimized by one of the screening methods to be discussed below.

SCREENING AND CLASSIFICATION OF PROJECTIONS

The molecules whose projections enter the reconstruction exhibit variations in staining, in the precise position of flexible components, and in orientation. These variations have the effect of smearing out the features of fine structural components in the reconstruction. In order to keep this undesired effect to a minimum, molecules with excessive variations need to be eliminated. There are three ways of achieving this screening that we would like to discuss: (i) to sort molecules on the basis of their appearance in the *untilted* view; (ii) to sort molecules on the basis of their appearance in the *tilted* view; and (iii) to eliminate any molecules whose tilted views deviate substantially from the corresponding projections of a first three-dimensional reconstruction computed without screening (self-consistency approach, cf. REFERENCE 25, where self-consistency is used to refine angles, whereas in our case, the azimuthal angles are accurately known). Sorting, employed in the first two approaches, is accomplished by multivariate statistical analysis methods.[26,27]

The first method of screening seems most logical at first glance, because the untilted view, which is collected for each molecule in the course of the reconstruction procedure, is a view common to all molecules and should be the same for all molecules in the absence of noise, deviations in orientation, and structural damage. The disadvantage of this screening method is that it is applied to the molecules at a stage where they have already been irradiated. The groupings found at this stage may not be relevant for judging the quality of molecules in their unirradiated state. However, this method may still be useful for prescreening the data at reduced resolution.

The second method, which is the most promising, employs multivariate statistical analysis of the entire conical tilt series after alignment. This method makes use of the fact that the images of the conical series are linked by a closed similarity pathway, which can be visualized by multivariate statistical analysis.[28] Since the individual low-dose projections contain a substantial amount of noise, the data path zigzags its way around the averaged path, which appears, when projected into one of the major factor maps, like a shadow of a twisted wire frame (FIGURE 5). It should now be possible to define a zone (in two dimensions) or a torus (in three or more dimensions) of allowed occupancy, which demarcates the boundary between "acceptable" and "unacceptable" data.

The third method is based on the assumption that the three-dimensional structure derived by indiscriminately including all experimental projections is a good estimate—albeit resolution-limited—for the true structure. This estimate can, therefore, be projected and compared with the input projections to check for consistency, and input projections with low correlations can be removed in a second reconstruction step.

SOFTWARE AND HARDWARE REQUIREMENTS

The successful reconstruction of the 50S ribosomal subunit enables us to define the requirements that must be met for reconstructing macromolecular assemblies of similar sizes.

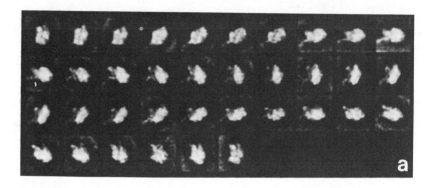

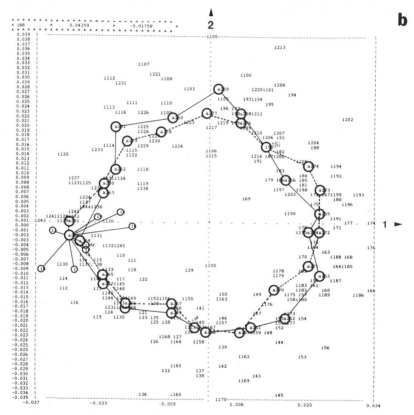

FIGURE 5. Screening of the projections by correspondence analysis. (a) Averages of groups of tilted projections that fall within a 10° range of one another. The number of images in each average is from 10 to 20. The 36 averages span the 360° conical tilt range with a tilt angle of 50°. (b) and (c) Correspondence analysis maps of images in (a). These images are marked with *large circles*. The original, unaveraged projections (*unmarked numbers preceded by "i"*) were inactively coprojected. (b) Factor 1 versus 2.

The implementation of the reconstruction procedure outlined above requires a large, sophisticated software system that makes available in modular form all operations required for preprocessing, alignment, multivariate classification, and three-dimensional back-projection and that supports a hierarchical macro calling structure. In addition, it must facilitate the bookkeeping associated with the storage and transfer of geometrical parameters and logical flags that characterize the role of each projection in the reconstruction process. The SPIDER system,[29] since 1981 supported in a VAX/VMS environment, meets these requirements. Our method uses

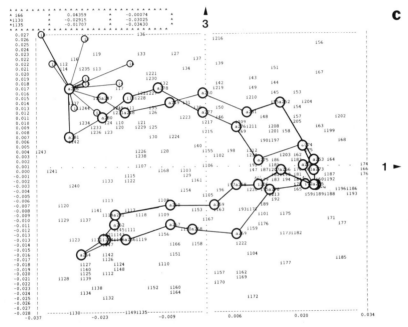

FIGURE 5. *Continued* (c) Factor 1 versus 3. The average images fall on a closed double loop, because projections 180° apart have similar features in the width-to-height ratio and in the size of the characteristic protrusions. Some separation occurs between the two loops along the third factorial axis. The constituent projections for one of the average images are marked with *small circles* to show the degree of scatter. By limiting the points to a toroid around the averaged data path, outliers of the projection series can be identified that should be excluded from the reconstruction.

about 25 high-level procedures written in the SPIDER command language. The bookkeeping is supported by the so-called document files, which are keyed, readable files generally used for saving and unsaving the values of important parameters such as rotation angles, average densities, etc.

As far as the hardware is concerned, at least a machine of the size and speed of the VAX11/780 (Digital Equipment Corp., Maynard, MA) is needed to keep the computation time within reasonable limits. Since numerous Fourier transformations are performed in the course of preprocessing, alignment, and weighting of projections, a large acceleration can be achieved by the use of parallel processing for computing the fast two-dimensional Fourier transform (FFT). We have recently acquired a MINI-

MAP (CSPI, Bellerica, MA) array processor for this purpose, which promises to speed up FFTs in the size range needed (between 64 × 64 and 128 × 128) by an estimated factor between 10 and 50 on a VAX11/750 (Digital Equipment Corp., Maynard, MA) equipped with floating point accelerator.

OUTLOOK

We believe that the new reconstruction method is general enough to be applicable to a large number of macromolecular assemblies for which the prospects of alternative methods of structure research, *e.g.,* by using X ray crystallography or electron microscopy of crystals, are currently remote.

Immediate candidates that meet the requirements of the method (suitable size range and existence of a preferred, stable view) are the 40S ribosomal subunit,[16,30] various dissociation products of arthropod hemocyanin molecules,[26,31,32] and fatty acid synthetase.[3]

Since, on the basis of angular statistics, a resolution of 20 Å is in principle attainable with 500 projections, we are now able to localize, in three dimensions, sites of immuno-label and factor binding, or entire proteins through partial reconstitution experiments. The use of negative stain limits the interpretation of three-dimensional information, in most cases, to the interpretation of the topography of the molecular envelope. Interior structural information will become accessible when the specimens are prepared and maintained in frozen-hydrated form. Preliminary experiments with 50S ribosomal subunits indicate that the signal-to-noise ratio in the micrograph of the tilted specimen is sufficient to accomplish the search for translation and rotation parameters.

REFERENCES

1. RADON, J. 1917. Über die Bestimmung von Funktionen durch ihre Integralwerte längs gewisser Mannigfaltigkeiten. Berichte über die Verhandlungen der königlich sächsischen Gesellschaft der Wissenschaft in Leipzig. Math Phys. Klasse **69:** 262–277.
2. SKOGLUND, U., K. ANDERSON, B. STRANDBERG & B. DANEHOLT. 1986. Three-dimensional structure of a specific pre-messenger RNP particle established by electron microscope tomography. Nature **319:** 560–564.
3. HOPPE, W., J. GASSMANN, N. HUNSMANN, J. SCHRAMM & M. STURM. 1974. Three-dimensional reconstruction of individual negatively stained yeast fatty-acid synthetase molecules from tilt series in the electron microscope. Hoppe-Seyler's Z. Physiol. Chem. **355:** 1483–1487.
4. FRANK, J. 1980a. Three-dimensional reconstruction of single molecules. *In* Three-Dimensional Ultrastructure in Biology. J. N. Turner, Ed. 31–51. Academic Press. New York, NY. (Methods in Cell Biology, Vol. 22).
5. OETTL, H., R. HEGERL & W. HOPPE. 1983. Three-dimensional reconstruction and averaging of 50S ribosomal subunits of *Escherichia coli* from electron micrographs. J. Mol. Biol. **163:** 431–458.
6. KNAUER, V., R. HEGERL & W. HOPPE. 1983. Three-dimensional reconstruction and averaging of 30S ribosomal subunits of *Escherichia coli* from electron micrographs. J. Mol. Biol. **163:** 409–430.
7. WILLIAMS, R. C. & H. W. FISHER. 1970. Electron microscopy of TMV under conditions of minimal beam exposure. J. Mol. Biol. **52:** 121–123.
8. UNWIN, P. N. T. 1974. Electron microscopy of the stacked disk aggregate of tabacco mosaic virus protein. II. The influence of electron irradiation on the stain distribution. J. Mol. Biol. **87:** 657–670.
9. BAKER, T. S., G. D. SOSINSKY, D. L. D. CASPAR, C. GALL & D. A. GOODENOUGH. 1985.

Gap Function Structures. VII. Analysis of connexion images obtained with cationic and anionic negative stains. J. Mol. Biol. **184:** 81–98.

10. TAYLOR, K. A. & R. M. GLAESER. 1974. Electron diffraction of frozen, hydrated protein crystals. Science **186:** 1036–1037.

11. ADRIAN, M., K. DUBOCHET, J. LEPAULT & A. W. MCDOWELL. 1984. Cryo-electron microscopy of viruses. Nature **308:** 32–36.

12. VERSCHOOR, A., J. FRANK, M. RADERMACHER, T. WAGENKNECHT & M. BOUBLIK. 1984. Three-dimensional reconstruction of the 30S ribosomal subunit from randomly oriented particles. J. Mol. Biol. **178:** 677–698.

13. RADERMACHER, M. & W. HOPPE. 1980. Properties of 3D reconstruction from projections by conical tilting compared to single-axis tilting. *In* Proc. 7th European Congress on Electron Microscopy. Vol. 1: 132–133.

14. FRANK, J., W. GOLDFARB, D. EISENBERG & T. S. BAKER. 1978. Reconstruction of glutamine synthetase using computer averaging. Ultramicroscopy **3:** 283–290.

15. FRANK, J. & W. GOLDFARB. 1980. Methods for averaging of single molecules and lattice fragments. *In* Electron Microscopy at Molecular Dimensions. W. Baumeister & W. Vogell, Eds. 261–269, Springer. Berlin.

16. FRANK, J., A. VERSCHOOR & M. BOUBLIK. 1981. Computer averaging of electron micrographs of 40S ribosomal subunits. Science **214:** 1353–1355.

17. RADERMACHER, M., T. WAGENKNECHT, A. VERSCHOOR & J. FRANK. 1986. A new reconstruction scheme applied to the 50S ribosomal subunit of *E. coli*. J. Microsc. **141:** RP1.

18. RADERMACHER, M., T. WAGENKNECHT, A. VERSCHOOR & J. FRANK. Three-dimensional reconstruction of the 50S ribosomal subunit from single-exposure, random conical tilt series. In preparation.

19. GORDON, R., R. BENDER, G. T. HERMAN. 1970. Algebraic reconstruction techniques (ART) for three-dimensional electron microscopy and X-ray photography. J. Theor. Biol. **29:** 471–481.

20. GILBERT, P. F. C. 1972. Iterative methods for the three-dimensional reconstruction of an object from projections. J. Theor. Biol. **36:** 105–111.

21. TISCHENDORF, G. W., H. ZEICHARDT & G. STOEFFLER. 1974. Determination of the location of proteins L14, L17, L18, L19, L22 and L23 on the surface of the 50S ribosomal subunit of *Escherichia coli* by immune electron microscopy. Mol. Gen. Genet. **134:** 187–208.

22. SAXTON, W. D. & W. BAUMEISTER. 1982. The correlation averaging of a regularly arranged bacterial cell envelope protein. J. Microsc. **127:** 127–138.

23. VERSCHOOR, A., J. FRANK & M. BOUBLIK. 1985. Investigation of the 50S ribosomal subunit by electron microscopy and image analysis. J. Ultrastruct. Res. **92:** 180–189.

24. RADERMACHER, M. & J. FRANK. 1984. Representation of three-dimensionally reconstructed objects in electron microscopy by surfaces of equal density. J. Microsc. **136:** 77–85.

25. VAN HEEL, M. 1984. Multivariate statistical classification of noisy images (randomly oriented biological macromolecules). Ultramicroscopy **13:** 165–184.

26. VAN HEEL, M. & J. FRANK. 1981. Use of multivariate statistics in analyzing the images of biological macromolecules. Ultramicroscopy **6:** 187–194.

27. FRANK, J. & M. VAN HEEL. 1982. Correspondence analysis of aligned images of biological particles. J. Mol. Biol. **161:** 134–137.

28. FRANK, J. & M. VAN HEEL. 1982. Averaging techniques and correspondence analysis. *In* Proc. 10th International Congress on Electron Microscopy. Vol. 1: 107–114.

29. FRANK, J., B. SHIMKIN & H. DOWSE. 1981. SPIDER—a modular software system for electron image processing. Ultramicroscopy **6:** 343–358.

30. FRANK, J., A. VERSCHOOR & M. BOUBLIK. 1982. Multivariate statistical analysis of ribosome electron micrographs. L and R lateral views of the 40S subunit from HeLa cells. J. Mol. Biol. **161:** 107–137.

31. LAMY, J., P.-Y. SIZARET, A. VERSCHOOR, R. FELDMANN & J. BONAVENTURA. 1982. Architecture of *Limulus polyphemus* hemocyanin. Biochemistry **21:** 6825–6833.

32. BIJLHOLT, M. M. C., M. G. VAN HEEL & E. F. J. VAN BRUGGEN. 1982. Comparison of 4x6-meric hemocyanins from three different arthropods using computer alignment and correspondence analysis. J. Mol. Biol. **161:** 139–153.

Actin and Flagellar Filaments: Two Helical Structures with Variable Twist

SHLOMO TRACHTENBERG,[a] DAVID STOKES,[b]
ESTHER BULLITT,[b] AND DAVID DeROSIER[a]

[a]*Rosenstiel Basic Medical Sciences Research Center*
and
[b]*Graduate Biophysics Program*
Brandeis University
Waltham, Massachusetts 02254

Helical assemblies of protein subunits are filamentous; that is, they are much longer than they are wide. It is natural for us to concentrate on this aspect of the structure in thinking about their biological functions. Microfilaments, microtubules, and intermediate filaments are correctly thought of as struts or cables, which are used, for example, to push or pull other subcellular structures. A less obvious aspect of helical structures is their twist and its biological role.

What is twist? Helical symmetry can be described by the product of two primitive operations: 1) a translation and 2) a rotation (FIGURE 1). The repeated pairwise application of these two operations generates a helical lattice; that is, it describes the spatial relationship between two equivalent subunits in the helical structure. In the case of actin, the translation, Δz, is about 27 Å and the rotation, $\Delta \phi$, about 167°. Twist is related to the angle of rotation. If a structure changes its twist, the angle of rotation between equivalent points changes.

In nature, there are a variety of structures that show a variation in twist: DNA (right- versus left-handed, for example).[2] There are two extremes of variable twist structures one can imagine. At one extreme, a structure might have a continuously varying twist along its length at any instant of time. At the other extreme, a structure would possess different twists at different times, but at any one time the twist would be essentially constant along the length. Actin is an example of the former, and the bacterial flagellar filament of the latter.

Let us begin by considering the bacterial flagellar filament. The macroscopic shape of this structure is a corkscrew, and it acts as a bacterial propellor connected via the hook structure to a rotary motor[3-6] located in the membrane of the cell: that is, the cell works like a motorboat. The filament, which consists of a helical array of 11 rows of flagellin subunits,[1] is not a simple helical structure in which all subunits are equivalently positioned. Indeed, it has been proposed by Asakura (1970),[7] Wakabayashi and Mitsui (1972),[8] and Calladine (1975)[9] that this structure normally consists of nonequivalent but chemically identical subunits, which can exist in one of two conformational states—let's call them R and L. The corkscrew shape represents strain or deformation resulting from the stress of having within a filament some rows in which all of the subunits are in the R state and the rest of the rows with all subunits in the L state.

In *Salmonella*, the filament is a dynamic structure. In response to a change in the direction of rotation of the motor additional rows switch from the L to the R state causing a change in twist and hand of the filament. Macnab and Ornston (1977)[10] suggested that this dynamic behavior is involved in the cell's ability to change swimming direction by causing the cell to tumble. When bacteria such as *Salmonella*

88

swim, several of their flagellar filaments come together to form a bundle. As Berg and Anderson (1973)[3] have shown, each filament within the bundle can continue to rotate so that the bundle itself, which of course has the same left-handed corkscrew shape of the component filaments, now acts as the propellor. When the motors driving each of the component filaments reverse, the filaments change their twist and hand. The altered filaments can no longer pack in the left-handed bundle and the bundle flies apart, causing the cell to tumble randomly. When the motors return to their original direction of rotation, the bundle reforms as the cell swims away in a new direction. The elegant details of these classic works are outside the bounds of this paper, but the point is that occurrence of the R and L states is thought to be responsible for the corkscrew shape of the filament and that, under physiological conditions, there is a change with twist and hand of the filament.

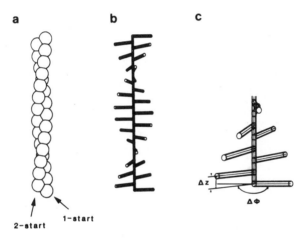

FIGURE 1. Geometry of the actin filament. A filament is schematically drawn in (**a**); actin monomers are represented as spheres. In (**b**), the spherical subunits have been replaced by thin rods to illustrate the angular orientation and translation of the subunits. In (**c**) the first few subunits have been magnified to show two parameters, which together completely describe the helix: the angle ($\Delta\phi$) and the axial rise (Δz) between subunits along the one-start helix.

How does the filament change twist? That is, what are the structure and organization of the subunit in the filament? How do they change in the R to L transition? To answer these questions, we have begun a study of filaments from two different mutants, which is advantageous because the "wild-type" helical filament is generally not the ideal candidate for structural studies. First, it is not straight but corkscrew-shaped, and hence the powerful methods of image reconstruction do not apply in a straightforward way. Second, the "wild-type" filament is a mixture of subunits in both the R and L conformations, so that in the process of subunit averaging, one may be obscuring or averaging out important differences. Kamiya *et al.* (1980)[11] have been able to produce mutant bacteria in which, in one case, all subunits are in the R form, and in another, all are in the L form. Both types of filaments are straight, showing no evidence of the corkscrew shape, and appear to be helically symmetric. The pair are ideally suited for image analysis and will provide images of the subunits in the two different states. The R form has a helical structure with a right-handed twist, while the L form is left-handed.

Mutant bacterial strains of *Salmonella* were kindly given to us by Dr. Kamiya. We prepared filaments from the mutants and examined them by electron microscopy in the frozen-hydrated state. The microscopy was carried out in the laboratory of Dr. Nigel Unwin, and we wish to acknowledge the substantial help of Drs. Paula Flicker and Nigel Unwin with the technique. From several hundred electron microscopic plates, we selected 6 filament images of the L form for averaging and three-dimensional reconstruction. A typical good image is rather disappointing to look at (FIGURE 2, top). It lacks the crisp detail often seen in negatively stained images. The diffraction patterns (FIGURE 3, left side), however, tell a different story. In them, we see that the disappointing images give rise to a set of four very strong, sharp layer lines, at $1/906$ Å^{-1},

SALMONELLA TYPHIMURIUM

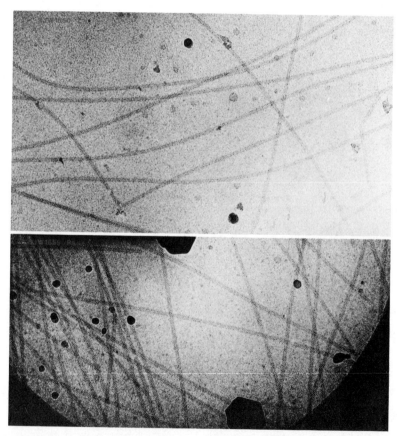

FIGURE 2. Electron micrographs of flagellar filaments of *Salmonella typhimurium*. *Top:* left-handed filaments from the mutant SJW1660. *Bottom:* right-handed filaments from the mutant SJW1655. The filaments, in 50 mM ammonium acetate, were applied to a holey carbon film and fast-frozen. The diameter of the filaments is 240 Å.

$\frac{1}{54}$ Å^{-1}, $\frac{1}{51}$ Å^{-1}, and $\frac{1}{26}$ Å^{-1}. The apparent discrepancy in quality between image and transform is due to the very low contrast and dose and the relatively high defocus of the images of the frozen-hydrated filaments.

The images (FIGURE 2, bottom) of the R filament and transform (FIGURE 3, right) look very similar to those of the L form, except that the $\frac{1}{26}$ Å^{-1} layer line is weaker or absent. It is possible that these filaments are somehow not as well preserved or ordered as the L form, but the strength and sharpness of the other layer lines argue against this. Another alternative is that the subunits have a different shape from those in the L filaments and that the reduction in intensity of the $\frac{1}{26}$ Å^{-1} layer line reflects this.

One of the primary reasons for studying this filament is to determine the nature of the R to L switch. The diffraction patterns present our first clue to the process by telling us how the geometry or helical lattice changes. Calladine (1976)[9] predicted that the R and L transition would include a shortening by 0.8 Å of the distance between subunits along the eleven-start rows and a transverse shift of 8 Å in the intersubunit bonding. The former change would be responsible for the corkscrew shape of the filament and the latter for the twist. The first attempt to study these changes was made by Kamiya *et al.* (1979)[12] using X ray diffraction. Their lattice results indicate a transverse shift of about 7 Å as predicted, but a much smaller shortening than predicted of only 0.2 Å. Our results (FIGURE 4), which are still preliminary, are similar regarding the transverse shift. We find, however, a 0.7-Å ± 0.3-Å shortening, which is more in line with Calladine's prediction. Thus, the measured changes in the helical lattice are in agreement with Calladine's predictions.

Salmonella typhimurium

SJW 1660   SJW 1655

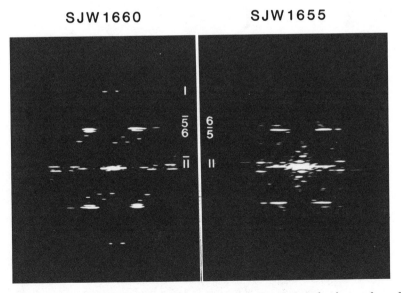

FIGURE 3. Diffraction patterns of images of the filaments in FIGURE 2. *Left side:* transform of a left-handed filament. *Right side:* transform of a right-handed filament. The order of each of the layer lines is indicated *to the side.* The spacing of the order 1 layer line is $\frac{1}{26}$ Å^{-1}.

Let us turn now to examine changes in the subunits themselves. One possibility is that the subunit maintains its intersubunit contacts while altering its conformation in the R to L transition. An alternative is that the conformation is not changed as much as are the points of intersubunit contact. In order to try to visualize the actual changes taking place, we have carried out three-dimensional reconstruction from the electron micrographs. The use of three-dimensional reconstruction allows us to visualize the

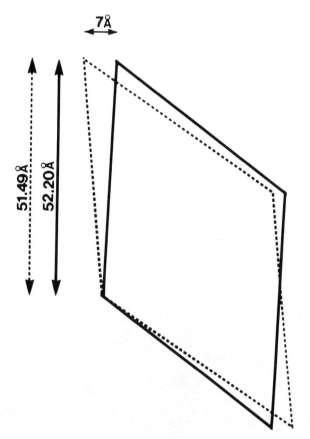

FIGURE 4. The unit cell of a filament. *Dashed line:* right-handed filament. *Solid line:* left-handed filament. The two unit cells, which are extracted from a helical net drawn at a radius of 70 Å, are very similar. The change in twist manifests itself in the 7-Å shift of the upper vertex. Each *arrow* denotes the length of the unit cell edge to the upper vertex.

morphology of the R and L filaments and perhaps to see what morphological changes accompany the transition. The structure of the L filament (FIGURE 5, top) is built around 11 rods of density, which lie along the eleven-start lattice lines. (These same rods are seen in reconstructions of the filament of *Rhizobium lupini.*) Protruding from these rods are knobs that may represent the antigenic part of the subunit. To the inside of the 11 rods are ringlike or tubelike morphological features, which, by their looped nature, may provide strength to the structure.

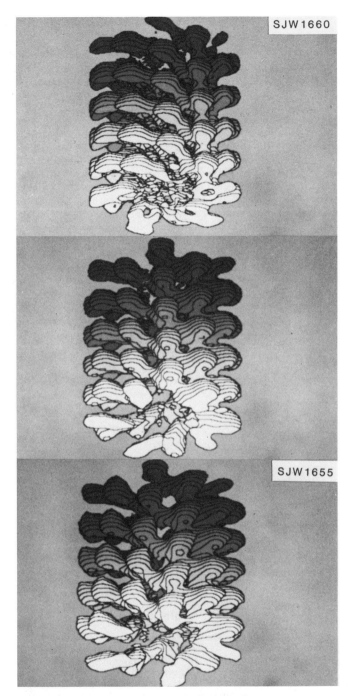

FIGURE 5. Three-dimensional reconstruction of the filaments. *Top:* left-handed filament. *Bottom:* right-handed filament. *Middle:* right-handed filament that has been twisted computationally to left-handed. The right- and left-handed forms are somewhat difficult to compare because of the slightly different geometries. The reconstruction of the right-handed filament was therefore twisted so that its geometry was the same as that of the left-handed one.

The reconstruction of the R filament (FIGURE 5, bottom), obtained very recently, is very similar to that of the L filament. In order to show the similarity, we have changed the twist of the R filament to that of the L filament in the computer. This change corresponds to a shearing deformation of the R form of the subunit and is not intended to or expected to be the same changes as are involved in the actual R to L transition. The point of the manipulation is that it allows us to directly compare the two structures. We aligned the retwisted R (R$_L$) form with the L form in order to compare them. At the outside of the structure it is clear that the knobs have less slew in the R (FIGURE 5, middle) than in the L structure (FIGURE 5, top). There are other subtler changes, but it is too early to decide which are real and which are artifactual.

Let us now turn from the flagellar filament, which has a fixed twist along its length, to actin, which has a twist that appears to vary along its length. How do we know its twist varies along the length, and how do we rule out other forms of disorder? Hanson (1967)[13] was the first to note variability in the helical symmetry, in this case in the pitch of the 2-start helices. It was not until much later that Egelman, Francis, and DeRosier (1982)[14] suggested that this effect was due to variable twist. The measurements of twist were made from Fourier transforms of the electron micrograph. From the transforms, the relative heights of the first (a ≈ 1/360 Å) and sixth (b ≈ 1/59 Å^{-1}) layer lines were determined and the twist S, which is derived from the ratio of the heights, a/b. In particular S = 360°/(2 + a/b).[15] Thus, the positions or heights of the layer lines can be used to measure the amount of twist for a particular stretch of treatment.

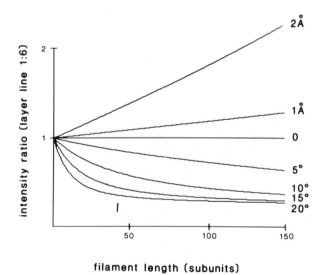

filament length (subunits)

FIGURE 6. Theoretical effects of cumulative angular and of cumulative axial disorder on intensity ratios from Fourier transforms. The intensity ratio of layer line 1: layer line 6 has been plotted as a function of filament length. The intensity ratio in the absence of disorder (Y° corresponding to the y-intercept) has arbitrarily been taken as 1. The *flat curve* corresponds to a constant intensity ratio in the absence of either type of cumulative disorder. *Upward curves* correspond to 1 Å and 2 Å of cumulative axial (as opposed to angular) disorder, and *downward curves* correspond to 5°, 10°, 15°, and 20° of cumulative angular disorder. The asymptotic limit at infinite filament length is 1/4 for angular and 36 for axial disorder.

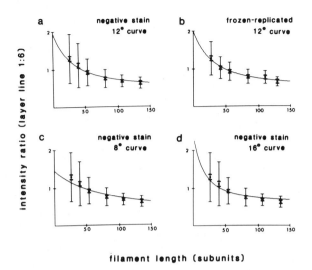

filament length (subunits)

FIGURE 7. Fitting of theoretical curves to the data both from negatively stained and from frozen-replicated actin filaments. The data have been grouped according to filament length; the mean, standard error and standard deviation of the mean are denoted by an x, the *smaller error bars*, and the *larger error bars* respectively. The *smooth lines* drawn through the data represent the theoretical curves that best fit the data. Two parameters characterize each curve: the y-intercept ($Y°$) and the amount of cumulative angular disorder (d_{rms}). The curves drawn in (**a**) and in (**b**) represent the best fit for negatively stained ($Y° = 1.95$, $d_{rms} = 12°$) and for frozen-replicated ($YH° = 2.0$, $d_{rms} = 12°$) filaments. The curves in (**c**) ($Y° = 1.45$, $d_{rms} = 8°$) and in (**d**) ($Y° = 2.4$, $d_{rms} = 16°$) do not fit the data for negatively stained filaments as well as the curve in (a), and represent what we consider to be the outer limits on the value of d_{rms}.

Egelman *et al.* (1982)[14] made a frequency plot for the occurrence of a particular twist. The variance or spread in the histogram provided a measure of the variation in angle between subunits. The estimate was about 10°.

Egelman *et al.* (1982)[14] made another measurement that more dramatically demonstrated the variable twist, since the result was peculiar to this form of (angular) disorder. They showed that with the increasing filament length, the strength of the first layer line grows less rapidly than that of the sixth. If there were no disorder but simply noise, the two intensities would grow at the same rate. If there were axial disorder (variation in height but not angle of the subunit), the effect would be reversed; that is, the first layer line would grow in strength more quickly than the sixth.

We decided to test this theory quantitatively and to examine the amount of variability in twist as a function of specimen preparation. We examined negatively stained filaments, fast-frozen, etched, replicated filaments (kindly made available to us by John Heuser); and we examined frozen-hydrated filaments (kindly supplied by Dr. Paula Flicker). To carry out the study, we obtained two independent measures of variable twist. First, we measured the distribution of twist in a population of filaments. Second, we determined the variation in layer line intensity as a function of filament length. To make the latter measurements, we selected images of straight filaments or regions of them at least 100 subunits (about 2,700 Å) in length. We digitized them and then divided up the regions into segments of 26 to 130 subunits in length. For each of

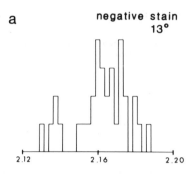

a   negative stain
     13°

2.12   2.16   2.20

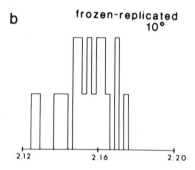

b   frozen-replicated
       10°

2.12   2.16   2.20

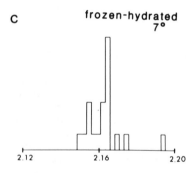

c   frozen-hydrated
        7°

2.12   2.16   2.20

helical twist
(subunits/turn)

FIGURE 8. Distributions of helical twist in the three preparations of actin filaments. The standard deviation of each of these distributions is linearly related to the amount of cumulative angular disorder. Panel (**a**) contains data from 33 negatively stained filaments, with an average length of 102 subunits. This distribution has a standard deviation of 0.0172 corresponding to $d_{rms} = 13°$. Panel (**b**) contains data from 21 frozen-replicated filaments with an average length of 88 subunits; the standard deviation of 0.0133 corresponds to $d_{rms} = 10°$. Panel (**c**) contains data from 15 frozen-hydrated filaments with an average length of 98 subunits; the standard deviation of 0.00961 corresponds to $d_{rms} = 7°$.

these segments we computed a transform and measured the integrated intensities of layer lines 1 and 6 and computed the ratio I_1/I_6. We plotted these ratios against length.

To provide an idea of what to expect, we plotted the theoretical curves for different amounts and kinds of disorder (FIGURE 6). Note that for angular variable twist (disorder) the ratio decreases with length to a value ¼ the intercept. By contrast, the ratio rises by a factor of 36 (at $\ell = \infty$) for axial disorder. Now let us turn to real data. In FIGURE 7 we show results for negatively stained filaments and freeze-etched, replicated filaments. On the former we have drawn in curves for 8°, 12°, and 16° of angular variation. The large error bars represent the standard deviation of the population at that length, and the small error bars are the standard error of the mean.

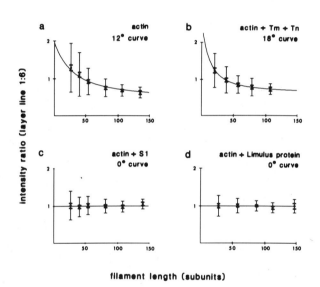

FIGURE 9. The effects of actin-binding proteins on measurements of angular disorder from plots of intensity ratio versus filament length. The data have been plotted as in FIGURE 7, and the curves represent the theoretical relationship between intensity ratio and filament length. The amount of disorder (d_{rms}) and the y-intercept ($Y°$) used to draw each curve are as follows: (a) actin, $d_{rms} = 12°$, $Y° = 1.95$; (b) actin + Tm + Tn, $d_{rms} = 18°$, $Y° = 2.5$; (c) actin + S1, $d_{rms} = 0°$, $Y° = 1.0$; (d) actin + *Limulus* protein, $d_{rms} = 0°$, $Y° = 1.0$.

It seems clear that both kinds of filaments have about 12° of angular variability and that 8° or 16° lie outside the expected errors, thus providing a measure of the precision of the result, *i.e.,* about 3–4°. The frozen-hydrated filaments could not be subjected to this analysis, since they have too low a signal-to-noise ratio. We analyzed these by examining the distribution of twist in the population.

To do so we plotted the distribution of twist for segments of filaments about 100 subunits in length (FIGURE 8). We did the same for the negatively stained and freeze-etched filaments. The value for angular variability obtained from the variation in twist for the last two agrees well with those obtained from the data on intensity ratios, *i.e.,* 13° versus 12° and 10° versus 12° respectively. The angular variation for the frozen-hydrated filaments, however, is less—about 7°. This figure probably represents a lower bound, since only those filaments with strong transforms and hence less angular variation could be included because of the lower signal-to-noise ratio.

Let me now add the clincher, which not only acts as a control but also shows how the variable twist can be modulated by accessory proteins. We compared data for actin with that for actin in association with three other proteins: scruin, a 55 kD bundling protein from the sperm of *Limulus;* the S1 head fragment of myosin; and troponin plus tropomyosin (FIGURE 9). We found that troponin and tropomyosin do not reduce and might even increase the angular variation, while the other two, scruin and S1, reduce it to near zero. Thus, we have detected and measured comparable amounts of variability in twist in three very differently prepared samples of actin. We conclude that this variation is a property of actin and is not an artifact produced by the preparation procedure. We find that the amount of variation in the angular position of the subunit is at least 7° and is more likely as large as 12°. We have further shown that the presence of auxiliary proteins can modulate this variation in twist.

What is the biological role of the variable twist in actin? The answer appears to be that it allows for the construction of higher order assemblies built from actin.[16] For example, one of the common forms of actin in cells, such as those in the brush border of the intestine or in the hair cell of the cochlea, is the actin bundle. In the bundle, actin filaments are hexagonally packed, held together by a bundling protein such as scruin, fascin, or fimbrin. In the bundle of filaments, the cross bridges must lie along hexagonal directions. If actin had strict symmetry, its subunits, to which the cross bridges bind, would not point along the hexagonal directions. Because of the angular variability, that is, the variable twist, the actin subunits can assume hexagonal positions. As the bundle forms, the cross bridges can find suitably oriented subunits, bind to them, and cross-link the filaments.

Thus, unlike the situation in flagellar filaments, the variation in twist of actin is used to build higher order structures such as bundles. How the variability in twist of actin is brought about is unknown. It is not known whether the angles between subunits can vary continuously or whether there are discrete states as there are in flagellar subunits. These puzzles remain to be solved.

In summary, we have presented the concept of variable twist in helical structures, have provided two extreme examples of its occurrence, and have shown how this property is useful biologically.

REFERENCES

1. O'BRIEN, E. J. & P. M. BENNETT. 1972. J. Mol. Biol. **70:** 133–152.
2. WANG, A. H.-J., S. FUJII, J. H. VAN BLOOM & A. RICH. 1982. Cold Spring Harbor Symp. Quant. Biol. **47:** 33–44.
3. BERG, H. C. & R. A. ANDERSON. 1973. Nature **245:** 380–382.
4. SILVERMAN, M. & M. SIMON. 1974. Nature **249:** 73–74.
5. LARSEN, S. H., R. W. READER, E. N. KORT, W.-W. TSO & J. ADLER. 1974. Nature **249:** 74–77.
6. BERG, H. C. 1974. Nature **249:** 77–78.
7. ASAKURA, S. 1970. Adv. Biophys. **1:** 99–155.
8. WAKABAYASHI, K. & T. MITSUI. 1972. J. Phys. Soc. Japan **33:** 175–182.
9. CALLADINE, C. R. 1976. J. Theor. Biol. **57:** 469–489.
10. MACNAB, R. M. & M. K. ORNSTON. 1977. J. Mol. Biol. **112:** 1–30.
11. KAMIYA, R., S. ASAKURA & S. YAMAGUCHI. 1980. Nature **286:** 628–630.
12. KAMIYA, R., S. ASAKURA, K. WAKABAYASHI & K. NAMBA. 1979. J. Mol. Biol. **131:** 725–742.
13. HANSON, J. 1967. Nature **213:** 353–356.
14. EGELMAN, E. H., N. FRANCIS & D. J. DEROSIER. 1982. Nature **298:** 131–135.

15. DeRosier, D. J. & R. Censullo. 1981. J. Mol. Biol. **146:** 77–99.
16. DeRosier, D. J. & L. G. Tilney. 1984. *In* Cell and Muscle Motility. J. W. Shay, Ed. Vol. 5:139–169. Plenum. New York, NY.

DISCUSSION OF THE PAPER

L. Makowski (*Columbia University, New York, New York*): You just showed an analysis indicating that the ratio of layer line 1 to layer line 6 depends on how long a stretch of the filament you average over. I haven't thought about this very much, but what's the right number for the X ray data?

DeRosier: I don't think it's known; we don't know it.

Makowski: How you interpret the X rays is directly dependent on what that number is.

DeRosier: Our numbers here are not significant, because the starting ratio varies in different filaments, and we scaled them all to the same height. So our absolute intercept is not relevant to what you'd measure in X ray.

A. C. Steven (*National Institutes of Health, Bethesda, Maryland*): A question on the binding of the two types of proteins, the S1 and the *Limulus* proteins that seem to abolish the angular disorder entirely, if I understood your slide correctly. In the first case, with S1, it doesn't seem to me you're really looking at much actin anymore if you've got saturating amounts of S1. The stain distribution is being dominated by the S1, so is it valid to draw conclusions about the state of the actin underneath it? And secondly, in the *Limulus* protein case, when you abolish angular disorder, do you get markedly superior helical diffraction patterns?

DeRosier: The answer to your second question is yes. We've got reconstruction of those that I didn't show. The answer to your first question is yes, most of the diffraction is from the S1. But what holds the S1 in place? Clearly it's some interaction between the S1 and the actin. If you mix S1 together, you don't get an actin structure. What you must assume is that when S1 binds to actin it may interact with neighboring S1 subunits to limit the angular disorder present in actin. Alternatively it may somehow cross-link a hinged region in actin to lock out the angular disorder. We don't know why, just that it does. The point was to show that you can modulate it.

E. Mandelkow (*Max-Planck-Unit for Structural Molecular Biology, Hamburg, Federal Republic of Germany*): Is this an effect that depends on the full occupation of the actin filament by myosin, or do you also think that you could just add a few myosin molecules, say 1% or so?

DeRosier: We haven't tried.

Mandelkow: What happens with calcium in your actin tropomyosin? Do you see a change there, or haven't you tried?

DeRosier: We haven't tried it. What was the state of those thin filaments? They were in rigor buffers, so there's calcium present.

The Three-Dimensional Structure of the Actin Filament Revisited[a]

UELI AEBI,[b,c] ROBERT MILLONIG,[b,c] HOPE SALVO,[b]
AND ANDREAS ENGEL[c]

[b]*Department of Cell Biology and Anatomy*
The Johns Hopkins University School of Medicine
Baltimore, Maryland 21205
and
[c]*M.E.-Müller-Institute for High-Resolution Electron Microscopy*
Biozentrum, University of Basel
CH-4056 Basel, Switzerland

INTRODUCTION

Ever since actin has been implicated in muscle contraction, and subsequently in cellular motility and cytoskeletal architecture, elucidation of its high-resolution three-dimensional (3-D) structure—both of its monomeric as well as of its filamentous form—has been a challenge to protein crystallographers and electron microscopists. Thirty years ago Bear and Selby[1] deduced certain features of the structure of actin-containing "thin" filaments from X ray diffraction patterns of intact dried muscle although they could not decide whether the scattering centers lie on a planar net or on a helix. A more complete picture emerged from electron microscopic (EM) analyses of negatively stained actin filaments by Hanson and Lowy[2] and by Huxley:[3] accordingly, actin-containing filaments—whether being natural (*i.e.* thin filaments) or synthetic (*i.e.* polymerized *in vitro* from pure actin—F-actin)—are about 80 Å in diameter (*i.e.* at their widest parts) and are made of two helically-wound strands (*i.e.* "long pitch" helices) of roughly spherical subunits, which cross each other about every 350 Å (see FIGURE 1). The spacing of the subunits along these two long-pitch helices is 55 Å, and the two helices are axially staggered by half a subunit (*i.e.* 27.5 Å). Alternatively, this subunit arrangement in the filament can be described by a one-start "genetic" helix having 13 subunits in 6 turns with a pitch of 59 Å (see FIGURE 1) and a "screw angle" (c.f. REFERENCE 4) of $-166°$.

It was first noted by Hanson[6] that the distance between the crossover points of the two long-pitch helices was rather variable in negatively stained filament preparations. More recently, Egelman *et al.*[7] have proposed a model for "cumulative angular disorder" in actin filaments in an attempt to quantitate Hanson's observation. According to this model, successive subunits along the genetic helix have a random variable twist of up to $\pm 10°$ from the average screw angle of $-166°$. The recent observation by Trinick *et al.*[8] that actin filaments exhibit variable crossovers even when embedded in a film of vitreous ice, without attachment to a substrate, supports the notion that this disorder is an *intrinsic* property of actin filaments and not just a

[a]Supported by National Institutes of Health (NIH) Grants GM 31940 and GM 35171 (to U.A.) and by Swiss National Science Foundation Grant 3.251.82 (to A.E.). R.M. was partly supported by NIH Training Grant 5T32 GM 24775. U.A. was also the recipient of a research award from the M.E.-Müller-Foundation of Switzerland.

specimen-preparation artifact. In addition, the apparent diameter along the length of isolated actin filaments can vary considerably—typically 70–100 Å —both in negatively stained preparations (c.f. REFERENCES 2, 9, 10), as well as when embedded in ice (REFERENCE 8; our own unpublished results). This phenomenon, at least in part, may be correlated with the variable crossovers of the filaments (see section on "The 'Lateral Slipping' Model of the Actin Filament").

The variabilities discussed above, together with the relatively low yield of sufficiently straight stretches (*i.e.* ≥4 crossovers), have discouraged many workers in the past from determining filament 3-D reconstructions from micrographs of isolated, undecorated actin filament preparations. Instead, they have masked out and reconstructed (c.f. REFERENCES 10–16) what appeared to be individual filaments from single- or multilayered actin filament paracrystals (for examples see FIGURE 2), which can easily be formed from either F-actin or thin filaments under a variety of induction conditions (c.f. REFERENCES 17–19). While the putative "filaments" in such paracrystalline arrays are highly ordered and straight over many crossovers, we are faced with

The Actin Helix

FIGURE 1. Schematic representations of the "actin helices" describing the geometrical arrangement of subunits in the actin filament. *Left*, the actin filament can be represented by two right-handed "long-pitch" helical strands[5] with 13 subunits in 1 turn, which have a 715-Å pitch and a 27.7° screw angle, and are angularly displaced by 180° and axially shifted by 27.5 Å relative to one another. As a consequence, the two long-pitch helices cross each other every 357.5 Å. *Right*, as an alternative, the actin filament may be represented by a left-handed 1-start "genetic" helix with 13 subunits in 6 turns, which has a 59-Å pitch and a −166° screw angle.

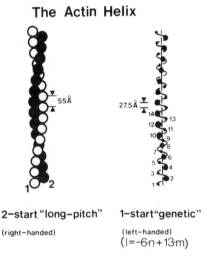

2-start "long–pitch"
(right–handed)

1-start "genetic"
(left–handed)
$(l = -6 \cdot n + 13 \cdot m)$

two potential problems: (1) In both single- and multilayered paracrystals, adjacent filaments may significantly laterally interdigitate, making clean separation of individual filaments difficult. (2) In multilayered paracrystals, what appear to be single filament projections actually represent superpositions of several filaments in the direction of the electron beam. Unless all superimposed filaments are in axial and angular register, and have the same polarity, the resulting filament projections may be highly artifactual. As a consequence, most filament 3-D reconstructions obtained from paracrystal data appear rather "compact" with overall diameters ranging from 60 to 75 Å, they exhibit strong axial connectivity and usually no significant axial polarity, and delineation of the subunits is mostly ambiguous (c.f. REFERENCES 10–13).

Two exceptions to the above are reconstructions determined from single-layered thin filament paracrystals, both of which appear more "open," with the axial connectivity being confined to distinct parts of the filament cross sections. At first glance, these density maps yield the impression that the major intersubunit contacts may be along the genetic helix.[14,16] While in both cases the interfilament distance within the single-layered filament arrays is about 75 Å—as is the interfilament distance within the single-layered pure F-actin filament arrays shown in FIGURE 2—in one case[14] the resulting filament reconstruction becomes about 75 Å in its widest part, whereas in the other case[16] it becomes almost 95 Å. The possible reasons for this apparent discrepancy are unclear.

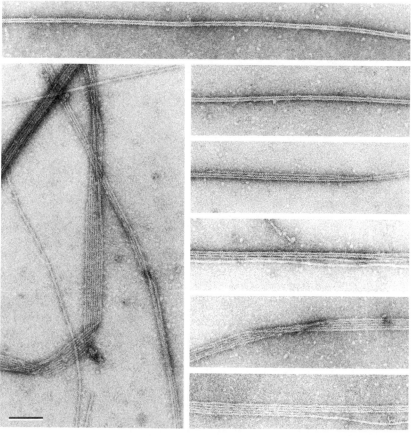

FIGURE 2. Examples of negatively stained paracrystalline F-actin filament arrays, which are mostly single-layered. G-actin (about 0.5 mg/ml) was polymerized with 25 mM MgCl₂ at 25° C for 30 min and was then dialyzed by "open" dialysis at 4° C against 5 mM MgCl₂ for various lengths of time before 5-μl aliquots were removed for electron microscopy. Grids were prepared as described,[18] using 0.75% uranyl formate, pH 4.25, for negative staining. *Scale bar:* 1,000 Å.

Recently, a 3-D reconstruction from isolated pure F-actin filaments embedded in a thin film of ice and recorded at a nominal defocus of 2–3 μm has been calculated to a resolution of about 40 Å.[8] It too is about 95 Å in its widest part, and most connections between successive subunits appear to be made along the genetic helix (*i.e.* close to the filament axis). The putative subunits are somewhat curved and resolved into a large and a small domain such that they appear elongated in a direction nearly perpendicular to the filament axis. Since this model has been obtained from material that was supposedly preserved in a more native state than e.g. negatively stained material (c.f. REFERENCE 20), it may reveal some gross features of subunit packing in the actin filament. On the other hand, it should be emphasized that, due to the excessive amount of defocus needed to generate enough contrast (c.f. REFERENCE 20), too much faith cannot be placed on the finer structural details of this comparatively low-resolution 3-D reconstruction.

Several investigators have obtained 3-D reconstructions from negatively stained isolated F-actin filaments or thin filaments stoichiometrically decorated with myosin subfragment-1 (S-1) (c.f. REFERENCES 11, 21–25). Most of the resulting "acto-(S-1)" models have yielded similar general mass distributions, but all of them suffer from the lack of clear-cut boundaries between the different molecular components. Thus the differences among them center primarily upon *interpretation* of what is actin and what is myosin S-1, a step which invariably relies on some *a priori* knowledge regarding the overall size and shape of the two constituents. As a consequence, an actin filament "carved out" from any of these acto-(S-1) reconstructions is only as good as the assumptions it is based upon.

Egelman and DeRosier[26] have modelled the F-actin filament by varying the orientation of a "synthetic" actin subunit—represented by two interpenetrating spheres of equal size—along the genetic helix until an optimal match between the model and actual filament diffraction patterns was reached. In the resulting filament model, the long axis (*i.e.* 68 Å) of the "two-sphere" model subunit is oriented nearly perpendicular to the filament axis giving rise to a filament that has a diameter of about 95 Å at its widest parts and in which all intersubunit contacts are along the genetic helix. Apparently, this model is consistent with the observed thin filament layer line intensities and positions from X ray diffraction patterns of live muscle[27,28] and with isolated unstained F-actin filaments embedded in a thin film of ice,[8] and it can easily be carved out from the acto-(S-1) 3-D reconstructions of Taylor and Amos.[21,24] Therefore Egelman feels that his new picture of the actin filament—which also seems to be consistent with his model for cumulative angular disorder in actin filaments[7]—is well enough founded to be referred to as the current "consensus" model of the actin filament.[29]

Here we present new data on isolated F-actin filaments that may necessitate some revision of the consensus model. Taken together, our data demonstrate that the physically stronger interfilament contacts are along the two long-pitch helices and not along the genetic helix. As a consequence, the observed variable crossover distances of the two long-pitch helices[6] may be due to "lateral slipping" of the two long-pitch helices past each other rather than to cumulative angular disorder of the genetic helix.[7] This lateral slipping, in turn, may explain the variable width of negatively stained and of unstained, isolated actin filaments. Finally, in 3-D reconstructions obtained from negatively stained single-filament stretches, actin subunits have been resolved, enabling us to orient molecular models of G-actin within F-actin filament models.

3-D RECONSTRUCTION OF ISOLATED F-ACTIN FILAMENTS

Using an improved method for preparing negatively stained, isolated F-actin filaments (c.f. FIGURE 3), we have been able to obtain significant numbers of straight and well-preserved filament stretches—typically 3–6 crossover distances long—suitable for 3-D helical reconstruction. Preparation involved the following steps: (1)

"injecting" 2–3 µl rabbit muscle actin filaments (about 0.5 mg/ml, polymerized with 50 mM KCl) into a 2–3-µl drop of polymerization buffer deposited on a lightly glow-discharged carbon-coated copper grid and allowing the filaments to adsorb for 1 min; (2) blotting off excess material with a piece of filter paper; (3) washing for 1 sec on a 100-µl drop of water; (4) staining for 30 sec in series on three 100-µl drops of

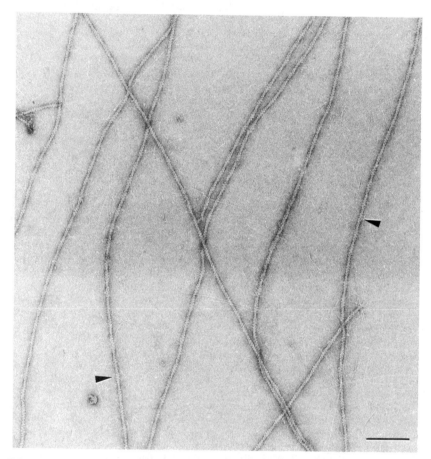

FIGURE 3. Micrograph of negatively stained isolated F-actin filaments of the type used for 3-D reconstructions (see FIGURE 4). Filaments were made by polymerizing rabbit muscle G-actin (about 0.5 mg/ml) with 50 mM KCl. Specimen preparation was performed as described in the text. The *arrowheads* point to filament stretches suitable for 3-D helical reconstruction. *Scale bar:* 1,000 Å.

0.75% uranyl formate, pH 4.25; (5) quickly blotting off excess stain with filter paper; (6) further removal of liquid via suction with a finely drawn-out Pasteur pipet gently moved around the edge of the grid; (7) air-drying the grid. Sometimes the negative-staining step was followed by a 1-sec wash on a 100-µl drop of 0.5% glucose before final blotting, suction, and air-drying of the grid.

For 3-D helical reconstruction, we looked for straight filament stretches with 3–6 evenly spaced "357-Å" crossovers (c.f. FIGURE 3, and FIGURES $4a_1$ and $4b_1$, arrowheads). The optimal helical repeat over a digitized filament stretch (c.f. FIGURES $4a_2$ and $4b_2$) was determined by varying it such as to minimize the power loss on averaging over the number of helical repeats contained in the stretch. Two examples of 4-helical-repeat long optimally averaged stretches are shown in FIGURES $4a_3$ and $4b_3$. With this procedure, the helical repeat could be determined to within a fraction of a 5.5-Å sample point. Averaged helical repeats were then $D(Z, k)$-filtered[30] to remove all the information inconsistent with the approximate integer helical selection rule $1 = -6n + 13m$.[10,15,31] The filtered data (REFERENCES 15, 31; see FIGURES $4a_4$ and $4b_4$) were used to subsequently compute the 3-D helical reconstruction (see FIGURES $4a_5$ and $4b_5$) and therefore represent a projection image normal to the filament axis of the 3-D reconstruction.

Two types of helical reconstructions could basically be distinguished, as illustrated in FIGURE 4: Type I (FIGURE $4a_5$) was revealed from filaments that were 90–100 Å at their widest parts (FIGURE $4a_1$) and had relatively short crossover regions (FIGURE $4a_3$). Type II (FIGURE $4b_5$) was obtained from filaments that were 80–90 Å at their widest parts (FIGURE $4b_1$) and had more extended crossover regions (FIGURE $4b_3$). Generally, type I filaments were "less helical" than type II filaments judged from their 5–10% bigger power losses upon $D(Z, k)$-filtration. Both types of filament reconstructions revealed radii of gyration[10] ranging from 23 to 26 Å, in agreement with values published earlier from various sources (c.f. REFERENCES 10, 28, 32).

At first glance, type I filament reconstructions seem to have no significant contacts along their long-pitch helices. Closer examination, however, reveals two "columns" of density, each following one long-pitch helical strand and being confined to relatively small filament radii (see arrows, FIGURE $4a_5$). This appearance can best be explained by a "2-domain" actin subunit with a relatively globular domain placed at large filament radii and making no contacts along the long-pitch helices, and a more elongated domain (*i.e.* parallel to the filament axis) at small filament radii making contacts along both one of the long-pitch helices and the genetic helix. This design gives the type I filament a relatively "open" appearance and accounts for the short but distinct crossover regions (see FIGURES $4a_{4,5}$).

Type II filament reconstructions (see FIGURE $4b_5$) appear more "compact" with a strong emphasis of the two long-pitch helices, and they exhibit a less distinct globular domain at high radii than type I filament reconstructions. The putative subunit consists of a large "base" with its longest dimension (*i.e.* 60–65 Å) approximately perpendicular to the filament axis and its shortest dimension (*i.e.* 30–35 Å) normal to a plane parallel to the filament axis, and a smaller "top," which is confined to smaller radii and connects adjacent subunits along the long-pitch helices. Type II filament reconstructions appear narrowest (*i.e.* 60–70 Å) in projection where their two long-pitch helices lie next to each other and widest (*i.e.* 80–90 Å) where they cross over each other.

To a first approximation, type I filament reconstructions resemble the 20-Å resolution model of the actin filament that was obtained by Egelman (see Fig. 2 in REFERENCE 29) upon positioning the actin subunit determined from actin-DNase I cocrystals[33] in a model filament to best fit the observed actin layer lines from X ray diffraction patterns of live muscle by Tajima *et al.*[27] However according to Egelman,[26,29] the actin subunits make no contacts along the two long-pitch helices in his "consensus" filament model. Furthermore, there is a striking similarity between type I filament reconstructions in projection (see FIGURE $4a_4$) and the Egelman-DeRosier model[26] when projected in the same direction (see Fig. 5 in REFERENCE 10).

Nevertheless, we favor the type II over the type I filament reconstructions for the

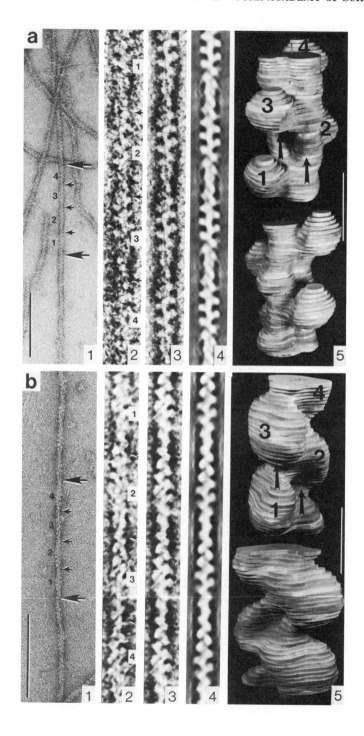

following reasons: (1) they are found more frequently in our preparations; (2) their helical symmetry is usually better preserved (see above); (3) in most cases their crossovers are more regularly spaced over straight filament stretches; (4) they appear less spread due to flattening on EM grids (see also REFERENCE 10); (5) based on the "lateral slipping" model proposed and discussed in the last section, type II filaments are preserved in a stabler state, in the sense that their two long-pitch helices make more extensive contacts with each other; (6) type II reconstructions are more similar to each other (see below) than are type I reconstructions; (7) an actin subunit model determined from negatively stained crystalline actin sheets[10,15] can be better fitted into type II (see next section) than into type I filament reconstructions.

We were struck by the reproducibility of type II filament reconstructions determined from different filament stretches as demonstrated by the "hybrid" filament model shown in FIGURE 5, where the bottom half (I) represents one reconstruction and the top half (II) another (the two coming from different micrographs). Not only was the type II reconstruction the most reproducible, but it also was the most predominant, found in about 35% of our filament 3-D reconstructions. Type I reconstructions accounted for roughly 10% of our reconstructions, and they were less similar among themselves than were the type II. The remainder of the reconstructions represented a sort of "continuum" between type I and type II, with a clear bias towards type II. This is a further reason that gives us more faith in type II over type I reconstructions. Therefore, for the remainder of this presentation, we will limit ourselves to type II filaments respectively type II reconstructions.

ALIGNING 3-D MODELS OF THE ACTIN MOLECULE WITHIN FILAMENT RECONSTRUCTIONS: TOWARDS A MOLECULAR MODEL OF THE ACTIN FILAMENT

The ultimate goal of our efforts is to obtain an actin filament model by high-resolution EM revealing enough detail to unambiguously trace the component

---

FIGURE 4. 3-D helical reconstruction of negatively stained, isolated F-actin filaments: (**a**) representative example of a type I filament respectively its 3-D reconstruction; (**b**) representative example of a type II filament respectively its 3-D reconstruction. (a_1, b_1) Micrographs of F-actin filaments made by polymerizing rabbit muscle G-actin (about 0.5 mg/ml) with 50 mM KCl. Specimen preparation was performed as described in the text. The *arrows* demarcate 4-helical-repeats long filament stretches used for 3-D reconstruction. The four helical repeats are numbered "1," "2," "3," "4." (a_2, b_2) The 4-helical-repeats long filament stretches demarcated in (a_1, b_1) are photowritten after digitization of the corresponding micrograph areas with a 25 μm raster yielding a 5.5-Å sampling of the data. The approximate crossovers of the two long-pitch helices are indicated by *arrowheads*, and the four helical repeats are numbered "1," "2," "3," "4." (a_3, b_3) The filament stretches shown in (a_2, b_2) after axial averaging over their optimal helical repeats (for details see text). Note that the four helical repeats in the averaged filament stretches are identical. (a_4, b_4) The axially averaged filaments shown in (a_3, b_3) after D(Z, k)-filtering[30] to remove all the information inconsistent with the approximate helical selection rule $1 = -6n + 13m$.[10,15,31] Note that the thus-filtered data represent projection images normal to the filament axis of the 3-D helical reconstructions shown in (a_5, b_5). (a_5, b_5) Two different angular views (*i.e.* rotated by approximately 90° about the filament axis relative to one another) of 3-D helical reconstructions computed from the filament projection data shown in (a_4, b_4). The filament reconstructions are presented in the form of balsa wood models, 4-subunit stretches (*i.e.* along the left-handed genetic helix $1 = -6n + 13m$) long, and contoured at a density level to include 100% mass (c.f. REFERENCES 15, 31). The subunits are numbered "1," "2," "3," "4," and the *arrows* point to the two long-pitch helical strands along which major intersubunit contacts are maintained. *Scale bars:* 1,000 Å (a_1, b_1); 50 Å (a_5, b_5).

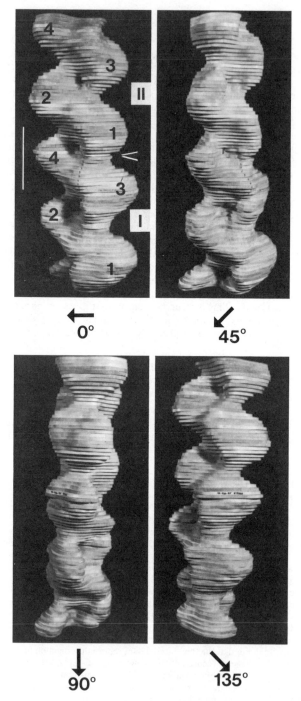

FIGURE 5. Four different angular views (*i.e.* rotated by approximately 45° about the filament axis relative to one another) of a type II "hybrid" filament model, where the *bottom* 4-subunit stretch (**I**) represents one type II filament reconstruction and the *top* 4-subunit stretch (**II**) another. The balsa wood model is contoured at a density level to include 100% mass (c.f. REFERENCES 15, 31). The subunits of each reconstruction are numbered "1," "2," "3," "4." *Scale bar:* 50 Å.

subunits. As shown in FIGURE 6, we believe that our type II filament 3-D reconstructions allow us to achieve this goal. We have contoured 10 sections perpendicular to the filament axis over a 27.5-Å long axial stretch corresponding to one axially repeating unit (*i.e.* half the axial extent of a 55-Å long subunit) of the $1 = -6n + 13m$ genetic

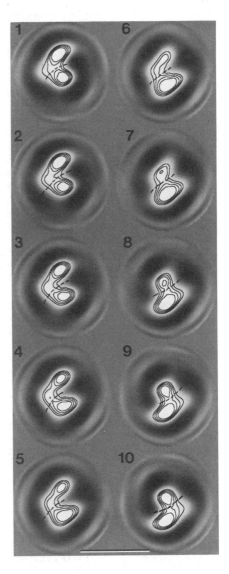

FIGURE 6. Ten equidistant sections perpendicular to the filament axis representing a 27.5-Å long axis stretch (*i.e.* half the axial extent of a 55-Å long actin subunit) of a type II filament 3-D reconstruction. Four contouring levels have been chosen to include 100%, 75%, 50%, and 25%, respectively, of the subunit mass. The contours are superimposed onto grey level representations of the sections. The *dashed line* in each section cuts the map approximately along its density minimum into two parts representing the contributions from two subunits that are axially staggered by 27.5 Å and belong to different long-pitch helical strands. *Scale bar:* 100 Å.

helix of a type II filament reconstruction. Each section contains contributions from two subunits that are axially staggered by 27.5 Å relative to each other and belong to different long-pitch helical stands. The 10 sections therefore represent exactly one subunit, one half contributed by one long-pitch helix and the other half by the other.

Our criterion for separating the contributions from the two subunits in each section was to make a "smooth cut" approximately following the density minimum separating the two major density maxima present in each section (see FIGURE 6, dashed lines). Furthermore, we made sure that these cuts defined a smooth interface between the two subunits, corresponding to the intersubunit contacts along the genetic helix. Four characteristic views of a thus "cut out" subunit are displayed in FIGURE 7b, with the view revealing the interface being in the bottom panel. The overall dimensions of the subunit are 55 × 34 × 64 Å, with the 55-Å spacing being parallel and the 64-Å spacing being perpendicular to the filament axis. The filament reconstruction from which the subunit was cut out is shown in FIGURE 7a.

Following this, we have tried to orient this subunit within a refined model of an actin subunit determined previously from negatively stained crystalline actin sheets.[15] Four corresponding views of this refined subunit are displayed in FIGURE 7d. After we compensated for the collapsed (i.e. by about 30%) thickness of the sheets used for the original subunit reconstruction,[10,15] the refined model yielded overall dimensions similar to those of the actual filament subunit (FIGURE 7b): 56 × 38 × 62 Å—cut out from a near-rectangular 56 × 65-Å dimer unit cell.[15] This subunit too consists of a massive "base" and a slender "top" and has slightly concave faces perpendicular to the 38-Å axis (compare FIGURE 7d with FIGURE 7b).

Finally, after having oriented the sheet-derived actin subunit (FIGURE 7d) relative to the filament-derived actin subunit (FIGURE 7b), we have built a "synthetic" actin filament (FIGURE 7c) using the 15-Å resolution sheet subunit and placing it the same way in the filament as is the actual filament subunit. In fact, the sheet subunits more or less "force" each other into the proper positions by utilizing "complementary" surface areas to form the intersubunit contacts along the genetic helix: a convex area of the slender top snugly fits into a concave area of the massive base of the subunit, with the two complementary contact areas being separated axially by almost exactly 27.5 Å (see FIGURE 7d, bottom). As can be seen from FIGURE 7c, the subunits also make significant top-to-base contact along the two long-pitch helices. The emerging filament measures about 66 Å in its narrowest part (i.e. where the two long-pitch helices lie side by side) and about 86 Å in its widest part (i.e. where the two long-pitch helices cross each other), and its radius of gyration is about 26 Å. Qualitatively, the 15-Å resolution synthetic filament model compares very favorably with the 27.5-Å resolution type II filament reconstruction shown in FIGURE 7a.

While in this model the 56-Å axis of the actin subunit is still oriented parallel to the filament axis, the angular orientation and radial distance of the subunit relative to the filament axis are different from those proposed earlier for the "best fit" model (see REFERENCE 15; FIGURE 8a). Instead our model is more similar to the "best contact" model—which was suggested as an alternative[15]—except that the two long-pitch helical strands have laterally slipped past each other compared to the best contact model (see REFERENCE 15; FIGURE 8b), giving the new model a more open appearance and making it wider in the parts where the two long-pitch helices cross each other in projection.

ACTIN FILAMENTS CAN BE PARTIALLY UNRAVELLED INTO THEIR TWO LONG-PITCH HELICES

As discussed and documented (see FIGURES 4–8) in the previous two sections, all our actin filament 3-D reconstructions exhibit significant intersubunit contacts both along their two long-pitch helices as well as along their genetic helix. However, it has been argued by Egelman[26,29] that the actin filament has no significant intersubunit

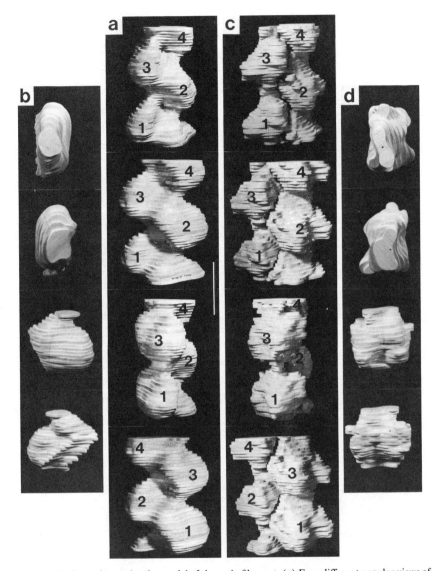

FIGURE 7. Towards a molecular model of the actin filament. (**a**) Four different angular views of a 4-subunit stretch of a type II filament 3-D reconstruction. (**b**) Four different views of an actin subunit "cut out" from the filament 3-D reconstruction displayed in (a) as illustrated in FIGURE 6. The view revealing the "cut" (*i.e.* the intersubunit contact along the genetic helix) is shown in the *bottom panel*. (**c**) Four different views of a 4-subunit-stretch of a "synthetic" actin filament built from the 15-Å resolution actin subunit model shown in (d). In this synthetic filament, the actin subunit model has been oriented the same way as the cut-out subunit (see b) in the actual filament 3-D reconstruction (see a). The four angular views of this filament model have been approximately matched with those of the actual filament 3-D reconstruction shown in (a). (**d**) Four different views of a 15-Å resolution refined model of the actin molecule determined from negatively stained crystalline actin sheets.[15] In this refined model, we have compensated for the collapsed thickness of the sheets used for the original subunit reconstruction.[10,15] The views shown here have been oriented relative to those of the cut-out subunit displayed in (b). All models shown have been contoured at a density level to include 100% of the actin subunit mass (c.f. REFERENCES 15, 31). In the 4-subunit stretch filament models shown in (a) and (c) the subunits have been numbered "1," "2," "3," "4." *Scale bar:* 50 Å.

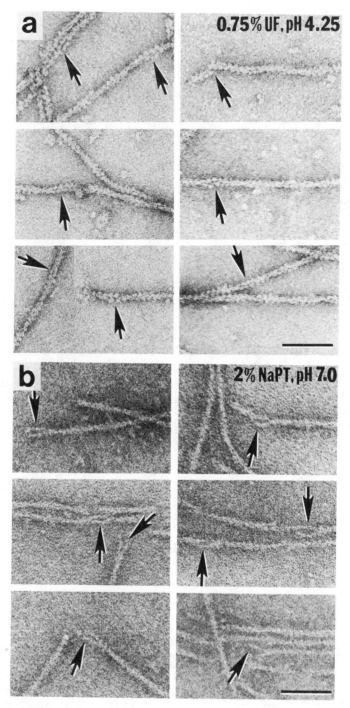

FIGURE 8. Examples of actin filament stretches that are partially unravelled into their two long-pitch helical strands. Filaments were made by polymerizing rabbit muscle G-actin (about 0.5 mg/ml) with 50 mM KCl, prepared for EM as described in the text, and negatively stained with 0.75% uranyl formate, pH 4.25, (**a**), or with 2% Na-phosphotungstate (NaPT), pH 7.0, (**b**). *Arrows* point to short "loops" along filaments, or to filaments "splaying into two strands" at their ends. *Scale bars:* 1,000 Å (a,b).

contacts along the two long-pitch helices as evidenced both by his consensus model of the filament[29] as well as by the cumulative angular disorder of the genetic helix.[7,29]

Being aware that the spatial extent of the different types of intersubunit contacts revealed by our filament reconstructions may not necessarily be a measure of the physical strength of these contacts—particularly when looking at 3-D reconstructions from negatively stained filament preparations—we have attempted to probe more directly for the *relative* physical strength of the intra- and inter-long-pitch helix intersubunit contacts. We have assumed that if the intra-long-pitch helix contacts are physically stronger than the inter-long-pitch helix ones, it should be possible to find conditions causing the two long-pitch helices to *partially unravel,* resulting in short "loops" along filaments or "splaying into two strands" at the ends of filaments.

As shown in FIGURE 8a, such partially unravelled F-actin filament stretches can be found quite easily in standard filament preparations (see FIGURE 3) upon systematic screening of the grids. As documented with representative examples in FIGURE 8b, even more striking examples of unravelling filament stretches were found in filament preparations after negative staining with 2% Na-phosphotungstate instead of 0.75% uranyl formate. Based on these results, we believe that the physically stronger intersubunit contacts are along the two long-pitch helices (*i.e. intra*-long-pitch helix contacts) rather than along the genetic helix (*i.e. inter*-long-pitch helix contacts). In fact, ours are by no means the first micrographs indicative of unravelling actin filaments (see *e.g.* REFERENCE 23; FIGURE 9), although no one has apparently stressed this phenomenon in the past. As will be discussed in the last section, this observation may have an important bearing on the filament architecture and, ultimately, on its mechanical and functional properties.

WHAT IS THE "TRUE" WIDTH OF THE ACTIN FILAMENT?

Published width measurements of negatively stained F-actin filaments range from 60 Å to 100 Å (c.f. REFERENCES 2, 3, 9, 10, 16, 34). The reasons for this variability have been ascribed to various sources, including the difficulty in measuring true widths from stain exclusion profiles (c.f. REFERENCES 10, 34), and the possibility of spread-flattening of the filaments on specimen preparation (c.f. REFERENCES 9, 10). In support of a 90–100-Å filament width are X ray diffraction analysis of thin filaments in live muscle[27,28] as well as width measurements of F-actin filaments embedded in a thin film of vitreous ice.[8] However, it is without question that the apparent width of negatively stained F-actin filaments may vary greatly (see *e.g.* REFERENCE 35, Fig. 7b; REFERENCE 10, Fig. 1d; our own unpublished results), an observation also being true for filaments embedded in ice (see *e.g.* REFERENCE 8, Fig. 2; our own unpublished results). Taken together, the possibility arises that the F-actin filament width exhibits a significant degree of *inherent* variability, which, in turn, may directly be correlated with the observed variability of the distance between crossover points of the two long-pitch helices (REFERENCE 6; see also next section).

Here we have simultaneously determined the "mass-per-unit-length" (MPUL) and the "full-width-half-maximum" (FWHM) of unstained and *in situ* freeze-dried F-actin filament stretches by scanning transmission electron microscopy (STEM) (c.f. REFERENCES 36–38). As can be gathered from FIGURE 9a, STEM dark-field images of filaments prepared this way appear quite variable in width. This rather qualitative observation is expressed more quantitatively by the FWHM histogram shown in FIGURE 9d, which is based on 140 measured filament segments. According to this histogram, the average width of these filament segments—*i.e.* as defined by their FWHM (c.f. REFERENCE 38)—is 72 Å, with about 75% of the values falling into a

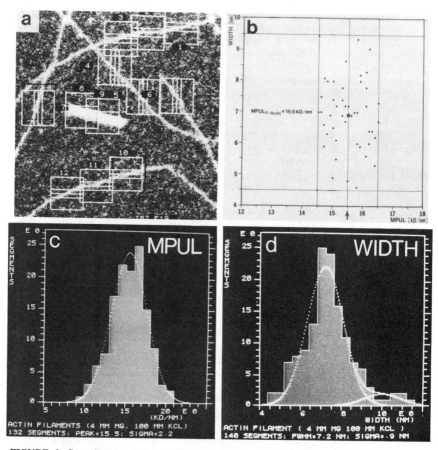

FIGURE 9. Scanning transmission electron microscopy (STEM) of unstained and *in situ* freeze-dried actin filaments to determine their mass per unit length (MPUL) and their full width half maximum (FWHM). Actin filaments were made by polymerizing rabbit muscle G-actin (about 0.5 mg/ml) with 4 mM MgCl$_2$/100 mM KCl and prepared for STEM as described.[37,38] (a) STEM dark-field electron micrograph of unstained and *in situ* freeze-dried actin filaments. The *boxes* enclose filament stretches (typically 1,200 Å long) used for MPUL/FWHM determination. For mass calibration, tobacco mosaic virus (TMV) rods were mixed into the sample (see boxes #8 and #9) (c.f. REFERENCE 38). (b) Plot of MPUL-FWHM pairs for filament segments with MPULs falling into the range 15.5 ± 1.0 kD/nm. According to this plot, there is no correlation between the MPUL and the FWHM of the measured filament segments. (c) MPUL histogram containing 132 measurements. The histogram has been fitted by a Gaussian curve (c.f. REFERENCE 38) whose peak corresponds to a MPUL of 15.5 kD/nm. (d) FWHM histogram containing the same set of measurements used for the MPUL histogram shown in (c). The histogram has been fitted by a set of two overlapping Gaussian curves, both having the same half-weight-width (c.f. REFERENCE 38). The peak of the major Gaussian corresponds to a FWHM of 72 Å.

range between 60 Å and 85 Å. The corresponding MPUL values are histogrammed in FIGURE 9c, yielding an average value of 15.5 kD/nm, which is almost exactly the value (*i.e.* 15.6 kD/nm) predicted for an F-actin filament with a subunit MW of 43 kD and a 55-Å axial repeat of the subunits along the two long-pitch helices.

In FIGURE 9b we have represented the MPUL-FWHM pairs for the filament segments with MPULs falling into the range 15.5 ± 1.0 kD/nm. According to this MPUL-FWHM plot, there is no correlation between these two values, meaning that the FWHM of filament segments having the same MPUL may vary considerably. We

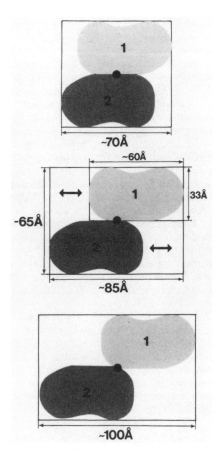

FIGURE 10. Schematic representation of the "lateral slipping" model of the actin filament in the form of cross sections normal to the filament axis. One long-pitch helical strand is shaded in *grey,* the other in *black.* Dimensions and cross-sectional shapes of the actin subunits are taken from actual filament 3-D reconstructions and from a 3-D model of the actin molecule (see FIGURES 6 and 7). Note that in projection the filament appears widest (*i.e.* 70–100 Å) where the two long-pitch helical strands come to lie on top of each other, and narrowest (*i.e.* 60–70 Å) where they lie side by side.

believe this to be ample evidence that the F-actin filament has by no means a unique width, but rather its diameter exhibits a significant degree of inherent variability, which may be due to its distinct helical architecture (see next section and FIGURE 10) and, ultimately, may explain the wide range of filament diameters (*i.e.* 60–100 Å) reported in the literature. While the FWHM represents a clearly defined and highly reproducible width measurement (c.f. REFERENCE 38), it does not necessarily provide us with an "absolute" width value. We guess that in the case of actin filaments, the

FWHM typically represents a 15–20% underestimate of their actual width, resulting in an average "true" width for F-actin of about 85 Å.

THE "LATERAL SLIPPING" MODEL OF THE ACTIN FILAMENT

Taken together, the following three observations have prompted us to revisit the molecular architecture of the F-actin filament: (1) that the spacings between crossovers of the two long-pitch helices are rather variable[6]; (2) that F-actin filaments can be partially unravelled into their two long-pitch helical strands (see FIGURE 8); and (3) that the apparent F-actin filament width is highly variable and typically ranges between 70 Å and 100 Å (see FIGURE 9).

Based on these observations we propose a model where *the physically strongest intersubunit contacts are along—and not across—the two long-pitch helical strands.* While this feature is not immediately apparent from the spatial extent of the intersubunit contacts in our 3-D reconstructions when contoured at a level to include 100% mass (see *e.g.* FIGURES 5 and 7), it becomes more obvious when looking at higher-density contours (*i.e.* those including less mass; see FIGURE 6): usually the contacts between the two long-pitch helical strands are lost first, although they appear spatially more extensive than those along the long-pitch helical strands when contoured at low-mass density. This intersubunit bonding situation, in turn, may allow "lateral slipping" of the two long-pitch helical strands past each other as illustrated in FIGURE 10, a feature of the actin filament that obviously has important implications for its structural and mechanical properties.

First, depending on the amount and direction of lateral slipping, the effective filament diameter may vary greatly—typically between about 70 Å and 100 Å (see FIGURE 10). That this observed width variation is an inherent property of the actin filament—and not merely a specimen preparation artifact (*i.e.* a variable amount of spread-flattening)—is supported by the finding that actin filaments exhibit variable diameters even when embedded in a film of vitreous ice spanning holes in the specimen support film (c.f. REFERENCE 8; our own unpublished observation).

Second, the observed variable crossover spacings (c.f. REFERENCE 6) may simply be explained by a variable amount and random direction of lateral slipping of the two long-pitch helical strands past each other rather than by a cumulative angular disorder of the genetic helix as proposed by Egelman and co-workers.[7] While both mechanisms may explain the variable crossover spacings, we believe that, due to the intersubunit bonding pattern described above, lateral slipping of the two long-pitch helices—and not a random variable twist of the genetic helix—is the *primary cause* of the observed disorder of the actin filament.

Third, in the extreme case, lateral slipping of the two long-pitch helical strands may lead to complete breaking of the intersubunit bonds between the two strands extending over several subunits, thereby giving rise to the observed loops and fraying of the filaments (see FIGURE 8). Consistent with this finding, 3-D reconstruction from tilted thin sections of rigor insect flight muscle has revealed a periodic untwisting and overtwisting of the two long-pitch helical strands of the thin filament.[39] Within the untwisted zones the two strands appeared widely separated, whereas in the overtwisted zones the two strands appeared more normally spaced. While variable long-pitch helix strand separation has as yet no recognized correlate in X ray studies of either native or preserved muscle, we believe that this is a *real* phenomenon exhibited by the actin filament under appropriate conditions and not simply an EM specimen preparation artifact. In particular, the lateral slipping mechanism allows in a straightforward way for periodic perturbation of the actin helix to accommodate the systematically varying interfilament bonding distances in paracrystalline actin filament arrays, as well as the wide range of cross-bridge structures and angles in muscle fibers consisting of layers of alternating thick and thin filaments.

CONCLUSIONS

In this report we have presented a *molecular* model of the F-actin filament obtained from 3-D reconstructions of negatively stained isolated filaments, and by orienting a 15-Å resolution 3-D model of the actin molecule determined previously from crystalline sheets within these filament 3-D reconstructions. In this filament model the physically strongest intersubunit contacts are along the two long-pitch helical strands and not along the one-start genetic helix. The resulting intersubunit bonding pattern is supported by images yielding filament segments with partially unravelled long-pitch helical strands giving rise to loops and frayed ends. Furthermore, this intersubunit bonding pattern allows for variable amounts and directions of lateral slipping of the two long-pitch helical strands past each other, a mechanism that can easily account for the variable crossover distances commonly observed with isolated filaments and the systematic perturbations of the actin helix found in actin-actin as well as in actin-myosin filament arrays. This actin filament model represents an *alternative* to the "consensus" model previously proposed by Egelman.[29] At first glance, it appears to be consistent with most of the phenomena and properties reported as yet about actin filaments. More rigorous testing respectively refinement of this molecular F-actin filament model is currently in progress in our laboratory.

ACKNOWLEDGMENTS

The help of Drs. P. R. Smith and E. L. Buhle, Jr. with some of the computing and image displaying is highly acknowledged. We would like to thank Drs. M. Kessel and R. Eichner for critically reading the manuscript and making many useful comments and suggestions.

REFERENCES

1. BEAR, R. S. & C. C. SELBY. 1956. The structure of paramyosin fibrils according to X-ray diffraction. J. Biophys. Biochem. Cytol. **2:** 55–69.
2. HANSON, J. & J. LOWY. 1963. The structure of F-actin and of actin filaments isolated from muscle. J. Mol. Biol. **6:** 46–60.
3. HUXLEY, H. E. 1963. Electron microscope studies on the structure of natural and synthetic protein filaments from striated muscle. J. Mol. Biol. **7:** 281–308.
4. SMITH, P. R. & U. AEBI. 1975. The determination of the helical screw angle of a helical particle from its diffraction pattern (Appendix). J. Mol. Biol. **106:** 271–275.
5. DEPUE, R. H., JR. & R. V. RICE. 1965. F-actin is a right-handed helix. J. Mol. Biol. **12:** 302–303.
6. HANSON, J. 1967. Axial period of actin filaments. Nature **213:** 353–356.
7. EGELMAN, E. H., N. FRANCIS & D. J. DEROSIER. 1982. F-actin is a helix with a random variable twist. Nature **298:** 131–135.
8. TRINICK, J., J. COOPER, J. SEYMOUR & E. H. EGELMAN. 1986. Cryo-electron microscopy and three-dimensional reconstruction of actin filaments. J. Microsc. **141:** 349–360.
9. FOWLER, W. E. & U. AEBI. 1983. A consistent picture of the actin filament related to the orientation of the actin molecule. J. Cell Biol. **97:** 264–269.

10. SMITH, P. R., W. E. FOWLER & U. AEBI. 1984. Towards an alignment of the actin molecule in the actin filament. Ultramicroscopy **13**: 113–124.

11. MOORE, P. B., H. E. HUXLEY & D. J. DEROSIER. 1970. Three-dimensional reconstruction of F-actin, thin filaments and decorated thin filaments. J. Mol. Biol. **50**: 279–295.

12. SPUDICH, J. A., H. E. HUXLEY & J. T. FINCH. 1972. Regulation of skeletal muscle contraction. II. Structural studies of the interactions of the tropomyosin-troponin complex with actin. J. Mol. Biol. **72**: 619–632.

13. WAKABAYASHI, T., H. E. HUXLEY, L. A. AMOS & A. KLUG. 1975. Three-dimensional image reconstruction of actin-tropomyosin complex and actin-tropomyosin-troponin I complex. J. Mol. Biol. **93**: 477–497.

14. SEYMOUR, J. & E. J. O'BRIEN. 1980. The position of tropomyosin in muscle thin filaments. Nature. **283**: 680–682.

15. SMITH, P. R., W. E. FOWLER, T. D. POLLARD & U. AEBI. 1983. Structure of the actin molecule determined from electron micrographs of crystalline sheets with a tentative alignment of the molecule in the actin filament. J. Mol. Biol. **166**: 641–660.

16. O'BRIEN, E. J., J. COUCH, G. R. P. JOHNSON & E. P. MORRIS. 1983. Structure of actin and the thin filament. *In* Actin: Structure and Function in Muscle and Non-Muscle Cells. C. dos Remedios and J. Bardeen, Eds. 3–15. Academic Press. Sidney, Australia.

17. HANSON, J. 1973. Evidence from electron microscope studies on actin paracrystals concerning the origin of the cross-striation in the thin filaments of vertebrate skeletal muscle. Proc. R. Soc. London, Ser. B. **183**: 39–58.

18. FOWLER, W. E. & U. AEBI. 1982. Polymorphism of actin paracrystals induced by polylysine. J. Cell Biol. **93**: 452–458.

19. RIOUX, L. & C. GICQUAUD. 1985. Actin paracrystalline sheets formed at the surface of positively charged liposomes. J. Ultrastruct. Res. **93**: 42–49.

20. STEWART, M. & G. VIGERS. 1986. Electron microscopy of frozen-hydrated biological material. Nature **319**: 631–636.

21. TAYLOR, K. A. & L. A. AMOS. 1981. A new model for the geometry of the binding of myosin crossbridges to muscle thin filaments. J. Mol. Biol. **147**: 297–324.

22. WAKABAYASHI, T. & C. TOYOSHIMA. 1981. Three-dimensional image analysis of the complex of thin filaments and myosin molecules from skeletal muscle. II. The multidomain structure of actin-myosin S1 complex. J. Biochem. **90**: 683–701.

23. VIBERT, P. & R. CRAIG. 1982. Three-dimensional reconstruction of thin filaments decorated with Ca^{2+}-regulated myosin. J. Mol. Biol. **157**: 299–319.

24. TAYLOR, K. A. & L. A. AMOS. 1983. Structure of actin in reconstructed images of S1 decorated filaments. *In* Actin: Structure and Function in Muscle and Non-Muscle Cells. C. dos Remedios and J. Bardeen, Eds. 25–26. Academic Press. Sidney, Australia.

25. TOYOSHIMA, C. & T. WAKABAYASHI. 1985. Three-dimensional image analysis of the complex of thin filaments and myosin from skeletal muscle. V. Assignment of actin in the actin-tropomyosin-myosin subfragment-1 complex. J. Biochem. **97**: 245–263.

26. EGELMAN, E. H. & D. J. DEROSIER. 1983. A model for F-actin derived from image analysis of isolated filaments (Appendix). J. Mol. Biol. **166**: 623–629.

27. TAJIMA, Y., K. KAMIYA & T. SETO. 1983. X-ray structure analysis of thin filaments of a molluscan smooth muscle in the living relaxed state. Biophys. J. **43**: 335–343.

28. EGELMAN, E. H. & R. PADRON. 1984. X-ray diffraction evidence that actin is a 100 A filament. Nature **307**: 56–58.

29. EGELMAN, E. H. 1985. The structure of F-actin. J. Muscle Res. Cell Motil. **6**: 129–151.

30. SMITH, P. R. & U. AEBI. 1974. Computer-generated Fourier transforms of helical particles. J. Phys. A: Gen. Phys. **7**: 1627–1633.

31. SMITH, P. R., U. AEBI, R. JOSEPHS & M. KESSEL. 1976. Studies of the structure of the bacteriophage T4 tail sheath. I. The recovery of 3-D structural information from the extended sheath. J. Mol. Biol. **106**: 243–271.

32. HARTT, J. & R. MENDELSON. 1980. X-ray scattering of F-actin and myosin subfragment-1 complex. Fed. Proc. **39**: 1729.

33. KABSCH, W., H. G. MANNHERZ & D. SUCK. 1985. Three-dimensional structure of the complex of actin and DNase I at 4.5 Å resolution. The EMBO J. **4**: 2113–2118.

34. EGELMAN, E. H., N. FRANCIS & D. J. DEROSIER. 1983. Helical disorder and the filament

structure of F-actin are elucidated by the angle-layered aggregate. J. Mol. Biol. **166:** 605–622.
35. BULLARD, B., J. BELL, R. CRAIG & K. LEONARD. 1985. Arthrin: A new actin-like protein in insect flight muscle. J. Mol. Biol. **182:** 443–454.
36. ENGEL, A. 1978. Molecular weight determination by scanning transmission electron microscopy. Ultramicroscopy **3:** 273–281.
37. DiCAPUA, E., A. ENGEL, A. STASIAK & T. KOLLER. 1982. Characterization of complexes between recA protein and duplex DNA by electron microscopy. J. Mol. Biol. **157:** 87–103.
38. ENGEL, A., R. EICHNER & U. AEBI. 1985. Polymorphism of reconstituted human epidermal keratin filaments: Determination of their mass-per-length and width by scanning transmission electron microscopy (STEM). J. Ultrastruct. Res. **90:** 323–335.
39. TAYLOR, K. A., M. C. REEDY, L. CORDOVA & M. K. REEDY. 1984. Three-dimensional reconstruction of rigor insect flight muscle from tilted thin sections. Nature **310:** 285–291.

DISCUSSION OF THE PAPER

D. J. DEROSIER (*Brandeis University, Waltham, Massachusetts*): Is the amount of slip that you see or what you would predict, comparable to the amount that Dr. Clarence E. Schutt at Princeton University finds in the variation in the C-axis along the crystals?

AEBI: I don't know.

DEROSIER: I think he finds that along one of the axes in the crystal the unit cell dimension can vary almost continuously by about 10 Å, which is a rather substantial amount. One of the reasons that it was such a slow project. I was just wondering, if that was at all similar to what you would find? Did you find as much as 15 Å?

AEBI: Yes. It was quite substantial. But I don't know whether it would be the same orientation.

DEROSIER: Have you determined what the effect on the transform would be, if you had such cumulative slippage, as you call it?

AEBI: Not particularly. Of course the most obvious thing is that you get the variable crossovers. Basically, it's as you illustrated in your first slide with the two plastic tubes you wound around each other.

K. A. TAYLOR (*Duke University Medical Center, Durham, North Carolina*): In some recent three-dimensional reconstructions on insect flight muscle that the Reedys and I did, we found that in regions where the cross bridges bound, the two actin strands were separated. At the time I thought that this must have been an artifact, but that observation would seem to fit in with your micrographs of the two separated actin strands.

AEBI: In fact, I would have loved to mention the finding you made some time ago in three-dimensional reconstructions of the *myac* layer (c.f. Taylor *et al.* 1984. Nature **310:** 285–291), since it so nicely supports the "lateral slipping" model I proposed in my talk. The reason that I did not is because at a recent meeting you tried to convince the audience with some newer reconstructions that this earlier data was most likely biased by a reconstruction artifact—so I did not want to embarrass you!

The Structure of Membrane Bound Cytochrome c Oxidase[a]

TERRENCE G. FREY[b] AND TAN CHANG[c]

Department of Biochemistry and Biophysics
University of Pennsylvania
Philadelphia, Pennsylvania 19104-6089

INTRODUCTION

Cytochrome c oxidase is the enzyme responsible for consumption of virtually all the molecular oxygen used by eukaryotes during aerobic metabolism. As the terminal component of mitochondrial electron transport, cytochrome oxidase accepts electrons from ferrocytochrome c in the intermembrane space to reduce oxygen in a reaction that consumes protons from the mitochondrial matrix. It also transports protons (*i.e.* cytochrome oxidase is a proton pump) from the matrix to the intermembrane space.[1] The electrochemical proton gradient created across the inner mitochondrial membrane by cytochrome oxidase and the other respiratory complexes is the form in which energy from electron transport is stored prior to synthesis of ATP by the mitochondrial F_0-F_1 ATP synthetase.[2] Cytochrome oxidase is a very complex integral membrane protein consisting of 7–12 different polypeptides,[3] 2 molecules of heme *a*, and 2 copper ions. The exact number of polypeptides in cytochrome oxidases isolated from mammals is controversial, but Buse *et al.* have isolated and sequenced 12 unique polypeptides present at unit stoichiometry with an aggregate molecular weight of 203,000.[4] The 3 largest polypeptides (I, II, and III) are coded by mitochondrial DNA, while all of the smaller polypeptides are coded by nuclear DNA (for reviews see REFERENCES 5–7).

TWO-DIMENSIONAL CRYSTALS

Monomers

Formation of two-dimensional crystals of cytochrome oxidase[d] was reported as early as 1968,[8] and two crystal forms have now been characterized. Crystalline sheets of cytochrome oxidase monomers are produced by extracting beef heart mitochondria with sodium deoxycholate to remove other membrane proteins and excess phospholipid.[9] The resulting preparation contains 12–15% phospholipid by weight with monomers of cytochrome oxidase packed into a rectangular unit cell with dimensions **a** = 68 Å and **b** = 174 Å and the symmetry of the two-sided group p12$_1$.[10] The three-dimensional reconstruction determined by Fuller *et al.* from images of crystals

[a]Supported by National Institutes of Health Grants GM-28750 to T. Frey and RRO-1747 for computer equipment.
[b]Present address: Department of Biology, San Diego State University, San Diego, CA 92182.
[c]Present address: Department of Biochemistry, University of Arizona, Tucson, AZ 85721.
[d]A two-dimensional crystal contains molecules periodically arranged on a two-dimensional lattice with no repeat in the third dimension.

that had been tilted by up to 60° shows that a cytochrome oxidase monomer is shaped somewhat like a lowercase letter "y" with a height of 110 Å and a center-to-center separation of the arms of 40 Å.[10] There is no evidence for the presence of a continuous lipid bilayer in the three-dimensional reconstruction, which is consistent with the low phospholipid content of these preparations. Although their three-dimensional model enclosed a volume corresponding to a molecular weight of only 102,000, the arrangement of structural domains agrees very well with the structure of cytochrome oxidase dimers observed in the other crystal form described below.

Dimers

The other two-dimensional cytochrome oxidase crystal form was first described by Vanderkooi *et al.,*[11] who observed it in preparations of membranous cytochrome oxidase produced by extracting beef heart mitochondria with Triton detergents (X-114 and X-100) in the presence of KCl. The Tritons remove other mitochondrial membrane proteins and excess phospholipid producing a near quantitative recovery of cytochrome oxidase in a membranous fraction containing 25–30% phospholipid by weight (FIGURE 1a). In projection, the crystals display the symmetry of plane group *pgg* with a rectangular unit cell with dimensions $\mathbf{a}$ = 95 Å and $\mathbf{b}$ = 125 Å;[11–12] this symmetry requires that molecules lie on crystallographic twofold axes indicating that, at least in these crystals, cytochrome oxidase is dimeric. Each unit cell contains two cytochrome oxidase dimers having opposite hands in projection (FIGURE 1b). Kim *et al.* have produced an apparently identical crystal form by reconstituting purified cytochrome oxidase with soybean phospholipids.[13] The similarity of lipid concentration in these crystals to the inner mitochondrial membrane suggests that the enzyme may be dimeric *in vivo,* but there is little evidence to support or contradict this hypothesis.

Henderson *et al.* determined the three-dimensional space group, molecular packing, and low-resolution three-dimensional structure from images of tilted, negatively stained specimens.[12] They proposed that the crystals form in collapsed vesicles with molecules from opposing sides interlocking on the inside to provide the forces necessary to stabilize the crystal. The two-sided space group is $p22_12_1$ with the twofold axes perpendicular to the membrane, while the 2_1 screw axes lie at the center of the crystal parallel to the $\mathbf{a}$ and $\mathbf{b}$ crystal axes. The two classes of dimers with opposite hands observed in filtered images of these crystals arise from the fact that molecules on one side of the collapsed vesicle are viewed in projection upside down relative to those on the opposite side (FIGURE 1b, c). Each unit cell contains one dimer from the upper membrane layer of the vesicle and one dimer from the lower layer. Although the structures of most of the crystals present in these preparations are considerably more complicated with multilayered vesicles,[11,14] the basic arrangement proposed by Henderson *et al.* fits all available data. The symmetry of space group $p22_12_1$ requires that the enzyme molecules all face the same direction with respect to the vesicle interior. This is indeed observed in three-dimensional reconstructions, which all show a markedly asymmetric distribution of protein protruding 50–60 Å beyond the bilayer surface on the insides of vesicle crystals but little (less than 10–15 Å) if at all on exterior surfaces.[12] This result was confirmed in the three-dimensional reconstruction determined by Deatherage *et al.*[15] An asymmetric protein distribution is also observed in membrane profile structures determined by X ray diffraction of membranous cytochrome oxidase.[16] Frey *et al.*[17] identified the exterior surfaces of vesicle crystals as corresponding to the matrix side (M-side) of the inner mitochondrial membrane by demonstrating the binding of antibodies against subunit IV—which bind only to the M-side of the mitochondrial membrane—to the outside of the crystals and the lack of

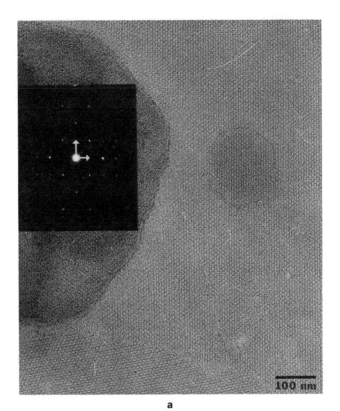

a

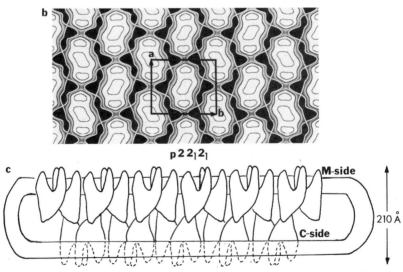

p 2 2₁ 2₁

FIGURE 1. (a) Electron micrograph of a negatively stained (1% uranyl acetate) crystal of cytochrome *c* oxidase dimers embedded in a lipid bilayer of a collapsed vesicle. The *inset* is an optical diffraction pattern of a subarray of this crystal with the *lattice vectors* of the reciprocal lattice indicated; the **a*** axis is vertical and the **b*** axis horizontal. (b) Noise-filtered image of the crystal in (a) with one unit cell outlined and real space *lattice vectors* indicated; **a** = 95 Å and **b** = 125 Å. Each unit cell contains a total of two dimers, which lie on crystallographic twofold axes; the two alternate columns of dimers have opposite hands as they appear upside down relative to one another. (c) Diagram of the molecular packing of cytochrome oxidase dimers in this crystal form. The two-sided plane group is p22₁2₁, and the exterior surfaces of the crystals correspond to the matrix side (M-side) of the inner mitochondrial membrane.

binding of antibodies against subunits II and III— which bind only to the cytoplasmic or C-side of the membrane.[18] Thus, cytochrome oxidase exposes the bulk of its mass to the intermembrane space (C-side) with relatively little mass exposed to the mitochondrial matrix (M-side).

QUATERNARY STRUCTURE

M-Side

Frey *et al.* determined the arrangement of y-shaped monomers in cytochrome oxidase dimers by selective contrast of three regions of vesicle crystals accomplished by different methods of specimen preparation.[19] The M-side structure is revealed by freeze-drying and heavy atom shadowing. Freeze-drying is employed to effect relatively gentle drying of specimens; heavy atom shadowing shows locations, sizes, and relative heights of protein domains on surfaces. By measuring the lengths of shadows cast by cytochrome oxidase dimers at the edges of crystalline arrays, they were able to calculate that the protein protrudes approximately 25–30 Å from the lipid bilayer surface on the exterior M-sides of the membrane.[19] This is somewhat more protein than is observed in any of the three-dimensional reconstructions[12,13,15] but is consistent with other data (see below). As expected, each unit cell contains only one cytochrome oxidase dimer, as the surface of only one membrane is contrasted by shadowing. In order to compensate for asymmetry in the images of unidirectionally shadowed crystals, the distribution of protein domains on the M-side of the membrane was determined from images of 12 crystals shadowed at different angles relative to the crystalline lattice. The 12 images were combined by cross-correlation alignment with twofold symmetry imposed on the resultant image, which is shown at the top of FIGURE 2; dark areas correspond to locations of protein domains. As seen from the M-side, a cytochrome oxidase dimer contains 4 protein domains (2 per monomer) aligned with two identical domains closely apposed at the twofold axis and two domains approximately 70 Å apart roughly along a unit cell diagonal from lower right to upper left.[19]

Intrabilayer Structure

Portions of membrane proteins lying within lipid bilayers are frequently revealed in freeze-fracture replicas as intramembrane particles (but see below) but only to low resolution and with many potential artifacts of the fracturing process. In the case of a two-dimensional crystal, the intrabilayer protein domains can be observed if the protein in solvent-accessible regions is contrast-matched by surrounding it with a medium of similar density. This is the situation that results when specimens are embedded in glucose by drying them down a thin layer of a 1% glucose solution; the glucose forms an amorphous solid with a density very similar to that of most proteins and also preserves the native protein structure in vacuum.[12,20–22] Such specimens have very low overall contrast and are very sensitive to electron irradiation, so that images must be recorded at low electron doses (1–5 electrons Å^2) with signal averaging over hundreds or thousands of unit cells to make the signal-to-noise ratio acceptable.[12,19,22]

An image of intrabilayer domains of cytochrome oxidase dimers is shown to the right in FIGURE 2. Protein appears dark with lighter areas corresponding to the low density hydrophobic core of the bilayer. As in images of negatively stained crystals, each unit cell contains 2 dimers of opposite hand, one from the upper layer and the other from the lower layer of the vesicle. The center of this image is chosen to

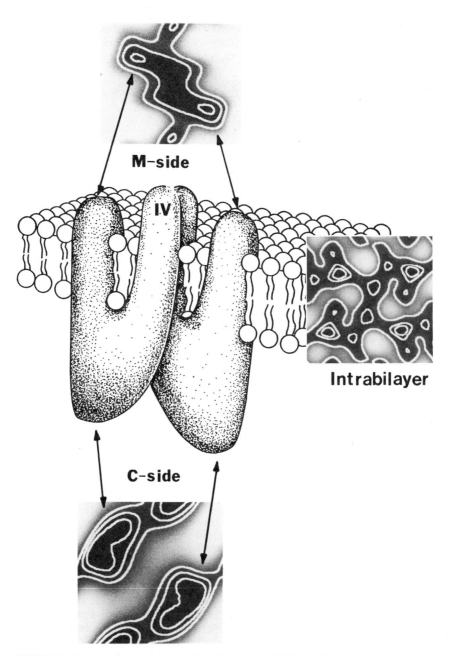

FIGURE 2. Drawing of a cytochrome oxidase dimer in a lipid bilayer. See text for an explanation of the surrounding images of M-side, intrabilayer, and C-side protein domains. The locus of subunit IV based upon decoration with anti-IV Fab antibody fragments is indicated.

correspond most closely with the M-side structure determined from shadowed specimens. Both M-side and intrabilayer images contain 4 domains per dimer located in exactly the same relative positions. This indicates that the "arms" of y-shaped monomers lie within and perpendicular to the lipid bilayer protruding 25–30 Å on the M-side of the membrane.[19] The same arrangement of protein domains within the bilayer was determined by Deatherage *et al.* based upon images of crystals embedded in media with densities intermediate between those of glucose and of negative stain.[22]

C-Side

The determination of M-side and of intrabilayer structures described above does not specify which of the two types of monomer arms (termed M1 and M2 by Fuller *et al.*, REFERENCE 10) is closer to the twofold axis. Fortunately, preparations of vesicle crystals exhibit an unusual pattern in freeze-fracture that reveals the structures of C-side surfaces. Unlike most membranous specimens, fracture planes travel along hydrophilic membrane surfaces in these crystals exposing both M-side and C-side surfaces.[14,19] Most freeze fracture replicas contain areas in which both types of surface have been exposed within a single crystal, and they can often be distinguished because of the similarity of one of them to freeze-dried and shadowed crystals, which contrast only M-side surfaces. An image of a C-side surface is shown at the bottom of FIGURE 2. Only 2 protein domains are observed per dimer in C-side images, and irrespective of the origin chosen for the center of the dimer, the two domains are well separated. Thus, cytochrome oxidase dimers must contain a large cleft on the C-side. This cleft is also observed in the three-dimensional reconstruction determined by Deatherage *et al.*,[15] and Capaldi *et al.* have suggested that it may contain the cytochrome *c* binding site.[23] There are only two ways to arrange the y-shaped monomers to form a dimer with a cleft on the C-side,[19] and the drawing in FIGURE 2 shows the one that agrees better with the three-dimensional reconstructions.[15]

Sectioned Grids

The distribution of protein across the membrane is clearly asymmetric in the case of cytochrome oxidase, and the question is a quantitative rather than a qualitative one, "How much protein protrudes beyond the lipid bilayer surface on the C-side and how much on the M-side of the mitochondrial inner membrane?" Three-dimensional reconstructions from images of tilted specimens provide only a semiquantitative answer, as there are technical problems arising from: a) The inability to tilt specimens by more than ± 60°, leaving a 60° cone of missing data, which reduces resolution along directions nearly perpendicular to the membrane, making it difficult to determine the extent of the structure along the membrane normal;[24] b) Uneven staining of upper and lower surfaces of the crystals, as one surface is in contact with the support film; and c) Possible disorder in protein domains exposed on exterior surfaces of the crystal. The volume enclosed by three-dimensional models of cytochrome oxidase would contain a protein of molecular weight less than 150,000;[15] apparently some portions of the 203,000 D molecule are not observed in the three-dimensional reconstructions.

One method for recording the missing information (a) and for determining the extent of uneven staining (b) and disorder (c) is to embed and section grids of negatively stained specimens. Jesior has elegantly demonstrated the efficacy of this approach by sectioning grids of negatively stained protein crystals in a predetermined orientation relative to the crystal axes.[25] We have applied his techniques to negatively stained vesicle crystals of cytochrome oxidase with very informative results.

Specimens were adsorbed to carbon/formvar coated finder grids and examined in the electron microscope to locate quality specimens; images of these specimens were recorded at high (around 40,000X) as well as intermediate magnifications to facilitate locating them with respect to grid fiducial marks. The grids were applied to thin layers of partly polymerized Epon 812 (Fluka) in an orientation that would allow sectioning along predetermined crystallographic directions. They were then covered with a layer of unpolymerized Epon and hardened by incubation at 70° C over one to two days. Blocks were carefully trimmed with glass knives, and then sections of 500–700 Å were taken perpendicular to the grid through the selected specimens; care was taken to pass the diamond knife first through the specimen and then through the carbon/formvar support film with the plane of the crystal parallel to the knife edge.

FIGURE 3a shows a vesicle crystal prior to embedding in Epon, and FIGURES 3a, b are sections perpendicular to the membrane along the (1, 1) crystal axis. Such sections typically show two 30-Å thick stain-excluding bands indicating the positions and extents of the top and bottom lipid bilayers. The fact that the bilayers are only 30 Å thick rather than 45–50 Å suggests penetration of the uranyl acetate into the headgroup regions of the phospholipid. A third stain-excluding band sandwiched between the two lipid bilayers is a region of high protein concentration where molecules of cytochrome oxidase from the top and bottom membranes meet at the center of the crystal. These features are seen more clearly in filtered images formed from segments of FIGURE 3b, c that have been linearly superimposed upon themselves along the plane of the membrane; these are shown in FIGURES 3d–g. Horizontal double-headed arrows indicate the positions of the lipid bilayers, and the support film is below the crystal. The magnification has been calibrated based upon the 200-Å thickness of these crystals to take into account possible compression along the membrane normal during sectioning.[26]

In projection down the (1, 1) axis molecules from the top and bottom membrane layers of the vesicle should line up vertically; this is diagramed in the filtered image in FIGURE 3h, where rows of particles perpendicular to the surfaces of the section alternate between the two hands. The vertical alignment of upper and lower molecules is exactly what is observed in the filtered images, and the white band in the center shows the region of highest protein concentration where the molecules interdigitate. Protein is clearly observed in and protruding through the thin layers of stain on the upper and lower surfaces of the crystal, and the two stain layers are seen to be uneven. The lower stain layer, which is in contact with the support film, is approximately 28 Å thick, while the upper is only 18 Å. These measurements correspond to the distance that protein protrudes beyond the hydrophobic core of the lipid bilayers. The measurement for the lower membrane in contact with the support film (28 Å) is in good agreement with measurements of the heights of protein domains on M-side surfaces observed in shadowed crystals.

The obvious uneven staining of top and bottom external surfaces of the vesicle is a partial explanation for the absence of significant structure there in three-dimensional reconstructions. These are usually calculated incorporating the full symmetry of the space group[12,15] and would, therefore, average the two surfaces, reducing the contribution of the well-contrasted surface in contact with the supporting film. The M-side protein domains observed in sectioned grids appear to be well-defined after linear superposition used to filter out random noise, arguing against disorder of the protein as a reason for their absence in three-dimensional reconstructions. We are now in the process of incorporating these images of sectioned grids into our three-dimensional dataset to see if this will improve the definition of M-side surfaces.

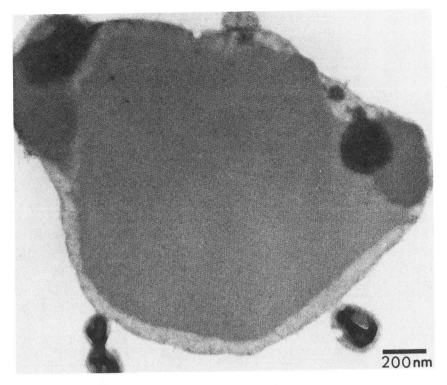

a

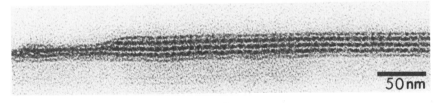

b

c

FIGURE 3a–c. (a) Image of a negatively stained vesicle crystal prior to embedding in Epon and sectioning. (b, c) Cross-sections of the crystal in (a) recorded from sections of the embedded grid. The carbon/formvar support film is the slightly darker area *beneath* the crystal; the thinner area to the *left* of (b) is a region at the edge of the crystal containing no protein molecules.

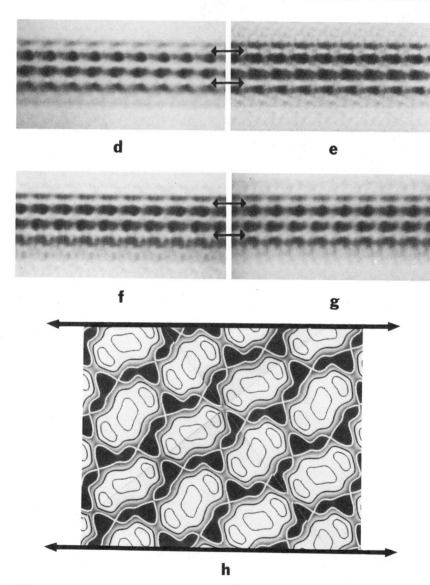

FIGURE 3d–h. (d–g) Filtered images of segments of (b,c) produced by linear superposition parallel to the membrane. The two lipid bilayers on the two sides of the collapsed vesicle are seen as two 30-Å thick stain-excluding light bands indicated by *horizontal double headed arrows*. The central stain-excluding band is a region of high protein concentration where molecules from the top and bottom of the vesicle interdigitate. In this orientation, molecules from the top and bottom align vertically. **(h)** FIGURE 1b oriented to illustrate the orientation of the crystal in the sections (b–g). The *double headed arrows* indicate the positions of the two surfaces of the thin section that would be oriented perpendicular to the plane of this image. The dimers in vertical columns alternate hand as they come from both the top and the bottom layers of the collapsed vesicle.

CONCLUSIONS

The drawing in FIGURE 2 summarizes our interpretation of the relevant data concerning the structure of cytochrome oxidase dimers, and their position in the phospholipid bilayer. A wealth of information on the low-resolution structure was obtained from three-dimensional reconstructions of negatively stained specimens based upon two-dimensional projection images of tilted specimens. Data obtained from shadowed specimens and from sectioned grids helps to define the limits of such three-dimensional reconstructions. The large C-side domains are well-defined in the three-dimensional models, which show them protruding 50–60 Å beyond the lipid bilayer, forming a large cleft in the dimer on the C-side.[12,15] The 25–30-Å M-side domains do not appear, however, in three-dimensional reconstructions but are clearly observed in shadowed specimens and in sections of grids of negatively stained specimens.[19] The grid sections provide the information missing from most three-dimensional datasets of this type (they correspond to a tilt angle of 90°), and addition of this data will hopefully improve the quality of three-dimensional reconstructions.

The available evidence suggests that subunit IV contributes a large portion of the protein mass on the M-side surface, as it is readily accessible to antibodies,[18] proteolytic enzymes,[27] and hydrophilic protein labeling reagents[28] from the M-side of the membrane. Frey *et al.* have used Anti-IV Fab fragments (monovalent antibody fragments) to locate subunit IV on the M-side domains nearest the twofold axis.[29,30] Analysis of the sequence of subunit IV predicts one transmembranous helix dividing it into a larger N-terminal domain and a smaller C-terminal domain. Proteolysis studies show that the N-terminal domain of subunit IV lies on the M-side of the membrane,[27] and calculation of its partial molecular volume indicates a diameter, if spherical, of 28 Å.[30] Thus, subunit IV could account for the 25–30-Å protein domains observed on the M-side of vesicle crystals in shadowed specimens and in sectioned grids.

ACKNOWLEDGMENTS

We would like to thank J. C. Jesior for many helpful suggestions.

REFERENCES

1. WIKSTROM, M. K. F. & H. T. SAARI. 1977. Biochim. Biophys. Acta **462:** 347–361.
2. MITCHELL, P. 1961. Nature **191:** 144–148.
3. KADENBACH, B., J. JARAUSCH, R. HARTMANN & P. MERLE. 1983. Anal. Biochem. **129:** 517–521.
4. BUSE, G., L. MEINECKE & B. BRUCH. 1985. J. Inorg. Biochem. **23:** 149–153.
5. WIKSTROM, M., K. KRAB & M. SARASTE. 1981. Cytochrome Oxidase: A Synthesis. Academic Press. London.
6. AZZI, A. 1979. Biochim. Biophys. Acta **549:** 177–222.
7. CAPALDI, R. A., F. MALATESTA & V. DARLEY-USMAR. 1983. Biochim. Biophys. Acta **726:** 135–148.
8. CRANE, F. L., J. W. STILES, K. S. PREZBINDOWSKI, F. J. RUZICKA & F. F. SUN. 1968. In Regulatory Functions of Biological Membranes. J. Jarnefelt, Ed. 21–56. Elsevier Press. Amsterdam.
9. SEKI, S., H. HAYASHI & T. ODA 1970. Arch. Biochem. Biophys. **138:**110–121.
10. FULLER, S. D., R. A. CAPALDI & R. HENDERSON. 1979. J. Mol. Biol. **134:** 305–327.
11. VANDERKOOI, G., A. E. SENIOR, R. A. CAPALDI & H. HAYASHI. 1972. Biochim. Biophys. Acta **274:** 38–48.

12. HENDERSON, R., R. A. CAPALDI & J. S. LEIGH. 1977. J. Mol. Biol. **112:** 631–648.
13. KIM, C. H., M. RADERMACHER, M. KESSEL, J. FRANK & T. E. KING. 1985. J. Inorg. Biochem. **23:** 163–169.
14. COSTELLO, M. J. & T. G. FREY. 1982. J. Mol. Biol. **162:** 131–156.
15. DEATHERAGE, J. F., R. HENDERSON & R. A. CAPALDI. 1982. J. Mol. Biol. **158:** 487–499.
16. JAYARAMAN, U., J. K. BLASIE, T. G. FREY & T. CHANG, submitted.
17. FREY, T. G., S. H. P. CHAN & G. SCHATZ. 1978. J. Biol. Chem. **253:** 4389–4395.
18. CHAN, S.H.P. & R. P. TRACY. 1978. Eur. J. Biochem. **89:** 595–605.
19. FREY, T. G., M. J. COSTELLO, B. KARLSSON, J. C. HASELGROVE & J. S. LEIGH. 1982. J. Mol. Biol. **162:** 113–130.
20. UNWIN, P. N. T. & R. HENDERSON. 1975. J. Mol. Biol. **94:** 425–440.
21. KUHLBRANDT, W. 1982. Ultramicroscopy **7:** 221–232.
22. DEATHERAGE, J.F., R. HENDERSON & R. A. CAPALDI. 1982. J. Mol. Biol. **158:** 501–514.
23. CAPALDI, R. A., V. DARLEY-USMAR, S. D. FULLER & F. MILLETT. 1982. FEBS Lett. **138:** 1–7.
24. HENDERSON, R. & P. N. T. UNWIN. 1975. Nature **257:** 28–32.
25. JESIOR, J. C. 1982. The EMBO Journal **1:** 1423–1428.
26. JESIOR, J. C. 1985. J. Ultrastruct. Res. **90:** 135–144.
27. MALATESTA, F., V. DARLEY-USMAR, C. DE JONG, L. J. PROCHASKA, R. BISSON, R. A. CAPALDI, G. C. M. STEFFENS & G. BUSE. 1983. Biochemistry **22:** 4405–4411.
28. LUDWIG, B., N. W. DOWNER & R. A. CAPALDI. 1979. Biochemistry **18:** 1401–1407.
29. FREY, T. G., M. J. COSTELLO & S. H. P. CHAN. 1984. Ultramicroscopy **13:** 85–92.
30. FREY, T. G., L. A. KUHN, J. S. LEIGH, M. J. COSTELLO & S. H. P. CHAN. 1985. J. Inorg. Biochem. **23:** 155–162.

DISCUSSION OF THE PAPER

G. C. RUBEN (*Dartmouth College, Hanover, New Hampshire*): It was reported before that cytochrome oxidase was about 110 to 120 Å high. What kind of figure are you getting for the height of cytochrome oxidase through the membrane?

FREY: We are just now in the process of determining a three-dimensional reconstruction incorporating the grid sectioning data. The only thing I can add to what was published previously is that we would add 10–20 Å to the matrix side domains of the molecule, which Deatherage *et al.* (1982)[22] reported to be less than 20 Å. I should point out that the molecular volume of their models is significantly smaller than expected from the 200,000 molecular weight which is now believed by many to be the true molecular weight of cytochrome oxidase in these preparations.

D. F. PARSONS (*New York State Department of Health, Albany, New York*): Did you dry-embed these grids before sectioning them, or were they processed through dehydration solvents?

FREY: They were not preprocessed in any way. They're just negatively stained grids. They were placed on partially polymerized resin and unpolymerized resin was layered over the grid.

PARSONS: What does careful examination of those cross sections tell you about the distribution and uniformity or heterogeneity of mass loss in relation to the electron dose that you use? Could it possibly be that the thin layer that you observe on top represents preferential mass loss?

FREY: I have no idea, I have not measured mass loss.

PARSONS: What was your dose?

FREY: Rather high, more than 100 electrons per Å^2. You need to give the negatively stained crystal a high enough dose to transform the uranyl acetate into uranyl oxide and uranium dioxide to make it stable during the sectioning process.

Structural Studies of Na,K-ATPase[a]

MANIJEH MOHRAZ, MOVIEN YEE, AND P.R. SMITH

Department of Cell Biology
New York University School of Medicine
New York, New York 10016

INTRODUCTION

Sodium and potassium ion activated adenosine-triphosphatase (Na,K-ATPase) is one of a group of transport ATPases that couple the hydrolysis of ATP to the active transport of ions across membranes. These include Ca-ATPases of the cell membrane and the sarcoplasmic reticulum, the gastric H,K-ATPase, and the H-ATPase of bacterial and fungal membranes. Na,K-ATPase and H,K-ATPase function as bidirectional pumps, since they transport ions in both directions across the cell membrane. It has been demonstrated that in red blood cells Na,K-ATPase uses the energy from the hydrolysis of one ATP molecule to transport 3 Na^+ out of the cell and 2 K^+ into the cell.[1,2] The enzyme consists of two polypeptide chains: α (M_r 110,000) is the catalytic subunit, and β (M_r 50–60,000) a glycoprotein whose exact function is not yet understood. The two subunits exist in a 1:1 molar stoichiometry.[3,4]

Purification of Na,K-ATPase from a number of tissues[5-8] has allowed extensive studies of its properties (see reviews in REFERENCES 9–12). A model, first proposed by Post and his collaborators,[13,14] has been elaborated to explain many of the existing kinetic and biochemical observations. According to this model the enzyme can exist in two major conformational states during the pumping cycle. In the E_1 form it has affinity for Na^+ and ATP and is stabilized by these ligands. In the E_2 form, the higher energy state, it binds K^+ and/or inorganic phosphate and is stabilized by these ions. E_1 and E_2 have different conformations: digestion by trypsin in the presence of Na^+ produces a different fragmentation pattern than proteolysis in the presence of K^+.[15,16]

Significant new structural information concerning the Na,K-ATPase has emerged in the past five years. The amino acid sequences of α[17,18] and β[19-21] have been determined. Furthermore, the discovery of a scheme to induce two-dimensional crystals of the enzyme in membrane fragments[22] has opened the way for structural studies by electron microscopy and image analysis. A number of investigators have reported structural analysis of the enzyme in projection[23-25] as well as preliminary data on its three-dimensional structure.[26,27] The oligomeric structure of the pump has, however, remained a subject of controversy. While some studies have produced data to suggest that the promoter ($\alpha\beta$) is capable of active transport,[28,29] others have pointed to the dimer ($\alpha\beta)_2$ as the functional unit.[30-34]

In our laboratory we have conducted structural studies of Na,K-ATPase and biochemical experiments designed to complement them. The combination of the two approaches has yielded interesting results that address the question of the oligomeric form of the functional unit and offer suggestions as to how the transport of Na^+ and K^+ might occur. From these results we propose a model for the functioning of the Na,K

[a]Supported by National Institute of General Medical Sciences Grants GM-35399 to M.M. and GM-26723 to P.R.S. and Biomedical Research Support Grant RR-05399.

pump that is consistent with much of the biochemical and kinetic information on the enzyme.

STRUCTURE OF Na,K-ATPase IN PROJECTION

In our initial structural studies of Na,K-ATPase we demonstrated that in addition to the intact enzyme, it was possible to crystallize preparations of a trypsin-digested form of it in which a fragment of M_r 20,000 had been removed from the N terminus of its α subunit.[24]

Computer filtration of negatively stained images of the intact, crystalline enzyme (FIGURE 1) showed that, in projection, the ATPase molecule was composed of a massive "body" with a less massive "hook" extending from it. Comparison of these images with those of the digested enzyme indicated that the mass lost by proteolysis had been removed from the body region of the intact ATPase molecule. This allowed the identification of at least a portion of the body with the α subunit. Furthermore, comparison of the shape of the molecule and the unit cell dimensions in the crystalline sheets of Na,K-ATPase with the corresponding parameters in the Ca-ATPase[35–37] indicated that the hook region might accommodate the β subunit. FIGURE 1 shows the domains in the projection view of the enzyme that we have identified as the α and β subunits.[24,38]

DIMERIC CRYSTALLINE SHEETS OF Na,K-ATPase

In an effort to produce better-ordered crystalline sheets of Na,K-ATPase we discovered that incubation of the enzyme with phospholipase A_2, while simultaneously dialyzing it against a buffer that contained crystal-inducing ions, caused extensive crystalline arrays to form overnight.[39] These phospholipase-induced sheets were

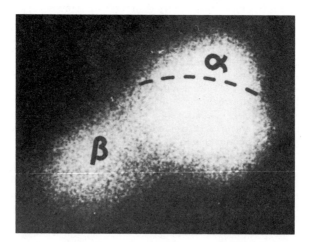

FIGURE 1. Computer-filtered image of Na,K-ATPase molecule derived from micrographs of negatively stained crystalline sheets of the enzyme. The view is perpendicular to the plane of the membrane. The regions corresponding to the α and β subunits are indicated in the figure.

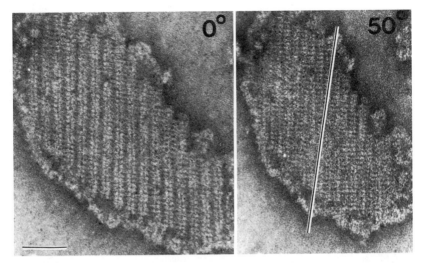

FIGURE 2. Zero and fifty degree tilted views of a negatively stained dimeric sheet of Na,K-ATPase. Crystallization was induced by phospholipase A_2 treatment and dialysis against a buffer containing 10 mM Tris-HCl, 1 mM NaVO$_3$, 5 mM MgCl$_2$, 5 mM CaCl$_2$, and 0.005% NaN$_3$ at pH 7.5. The *diagonal line* shows the orientation of the tilt axis. *Scale bar:* 100 nm.

exclusively dimeric with a $(\alpha\beta)_2$ structure in the unit cell (FIGURE 2). By contrast, the sheets produced by the previous methods of crystallization were nearly always monomeric (one $\alpha\beta$ structure in the unit cell) with dimeric patches only rarely seen in suspensions of the intact enzyme and never seen in the digested ATPase. Examination of the crystallization process showed that pairs of ATPase molecules initially interacted to form long ribbons. These ribbons, which formed as soon as incubation with phospholipase started, subsequently associated laterally to form the extended arrays. Inspection of the ribbons revealed that the initial interaction in the pair of enzyme molecules was between the domains that we had identified as α subunits (FIGURE 1).

THREE-DIMENSIONAL STRUCTURE OF Na,K-ATPase

Tilt series were recorded at 60, 55, 50, 45, and 30 degrees from negatively stained, phospholipase A_2-induced dimeric sheets of the ATPase similar in quality to the sheet shown in FIGURE 2. Side views of the sheets were obtained when they were observed to curl up at their edges. A total of seven tilt series were computer-filtered to yield 42 independent projection images of the dimer sheets. In each tilt series identical areas of at least 100 unit cells were averaged in the filtration process to yield a single filtered unit cell for each set of tilt parameters.

Synthesis of the images of the tilted molecule into a three-dimensional reconstruction was performed according to the basic scheme of Fuller *et al.*,[40] using methods we employed previously.[41,42] The three-dimensional Fourier transform was filled with significant information to an overall resolution limit of about 2.5 nm with data from the side view being added into the transform on the Z* axis only. Reverse Fourier transformation yielded a reconstruction that was represented on a series of 26 slices parallel to the plane of the membrane and separated by 0.577 nm.

The final reconstruction showed strong structure in only a limited volume. This was a region of approximately 6.0 nm in depth, abutting one side of the membrane as determined from the side view. While some structure could be seen on the other side of the membrane, its contrast was low, and it was therefore impossible to be confident of the information presented by it. In FIGURE 3 we show the high-contrast region on one side of the membrane. This is most probably the cytoplasmic side, since most of the mass of the molecule is known to protrude from this side with little mass being found on the exoplasmic surface.[9–12] The volume for a unit cell of the reconstruction shown in FIGURE 3 was calculated to be 170 cubic nm. This represents a dimer mass of 144,000 daltons, which is approximately 70% of the mass of the α subunit.

The reconstruction shows a complex mass distribution consisting of ribbons of

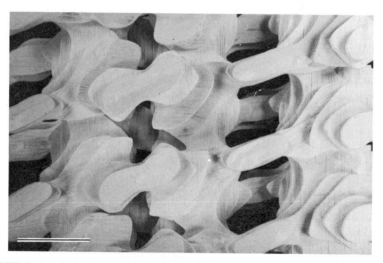

FIGURE 3. A balsa wood model of the three-dimensional reconstruction of Na,K-ATPase viewed from inside the cell and perpendicular to the membrane. It shows the protein mass on the cytoplasmic side of the plasma membrane. *Scale bar:* 4 nm.

paired stain-excluding columns attached to the membrane and sets of bridges interconnecting them. At the level closest to the membrane, bridges are seen to run parallel to the 14.5-nm lattice vector connecting the bases of the paired stain-excluding regions in each ribbon and continuing to connect adjacent ribbons. Since the β subunit has been shown to protrude only slightly from the cytoplasmic side of the membrane, we think that the bridges between ribbons are most likely associated with the β subunit. Each of the stain columns can be seen to be tilted along the ribbon and inwards towards its center. At a level of about 3.0 nm above the membrane a new set of bridges are formed, which connect molecules parallel to the axis of the ribbon. In the upper levels of the reconstruction new sets of bridges are formed between the α subunits, reflecting the strong and specific interactions needed to ensure dimer formation. A more detailed description of these results is presented elsewhere.[43]

Brief reports of reconstructions of the Na,K-ATPase by Hebert *et al.*,[26] and Ovchinnikov *et al.*[27] have been published. The specimens used were nominally of dimer sheets of the ATPase, but the lattice parameters of the sheets and their appearance in the electron microscope appear quite different from each other, and also different from the sheets we employ. Indeed the sheet used by Ovchinnikov *et al.*[27] shows little of the dimer ribbon structure common to the specimens studied by Hebert *et al.*[23,26] and ourselves.[24,38]

Both groups report reconstructing an ATPase molecule that spans the membrane, and both reconstructions are approximately 10.0 nm in depth, in contrast to our result, which shows significant structure confined to one side of the membrane only. Insufficient data is available to assess the results of Ovchinnikov *et al.* However, the result of Hebert *et al.* for the thickness of the structure can be checked by reverse Fourier transformation of the data they present for their (1,0) and (2,1) lattice lines. This procedure shows that the contribution of these lattice lines is significant in only a very thin slice of their final reconstruction, and is inconsistent with a 10.0-nm thick model.

CONFORMATION OF Na,K-ATPase IN THE DIMERIC SHEETS

Jørgensen[15] showed that limited digestion of membrane-bound Na,K-ATPase by trypsin in the presence of Na^+ and in the presence of K^+ produced different cleavage patterns, indicating that the enzyme assumed significantly different conformations depending on whether it had bound Na^+ (E_1) or bound K^+ (E_2). He termed the two conformations "Na form" and "K form." Absence of ions in the proteolysis buffer produced a pattern that corresponded to the Na form. The gel in FIGURE 4 shows the pure ATPase in lane A and its trypsin-digested patterns in lanes B (Na form) and C (K form). In the Na form the α subunit is cleaved to a fragment of M_r 77,000, with no apparent digestion of the β subunit. The 77,000 piece is digested further if the proteolysis is allowed to proceed too long (lower bands in FIGURE 4, lane B). In the K form the α subunit is digested to produce two fragments of M_r 58,000 and 41,000. These fragments lose a further 2,000 piece each to produce the doublets[16] seen in FIGURE 4, lane C.

We exploited the specificity of trypsin digestion to investigate the conformational state of the ATPase molecule in the dimeric sheets. Crystalline sheets of the enzyme, induced by phospholipase A_2 and various inducing ions (FIGURE 5A,B), were digested with trypsin. Na^+ and K^+ were excluded from the proteolysis buffer so that the conformation of the ATPase molecule in the crystalline state would not be disturbed. FIGURE 4, lane D, shows the fragmentation pattern of the dimeric sheet of FIGURE 5A. The pattern corresponds to a combination of the Na and the K forms. In order to be sure that this unexpected cleavage pattern was not caused by phospholipase A_2, but rather was the result of a natural association in the dimers, we treated the purified enzyme with phospholipase A_2 under ionic conditions that did not induce dimeric sheets (FIGURE 5C; see figure caption for experimental conditions). Trypsin digestion of vesicles shown in FIGURE 5C produced only the expected Na form. Subsequent extensive proteolysis studies of dimeric and "nondimeric" sheets induced under various conditions have confirmed that trypsin digestion of dimeric sheets always gives a pattern that is a mixture of Na and K forms.[44]

A MODEL FOR THE FUNCTIONING OF THE Na,K PUMP

The results we have presented here have implications for the oligomeric structure as well as for the functioning of the Na,K pump. They suggest that the dimers are formed by interactions of two ATPase molecules, one in the E_1 (Na-binding) and the other in the E_2 (K-binding) conformation. Although there is no direct evidence that the dimer is indeed the functional unit, formation of very stable, extensive crystalline arrays in which a dimer of conformationally distinct molecules constitutes the building block does imply functional significance.

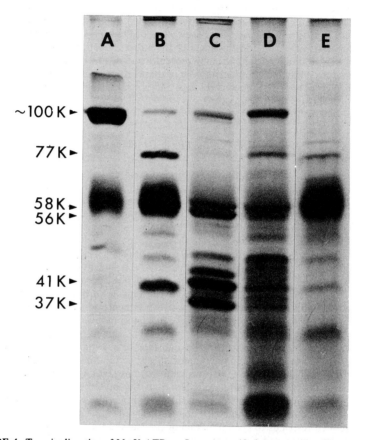

FIGURE 4. Trypsin digestion of Na,K-ATPase. Lane **A,** purified enzyme; **B,** ATPase digested for 30 minutes at 37° C in 25 mM imidazole and 1 mM EDTA at pH 7.5 with a trypsin-to-ATPase weight ratio of 1:100; **C,** ATPase digested for 15 minutes at 37° C in 150 mM KCl, 25 mM imidazole, and 1 mM EDTA at pH 7.5 with a trypsin-to-ATPase weight ratio of 1:10; **D,** ATPase dimeric sheets induced by phospholipase A_2, Mg^{2+}, and PO_4^{3-} digested for 30 minutes at 37° C in 25 mM imidazole and 1 mM EDTA at pH 7.5 with a trypsin-to-ATPase weight ratio of 1:100; **E,** ATPase treated with phospholipase A_2 and NaCl digested for 30 minutes at 37° C in 25 mM imidazole and 1 mM EDTA at pH 7.5 with a trypsin-to-ATPase weight ratio of 1:100. The gel was 10% acrylamide stained with silver nitrate.

We propose a model for the operation of the Na,K pump that is consistent with these findings. According to this model, which is shown schematically in FIGURE 6, the functional unit is a dimer $(\alpha\beta)_2$ of Na,K-ATPase formed through the interaction of the α subunit. During the pumping cycle the two halves of the unit operate 180° out of

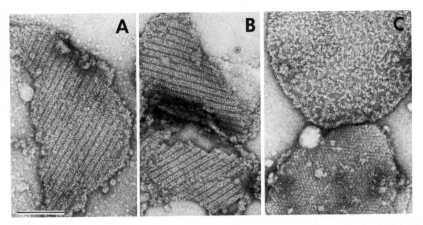

FIGURE 5. Sheets of phospholipase A_2 treated ATPase. 10 μg of ATPase was mixed with 0.8 units of phospholipase A_2 in 10 mM Tris at pH 7.5 and dialyzed at 4° C against: (**A**) 10 mM Tris, 5 mM H_3PO_4, 1 mM $MgCl_2$, 1 mM $CaCl_2$, 0.005% NaN_3 at pH 7.3; (**B**) 10 mM Tris, 50 mM NaCl, 50 mM KCl, 1 mM $CaCl_2$, 0.005% NaN_3 at pH 7.0; (**C**) 10 mM Tris, 150 mM NaCl, 1 mM $CaCl_2$, 0.005% NaN_3 at pH 7.5. *Scale bar:* 100 nm.

phase, so that while one enzyme molecule is in the E_1 conformation and transports Na^+ out of the cell, the other is in the E_2 state and pumps K^+ into the cell. By the end of the cycle the two would have exchanged conformation so that the former would now transport K^+ and the latter, Na^+.

Support for the dimeric structure comes from numerous studies. These include radiation inactivation of the native enzyme,[30] chemical cross-linking of the membrane-bound enzyme,[31,32] molecular weight determination of the active complex solubilized in nonionic detergents,[33] and low-angle neutron scattering of the solubilized enzyme.[34] Additionally, the mechanism that we propose for the functioning of the pump provides clarification for some of the experimental observations that cannot be readily explained by a protomeric pump. These include "half-of-the-sites" behavior, high- and low-affinity ATP binding sites, and simultaneous versus sequential transport of Na^+ and K^+ (for a review, see REFERENCE 12).

Models similar to ours have been proposed previously;[45-47] however, they have been largely ignored due to lack of direct evidence. The existence of a dimer of conformationally distinct molecules represents the first structural result in support of a

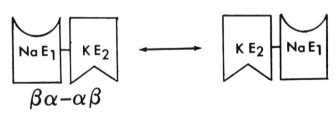

FIGURE 6. A model proposed for the function of Na,K-ATPase ion exchange pump. The two ATPase molecules are associated through their α subunits. During the pump cycle the two (αβ) units function 180° out of phase with one in the **Na**-binding (E_1) and the other in the **K**-binding (E_2) state. ATP hydrolysis causes ion exchange across the membrane, and the two (αβ) units change roles, ready to pump the other ion across the membrane.

mechanism of transport where the two halves of the dimeric pump function out of phase.

ACKNOWLEDGMENTS

We are grateful to Marcia Simpson for her excellent electron microscopy, to Billy DeLeon for technical assistance, and to Dr. E.L. Buhle for help with image digitization.

REFERENCES

1. SEN, A. K. & R. L. POST. 1964. J. Biol. Chem. **239:** 345–352.
2. GARRAHAN, P. J. & I. M. GLYNN. 1967. J. Physiol. **192:** 217–235.
3. LIANG, S. M. & C. G. WINTER. 1977. J. Biol. Chem. **252:** 8278–8284.
4. CRAIG, W. S. & J. KYTE. 1980. J. Biol. Chem. **255:** 6262–6269.
5. KYTE, J. 1971. J. Biol. Chem. **246:** 4157–4165.
6. JØRGENSEN, P. L. 1974. Biochim. Biophys. Acta **356:** 36–52.
7. HOKIN, L. E., J. L. DAHL, J. D. DEUPREE, J. F. DIXON, J. F. HACKNEY & J. F. PERDUE. 1973. J. Biol. Chem. **248:** 2593–2605.
8. DIXON, J. F. & L. E. HOKIN. 1974. Arch. Biochem. Biophys. **163:** 749–758.
9. ROBINSON, J. D. & M. S. FLASHNER. 1979. Biochim. Biophys. Acta **549:** 145–176.
10. CANTLEY, L. C. 1981. *In* Current Topics in Bioenergetics. Vol. 2: 201–237. Academic Press. New York, NY.
11. JØRGENSEN, P. L. 1982. Biochim. Biophys. Acta **694:** 27–68.
12. GLYNN, I. M. 1985. *In* The Enzymes of Biological Membranes. A. N. Martonosi, Ed. Vol. 3: 35–114. Plenum Press. New York and London.
13. POST, R. L., C. HEGYVARY & S. KUME. 1972. J. Biol. Chem. **247:** 6530–6540.
14. POST, R. L., G. TODA & F. N. ROGERS. 1975. J. Biol. Chem. **250:** 691–701.
15. JØRGENSEN, P. L. 1975. Biochim. Biophys. Acta **401:** 399–415.
16. CASTRO, J. & R. A. FARLEY. 1979. J. Biol. Chem. **254:** 2221–2228.
17. SCHULL, G. E., A. SCHWARTZ & J. B. LINGREL. 1985. Nature **316:** 691–695.
18. KAWAKAMI, K., S. NOGUCHI, M. NODA, H. TAKAHASHI, T. OHTA, M. KAWAMURA, H. NOJIMA, K. NAGANO, T. HIROSE, S. INAYAMA, H. HAYASHIDA, T. MIYATA & S. NUMA. 1985. Nature **316:** 733–736.
19. NOGUCHI, S., M. NODA, H. TAKAHASHI, K. KAWAKAMI, T. OHTA, K. NAGANO, T. HIROSE, S. INAYAMA, M. KAWAMURA & S. NUMA. 1986. FEBS Lett. **196:** 315–320.
20. KAWAKAMI, K., H. NOJIMA, T. OHTA & K. NAGANO. 1986. Nuc. Acid Res. **14:** 2833–2844.
21. SCHULL, G. E., L. K. LANE & J. B. LINGREL. 1986. Nature **321:** 429–431.
22. SKRIVER, E., A. B. MAUNSBACH & P. L. JØRGENSEN. 1981. FEBS Lett. **131:** 219–222.
23. HEBERT, H., P. L. JØRGENSEN, E. SKRIVER & A. B. MAUNSBACH. 1982. Biochim. Biophys. Acta **689:** 571–574.
24. MOHRAZ, M. & P. R. SMITH. 1984. J. Cell Biol. **98:** 1836–1841.
25. ZAMPIGHI, G., J. KYTE & W. FREYTAG. 1984. J. Cell Biol. **98:** 1851–1864.
26. HEBERT, H., E. SKRIVER & A. B. MAUNSBACH. 1985. FEBS Lett. **187:** 182–186.
27. OVCHINNIKOV, YU. A., V. V. DEMIN, A. N. BARNAKOV, A. P. KUZIN, A. V. LUNEV, N. N. MODYANOV & K. N. DZHANDZHUGAZYAN. 1985. FEBS Lett. **190:** 73–76.
28. BROTHERUS, J. R., J. V. MOLLER & P. L. JØRGENSEN. 1981. Biochem. Biophys. Res. Commun. **100:** 146–154.
29. CRAIG, W. S. 1982. Biochemistry **21:** 5707–5717.
30. KEPNER, G. R. & R. I. MACEY. 1968. Biochim. Biophys. Acta **163:** 188–203.
31. KYTE, J. 1975. J. Biol. Chem. **250:** 7443–7449.
32. PERIYASAMY, S. M., W.-H. HUANG & A. ASKARI. 1983. J. Biol. Chem. **258:** 9878–9885.
33. ESMANN, M. 1984. Biochim. Biophys. Acta **787:** 81–89.

34. PACHENCE, J. M., B. P. SCHOENBORN & I. S. EDELMAN. 1983. Biophys. J. **41:** 370a.
35. BUHLE, E. L., B. E. KNOX, E. SERPERSU & U. AEBI. 1983. J. Ultrastruct. Res. **85:** 186–203.
36. CASTELLANI, L. & M. D. HARDWICKE. 1983. J. Cell Biol. **97:** 557–561.
37. TAYLOR, K., L. DUX & A. MARTONOSI. 1984. J. Mol. Biol. **174:** 193–204.
38. MOHRAZ, M., C. A. RINDER & P. R. SMITH. 1985. *In* The Sodium Pump, 4th International Conference on Na,K-ATPase. I. M. Glynn & J. C. Ellory, Eds. 45–49. The Company of Biologists, Ltd. Cambridge, England.
39. MOHRAZ, M., M. YEE & P. R. SMITH. 1985. J. Ultrastruct. Res. **93:** 17–26.
40. FULLER, S. D., R. A. CAPALDI & R. HENDERSON. 1979. J. Mol. Biol. **134:** 305–327.
41. SMITH, P. R. 1981. Ultramicroscopy **7:** 155–160.
42. SMITH, P. R., W. E. FOWLER, T. D. POLLARD & U. AEBI. 1983. J. Mol. Biol. **167:** 641–660.
43. MOHRAZ, M., M. V. SIMPSON & P. R. SMITH, in preparation.
44. MOHRAZ, M., M. YEE & P. R. SMITH, in preparation.
45. STEIN, W. D., W. R. LIEB, S. J. D. KARLISH & Y. EILAM. 1973. Proc. Natl. Acad. Sci. USA **70:** 275–278.
46. REPKE, K. R. H. & R. SCHON. 1973. Acta Biol. Med. Ger. **31:** K19–K30.
47. REPKE, K. R. H. & F. DITTRICH. 1979. *In* Na,K-ATPase: Structure and Kinetics J. C. Skou & J. G. Nørby, Eds. 487–500. Academic Press. London.

DISCUSSION OF THE PAPER

O. H. GRIFFITH (*University of Oregon, Eugene, Oregon*): How many sodiums get pumped per potassium?

MOHRAZ: In red blood cells, there are three sodiums pumped out of the cell for every two potassiums pumped in.

GRIFFITH: And how do you reconcile that with your model?

MOHRAZ: Well, I don't think this is different from any other model. Some data suggest that there are three identical sodium sites on the molecule inside the cell and two identical potassium sites on the outside; so the pumping could be three sodiums out of the cell simultaneously, rather than consecutively.

A. P. SOMLYO (*University of Pennsylvania School of Medicine, Philadelphia, Pennsylvania*): If you use vanadate rather than phosphate, do you still get a mixed E_1 and E_2 form?

MOHRAZ: Yes.

Electron Microscopic Studies
of Chromosomal Proteins[a]

MICHAEL BEER, JACOB VARKEY, JOHN BRANTLEY,
AND KRISHNA NIYOGI

The Thomas C. Jenkins Department of Biophysics
Johns Hopkins University
Baltimore, Maryland 21218

INTRODUCTION

A central theme in current biological research is the clarification of the control mechanism of genes during development. Many groups are actively investigating this question by the powerful techniques of molecular biology. It is becoming increasingly clear that, in addition to the stretches of DNA that code for any particular protein, there are, for every gene, also control sequences. Some of these are adjacent to the coding sequences on the 5′ side and are sometimes even part of them. Proteins are synthesized in the cell, bind to various parts of these control regions, and so modulate the rate of RNA synthesis. There is no clear eukaryotic example of a trigger that activates a gene at the appropriate stage in development. Sensitive sequences are being recognized and proteins purified, which, as they bind to these DNA regions, participate in the regulation. The ultimate aim of this work is to recognize the proteins that are relevant to the rate of transcription, to determine their sites of binding on the DNA, and to determine the various molecular interactions and the structure of the nucleoprotein complex. We hope to show in this paper several ways in which the electron microscopist can contribute to this task.

LABELING WITH CLUSTER COMPOUNDS

First we shall describe some experiments designed to determine the arrangement of λ-repressor molecules when bound to DNA. This protein is a key component in what is probably the best understood genetic control: the switching between lytic and lysogenic growth in λ-phage. The work on this system was recently reviewed in an elegant monograph by Ptashne.[1] λ-repressor is a protein composed of 236 amino acids; the first 92 compose the amino terminal domain, which contains the binding site for DNA, while those from 132 to the end form the carboxy terminal domain, which accounts for the dimerization of the protein. Joining the two is a connector that is sensitive to proteolytic digestion.[2] Repressor exerts control through the binding of its dimers to three operators, O_R1, O_R2, and O_R3. The affinity is greatest to O_R1, and occupancy of that site greatly enhances binding to O_R2. This effect is called co-operativity.

The arrangement of the co-operatively binding proteins is of interest. To study it, λ-repressor was labeled with undecagold prior to microscopy. This reagent was first synthesized and proposed for electron microscopy by Bartlett *et al.*[3] It consists of eleven

[a]Supported by National Institutes of Health Grants RR01214 and RR01777 and Biomedical Research Support Grant S07 RR07041.

BEER *et al.*: CHROMOSOMAL PROTEINS

gold atoms held together in a compact structure by phosphines. The compound as originally synthesized has 21 reactive primary amines. Its conversion to a reagent with a single SH specific maleimide group and the amino groups blocked was worked out independently by Yang *et al.*[4] and by Safer.[5] Wall *et al.*[6] showed that the undecagold clusters give high visibility in the scanning transmission electron microscope (STEM) and that they are stable enough in the electron beam to give only slight decrease in the apparent molecular weight after an irradiation dose of 10^5 electrons/nm^2. Because of

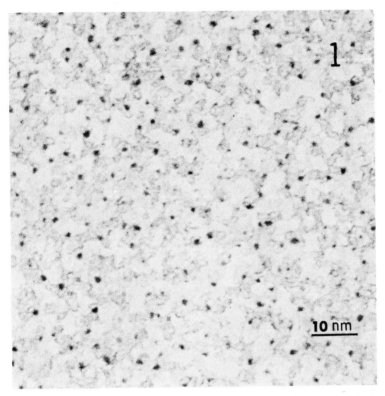

FIGURE 1. Bright-field image of undecagold clusters recorded using a Philips 420 CTEM with a super-twin lens operating at 40 kV. The sizes of the individual gold clusters varies from 8 to 12 Å.

the widespread use of transmission electron microscopes (TEMs) and the relative inaccessibility of STEMs, it seemed of interest to determine how readily the undeca-gold clusters could be imaged in a TEM. Both bright-field and dark-field electron micrographs were recorded and gave good visibility and high contrast in the TEM; a bright-field micrograph is shown in FIGURE 1. Thus the arrangement of proteins might be determined by examining a labeled system in a TEM.

Recently work has begun on another family of heavy atom containing cluster compounds with a view to developing labels for electron microscopy. These are the cluster compounds that contain 17 atoms of tungsten. These have now been synthesized with organic groups coupled to them. Their visibility in the conventional transmission electron microscope (CTEM) has been established and is shown in FIGURE 2.[9]

The known amino acid sequence of λ-repressor[7] suggests that up to three moles of SH reagents might be bound per mole of protein at cysteine residues 180, 215, and 219. When the maleimide derivative of the undecagold cluster was reacted with λ-repressor a product was obtained that on SDS gel electrophoresis gave a single new band with lower mobility and was shown by neutron activation analysis to contain gold. Under suitable conditions virtually all of the unreacted λ-repressor could be converted into this product. However, no second, slower band was ever observed, suggesting that of the three cysteines only one is reactive. In any case all three cysteines are in the carboxy end of the repressor molecule. The gold-labeled protein retained its ability to bind to DNA, as was shown by a filter binding assay. In any structural investigation involving labeled components it is, of course, essential to ascertain whether the modified component behaves normally; this seems to be the case with the λ-repressor.

To study the configuration of the co-operatively bound repressors, undecagold-labeled protein was allowed to interact with a restriction fragment of λ-DNA containing the three operators, O_R1, O_R2, and O_R3. This was spread on hydrophilic carbon films by the wet method: carbon films were floated off mica and lifted from

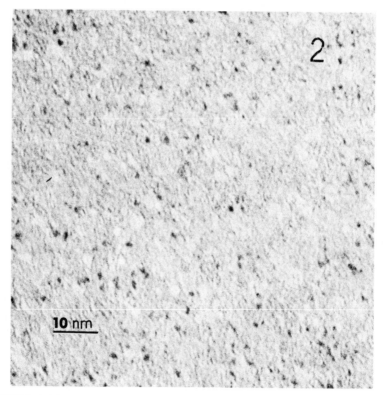

FIGURE 2. Bright-field image of W_{17} clusters recorded using a Philips 420 CTEM operating at 40 kV.

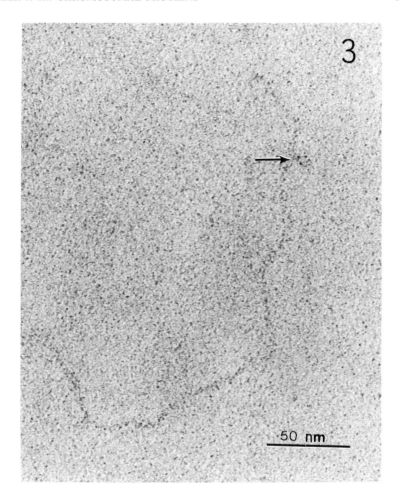

FIGURE 3. Undecagold-labeled λ-repressor-DNA complex. The DNA is 1,295 base pair restriction fragment from pRLM4. *Arrow* indicates the operator site, and the *four dots* presumably represent the four repressor molecules.

above, and the preparation was applied to the carbon surface while still wet. The sample was stained with 0.01% uranyl acetate. High-resolution electron micrographs of the DNA-protein complex showed clusters of four dots presumably due to the undecagold labels on four bound repressor monomers (FIGURE 3). This arrangement appears characteristic. In a recent paper Hochschild and Ptashne[8] showed that λ-repressor molecules bind co-operatively even when the operators are separated from each other by the insertion of a stretch of DNA, provided it comprises an integral number of helical turns. They also proposed three different arrangements for the co-operatively binding repressors. Our results appear most consistent with their model "C," in which the carboxy domains of the four repressor molecules are in each others' proximity. Experiments are in progress on mutants where it is known that the co-operativity is broken by an insertion, to test if the characteristic pattern is also eliminated.

SINGLE ATOM LABELING

The structure of chromatin has interested biologists for many years. The work prior to 1970 was discussed by DuPraw[10] in an interesting monograph. In 1969 Miller and Beatty[11] developed a procedure for gently centrifuging onto a grid the contents of a cell nucleus. Their micrographs of actively transcribing ribosomal genes gave a dramatic glimpse of transcription, led to the recognition of nontranscribed spacers, and indicated the rapid deposition of proteins onto nascent RNA. Then in 1974 Olins and Olins[12] showed that chromatin fibers were beaded. These beads, now called nucleosomes, were also recognized biochemically and were intensively studied in a number of laboratories (for a review see McGhee and Felsenfeld, REFERENCE 13). They are known to consist of two each of the four core histones: H2A, H2B, H3, and H4. In inactive chromatin the string of beads is wound into a supercoil stabilized by histone H1.

X ray diffraction, neutron diffraction, and electron microscopy indicated the gross morphology of the nucleosomes and established that the DNA was wrapped around the outside of an apparently cylindrical core. Considerable information became available through biochemistry on the affinities of the individual proteins for each other during the assembly. However the positions of the four histones within the nucleosome were unknown in 1977. It was therefore decided that an attempt would be made, using electron microscopy, to establish these locations.

The basis of the procedure was to form nucleosomes in which particular histones carried heavy atom labels, which then could be detected in the electron microscope.[14] For labeling, assembled nucleosomes were reacted with the amine reagent methyl (methylthio) acetimidate hydrochloride. This reaction had little effect on the pattern of digestion of nucleosomal DNA with micrococcal nuclease, indicating that the structure was not drastically altered. The histones were next separated and nucleosomes assembled using one or two of the labeled histones and the remainder unlabeled. These nucleosomes, modified in specific histones, were reacted with a platinum compound chloroglycyl-L-methioninatoplatinum (II),[15] in which only one bond is available for reaction; the other three of the four bonds are blocked by a chelating group. Such preparations were examined in the dedicated STEM, which is capable of imaging individual heavy atoms.[16,17]

The micrographs so obtained were noisy, and several particles had to be examined and the results treated statistically. The necessary correlation was made possible by two suppositions. First, that nucleosomes are flattened cylinders, and in preparations for electron microscopy they tend to be oriented with their axes perpendicular to the supporting film. Second, that the entry and exit points of the DNA are fixed in the nucleosome, and that a rotational alignment can be based on that observable point. The positions of the images of the Pt atom labels indicate the sites occupied by the histones that carry the labels. When the nucleosome was divided into a square array of 25 resolution elements, statistically significant accumulations of labels were observed; however, there appeared to be no higher-resolution information in the data. The deduced histone locations are consistent with other data obtained biochemically and crystallographically.

It is important to understand the limitation on resolution. Imaging single heavy atoms requires approximately 10^5 electrons/nm^2. Such heavy irradiation is known to result in considerable damage to the biological specimen. For example Unwin and

Henderson[18] in their classic studies of the structure of the membrane of *Halobacterium halobium* found that the irradiation had to be less than 10^2 electrons/nm^2. At present, studies that depend on the imaging of individual heavy atom labels in isolated particles, require dosages that are too high for resolution better than about 2 to 3 nm. Higher resolution requires crystalline specimens and analysis either by X ray diffraction or electron crystallography. (For modern illustrations of this approach see REFERENCE 19). Indeed, recently, excellent crystals of the histone octamer (nucleosome minus the DNA) suitable for high-resolution structure determination by X ray diffraction, were obtained by Burlingame *et al.*[20] and these promise to yield far more precise structure than the approach used by Stoeckert *et al.*[14]

GENE SPECIFIC LABELING

We shall now discuss a third area where the electron microscopist might make a contribution to our understanding of gene control in eukaryotic cells. Up to the present most electron microscopic studies of chromatin either examined general features without attempting to analyze a particular gene or were able to focus on particular genes, but only at times of vigorous transcriptional activity. There is a need for techniques for comparing—both biochemically and structurally—the chromatin of a particular gene at various times in the life cycle of a cell. We are developing procedures for labeling a particular gene specifically. Then it can be recognized in preparations for electron microscopy, and changes accompanying shifts in transcriptional activity can be monitored.

For initial studies the ribosomal 5S RNA genes in *Xenopus laevis* were chosen. In this system the genes exist in two major forms, the oocyte genes active during oogenesis and the somatic genes active in somatic cells. Both these genes are present in highly repeated tandem sequences, 20,000 per haploid genome for the oocyte and 800 for the somatic. In both cases each gene contains one Hind III cleavage site per repeat, and this is in the nontranscribed spacer region. We are attempting to compare the structure of the inactive oocyte gene and the active somatic gene from *Xenopus* erythrocytes.

The gene specific labeling to be used here is a modification of the chromatin purification proposal of Langmore and his collaborators.[21] Nuclei are isolated from suitable cells, chromatin is prepared and digested with a restriction enzyme, and single strand tails are produced by digestion with an exonuclease. An oligonucleotide complementary to the end of the single strand tail is synthesized, and a biotinylated nucleotide is coupled to it. This is annealed to the chromatin where it will bind efficiently only to the correct fragments, since in general no others will have terminal sequences complementary to the same oligonucleotides.

Some of the required steps have now been tested in our laboratory. It has been shown that chromatin, isolated from *Xenopus* erythrocytes, can be digested with the restriction endonuclease Hind III, which is known to cut genomic DNA once per tandemly repeating 5S RNA gene, to yield gene size fragments. We have confirmed the results of Reynolds *et al.*[22] that a similar digestion can be carried out on chromatin. Of course, if the molecules important in the control of gene activity are to be studied by examining these fragments, it is necessary to ensure that they have not been lost during the digestion. Four experiments indicate that the critical proteins survive. First, the template activity of chromatin is retained during digestion, indicating that the control proteins are not disturbed in a crucial way. Second, the morphology of chromatin appears the same before and after digestion. Third, using transcription as a criterion, the transcription complex formed *in vitro* appears to survive incubations designed to mimic those required for digestion by restriction enzyme. Fourth, gel electrophoresis of restriction-enzyme digested chromatin shows the histone bands with no evidence of proteolytic breakdown. These results strongly suggest that the control apparatus survives digestion, though they do not rigorously prove it.

Experiments have also been done to show the feasibility of labeling specifically the biotinylated nucleotides with streptavidin bound to colloidal gold. Linear plasmid

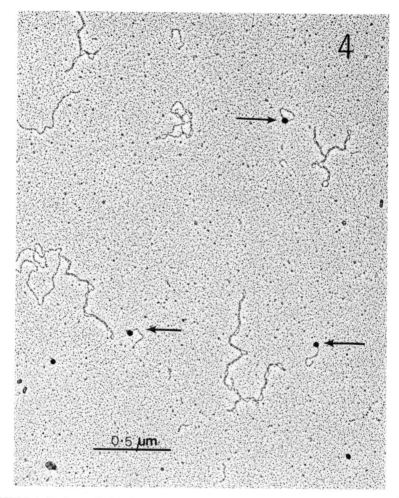

FIGURE 4. Biotin specific labeling of DNA. For details see text. *Arrows* indicate the terminally labeled 5S RNA genes.

DNA was terminally labeled with biotinylated nucleoside triphosphates. This was achieved as follows. A plasmid, pXLO8, was cut with restriction enzyme Hind III to give a 700-base pairs long 5S RNA gene and a 5,000-base pairs long bacterial DNA and was digested with lambda exonuclease to leave protruding single strand 3' tails.

Oligonucleotide primers specific for the 5S genes were annealed to these tails, and synthesis with biotinylated nucleotides was initiated to fill in the gap. The product was mixed with circular 0X-174 RF DNA. On adding streptavidin-gold, labeling was observed only on the 5S RNA gene size linear molecules as seen in FIGURE 4. This shows that streptavidin labeling was specific for the biotin.

CONCLUSIONS

Cluster compound labeling appears to be an effective method for the analysis of the relative arrangement of several proteins bound to DNA. Single atom labeling can yield low-resolution structural information, but the high irradiation required leads to too much damage for resolutions better than perhaps 2 to 3 nm. Gene specific labeling— though at present not fully worked out—promises to provide insights into the structural changes that accompany gene regulation.

REFERENCES

1. PTASHNE, M. 1986. A Genetic Switch: Gene Control and Phage. Cell Press and Blackwell Scientific Publications. Palo Alto, CA.
2. PABO, C. O., R. T. SAUER, J. M. STURTEVANT & M. PTASHNE. 1979. The λ-repressor contains two domains. Proc. Natl. Acad. Sci. USA **76**: 1608–1612.
3. BARTLETT, P. A., B. BAUER & S. J. SINGER. 1978. J. Am. Chem. Soc. **100**: 5085–5089.
4. YANG, H., J. E. REARDON & P. A. FREY. 1984. Synthesis of undecagold cluster molecules as biochemical labeling reagents. 2. Bromoacetyl and maleimide undecagold clusters. Biochemistry **23**: 3857–3862.
5. SAFER, D., L. BOLINGER & J. S. LEIGH, JR. 1986. Undecagold clusters for site-specific labeling of biological macromolecules: simplified preparations and model applications, J. Inorg. Biochem. **26**: 77–91.
6. WALL, J. S., J. F. HAINFELD, P. A. BARTLETT & S. J. SINGER. 1982. Observation of an undecagold cluster compound in the scanning transmission electron microscope. Ultramicroscopy **8**: 397–402.
7. SAUER, R. T. & R. ANDEREGG. 1978. Primary structure of λ-repressor. Biochemistry **17**: 1091–1100.
8. HOCHSCHILD, A. & M. PTASHNE. 1986. Cooperative binding of λ-repressors to sites separated by integral turns of the DNA helix. Cell **44**: 681–687.
9. KEANA, J. F. W., M. D. OGAN, Y. LÜ, M. BEER & J. VARKEY. Functionalized heteropolytungstate anions possessing a modified Dawson structure: small, individually distinguishable labels for conventional transmission electron microscopy. 1985. J. Am. Chem. Soc. **107**: 6714–6715.
10. DUPRAW, E. J. 1970. DNA and Chromosomes. Holt, Rinehart Winston, Inc. New York, NY.
11. MILLER, O. L. & B. R. BEATTY. 1969. Visualization of nucleolar genes. Science **164**: 955–957.
12. OLINS, A. L. & D. A. OLINS. 1974. Spheroid chromatin units (*v*-bodies). Science **183**: 330–332.
13. McGHEE, J. & G. FELSENFELD. 1980. Annu. Rev. Biochem. **49**: 1115–1156.
14. STOECKERT, C. J., JR., M. BEER, J. W. WIGGINS & J. C. WIERMAN. 1984. Histone positions within the nucleosome using platinum labeling and the scanning transmission electron microscope. J. Mol. Biol. **177**: 483–505.
15. KUHN, E. M., K. D. MARENUS & M. BEER. 1984. Discrimination of collagen types by methionine-specific stain, J. Ultrastruct. Res. **87**: 172–179.
16. CREWE, A. V., J. WALL & J. LANGMORE. 1970. Visibility of single atoms. Science **168**: 1338–1340.

17. WIGGINS, J. W., J. ZUBIN & M. BEER. 1979. High-resolution scanning transmission electron microscope at Johns Hopkins. Rev. Sci. Instrum. **50:** 403–410.
18. UNWIN, P. N. T. & R. HENDERSON. 1975. Molecular structure determination by electron microscopy of unstained specimens. J. Mol. Biol. **94:** 425–440.
19. GLAESER, R. M. 1984. Symposium on electron crystallography of macromolecules. Ultramicroscopy **13:** #1–2.
20. BURLINGAME, R. W., W. E. LOVE, B. -W. WANG, R. HAMLIN, N. -H. XUONG & E. N. MOUDRIANAKIS. 1985. Crystallographic structure of the octameric histone core of the nucleosome at a resolution of 3.3 Å. Science **228:** 546–553.
21. WORKMAN, J. L. & J. P. LANGMORE. 1985. Nucleoprotein hybridization: a method for isolating specific genes as high molecular weight chromatin. Biochemistry **24:** 7486–7497.
22. REYNOLDS, W. F., L. S. BLOOMER & J. M. GOTTESFELD. 1983. Control of 5S DNA transcription in *Xenopus* somatic cell chromatin: activation with an oocyte cell extract. Nuc. Acid Res. **11:** 57.

DISCUSSION OF THE PAPER

A. C. STEVEN (*National Institutes of Health, Bethesda, Maryland*): How does your provisional map of H2, H2B, and so on in the nuclear zone compare with the various possibilities that have been proposed recently by Burlingame on the one hand and by Klug's group on the other. Do they match up reasonably well?

BEER: They match up reasonably well at the crude resolution that we were able to gain.

Potential for High-Resolution Electron Crystallography at Intermediate High Voltage[a]

WAH CHIU,[b,c] TZYY-WEN JENG,[b] LAURA L. DEGN,[c]
AND B. V. VENKATARAM PRASAD[b]

[b]*Department of Biochemistry*
[c]*Department of Molecular and Cellular Biology*
University of Arizona
Tucson, Arizona 85721

INTRODUCTION

Electron crystallography is a unique technique for three-dimensional structure determination of macromolecules. Differing from X ray crystallography, this technique uses thin, small crystals and can retrieve the structural phases directly from electron images.[1] Henderson and Unwin demonstrated the usefulness of this technique with their determination of the structure of the purple membrane. They preserved the two-dimensional crystals of bacteriorhodopsin by embedding them in a thin layer of glucose, and reduced the electron radiation damage of the specimen by using a very low electron exposure.[2] Computer processing techniques were subsequently used to obtain the three-dimensional structure of bacteriorhodopsin to 7-Å resolution. The seven rods of mass density in their reconstruction were interpreted as seven helices.[3] Since then, an increasing number of two-dimensional crystals, particularly of membrane proteins, have been studied by the electron crystallographic technique. However, most of the three-dimensional structures have been limited to about 20-Å resolution primarily due to the poor crystallinity of these specimens or to specimen preparation techniques.[4] We have made a number of thin crystals of soluble proteins that are well suited for high-resolution electron crystallographic analysis.[5–8]

In order to record high-resolution electron images the specimens must be kept at low temperature to minimize radiation damage. The electron microscope must therefore be equipped with a mechanically stable cold stage. Images of the crotoxin complex, purple membrane, and paraffin crystals have been recorded to 3.5 Å or better in cryomicroscopes.[9–12] However, because of the low contrast in the structure factors for reflections beyond 7 Å, it is difficult to detect them in the optical diffraction patterns of the images.[13] Hayward and Stroud enhanced the detectability of the high-resolution structure factors by superimposing small image patches of purple membrane, and justified the validity of the high-resolution structural data by evaluating the probability distribution function of the phases and their figures of merit.[9] Henderson and co-workers recently used correlation analysis to define the extent of lattice distortion in images of large patches of purple membrane, and compensated for the distortion by a real space interpolation procedure prior to

[a]Supported by U.S. Public Health Service Grants NS16047 and RR02250.

conventional Fourier averaging.[11] They also found it necessary to correct for the phase errors due to the beam tilt. Their analysis demonstrated quite convincingly that high-resolution phases can be retrieved from low-contrast images. We also achieved a significant improvement in enhancing the signal-to-noise ratio in the Fourier transform of some of our images of the crotoxin complex crystals by using a similar procedure.[14] Independently, Grant and co-workers have developed a computer procedure to enhance the signals from multiple image areas by aligning and merging them in reciprocal space by vector summation of the complex structure factors. The resultant phases are evaluated in terms of a phase coherence factor (Q-factor). Improvement of this Q-factor and the rms phase residuals are observed as more images are added.[15]

In order to make the electron crystallographic technique practical for high-resolution three-dimensional structure determination, the efficiency of data collection needs to be improved. It has been suggested that the difficulty of recording high-resolution images routinely is due to the beam-induced movement of the specimen.[13] A preliminary result of Downing suggests that this problem may be solved by using a small beam spot to reduce the mechanical stress on the specimen. His experiment was done with a highly coherent electron source of a field emission gun.[16] The suitability of this procedure with the LaB_6 gun remains to be tested.

All of these advances encourage optimism for the determination of three-dimensional structures of biological and organic crystals to high resolution with electron crystallographic techniques. This paper addresses the potential advantages of the newly available intermediate high voltage (400-kV) electron microscopes for electron crystallographic study of macromolecules. Four factors that affect image reconstruction at varying accelerating voltages are discussed, and experimental evidence for the advantages of higher accelerating voltages is presented.

ELECTRON SCATTERING AND IMAGE FORMATION

The goal of image processing is to determine the scattering potential function of the object.[17] All current algorithms for three-dimensional reconstruction from electron micrographs assume the validity of the weak phase approximation. This approximation results in a simple linear relationship between the image intensity and the scattering potential function. In terms of physical optics, this means that the object changes the phases but not the amplitudes of the transmitted electron waves. According to scattering theory, this means that electrons are scattered no more than once by the object. The assumptions upon which the weak phase approximation is based are as follows: (i) only elastically scattered electrons are considered, (ii) the shape transform of the specimen thickness is ignored, (iii) Fresnel diffraction throughout the specimen thickness is negligible, and (iv) the curvature of the Ewald sphere is negligible. These assumptions are generally thought to be valid if the specimen is sufficiently thin and the image resolution is relatively low.[18] Some of these factors will be discussed separately below.

The most rigorous approximation relating the scattering potential function and the image intensity is the multislice dynamical approximation.[19] The complexity of this approximation makes it impractical from the computational point of view for retrieving the object structure from the experimentally observed images except perhaps for simple structures. Therefore, it is desirable to use experimental conditions in which the weak phase approximation is applicable, in order to simplify the computations required to reconstruct the scattering potential function of the object. Jap and Glaeser carried out a numerical simulation to compare conditions under which the weak phase or multislice dynamical approximations are valid in small organic crystals as a function of resolution, crystal thickness, and electron energy.[18] Their results suggest that, as electron energy is increased, the weak phase approximation is valid over a larger range of crystal thickness and resolution. These studies were extended to a protein crystal and similar trends were shown up to energies of 500 kV.[20] Therefore, it seems reasonable to expect that the use of an intermediate high voltage electron microscope will facilitate the structure determination of protein crystals by allowing the use of the weak phase approximation.

DEPTH OF FIELD

The three-dimensional structure of an object can be reconstructed by combining two-dimensional electron images of the object at different tilt views.[21,22] The current method of three-dimensional reconstruction assumes that each electron image represents a true projection of the three-dimensional object. This assumption is valid if the defocus difference between the top and the bottom of the object is less than the depth of field (D) of the electron microscope. The depth of field can be expressed as $D = d^2/2\lambda$ where d is the resolution and λ is the electron wavelength.[4] For example, if the expected resolution of the reconstruction is 4 Å, the depth of field would be 216 Å at 100 kV ($\lambda = 0.037$ Å). If the specimen's thickness exceeds 216 Å, the image detail finer than 4 Å must be treated as a superposition of images representing different depths of the object with different defocus values. This is equivalent to violating the assumption that Fresnel diffraction can be neglected throughout the specimen thickness. Consequently, beyond 4 Å in the Fourier transform, phases cannot be retrieved accurately merely by correcting for the phase reversal in the contrast transfer function. This problem becomes more pronounced for a tilted specimen, because tilting increases the effective thickness of the specimen. In our study of the crotoxin complex crystal, which has a unit cell thickness of 256 Å,[5] the depth of the field may affect the reliability of phases obtained from a 100-kV electron image at resolutions better than 4.5 Å. However, at 400 kV, the depth of field becomes a limiting factor for this crystal only beyond 2.9-Å resolution.

CURVATURE OF THE EWALD SPHERE

The Ewald sphere is a geometrical construction for describing the diffraction conditions. The radius of the sphere is $1/\lambda$. If λ is sufficiently small, and the resolution sufficiently low, the curvature of the sphere is small enough to be approximated by a plane. The consequence of such an approximation is that the electron image would be formed by a set of equivalent reflections $F(h \ k \ s_z)$ and $F(-h-k-s_z)$, which are related by Friedel symmetry. Under these conditions the image represents the projection of the object down the optical axis of the microscope. Cohen and co-workers have analyzed theoretically the validity of this projection approximation as a function of specimen thickness, tilt angle, and resolution.[23] When this approximation breaks down, the image becomes a mixture of two different "single side band" images. Such an image differs from a bright-field image, which by definition is formed by Friedel related reflections. Conventional methods used in image processing were developed for bright-field images. Application of these methods to images for which the curvature of the Ewald sphere is not negligible will result in the inaccurate retrieval of phases. It is conceivable that more sophisticated image processing procedures could be developed to take this into account. Alternatively, one can collect the electron microscopic data with higher energy electrons to reduce the curvature of the Ewald sphere and ensure that the data can be treated by the simple approximation. In the case of the crotoxin complex crystal, we estimate that the three-dimensional reconstruction can be reliably done to 7 Å at 100 kV and to 4 Å at 400 kV for 256-Å thick crystals.[23]

RADIATION DAMAGE

Radiation damage has been recognized as the most severe limitation preventing the collection of high-resolution microscopic data from beam sensitive materials. The mechanism of damage involves primary and secondary processes.[24] The primary damage results from ionization and bond breakage. Secondary damage results from mass migration, mass loss, free radical formation, and recombination of damaged species. Most recent efforts to reduce radiation damage have been directed at minimizing the destructive effects of the secondary damage processes by keeping the specimen at low temperature. Significant reduction of radiation damage is attained at temperatures around $-120°$ C.[25-28] This conclusion is further supported by the results of a set of collaborative experiments by several laboratories in the U.S. and in Europe.[29]

Another way of minimizing the radiation damage is to reduce the primary damage processes by using higher electron energy.[30] The stopping-power theory predicts that the energy loss due to the inelastic scattering of ionizing radiation is proportional to the thickness of the specimen and inversely proportional to β^2 where β is the ratio of the velocity of the radiation to the velocity of light. The primary damage process can therefore be reduced by a factor of 2 by increasing the electron energy from 100 kV ($\beta = 0.55$) to 400 kV ($\beta = 0.82$). For the same extent of damage, at 400 kV one can use twice the specimen thickness as at 100 kV. Results of a radiation damage study of l-valine crystals indicate that the critical dose determined from the fading of electron diffraction intensities increased by a factor of 2 from 100 kV to 400 kV.[31] A similar improvement can be anticipated for thin protein crystals. Experimental verification of this expectation is needed.

While inelastic scattering is reduced at higher electron energy, the signal from elastically scattered electrons is also reduced to an equivalent extent.[32] It has been shown that for a sufficiently thin specimen, the electron image can be interpreted in terms of elastically scattered signals. The inelastically scattered electrons are assumed to contribute incoherently to the image intensity and hence can be regarded as part of the background noise. In the weak phase approximation, the image intensity from elastic scattering is proportional to the electron diffraction amplitude, not to its intensity. The image contrast would thus be improved by a factor of 1.4 at 400 kV over that at 100 kV for an equivalent level of specimen damage.[4]

EXPERIMENTAL COMPARISON OF ELECTRON DIFFRACTION INTENSITIES OF PROTEIN CRYSTALS AT DIFFERENT VOLTAGES

From the above discussion, it is apparent that it is important to consider the potential failure of the weak phase approximation while determining the high-resolution structure of a protein crystal. Experimentally, it is difficult to evaluate quantitatively the suitability of the weak phase approximation in analyzing the electron image data of a protein crystal. We made a comparison of the electron

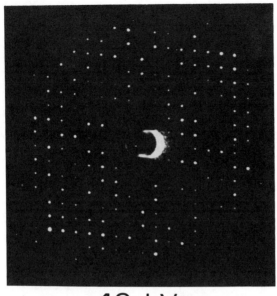

40 kV

100 kV

FIGURE 1. Computer-processed electron diffraction patterns of crotoxin complex crystal embedded in glucose recorded successively at **40 kV** and **100 kV** in a JEOL 100CX electron microscope operated at −145° C. The specimens were tilted 30 degrees.

diffraction intensities of the crotoxin complex crystal at 40 and 100 kV. The result may provide a qualitative indication of whether there is an advantage in using higher energy electrons for structure determinations.

The crotoxin complex crystal was embedded in glucose and examined in the JEOL 100CX electron microscope operated with a cold stage at $-145°$ C. A fixed angle holder of 30 degrees tilt was used to hold the specimen. The search for a suitable crystal was done with a low exposure rate (0.01 e/Å^2/s) at 40 or 100 kV. Low-dose electron diffraction patterns were initially recorded at 40 kV and then at 100 kV from the same crystal. The accumulated exposure of each crystal was less than 3 electron/Å^2. The electron diffraction patterns were digitized and the radial background was subtracted to improve the visibility of the low-resolution reflections, then photographed from the screen of an AED 512 graphics terminal. FIGURE 1 is an example of such a pair of processed diffraction patterns. The diffraction intensity of each reflection was calculated according to Baldwin and Henderson.[33] The symmetry reliability (R) factor for the apparent Friedel related reflections was calculated according to the following

TABLE 1. Six Sets of Symmetry Reliability (R) Factors for Apparent Friedel Related Pairs of Reflections Calculated from Diffraction Patterns Recorded Consecutively at 40 kV and 100 kV

Set	Tilt	keV	(%) R Factor	No. of Reflection Pairs
1	30°	40	26	74
		100	22	76
2	28°	40	44	78
		100	29	75
3	28°	40	30	73
		100	18	78
4	30°	40	28	76
		100	27	77
5	27°	40	28	77
		100	25	80
6	25°	40	21	72
		100	18	78

equation:

$$R = \frac{\sum\limits_{N} |I_1 - I_2|}{\sum\limits_{N} |I_1 + I_2|} \times 2$$

where I_1 and I_2 are the integrated intensities for the apparent Friedel pairs, and N is the total number of pairs of reflections. The R factor is a measure of the equivalence in intensities of apparent Friedel pairs. TABLE 1 shows six data sets where the diffraction intensities were analyzed out to 5-Å resolution. In all cases, the R factor for the 40 kV data is worse than the 100-kV data. In two cases, the difference in R factor is not as pronounced as the other data, probably because those particular crystals were very thin. In this experiment we did not measure the thickness of the crystal from which the data was recorded. The cause of the R factor difference between the two voltages may be a combination of effects from the Ewald sphere curvature and dynamical scattering. The trends suggest a potential advantage of using higher electron energy.

CONCLUSION

Cryoelectron microscopy has been demonstrated to be feasible for collecting electron diffraction patterns and images from thin crystals of organic and biological macromolecules with resolution identical to that of the 100-kV microscope. The ability to retrieve phases for reconstructing the three-dimensional structure of thin crystals to that resolution remains to be proved. Both the theoretical and experimental evidence suggest there are advantages to using higher energy electrons for high-resolution structure analysis. The availability of a new generation of intermediate high voltage electron microscopes opens up a new opportunity for structure research on thin crystals of macromolecules.

ACKNOWLEDGMENTS

We thank Robert A. Grant and Michael F. Schmid for their comments on the manuscript.

REFERENCES

1. KLUG, A. 1979. Chem. Scr. **14:** 245–256.
2. UNWIN, P. N. T. & R. HENDERSON. 1975. J. Mol. Biol. **94:** 425–440.
3. HENDERSON, R. & P. N. T. UNWIN. 1975. Nature **257:** 28–32.
4. GLAESER, R. M. 1985. Annu. Rev. Phys. Chem. **36:** 243–275.
5. JENG, T. W. & W. CHIU. 1983. J. Mol. Biol. **164:** 329–346.
6. COHEN, H., W. CHIU & J. HOSODA. 1983. J. Mol. Biol. **169:** 235–246.
7. CHIU, W. 1982. High resolution electron microscopy of unstained, hydrated protein crystals. *In* Electron Microscopy of Proteins. J. R. Harris, Ed. Vol. 2: 233–259. Academic Press. London.
8. GRANT, R. A., L. L. DEGN, W. CHIU & J. P. ROBINSON. 1983. Proc. 41st Annu. Mtg. Electron Microscopy Soc. Amer. 730–731.
9. HAYWARD, S. B. & R. M. STROUD. 1981. J. Mol. Biol. **151:** 491–517.
10. JENG, T. W., W. CHIU, F. ZEMLIN & E. ZEITLER. 1984. J. Mol. Biol. **175:** 93–97.
11. HENDERSON, R., J. M. BALDWIN, K. H. DOWNING, J. LEPAULT & F. ZEMLIN. 1986. Ultramicroscopy, **19:** 147–178.
12. ZEMLIN, F., E. REUBER, E. BECKMAN, E. ZEITLER & D. L. DORSET. 1985. Science **229:** 461–462.
13. HENDERSON, R. & R. M. GLAESER. 1985. Ultramicroscopy **16:** 139–150.
14. CHIU, W., M. F. SCHMID & T. W. JENG. 1986. Proc. 44th Annu. Mtg. Electron Microscopy Soc. Amer. 2–5.
15. GRANT, R. A., M. F. SCHMID, W. CHIU, J. DEATHERAGE & J. HOSODA. 1986. Biophys. J. **49:** 251–258.
16. DOWNING, K. H. 1986. Proc. 44th Annu. Mtg. Electron Microscopy Soc. Amer. 14–17.
17. GLAESER, R. M. 1982. Electron microscopy. *In* Methods of Experimental Physics. G. Ehrenstein & H. Lecar, Eds. Vol. 20: 391–444. Academic Press. New York, NY.
18. JAP, B. K. & R. M. GLAESER. 1980. Acta Crystallogr. Sect. A **36:** 57–67.
19. COWLEY, J. M. & A. F. MOODIE. 1957. Acta Crystallogr. **10:** 609–610.
20. HO, M. H., B. K. JAP & R. M. GLAESER. 1981. Proc. 39th Annu. Mtg. Electron Microscopy Soc. Amer. 30–31.

21. DeRosier, D. J. & A. Klug. 1968. Nature **217:** 130–134.
22. Hoppe, W. 1983. Agnew. Chem. Intl. Ed. Engl. **22:** 456–489.
23. Cohen, H. A., M. F. Schmid & W. Chiu. 1984. Ultramicroscopy **14:** 219–226.
24. Glaeser, R. M. 1971. J. Ultrastruct. Res. **36:** 466–482.
25. Glaeser, R. M. & K. A. Taylor. 1978. J. Microsc. **112:** 127–138.
26. Hayward, S. B. & R. M. Glaeser. 1979. Ultramicroscopy **4:** 201–210.
27. Jeng, T. W. & W. Chiu. 1984. J. Microsc. **136:** 35–44.
28. Chiu, W., E. Knapek, T. W. Jeng & I. Dietrich. 1981. Ultramicroscopy **6:** 291–295.
29. Chiu, W., K. H. Downing, J. Dubochet, R. M. Glaeser, H. G. Heide, E. Knapek, D. A. Kopf, M. K. Lamvik, J. Lepault, J. D. Robertson, E. Zeitler & F. Zemlin. 1986. J. Microsc. **141:** 385–391.
30. Cosslett, V. E. 1969. Q. Rev. Biophys. **2:** 95–133.
31. Howitt, D. G., R. M. Glaeser & G. Thomas. 1976. J. Ultrastruct. Res. **55:** 457–461.
32. Isaacson, M. S. 1977. Specimen damage in the electron microscope. *In* Principles & Techniques of Electron Microscopy. M. A. Hayat, Ed. Vol. 7: 1–78. Van Nostrand Reinhold Co. New York, NY.
33. Baldwin, J. M. & R. Henderson. 1984. Ultramicroscopy **14:** 319–336.

DISCUSSION OF THE PAPER

D. A. Agard (*University of California Medical Center, San Francisco, California*): I have a question about the unbending of the lattice. By it's very nature that has to put more intensity on the lattice point. How do you know that what it's doing to the phases is, in fact, correct?

Chiu: We also have another crystal that has a centrosymmetry. In fact, we do see improvement in the phase residual deviated from zero or π.

Agard: How much of an improvement is there?

Chiu: About ten degrees.

The Usefulness of the High Voltage Electron Microscope in Biomedical Ultrastructure Analysis

DONALD F. PARSONS

Wadsworth Center for Laboratories and Research
New York State Department of Health
Empire State Plaza
Albany, New York 12201

INTRODUCTION

Until recently uncertainty persisted about the role and usefulness of high accelera-tion voltage (200–3,000-kV) electron microscopy in biomedical research. However, four and a half years of collaboration between staff and a variety of users at our National Institutes of Health High Voltage Electron Microscope (HVEM) Resource have given us a clearer picture about the usefulness of the HVEM. During this time, interest has increased in the USA in obtaining three-dimensional images from thicker specimens as reflected by an increase in users of our HVEM installation (we now operate two shifts a day). Electron microscope manufacturers have responded by introducing objective lenses of lower chromatic aberration coefficients and by provid-ing correction of the off-axis chromatic blurring produced by projector lenses at low magnification. These improvements in lens design raise important new questions about what fraction of the biologist's thick-specimen needs can be met by using the new, smaller intermediate voltage electron microscopes at 300–400-kV acceleration voltage rather than the 1.0 MV HVEMs. We also see more extensive use of the 200-kV analytical electron microscopes now commonly available. Since these are usually fitted with the scanning transmission electron microscope (STEM), we can expect to see more comparisons of STEM with the fixed-beam electron microscope at different voltages. Some electron microscopes fitted with STEM also now have electron energy-loss spectroscopy (EELS) available. Exploration of enhancement of the contrast and resolution of thick specimens by STEM with energy filtering is now a practical possibility.

The theoretical exploration of the benefits of all these different energies and image modes does not (so far) have a good record of agreeing with experiment for biological specimens. Also, we know that some electron interactions at relativistic voltages near 1.0 MV cannot be the same as for lower, nonrelativistic electron energies.

Hence, we are entering a period of trial and reassessment of thick specimen imaging and must await quantitative evaluation of these different approaches before deciding where each type of electron microscope plays its optimum role.

OBSERVATION SAMPLING STATISTICS IN ELECTRON MICROSCOPY

Good evaluation of the statistical significance of observations was never a strong point of biomedical light microscopy and is no better for electron microscopy. For example, pathology reports give the size and number of specimens sampled but often no quantitation of the volume of tissue microscopically examined in relation to these bulk volumes. Conventional thin-section electron microscopy can be even more seriously at fault than light microscopy. Bernhard[1] emphasized that a thin section transects only a small fraction (about 1/300th) of the volume of a single cell and that random thin sections may give a poor indication of the size, shape, and distribution of organelles. Weeks of work are required to prepare and photograph the 300 *random* (spaced-apart) cell sections equivalent in volume to a single cell.[2] Because of this inconvenience, images of adjacent thin sections are often recorded and portrayed together without it being pointed out that they differ little from each other in information content. At this late date we recognize that many false interpretations of cell shape and structure have arisen from statistically inadequate data. Some of these interpretations have been subsequently corrected by serial thin sectioning, but this has proved too tedious a procedure for general use.

The introduction of increased penetration for electron microscopes has enabled one to examine thicker sections as stereo pairs and carry out serial sectioning in a shorter time. The emphasis has remained with serial thick sectioning rather than with viewing critical-point dried whole cell mounts because of suspicions about artifacts in critical-point dried specimens and the restriction of whole cell mounts to cultured cells. Recently better cytoskeleton preservation has been obtained by use of very dry CO_2 in critical-point drying,[3] so that an increase in use of whole cell mounts may be imminent.

Most problems of interpretation associated with using a few random sections are removed by complete serial thick sectioning or whole cell mount examination. Some accepted concepts about the size, shape, and distribution will probably need to be revised.

EXAMPLES OF REVISED CONCEPTS OF CELL STRUCTURE AND THE USE OF MORPHOMETRY AS A RESULT OF SERIAL SECTIONING

It has been assumed that with few exceptions eukaryotic species have numerous single mitochondria per cell. However, serial thin sectioning of the yeast *Saccharomyces cerevisiae* showed one giant mitochondrion per cell.[4] Previous random thin section work on *Candida utilis* also suggested single vesicles for the mitochrondria. Serial thick sectioning using the HVEM showed that some of these species (in particular energy states) contained a small number of highly branched mitochondria.[5] Surprisingly this kind of work has not been extended very much and we do not know how common macromitochondria are and how they are affected by metabolic state or stage of cell cycle in most cell types.

Traditionally, the Golgi apparatus is considered one of the most irregular and variable of cell organelles despite identification of a multitude of specific synthetic functions in the organelle. Computer-aided reconstruction of serial thin sections has yielded the complete ultrastructure of the Golgi apparatus in milk-secreting bovine mammary epithelial cells. The Golgi had a bowllike shape (the *cis*-face was oriented towards the periphery of the cell). The structure showed a channel that transversed the dictyosome and contained secretory vesicles.[6] In this cell type at least, the Golgi is more organized than was previously suspected.

Recently it has been shown in several cases that migrating cells are polarized, with the cytoplasmic vesicular structures and associated microtubules more or less oriented in front of the nucleus and facing the average direction of motion. This was usually not evident from random single sections or whole cell mounts, while serial sectioning was required to demonstrate it. Such polarization was demonstrated by Malech et al.[7] for polymorphonuclear leukocytes migrating along a chemotactic gradient. In tissue culture, fibroblasts migrating out of chick heart explants show a tight clustering of mitochondria near the nucleus while in motion, but the mitochondria (or macromitochondria) disperse more evenly when the cells come to rest in confluent growth.[8]

An interesting bonus from three-dimensional graphics reconstruction of profiles of serial sections is the availability of volumes and surface areas of whole cells or their parts. In a computer-aided reconstruction of serial thin sections of migrating fetal monkey neurons, cells in motion were seen to be polarized. The proliferative state of the migrating cells was also suggested by the increase in nuclear volume of some of these cells, presumably due to entering the G2 phase of the cell cycle.[9]

Recently, three-dimensional graphic reconstruction has shown us that some tumor cells (and also leukocytes) squeezing through small gaps in the elastic-collagen reticulum of the stroma of mouse peritoneum were polarized[10] (something we had not noticed in two years of random single thick section work on the same system!). We also noted that many such cells, in the process of squeezing through a barrier, were shrunken in volume.

Current work is directed at quantitating the polarization of organelles, the volume changes, and the areas of adhesion of the invasive cells to the connective tissue fibrils and to other cells. We hope that this data can be used to prepare mechanisms for the motor power of these invasive cells. To do this, we are upgrading our computer profiling methods so that we can obtain profiles from multiple levels in each section by directly digitizing contours in the stereo image of each section. Details of the approach will be reported shortly.

SUMMARY

We now have the tools to quantitate and correctly interpret the size, shape, and distribution of organelles in cells. It appears likely that the results will require some changes in some basic assumptions of ultrastructural cytology. This should also result in some new biochemical hypotheses about control of cell metabolism.

ACKNOWLEDGMENT

The author particularly thanks Dr. W. Tivol, Dr. M. Song, D. Barnard, and P. McCauley for technical support in running the HVEM facility and M. Marko for technical assistance with the author's research. The graphic reconstruction is a joint effort of J. Frank, R. Banerjee, M. Marko, D. Lieth and M. Radermacher. Much of our understanding of the new multilevel approach to profiling in stereo views is due to Dr. Murray V. King.

REFERENCES

1. BERNHARD, W. 1968. The detection and study of tumor viruses with the electron microscope. Cancer Res. 20: 712–715.
2. PARSONS, D. & D. A. WHALEY. 1976. Minimizing false negatives in electron microscope searches for virus or other specific features of cancer cells. Cancer Biochem. Biophys. 1: 299–302.
3. RIS, H. 1985. The cytoplasmic filament system in critical-point dried whole mounts and plastic-embedded sections. J. Cell Biol. 100: 1474–1487.

4. HOFFMAN, H.-P. & C. J. AVERS. 1973. Mitochondria of yeast: Ultrastructural evidence of one giant, branched organelle per cell. Science **181:** 749–751.
5. DAVISON, M. T. & P. B. GARLAND. 1975. Mitochondrial structure studied by high voltage EM of thick sections of *Candida utilis.* J. Gen. Microbiol. **91:** 122–138.
6. DYLEWSKI, D. P., R. M. HARALICK & T. W. KEENAN. 1984. Three-dimensional ultrastructure of the Golgi apparatus in bovine mammary epithelial cells during lactation. J. Ultrastruct. Res. **87:** 75–85.
7. MALECH, H. L., R. K. ROOT & J. I. GALLIN. 1977. Structural analysis of human neutrophil migration. Centriole, microtubule and microfilament orientation during chemotaxis. J. Cell Biol. **75:** 666–693.
8. COUCHMAN, J. R. & D. R. REES. 1982. Organelles-cytoskeleton relationships in fibroblasts: mitochondria, Golgi apparatus and endoplasmic reticulum in phases of movement and growth. Eur. J. Cell. Biol. **27:** 47–54.
9. RAKIC, P., L. J. STENSON, E. P. SAYRE & R. L. SIDMAN. 1974. Computer-aided three-dimensional reconstruction and quantitative analysis of cells from serial electron microscope montages of fetal monkey brain. Nature **250:** 31–34.
10. PARSONS, D. F., M. MARKO, M. RADERMACHER & J. FRANK. 1985. Shape changes and polarization of cells migrating through tissue. A high voltage electron microscope and computer graphics study of serial thick sections. Tissue Cell **17:** 491–510.

DISCUSSION OF THE PAPER

J. B. PAWLEY (*University of Wisconsin, Madison, Wisconsin*): I'm glad that you are again trying to access profile information from fixed sections viewed in stereo. I don't know if you are aware, but Tom Bower, a student of Bob Glaser's, did that some years ago in Berkeley. He didn't have the very best computer graphics to show the results, and so they were a little bit of a headache to look at. However, one of the great advantages of making several profiles of different depths in a stereo pair is that the alignment problem doesn't exist. Since at least the stereo pairs are aligned to start off with, you can extrapolate a little bit to see where the next set of three or four profiles should be. It's quite tedious. Tom Bower used the National Geodatic survey machines for making topographic maps from aerial photographs. He replaced the optics a little bit so he could put two micrographs in it and used the same techniques for tracing out profiles of a given height.

PARSONS: Yes, our equipment is much simpler than that, and we've recently found a way to do it. As you profile, you see the new profile being generated in depth. It's one-sided, but it's in depth. This is something that we didn't believe at first. So that simplified the equipment greatly. Our program also provides for the inclusion of absences due to sectioning along the material and also of essential absences due to lack of penetration of the stain from the outside. This is all accounted for in the reconstruction.

The Extraction of Three-Dimensional Information from Stereo Micrographs of Thick Sections Using Computer Graphics Methods[a]

LEE D. PEACHEY

Department of Biology-G7
University of Pennsylvania
Philadelphia, Pennsylvania 19104-4288

High voltage transmission electron microscopes (HVEM) operating with accelerating voltages of 1,000 kV and higher have been available for use by biologists and medical scientists for more than a decade.[1-4] A reduction in specific electron scattering (average electron energy loss per unit mass thickness) by the specimen as electron beam voltage increases allows the beam in these instruments to penetrate relatively thick preparations with a level of image quality that is attainable at 100 kV only with very thin specimens, 0.1 μm in thickness or less.[5] This has made possible the useful examination by electron microscopy both of dry mounts of whole cells grown in tissue culture[6] and of slices of embedded cells and tissues up to at least 5 μm in thickness.[7]

More recently, instruments operating at accelerating voltages intermediate between 100 kV and 1,000 kV, specifically at 300 to 400 kV (IVEM), have become commercially available. These newer instruments offer the advantage of increased accelerating voltage, state-of-the-art electronics and optical design, and a moderate size (compared to HVEMs) that can be accommodated in an average size laboratory. Though not having the full penetrating power of 1,000-kV instruments, these IVEMs have the potential for imaging specimens of embedded biological tissue several μm thick, which is sufficient for many purposes (Peachey, L.D., unpublished).

The usefulness of HVEM and IVEM images has been enhanced greatly by the development and use of stains that selectively increase the electron scattering of structures of particular interest in the specimens relative to that of other structures and the background.[7-9] This development has been crucial in the extension of the thickness of specimens that can usefully be examined up to 5 μm or more for slices of embedded cells and tissues while avoiding excessive confusion from overlying structures in the image. The result has been an unprecedented ability to visualize the three-dimensional structure and arrangement of cell organelles at high resolution that overcomes the severe restriction of two-dimensionality inherent in the use of very thin sections for structural study.

FIGURE 1 is a stereo pair of 400-kV electron micrographs of a 1.5-μm thick section of a frog skeletal muscle fiber stained selectively for the transverse tubular system using the Golgi method as modified by Franzini-Armstrong and Peachey.[9] Although all cellular structures in the muscle fiber are present in the section thickness, all except the selectively stained transverse tubules are seen only faintly in the background. The transverse tubules are seen throughout the depth of the section. This would be difficult

[a]Supported by National Institutes of Health Grants RR-2483 and HL-15835 and by the Muscular Dystrophy Association (Henry M. Watts Neuromuscular Disease Research Center).

to visualize in a single image and without selective stain, showing the power of this method.

Traditionally the three-dimensionality of thick specimens examined by high and intermediate voltage electron microscopy has been studied by viewing stereo pairs of images. Such images are produced by tilting the specimen appropriately in a special holder in the electron microscope between successive exposures.[10] The resulting stereo pairs of micrographs must be viewed in such a way that the two images are presented

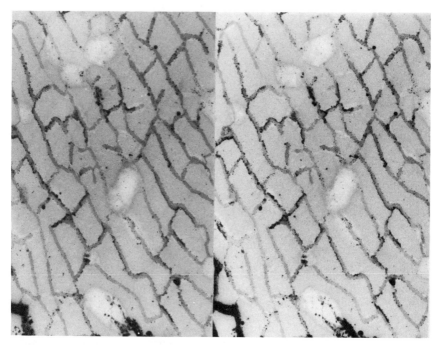

FIGURE 1. Stereo pair of electron micrographs taken at 400 kV on the JEOL JEM 4000-EX transmission electron microscope in the laboratory of Prof. Hatsujiro Hashimoto at Osaka University, Osaka, Japan. The specimen was a 1.5-μm thick transverse section of a plastic-embedded frog skeletal muscle stained selectively with a modified Golgi method. Over most of the area of this micrograph, the transverse tubular system, which is a network of tubules connecting from the surface of the muscle fiber to its interior, is stained. Myofibrils, which pass through the openings in the transverse tubular network, are relatively unstained. In some regions, the stain is particularly dense against one side of the transverse tubule. In two places near the bottom of the figure, the stain has also filled the sarcoplasmic reticulum resulting in a very dense precipitate. The very light, ovoid regions are unstained lipid droplets. 10,000X.

separately to the viewer's right and left eyes, either with or without optical aids. Parallax (horizontal displacement) differences between the two images are perceived by the viewer as "depth" or three-dimensionality in the image. Presentation of stereo depth in electron micrographs to a larger audience, through projected slides or in publications, has presented its own problems, though some useful methods have been devised.[10] (It should be noted that individual observers vary considerably in their ability to perceive parallax as depth, with a fraction of the population unable to do so at all).

Lacking in all this has been a convenient way to make measurements or otherwise extract quantitative information on the depth parameter in pairs of stereo micrographs. The visual perception of depth usually is the beginning and the end for those persons who can visualize micrographs stereoscopically, and even this is not possible for those who cannot. Therefore I have been seeking ways to extract three-dimensional information precisely and quantitatively from stereo electron microscopic images of thick specimens, taking advantage of recent developments in computer graphics and of the possibility of using more than one pair of images to increase the precision of the results. The procedures used are based on simple geometry, and no new or complex mathematics is needed or used. The goal has been to develop a system that is simple to use and that enhances the ability of a skilled observer to analyze images quantitatively, precisely, and objectively. The system under development is strongly interactive, with the computer and the human operator performing those parts of each task best suited to their individual abilities.

The electron microscope negatives that are to be analyzed first are digitized. Currently this is done with an Optronix International Inc. (7 Stuart Rd., Chelmsford, MA 01824) Photoscan P-1000 scanning digitizer to a spatial resolution of 650 by 800 pixels and to 256 grey levels. The resulting digital images are transferred via magnetic tape to a fixed magnetic disk on a DEC VAX 11/750, which is the host computer for the graphics/reconstruction system. All subsequent manipulation of the images is done digitally.

The graphics computer and display system used is the Raster Technologies, Inc. (9 Executive Park Drive, North Billerica, MA 01862) Model One/380 Graphics System, specially configured with dual display memories and drivers connected to two high-resolution color video raster displays, so that stereoscopic images and reconstructions can be displayed. The displays are viewed stereoscopically using an optical binocular viewing system.

Digital images are loaded from the host computer via a direct memory access interface into the display memories of the graphics system. Each display memory is 1,280 pixels wide, 1,024 pixels high, and 8 bits deep. The eight bit planes are divided so that some of the bits (usually 6) are used to display the digitized image in various grey levels, and the remaining bits (usually 2) are used to display graphics cursors and points or vectors used in reconstructing structures or as depth markers. Three colors are possible in the reconstruction when 2 bits are assigned to it, and 6 bits allow 64 grey levels in the image. These have been found to be adequate.

The Raster Technologies System One/380 incorporates a three-dimensional display list memory into which three-dimensional coordinates of points, ends of vectors, corners of polygons, etc. are loaded from the host computer. The host computer also chooses viewing angles and other display parameters and passes these to the graphics processor such that a satisfactory stereoscopic display of the graphics reconstruction is superimposed on the stereo image pair on the two display monitors. An X-Y digitizing tablet is used by the operator to input positional information during an interactive reconstruction and measurement session, and a set of push buttons on the tablet's cursor allows the operator to control program flow and select program options.

The process of generating and editing a reconstruction involves displaying stereo pairs of micrographs of the same specimen, and interactively positioning and moving vectors to correspond to the visualized position, shape, etc. in the specimen. The operator's judgment is crucial in positioning the vectors in relation to structures seen in the micrographs. The computer provides a high level of precision of display and a convenient way to view the images and to generate the desired reconstruction. Through this procedure, a three-dimensional, graphics model of the structure as visualized by the operator gradually is built up on the viewing screens and in the memories of the host computer and the graphics system.

When multiple stereo views of the same structures are available, they can be presented sequentially and used in generating a single reconstruction, using spatial information from all of the views. This use of multiple stereo views from different central viewing angles improves the validity of the reconstructions, since the direct line of sight view gives the least precision for establishing positions along the line of sight. In other words, "side" views help in editing the position of vectors that were laid down while viewing a structure "head on" in a previous stereo pair. This is a major advantage of this approach to three-dimensional reconstruction: more than one stereo pair can be used to provide information on the three-dimensional location of an individual feature in the specimen.

A typical reconstruction session proceeds as follows. A stereo pair of digitized electron micrographs is loaded from the host computer into the two display memories of the graphics system, leaving two bits set to zero for the graphics reconstruction data. Look-up tables available to the display processors are loaded with appropriate display intensity values so that a half-tone image is displayed on each monitor. These half-tone images must have been taken in the electron microscope with suitable tilt angles for stereo viewing, and are viewed stereoscopically on the display screens by the operator. The image appears to the observer much the same as does a pair of stereoscopic prints of the micrographs viewed with a standard mirror stereoscope. Errors in alignment of the two images with respect to the fixed coordinates of the display frames are ascertained at this point using a brief interactive procedure, following which the images are translated digitally in the host computer and reloaded to the graphics system correctly positioned. The position correction data used are stored in a data file associated with the particular image pair for use in any future editing sessions.

The graphics system then displays a single illuminated vector on each display. The three-dimensional coordinates of the ends of this vector are loaded from the host computer into the three-dimensional display list of the graphics system, which generates a pair of two-dimensional projections of this vector according to the current display parameters. These projections are loaded by the graphics processor into those bit planes of the two display memories assigned to the graphics reconstruction display, and thus appear on the two viewing screens in stereo. One end of this first vector, its "fixed" end, arbitrarily starts at the center of the viewing screen. The other, "free", end of this vector is controlled by the cursor of the digitizing tablet, which the operator uses to move that end of the vector at will. This vector, the current "active" vector, blinks on and off at a rate of about twice per second. In addition to moving the free end of the active vector parallel to the plane of the viewing field using the digitizing tablet, the operator can vary its apparent depth or Z-position perpendicular to the viewing field. This is accomplished by digital changes in the parallax between the two displays of the free end and of the lines connecting the free end to the fixed end of the active vector. The result is that the operator can move the free end of the active vector in three dimensions within the perceptual volume being observed, which also contains the three-dimensional image.

The immediate goal is to position vectors "inside" the space occupied by the stereoscopic image such that these vectors in some way model or reconstruct the objects of interest. This is accomplished through sequential manipulation of active vectors in various ways. For example, having positioned the free end of an active vector in some particular relationship with respect to objects seen in the image, the operator has several choices. One choice is to move the fixed end of the active vector to that point, effectively eliminating the previous active vector. A new active vector, whose fixed end is at the location of the free end of the previous vector when the change was made, is then displayed. Generally this would be the choice made at the beginning of a session, since the original fixed point was only arbitrarily placed with respect to the viewed structures in the image. The operator now has an active vector the fixed end of which bears some special relationship with respect to the structure being studied, *i.e.* it may lie at one end of the structure, at a branch point, on its surface, etc. The free end of the active vector then can be moved in three dimensions to another location with respect to the structure, under the control of the operator.

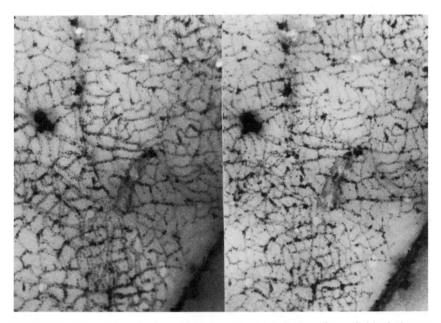

FIGURE 2. Photographs taken from a stereoscopic video display of two digitized electron micrographs of a transverse section 5 μm thick of a frog skeletal muscle fiber stained with the Golgi method and embedded in plastic. A network of transverse tubules is visible by virtue of the dense and selective staining of these structures. The network undulates considerably while retaining a generally transverse orientation. In a thick section such as this, where more than one layer of tubular network is found within the thickness of the section, graphic reconstruction is a valuable aid to understanding the geometric arrangement of the networks. The micrographs were taken at 1,000 kV in a JEOL JEM-1000 transmission electron microscope with a stereo tilt angle of 15 degrees total. 5,000X.

A second choice that can be made to dispose of an active vector is to "deposit" it, making it part of the visible reconstruction. This stops its blinking. It also creates a new active, blinking vector with the previous free point as its fixed end and a new free end following the digitizer and Z-control. A series of vectors can be deposited in three dimensions in this way, gradually building up a three-dimensional trace that in some way represents the structure being studied as seen by the operator. It is important to note that the image and the developing reconstruction coexist within the same three-dimensional perceptual space seen binocularly and stereoscopically by the operator.

Polygons and networks of vectors in three dimensions can be created by a third choice the operator has while building a reconstruction. If the free end of the currently active vector is placed near a previously deposited point (any end of a permanent vector), and a particular button on the digitizing tablet cursor is pressed, the free end of the active vector will jump to that point, *i.e.,* that point is "captured" and becomes the free end of the active vector. The position of the digitizing cursor becomes irrelevant at this time, until the operator makes one of several choices available by pressing cursor buttons. The active vector can be deposited, and the captured point becomes the fixed end of a new active vector. This allows, for example, for a polygon to be closed to its origin, or for several vectors to be attached to a single point, building up a network. Alternatively, and useful when editing complex reconstructions, the captured point can become a freely moving point, following the coordinate input by the operator through the digitizing tablet and the Z-control. In this case, all vectors connected to the free point now become active, that is, they move in three dimensions as the free point is moved. All these vectors blink in sequence to indicate their active status. If this free point subsequently is deposited at some location, then all the active vectors become permanent, and have that point as one of their ends. This procedure allows a reconstruction to be edited, by capturing previously deposited points and moving them, regardless of how many vectors are connected to them.

A further option for the fate of active vectors, either single or multiple, is deletion. Again this is useful for editing a reconstruction in progress or correcting a previously generated reconstruction at a later date, or simply for moving the free point around within a completed or partially completed reconstruction without depositing a vector, as usually is done at the start of a session.

FIGURES 2 to 4 show the results of a preliminary test of the system described here. FIGURE 2 is a stereo pair of high voltage electron micrographs of a selectively stained

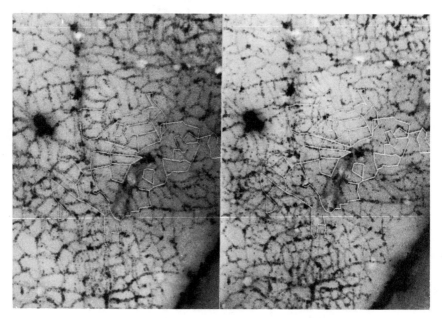

FIGURE 3. Same stereo micrographs as in FIGURE 2, with a partially completed vector reconstruction of a portion of the transverse tubular network.

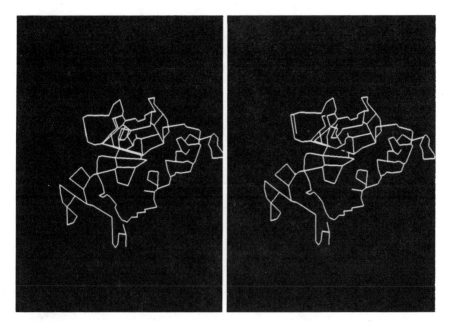

FIGURE 4. Stereo display of the reconstruction shown in FIGURE 3 without the image.

skeletal muscle fiber photographed directly from the video displays, and shows the image before any reconstruction is performed. A set of vectors then was generated to coincide with a portion of the transverse tubular network, which is densely stained in these preparations. FIGURE 3 shows the same stereo image with a partially completed reconstruction superimposed, again photographed directly from the display screens. The three-dimensional shape of the portion of the network that has been reconstructed is more clearly apparent with the reconstruction than without. In FIGURE 4, the reconstruction has been displayed and photographed without the image.

Extensions of this system that are planned or in progress include programs to perform calculations of three-dimensional parameters, such as length, surface area, and volume, from the three-dimensional coordinates of the data generated during the reconstruction, and the generation of solid surface displays of reconstructed closed objects. The latter will simplify the presentation of completed reconstructions to a larger audience, as discussed above, and will allow their three-dimensional nature to be appreciated by persons without strong parallax stereopsis. Dynamically rotating and zooming displays will also help to increase the apparent three-dimensionality of the reconstructions.

In summary, a system using advanced computer graphics to build reconstructions of objects viewed stereoscopically in electron micrographs of thick preparations is being constructed and programmed. It has the advantage that the reconstruction is built within the perceptual stereoscopic view space containing the original micrographs being reconstructed, and the reconstructions can be edited repeatedly while comparing them to different stereo views of the same objects, resulting in greater certainty and precision in the three-dimensional localization and shape of structures than can be obtained from a single stereo pair of micrographs. Once the reconstructions have been completed, they provide readily visible stereoscopic impressions of the structures that have been reconstructed, without the hindrance of other objects in the fields of view.

Quantitative three-dimensional information is readily available once the three-dimensional coordinates of structures have been determined during the reconstruction. In the future more realistic displays giving the impression of solid objects will be generated from the reconstruction data.

ACKNOWLEDGMENT

The specimens illustrated here are from a collaborative study with Dr. Clara Franzini-Armstrong, and the author gratefully acknowledges her contribution.

REFERENCES

1. HAMA, K. & K. R. PORTER. 1969. An application of high voltage electron microscopy to the study of biological materials. High voltage electron microscopy. J. Microscopie **8:** 149–158.
2. FAVARD, P., L. OVTRACHT & N. CARASSO. 1971. Observations de specimens biologiques en microscopie électronique à haute tension. I. Coupes épaisses. J. Microscopie **12:** 301–316.
3. RAMBOURG, A., Y. CLERMONT & A. MARRAUD. 1974. Three-dimensional structure of the osmium-impregnated Golgi apparatus as seen in the high voltage electron microscope. Am. J. Anat. **140:** 27–46.
4. PEACHEY, L. D., M. FOTINO & K. R. PORTER. 1974. Biological applications of high voltage electron microscopy. *In* High Voltage Microscopy. P. R. Swann, C. J. Humphries & M. J. Goringe, Eds. 405–413. Academic Press. New York, NY.
5. GLAESER, R. M. 1982. A critique of the theoretical basis for the use of HVEM in biology. Proc. 40th Annu. Mtg. Electron Microscopy Soc. Amer. 2–3.
6. WOLOSEWICK, J. J. & K. R. PORTER. 1976. Stereo high-voltage electron microscopy of whole cells of the human diploid line, WI-38. Am. J. Anat. **147:** 303–324.
7. PEACHEY, L. D. 1982. High voltage electron microscopy of muscle using thick slices of embedded tissue and selective stains. Proc. 40th Annu. Mtg. Electron Microscopy Soc. Amer. 14–17.
8. COUTEAUX, R., N. CARASSO & P. FAVARD. 1975. High and low voltage electron microscopical observations of the neurofibrillar bundles of the leech after silver impregnation. J. Microscopie Biol. Cell. **24:** 283–294.
9. FRANZINI-ARMSTRONG, C. & L. D. PEACHEY. 1982. A modified Golgi black reaction method for light and electron microscopy. J. Histochem. Cytochem. **30:** 99–105.
10. PEACHEY, L. D. 1978. Stereoscopic electron microscopy: principles and methods. Bull. Electron Microscopy Soc. Amer. **8:** 15–21.

DISCUSSION OF THE PAPER

J. JOHNSON (*Johns Hopkins University, Baltimore, Maryland*): Which voltages did you use to take the pictures? Were they all 1 MEV?

PEACHEY: Those were 1 MEV.

JOHNSON: You've shown us some rather remarkable micrographs from specimens that are 5 μm thick in which you've stained for a single cellular component. What, in your estimation, are the limitations in terms of thickness in specimens that are conventionally treated with gluteraldehyde and osmium and embedded and stained with uranyl acetate and lead?

PEACHEY: That question doesn't have a simple answer, because it depends on the nature of the specimen. If the specimen contains a lot of very small structures all equally well stained, then confusion can set in. For example, in a cell like a liver cell or a muscle cell that is pretty much packed with organelles, a quarter-μm thick section gives a pretty confusing image, if you've stained everything. If you do get the selective stain right, and you stain a single structure, such as we've done with the T system, then other limits apply. Ultimately you reach a limit of penetrating power. At one MeV with 5 μm, it is not difficult to do. At 8 μm, it is harder to focus and to have enough beam for a good exposure. The thickest I ever did was 12 μm, and that was pretty hard work.

D. F. PARSONS (*New York State Department of Health, Albany, New York*): I'm wondering if you tried the single monitor method. There are a number of people who flash between the two stereo images to view them on a single monitor using 1/20th of a second persistence of vision. Have you done this and, if so, what do you find that's useful?

PEACHEY: We haven't tried it. We considered a whole variety of ways of looking at stereo pairs and decided to go with this sort of exceptionally simple and somewhat brute-force way of using two separate displays. Though many of these other systems could be used, the hardware is more complicated, because you may have to wear glasses and blink your eyes on and off and things such as that. But you could do with a single monitor.

What is important in our experience is displaying the image and the reconstruction on the same tube but not trying to combine these in some optical way, because it's very difficult to build an optical system that does not deteriorate when you move your eyes slightly. What we found when we tried to do it with mirrors was that we couldn't keep the reconstruction and the image stereo, what I call perceptual space, coinciding very well; so then we went to the method of displaying both on the same tube.

PARSONS: Yes, I agree with you that that is a better approach.

W. CHIU (*University of Arizona, Tucson, Arizona*): Dr. Peachey, I think you have one of the first intermediate high voltage microscopes for biological applications. Can you comment at this point on the relative advantages of the intermediate high voltage and the high voltage microscope in terms of resolution and the thickness of the specimen?

PEACHEY: Our intermediate voltage microscope will be delivered in October, so we don't have it yet. However, I have used a couple of them. In my experience, for the kind of specimen that I showed you that is fixed in an ordinary way and embedded in epoxy resin and perhaps selectively stained, up to 3-μm sections viewed at 400 KV, which will be the voltage of our intermediate voltage microscope, is relatively easy to do. Three μm seems to be about as easy to use at 400 kV as, say, 5 μm is at 1,000 kV. The resolution that we get in the images seems to be comparable at those thicknesses for the two voltages in terms of resolution loss due to chromatic abberation, and that's about what you expect on theoretical grounds.

The other advantage of the intermediate voltage microscope, I suppose, is that it's a smaller, more reasonable microscope to live with, which will allow us to have one in our lab, not somewhere else.

CHIU: You were talking about the other uses.

PEACHEY: Yes. As long as I'm being pushed for a commercial here, our intermediate voltage microscope and the three-dimensional reconstruction facility are both supported by the National Institutes of Health as research resources, and they will be available to all of you on the same basis as are the high voltage facilities in Albany, Madison, and Boulder, at no cost.

B. P. M. MENCO: (*Northwestern University, Evanston, Illinois*): Can you get the same kind of reconstruction possibilities by simply rotating your specimen with a rotating stage in the microscope and viewing it that way from another angle?

PEACHEY: Yes, but you can't do this quite as dynamically as you would want to. A challenge to you people who are good at this sort of instrumentation would be to design a system so that we could sit at the microscope and look around and focus and move the specimen in stereo view all the while. The reason you can't do this is that in tilting the specimen, of course, the specimen moves laterally and it changes the focus and other things. Even in the best eucentric stages, I don't think you could do this. Hama, in Japan, at Okasaki, has built a system that I think is very useful. He tilts the specimen in one direction and records a video frame in a frame buffer; then he tilts the specimen into the other tilt direction and has a live video signal from that. He then displays the stored image at Tilt-I and the live signal at Tilt-II using two colors, red and green, on a color monitor and looks at it with red and green glasses. My expectation was that this wouldn't be very good, but I was wrong. It really is quite a good stereo display, letting you look at the image in stereo while sitting at the microscope. It's a big disadvantage not to know what your image looks like in stereo while you are sitting there. Traditionally, we don't know whether it's interesting or not until we've been to the dark room. So while we're scanning a fixed specimen, we may go by the most brilliant discovery we could make in our lives, and we don't even see it.

J. B. PAWLEY (*University of Wisconsin, Madison, Wisconsin*): On the subject of sample thickness and the comparative usefulness of different voltage microscopes, I agree regarding sectioned material. But there is, of course, quite a lot of interest in material that has been critical-point dried and is a whole mount. This is very different in that it is unsupported, fragile, and susceptible to electrostatic forces from charge buildup. It's also susceptible to radiation damage, causing macrochanges in the length of the fibers, and the motion of the sample. The reason that this is so important, is that we almost always make stereo pairs. If there is a motion of the sample between the first and the second stereo pair, of course, you end up with a distorted stereo pair that is not understandable. The million-volt microscope does have a definite advantage for looking at whole mounts, because dose deposited in the sample to get a given micrograph no matter which way you go about organizing it is less at a million volts than needed for 300 KV.

CHIU: You can use the low temperature.

PAWLEY: I'm not sure you really can. I don't see how you could define the temperature of something as fibrous as this in an environment of an electron beam irradiating it at 10^8 rads. Also the stereo system that you mentioned from Japan in fact was demonstrated in Madison in 1980 and was published at the Euro meeting. We find that it does work for very high-contrast objects such as yours but needs samples that are entirely stained. In normal staining, the contrast is not enough to make it go very well.

PEACHEY: I'm sorry, I forgot your contribution on that. I have two comments. One has to do with the high-contrast and the low-contrast objects. One of the things we're finding in this reconstruction study is that when the object does not have high contrast and very sharp boundaries and, with our floating vectors, when we try to position the vector with respect to the object, we find that we don't know exactly where the object is. It's like trying to put a point at the center of a cotton ball. It's very difficult. When we look at these objects that are subtle in their density distribution, we think we know what they look like in three dimensions and where they are; but when we have to put our finger on them and say that's where they are, it's very, very difficult to do. We may be fooling ourselves a little bit in thinking we know how to localize these objects. With respect to the voltages, I certainly am not one to say one voltage is better than another. If you are going to look at fixed and embedded specimens, 200 KV is better than 100, 400 is better than 200, 1,000 is better than 400; but the voltage is not always the limit. Sometimes the nature of the specimen, your ability to perceive the depth of the specimen, limits you to a thickness that's less than the voltage would limit you.

Four-Dimensional Microscopy of *Ascaris* Sperm Motility[a]

JAMES B. PAWLEY, SOL SEPSENWOL,[b] AND HANS RIS

Madison HVEM Facility
1675 Observatory Drive
University of Wisconsin
Madison, Wisconsin 53706
and
[b]*University of Wisconsin*
Stevens Point, Wisconsin 54481

INTRODUCTION

Powerful though electron microscopy is as a tool for ultrastructural investigation, it is seriously limited by the fact that it cannot be used to image living cells. This limitation is particularly important when the structure of interest is the motile system of a cell. As studies of the cytoskeleton and motility in general have constituted a major emphasis at the Madison High Voltage Electron Microscope (HVEM) Facility for the past several years, we recently installed a sophisticated video-enhanced light micros-copy (LM) system. This permits us to complement the three-dimensional images of carefully dried whole mounts obtained from fixed cells in the HVEM with information in the time dimension as seen in video recordings of the same or similar cells, made using the LM, while they were still alive and moving. Of course, the spatial resolution of the LM is limited to 0.2 μm by diffraction, but, as others have shown, structures smaller than this can sometimes be detected if the contrast of the image is increased by television-based electronic image enhancement.[1]

Earlier complementary HVEM/LM studies of this type involved motion of and contact between the growth cones of cultured nerve cells[2] and the activation of platelets.[3] The present contribution concerns pseudopodal motility in the sperm of the nematode *Ascaris suum*. Nematode sperm do not possess flagella, but creep over surfaces somewhat like amoebae. Recent studies on sperm of *Caenorhabditis elegans* and *Ascaris* species, however, have made it clear that this pseudopod motility is very different from the motion of amoebae, leucocytes, or cultured cells.[4] Nematode sperm contain only traces of actin and no detectable myosin.[5,6] Instead, there is a 15-KD protein called the major sperm protein that makes up 10–15% of the sperm and that in motile sperm has been localized in the pseudopod.[7] Electron microscopy of thin sections has revealed a mass of thin filaments throughout the pseudopod, but nothing definitive was revealed about their role in pseudopod motility. Pseudopod movement appears to be accomplished by short, fingerlike projections or villi that originate at the forward edge of the pseudopod and then move toward the cell body, where they disappear into the cell interior. This movement of villi can be observed in the LM if sperm are activated on a glass slide. When the cell body is attached to the glass but the pseudopod is not, such villi continuously form at the edge and move centripetally to the cell body.

[a]Supported by Division of Research Resources, National Institutes of Health, Biomedical Research Technology Program Grant P-41-RR00570.

In sperm that are actually moving, some of the villi attach to the glass surface, so that their centripetal movement relative to the cell body now moves the whole cell forward.

We recently studied activated sperm in critical-point dried (CPD) whole mounts and in thick sections viewed in the HVEM, and found that the pseudopodal filaments have a very characteristic arrangement in relation to the villi.[8] Filaments are arranged in complexes or bundles that connect each villus to the cell body. It appeared obvious that these filament complexes were somehow involved in the movement of the villi and thus in pseudopodal motility, but it was not clear how these fiber complexes actually produced motion. So we extended our investigations to the observation of living cells to see if the complexes were visible. Indeed, video-enhanced LM clearly reveals these complexes even though they are only about 0.2 μm thick. The combination of structural analysis by stereomicroscopy with HVEM and the changes in time obtained from videotapes is a powerful tool for observational and experimental analysis of this novel motility system.

MATERIALS AND METHODS

High Voltage Electron Microscopy

Spermatids from adult male *Ascaris suum* were collected from the seminal vesicle into 8–15 ml of HEPES-saline buffer (pH 7.4), preheated to 40° C, and gassed with 15% CO_2 in N_2. The spermatids were activated by adding 50 μl of a 20,000 g

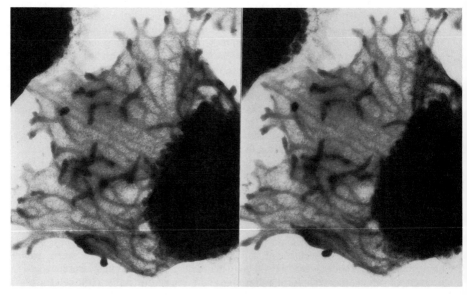

FIGURE 1. Stereomicrograph of a whole mount of an *Ascaris* sperm activated on a gold grid, fixed in 2% glutaraldehyde in 0.1 M HEPES buffer (pH 7.4) containing 0.1% Triton X-100. Postfixed in 0.1% OsO_4, stained with 1% uranyl acetate, and critical-point dried. 3,700X, tilt angle 12°.

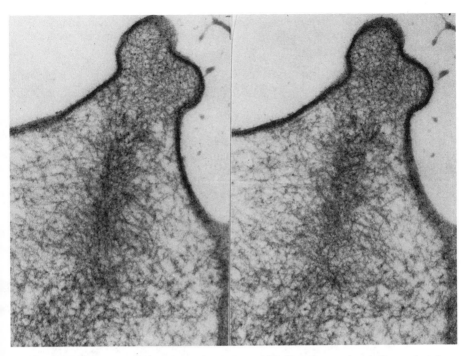

FIGURE 2. Stereomicrograph of a section 0.13 μm thick through pseuodopod of an activated sperm. Fixation 2.5% glutaraldehyde in 0.1 M HEPES buffer (pH 7.4) containing 0.05% saponin and 0.2% tannic acid. Postfixed in 0.1% OsO_4 followed by 1% uranyl acetate. The sections were stained with 7.5% uranyl magnesium acetate and lead citrate. Longitudinal section of a fiber complex and its termination in a short villus. 40,000X, tilt angle 30°.

supernatant of *vas deferens* homogenate (Sepsenwol, Taft, and Nguyen, unpublished). For whole mounts, formvar-carbon coated gold grids that had been made hydrophilic by glow discharge were placed at the bottom of a 60-mm plastic dish containing the sperm. After 15–20 minutes in a 40°-C incubator with 15% CO_2 in N_2, the grids were fixed in 2% glutaraldehyde and 0.1% Triton X-100 in 0.1 M HEPES buffer (pH 7.4). After washing in buffer the grids were postfixed in 0.1% OsO_4, stained with 1% uranyl acetate, and carefully critical-point dried. To preserve the structure during critical-point drying, it is essential to remove all traces of water or ethanol from the CO_2 in the pressure chamber.[9] To accomplish this, a molecular sieve was used to dry both the ethanol and the CO_2, and the mixture in the pressure chamber was stirred during the process of replacing the former with the latter.

For plastic embedding, activated sperm were pelleted and fixed in 2.5% glutaraldehyde in 0.1 M HEPES buffer (pH 7.4) containing 0.05% saponin and 0.2% tannic acid. After postfixation in 0.1% OsO_4 and staining in 1% uranyl acetate, the pellets were dehydrated and embedded in epon-araldite. Sections 0.13 μm thick were stained with 7.5% uranyl magnesium acetate in water (2 hours at 50° C) and Reynold's lead citrate (20 minutes at room temperature).

All HVEM samples were viewed in an AEI EM-7 microscope at 1 MeV, and stereo pairs were made using an axis-centered, side-entry tilt stage. To avoid damaging the delicate whole mount preparations, a sensitive television imaging system attached to the HVEM[10] was used for searching and focusing in order to minimize exposure of these samples to the beam.

Light Microscopy

A suspension of activated *Ascaris* sperm cells was prepared as for HVEM, and a 100-μl aliquot of the suspension was transferred to a well on a microscope slide formed by a cutout in an aluminum sheet sealed to the glass with silicone grease. After the addition of a grease-sealed #1 coverslip, the entire assembly was inverted and placed on the stage of an inverted Carl Zeiss (Thornwood, NY) Axiomat microscope and viewed in Nomarski differential interference contrast (DIC) microscopy using an oil immersion 100X, NA 1.3 plan Apochromat lens. The lens and stage area were insulated and maintained at 40–41° C with a hot air blower. The image was routed via the 0.8–3.2X zoom system of the Axiomat observation module to the side television port, where it was transmitted by a 1.6X coupling to a Sierra Scientific (Mountain View, CA) LSN-1, Newvicon television camera. The television signal was processed using a Quantex 9200 two-plane, real-time image memory system. The first memory was used to produce a rolling average of the image with a characteristic decay parameter of 1 or

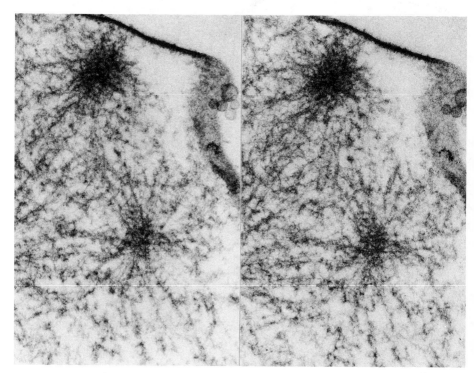

FIGURE 3. Stereomicrograph of a section 0.13 μm thick showing cross sections of two fiber complexes. Fixation and staining as in FIGURE 2. 40,000X, tilt angle 30°.

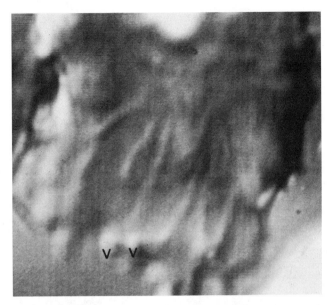

FIGURE 4. Light micrograph showing the pseudopod of a living activated sperm. The cell body is toward the *top*. Several villi (**V**) and fiber complexes are visible. (Compare with FIGURE 1.) Nomarski DIC, Axiomat with 100X oil immersion objective. Photographed from video tape. (See methods section for details.)

2 frames, and the second memory plane was used to store and then digitally subtract from succeeding frames an out-of-focus image in order to reduce the effect of the inhomogeneities of the imaging system, referred to as mottle.[1] A look-up table was used to increase the contrast of the difference image, and this signal was then recorded on a ¾-inch video cassette recorder. Photographs were made subsequently by storing the played-back signal in the Quantex and then photographing this image from a monitor screen.

RESULTS

High Voltage Electron Microscopy

A whole mount of an activated cell is shown in FIGURE 1. When viewed in stereo using a 2X stereo viewer, the villi arising from the upper surface of the pseudopod are clearly visible. They are filled with a fiber complex that connects each villus to the cell body, which can be seen on the right. Some fiber complexes branch near the periphery ending in two or more separate villi.

The arrangement of the fibers within the complexes is seen in FIGURES 2 and 3, which are stereomicrographs of sections 0.13 μm thick. FIGURE 2 illustrates a longitudinal section of a fiber complex and its termination in a short villus. In FIGURE 3 we see a cross section through two fiber complexes. Each fiber complex consists of an axial area where the fibers are more tightly packed, which is about 0.2 μm in diameter. From this axial area fibers radiate at right angles connecting to adjacent complexes and to the cell membrane. The complex terminates at the tip of the villus, where the filaments form a branching network that terminates at the cell membrane. The fibers themselves are variable in thickness (5–10 nm) with a fuzzy outline, and some appear to be composed of a thinner (2–3-nm) subfiber. (These fiber complexes will be described elsewhere in greater detail.)

Observations on Living Cells

The arrangement of the fiber complexes in the pseudopod suggests that they are somehow involved with the movement of the villi from its leading edge to the cell body. Routine phase contrast or DIC microscopy does not show these structures in the living cell, but they are visible if video-enhanced contrast DIC is used. FIGURE 4 shows a pseudopod in action as recorded on video tape and photographed from the video monitor. Several villi (V) and long fiber complexes connecting them to the cell body at the top of the picture are visible. Movement of the fiber complexes as the villi progress over the surface of the pseudopod can be seen in FIGURE 5a–d, which shows four consecutive images of a living pseudopod photographed from the videotape at 5-second intervals. Three generations of villi (numbered 1, 2, and 3 on FIGURE 5a–d) originating at the leading edge of the pseudopod are pictured as they move towards the cell body. The arrows point to a V-shaped branch in a fiber complex. Comparing FIGURE 5a, b and c it is apparent that the fiber complexes move centripetally without much change in shape and at the same speed as the villi. The way in which this movement is accomplished is still a mystery.

DISCUSSION

The HVEM has been closely identified with early attempts to view entire cells at a resolution higher than that possible in the LM. The structure of the cytoskeleton has been a principle area of interest in these studies, because the shape and topology of the fibrous elements that make up this structural network only become evident when it is viewed in slices substantially thicker than are present in ordinary EM thin sections. The utility of this approach is well demonstrated by the bundles of filaments making up the motile system of *Ascaris* sperm. The structural nature of and interconnections between these bundles only become evident when they are viewed in whole mounts and thick sections. It is important to use these two preparative techniques together, both because they are a check against the introduction of preparative artifacts and because the former makes it possible to see the overall structure of the network while the latter permits more detailed views relatively free from the complications of excessive structural overlap. These advantages should be readily evident in FIGURES 1–3, where the similarity of the structures made visible by the two preparative techniques and the different levels of structural organization can be seen.

The HVEM studies overlap well with those using the LM because, while both can image whole cells, the LM can also image living cells. The recent addition of video systems to increase the contrast of the LM image makes this interaction even more mutually supportive, because it permits the motion of smaller intracellular structures such as the fiber bundles to be easily visualized and recorded. Unfortunately, a series of halftone images on the printed page is a pale substitute for the impression one obtains from watching the playback of a video tape recording, but the motion of both villi and filament bundles can be readily inferred from FIGURE 5.

What should be evident in this discussion is the large extent to which information obtained by all three techniques is complementary. This is true not only in the sense that each provides information that cannot be obtained from the others, but also in the sense that information from each assists in the proper interpretation of the results obtained from the others. The fiber bundles are visible by all three techniques, but only

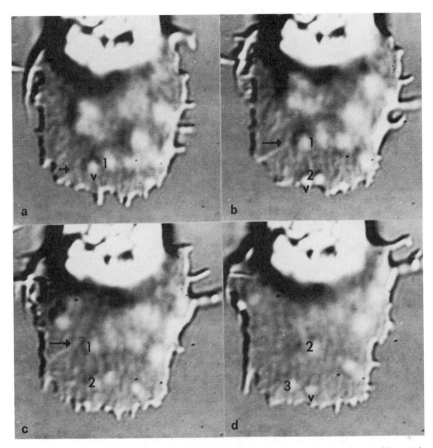

FIGURE 5. Light micrographs (Nomarski DIC) of living sperm taken at 5-second intervals. Photographed from video tape as described in methods section. Three generations of villi are shown as they form at the forward edge of the pseudopod and move upwards toward the cell body (labeled **1, 2,** and **3**). The *arrows* point to a V-shaped branch of a fiber complex that is seen to move centripetally without changing shape and at the same speed as the villi.

by using all of them can we determine both the fine structure of the fiber bundles and the direction in which these bundles move. Clearly, this integrated approach is important to the understanding of the three-dimensional morphological correlates of pseudopodal movement in *Ascaris* as well as to the understanding of changes in these structures in the fourth dimension, time.

REFERENCES

1. ALLEN, R. D. & N. S. ALLEN. 1983. Video-enhanced microscopy with computer frame memory. J. Micros. **129:** 3–17.
2. TSUI, H.-C. T., K. L. LANKFORD & W. L. KLEIN. 1985. Differentiation of neuronal growth cones: specialization of filopodial tips for adhesive interactions. Proc. Natl. Acad. Sci. USA **82:** 8256–8260.
3. ALBRECHT, R. M., J. A. OLIVER & J. C. LOFTUS. 1986. Observations of colloidal gold labelled platelet surface receptors and the underlying cytoskeleton using HVEM and SEM. *In* The Science of Biological Specimen Preparation for Microscopy and Microanalysis. M. Müller, R. Becker, A. Boyde, and J. J. Wolosewick, Eds. 185–193. SEM, Inc. Chicago, IL.
4. ROBERTS, T. & M. CRAWLING. 1983. *Caenorhabditis elegans* spermatozoan contact the substrate only by their pseudopods and contain 2 nm filaments. Cell Motil. **3:** 333–347.
5. NELSON, G. A., T. M. ROBERTS & S. WARD. 1982. *Caenorhabditis elegans* spermatozoan locomotion: amoeboid movement with almost no actin. J. Cell Biol. **92:** 121–131.
6. NELSON, G. A. & S. WARD. 1981. Amoeboid motility and actin in *Ascaris lumbricoides* sperm. Exp. Cell Res. **131:** 149–160.
7. WARD, S. & M. KLASS. 1982. The location of the major protein in *Caenorhabditis elegans* sperm and spermatocytes. Dev. Biol. **92:** 203–208.
8. SEPSENWOL, S. & H. RIS. 1985. Fibercomplexes associated with pseudopod motility in *Ascaris* sperm. J. Cell Biol. **101:** 401a.
9. RIS, H. 1985. The cytoplasmic filament system in critical-point dried whole mounts and plastic-embedded sections. J. Cell Biol. **100:** 1474–1487.
10. PAWLEY, J. B. 1984. An HVEM for high resolution low-dose studies of biomacromolecules. Ultramicroscopy **13:** 387–406.

DISCUSSION OF THE PAPER

D. F. PARSONS (*New York State Department of Health, Albany, New York*): Can you say a little more about this major sperm protein; does it have any homology with actin at all? And is it present in other cells? And were those the filaments you were showing us?

PAWLEY: Yes, we believe they are the filaments. In fact Thomas Roberts and Hans Ris have done some unpublished work with labeled antibodies that I didn't show you, and it appears that the filament bundles are composed of the major sperm protein. So it is the sperm protein.

G. J. BRAKENHOFF (*University of Amsterdam, Amsterdam, The Netherlands*): What kind of substrates did you use to restrict the mobility of the cells?

PAWLEY: It was more a matter of not treating it with various adhesive materials that would make it adhere well. We didn't put it on BSA or some other protein coating that might make it stick well. You can use terrible things like polylysine, which sticks it down well, but they're not happy with that. They can't move. So it's more a matter of leaving out. These were just glow-discharged grids.

J. W. BUCHANAN (*University of Pennsylvania School of Veterinary Medicine, Philadelphia, Pennsylvania*): In response to a previous question about using a rotating stage with a photomicroscope, this does work well for three-dimensional histology, and orientation is accomplished with the rotating stage. You can line up your specimens with embedded vertical hairs or use the edge of the serial section to show that up. The method was published in the *Journal of Microscopy* in 1973.

A. BOYDE (*University College London, London, England*): I just have two comments about this stereo projection. First, the point is the gentlemen who are unable to see through black sections by high voltage microscopy insist on showing us black pictures instead of just reversing the contrast. You get more light out of these low-intensity projectors if you try to shine it through the negative from which you made the slide, and we'd all see rather more detail in stereo under those circumstances.

The other comment is to Dr. Peachey regarding his method of determining orientations by laying down large poles in three dimensions rather than single points. I wondered whether it would be better to determine those intersections in your array, but with a single stereo cursor rather than the very large barlike object?

G. C. RUBEN (*Dartmouth College, Hanover, New Hampshire*): I wanted to ask Dr. Peachey about the diffuseness of the network and how accurate eventually he thinks that he's going to be able to locate things in space? By very high magnification stereo microscopy of molecules and freeze-etch replicas, we think we can get down to 5 to 10 Å. But with very, very thick specimens like you've shown, it must be very difficult to get within a few hundred?

L. D. PEACHEY (*University of Pennsylvania, Philadelphia, Pennsylvania*): I can't answer that very precisely, because we don't know. There are two things that I can say. One is I hope it's going to be better than a few hundred, but it's not going to be 3 to 5 Å. We're hoping we might be able to get down to something like 40 to 50 Å, which is fine for these networks that have a mesh size of about half a micron or several thousand angstroms.

The accuracy of things in three dimensions is not isotropic. We can position left and right in the field much more precisely than along the line of sight, called the Z. That's why we wanted to be able to use multiple tilts. In our intermediate voltage microscope we have full specimen rotation, so we can rotate the grid continuously. We're going to try to go to very high tilt, so that we effectively get side views. When we have a side view, then we can correct the Z-position in the way I described, because what was originally the Z-position now is a very large X and Y component that we can set more precisely. I'm afraid I can't give a more precise answer than that. At present, tests indicate that we can only position something in the Z-axis to about 1/20th of the resolution that we can position things laterally.

RUBEN: One of the difficulties that you are going to run into at very high tilts is that your specimen will get quite a bit thicker and your uncertainties greater. So I guess there's a trade-off here.

PEACHEY: There are two parts I think to that question. One was, how well can you image it? There's also the matter of the stereo imaging in the computer. Now I remember using 512 square arrays. Is that right?

UNIDENTIFIED SPEAKER: We can use up to 1024.

PEACHEY: All right, even 1024. That was 512 that we saw. It gets more complicated, because to read that out at 60 Hertz, 1024 in alternate frames is a problem. Stereo is a difference technique: you are relying on your perception of stereo. Usually you can not really fuse a stereo pair if the parallax is more than maybe 10 or, at most, 20% of the field of view. So even if we start off with 1024, 10% would be 100 pixels, so the most different levels you could imagine would be 100 levels. That's a lot, but it's not so bad. In fact it's a lot less than that, because of the fact that it's digitized. You've got to cut that down by the sampling factor and you are down to about 30 levels. So there is a limit even with the larger array size, imposed just because it has been digitized.

UNIDENTIFIED SPEAKER: Your last point was correct, but the one before was not, because when you are positioning a stereo cursor you are in fact only abstracting a small part of the three-dimensional space. So the limit of 10 to the largest parallax you mentioned is the limit that is in fact not enjoyed in practice. If you have any experience in doing stereo measurements you can accommodate, you can make measurements within a field that has very much more parallax than that.

E. MANDELKOW (*Max-Planck-Unit for Structural Molecular Biology, Hamburg, Federal Republic of Germany*): We should perhaps learn from the X ray crystallographers who, about a dozen years ago, enjoyed showing a lot of stereo pictures at conferences. They more or less stopped doing that because, although it is great to be working with stereo pictures while you're doing your analysis, it's notoriously difficult to show it to an audience, and it's usually distractive. I think computer methods will allow one to show pictures in a two-dimensional way that may be more illustrative than the three-dimensional way.

PEACHEY: That's probably true if you look at something like a microtubule. It's not so easy to abstract the information into a computer image from something like these cells.

PAWLEY: I agree with Dr. Peachey. Two-dimensional computer graphics are only useful for imaging surfaces that can easily be specified mathematically. We are a long way from being able to do this with higher-order biological structures. At present, the work has to be shown in three dimensions using stereo projection methods.

Mapping the Domains of Molecules and Complexes by Mass and Heavy Atom Loading[a]

JAMES F. HAINFELD AND JOSEPH S. WALL

Biology Department
Brookhaven National Laboratory
Upton, New York 11973

MASS MAPPING

The Scanning Transmission Electron Microscope (STEM) has proved valuable for determining the molecular weight of isolated molecules and complexes, because the fraction of elastically scattered electrons is in linear proportion to mass thickness for thin samples (<500 Å), and this signal can easily be corrected for multiple scattering point by point to measure thick samples as well. Data is digitized from the photomultiplier signal, so that the film and densitometry steps that are required with a conventional electron microscope (CEM) are eliminated. The errors and transfer functions of these additional steps are therefore also precluded. The STEM is very efficient in dark-field collection, where it measures about 80% of the signal. Thereby it uses lower doses than the CEM, whose dark-field collection efficiency is about 5%.[1] The Brookhaven STEM has two concentric elastic scattering detectors—15–40 mrad and 40–200 mrad half angle—though only the larger is used for mass analysis. Since on the smaller some diffraction effects are evident that could give rise to incorrect mass values in crystalline areas, one can employ only about 40% of its elastically scattered signal. The transfer function of the STEM for in-focus images is monotonic and does not oscillate at higher spatial frequencies as it does in the CEM. The interpretation of high-resolution details (<15 Å) from STEM images in terms of true mass distribution, therefore, does not require extensive computer correction, as do CEM micrographs. There are no zeros in the STEM transfer function, as in the CEM. Therefore, the STEM could complement CEM work in frozen-hydrated samples, for example, where at a high value of defocus low frequency data interpretation (due to the transfer function) is ambiguous and has a poor signal-to-noise ratio.

Mass mapping can be used in various ways. Not only can the native mass be determined, but mass can be added at specific sites, as for instance with antibodies or with components of a multienzyme complex. It can also be removed by dissociating components or by enzymatic digestion. With fibers such as DNA, viruses, and intermediate filaments, mass per unit length can be determined. Radial mass profiles of fibers have also been calculated from STEM data, as discussed in another paper in this volume. The usual purpose of such mass mapping is to discover the position of certain components in a structure or to find the stoichiometry based on known component weights. Such information is often easily obtained by these means and may help to distinguish between several proposed models of the structure.

[a]Supported by the Office of Health and Environmental Research of the U.S. Department of Energy and National Institutes of Health Grant GM-31975.

Most samples for STEM mass analysis are prepared by injecting the sample into a drop of buffer on the grid, so that one may avoid viewing protein denatured on the air-water interface of the applied drop.[2] After being attached for about one minute the samples are washed with several changes of vacuum-volatile buffer (usually ammonium acetate), quick-frozen in liquid N_2 slush, and freeze-dried overnight. The unstained results on 20–40-Å thick carbon films are reasonably well preserved, although in a few cases they suffer some shrinkage.

Mass measurement techniques have occasionally been applied to negatively stained samples, for which the stain must be thick enough to form a uniformly flat embedment. Otherwise distortion from stain draping occurs. Point by point correction for multiple scattering must also be used. The advantage of stain is that more subunit detail is often visible with it than without it, making alignments easier. Sometimes, too, shrinkage is less with stain. Although mass measurement of frozen-hydrated samples has not yet been tried with the STEM, it is entirely possible. It may also be quite useful, since proved specimen preservation at low doses could further enhance the resolution of mapping as well as avoid some of the problems with the CEM noted above.

Once the STEM data is collected, it is stored on magtape. The images are then analyzed in three steps. First, the background is measured in segments in a way that reasonably deals with gradients or local variations. Second, TMV, which was added to the grid during preparation, is measured as an internal mass calibration. Third, areas of molecules to be measured are interactively selected and their mass calculated, and a statistical study is made.[3,4]

The accuracy of STEM mass measurement depends on two factors—beam noise (counting statistics) and carbon film (substrate) noise, assuming that the sample is pure with no background material.[5] In summary form, at 10 el/$Å^2$ the accuracy (standard deviation) on single particles is approximately:

Mass (daltons)	% Error
1×10^3	100
10×10^3	30
100×10^3	6
1×10^6	1.5
10×10^6	.5

Of course, greater accuracy is attained by counting more particles. If particles are aligned either in thin crystals or by using correlation techniques for single particle alignment, resolution and accuracy can be increased. Engel, Baumeister, and Saxton used an unstained, ordered, bacterial-wall HPI layer to give a 30-Å resolution map showing features with 1.3 kD weight.[6]

Domainal mapping was done on single fibrinogen molecules. The intact molecules (I-4) had a molecular weight of 340 kd, and a proteolytically cleaved form (I-9, with the Aα chain cut) was 300 kd. The missing mass was localized to the central E domain of the molecule. Fab fragments directed toward the D and E domains helped to establish where they are in the intact molecules.[3,7]

Intermediate filaments, both native and reconstituted, were examined from various organisms and tissues and their mass per unit length measured.[8] Although several categories were found, they all had a dense core with diffuse edges, which implied a model consistent with sequence information.[9,10]

The E. coli pyruvate dehydrogenase complex (5.7 $\times$ 10^6 d) can be reconstituted in stages. From plots of projected radial mass it was found that the 6 E_3 (dihydrolipoyl dehydrogenase) dimers add to the faces of the E_2 (dihydrolipoyl transacetylase) cubic core structure, and the twelve E_1 dimers (pyruvate dehydrogenase) attach along the twelve edges of the cubic core.[11]

Dynein ATPase images showed three globular units connected by strands. Molecular weights of each of these heads (450, 450, and 550 kd) corresponded to subunits that have these molecular weights. Analysis of the mass at the periphery of this ATPase bound to microtubules determined the polarity of attachment of this complex.[12]

The number of various proteins per vesicular stomatitis virus (VSV) was determined from mass analysis and known stoichiometry.[13] The number of subunits per unit cell in bacterial flagellae was discovered by studies of mass per unit length.[14]

With respect to DNA studies, freeze-dried strands with proteins attached indicated that positions measured along the DNA are accurate to within a few base pairs.[15] T antigen (82 kd) was found to bind as monomers through tetramers at specific sites on SV40 DNA.[16] The number per site can easily be found by STEM mass measurements.

Mass mapping on helical or cylindrical structures is of interest since data along the axial direction can be combined to improve signal-to-noise. Radial reconstruction algorithms have been derived that permit a density vs. radius calculation for the structure. This technique has been used to find the RNA in TMV (a higher density annulus at 40 Å radius)[17] and to quantitate the distribution of a protein inside T4 tail tubes (probably gp48) that seems to act as a ruler that determines the length of the tail.[18] Further applications of this reconstruction technique are underway in fd virus and actin filaments and the method is being adapted to yield radial density plots from spherical structures as well. Steven et al. discuss this technique more fully in another article in this volume.[30]

CLUSTER LABELING

The high-resolution STEM can easily image single atoms with $Z \geq 47$ (silver). Previously single heavy atom reagents such as p-chloromercuribenzoate (PCMB) were used to label specific sites on proteins. However, the dose necessary to image single atoms clearly is from 1,000 to 10,000 el/$Å^2$. The lower of these is only applicable to atoms on thin substrates (about 20 Å) and not on top of large protein masses. Nevertheless both doses are too high to preserve high resolution in the native biological structures. Image averaging of ordered arrays or aligned molecules to achieve higher resolution and visibility of single heavy atom stains has been limited to special cases.[19,20] Therefore, a compromise for single molecule labeling was needed, namely a larger, more visible label at lower doses but one that was smaller than the usual markers such as ferritin.

Several years ago Singer found that the small size of a well-known inorganic compound that had eleven gold atoms at its core made it distinguishable from other conventional labels. Since the complex was only soluble in organic solvents, it had to be derivatized to make it water soluble. Bartlett did this with an undecagold complex that had the dense core of eleven gold atoms and an organic shell with twenty one amines at the periphery. Though this compound could not be seen in the bright-field CEM, it was easily visible in the more efficient dark-field STEM. While CEM applications temporarily ceased, therefore, STEM use was pursued.

The visibility and stability of these clusters were found to lose intensity as beam dose increased,[21] a behavior similar to that of proteins. About 40% of their intensity was lost after a dose of 1,000 el/$Å^2$. The signal-to-noise ratio on the cores was 6.2 at 100 el/$Å^2$, indicating good visibility.

The amines surrounding the undecagold cluster can be cross-linked to proteins or other molecules. An early success with this compound was to attach two or perhaps more modified biotin molecules to the gold cluster. Then avidin, which is a tetramer with four binding sites for biotin, was added. Resultant polymers showed gold clusters at the biotin binding sites with about 10-Å resolution thus delineating relative positons of the four sites on each avidin molecule. This was the first time that such a high resolution had been achieved on single molecules.[22]

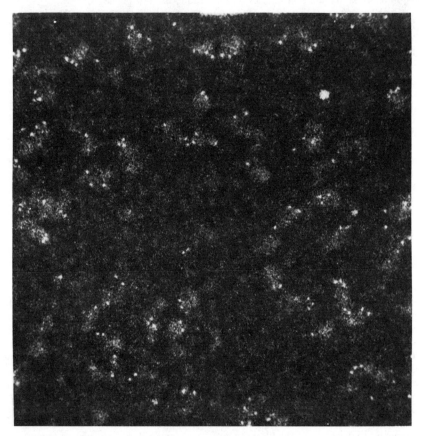

FIGURE 1. Fab′ fragments from rabbit IgG labeled with monofunctional undecagold complexes. Full width: 1,280 Å; dose: 60 el/Å².

Another binding scheme for the undecagold cluster was to oxidize carbohydrate moieties with periodate on a glycoprotein, react the aldehydes with the gold cluster amines to form a Schiff′s base, and stabilize with borohydride. This technique, which occurs in nature, localized the sugar moieties on the heavy chain of human haptoglobin in a haptoglobin-hemoglobin complex.[23]

More recently the gold complex was made monofunctional, since the twenty-one amine residues often led to unwanted aggregation during labeling. Frey *et al.,* working with the existing complex, blocked twenty of the twenty-one reactive groups, chromatographically separating the desired specie.[24] Safer, on the other hand, beginning with the Au_{11} synthesis, used a stoichiometric mixture of phosphines, so that only one reactive group would be on each undecagold.[25] Both of these monofunctional complexes have been synthesized with reactivity towards free sulfhydryls.

These labels have marked cysteine 374 on F-actin[26] and the β-93 CYS on hemoglobin in the haptoglobin-hemoglobin complex. The undecagold has also been attached to promoter proteins on DNA, S-1 myosin heads, and human fibrinogen. Labeling Fab' fragments of IgG makes the smallest antibody label yet that is clearly visible in the electron microscope. The distance from the gold to the antigen is about 45 Å, which is a five to ten times better resolution than that of previous antibody labels that attach whole antibodies to ferritin, gold colloids, or other electron-dense materials. An image of Fab' molecules (MW = 50 kd) with undecagolds attached is shown in FIGURE 1. This technique, which should have many biological applications, will be described in greater detail in a forthcoming publication.

There are many stable complexes of heavy metals in addition to the gold, such as Au_{55}[27] and W_{12}. The latter, of which the common E.M. stain phosphotungstic acid (PTA) is an example, has, apart from its charge (4-, a polyanion), little reactivity with biological samples. However, organic side chains can be synthetically attached to this cluster. A long chain hydrophobic tail was attached to create a fatty-acid-like molecule with the tungstate cluster as the charged end group. These when inserted into phospholipid bilayer vesicles could be seen in the microscope.[28] Compounds with C_4 through C_{22} chain tails were synthesized, and could be used as tags for membranes of cells. No other E.M. lipid labels exist except radioactive compounds, which lack high resolution and often require months to expose the film.

The W_{11} clusters, unlike the Au_{11} ones, do not deteriorate with beam dose[29] and appear to be almost indestructible. Perhaps in a double-label experiment with them a subsequent high-dose image could be used to distinguish the gold from the tungsten clusters. Work is proceeding to prepare a monofunctional derivative of the W_{11} cluster that is reactive towards sulfhydryls or other groups.

Future cluster labeling will include synthesis of larger clusters with the same specific reactivity but more easily seen in CEMs or in thin sections. The high scattering from these clusters may make them useful in X ray or electron diffraction studies as isomorphous replacements; this is valuable when with small crystals single atom derivatives give too small a signal.

In summary, the high-resolution STEM has made it possible to map domains of molecules or complexes by both quantitative mass distributions and ultra-small heavy atom cluster labels covalently attached to sites of interest.

REFERENCES

1. WALL, J. S., M. ISAACSON & J. LANGMORE. 1974. The collection of scattered electrons in dark field electron microscopy. Optik **39**: 359–374.
2. WALL, J. S., J. F. HAINFELD & K. D. CHUNG. 1985. Films that wet without glow discharge. *In* Proc. 43rd Annu. Mtg. Electron Microscopy Soc. Amer. 716–717.
3. MOSESSON, M. W., J. HAINFELD, R. H. HASCHEMEYER & J. S. WALL. 1981. Identification and mass analysis of human fibrinogen molecules and their domains by scanning transmission electron microscopy (STEM). J. Mol. Biol. **153**: 695–718.
4. HAINFELD, J. F., J. S. WALL & E. J. DESMOND. 1982. A small computer system for micrograph analysis. Ultramicroscopy **8**: 263–270.
5. WALL, J. S. & J. F. HAINFELD. 1986. Mass mapping with the scanning transmission electron microscope. Annu. Rev. Biophys. Biophys. Chem. **15**: 355–376.

6. ENGEL, A., W. BAUMEISTER & W. O. SAXTON. 1982. Mass mapping of a protein complex with the scanning transmission electron microscope. Proc. Natl. Acad. Sci. U.S.A. **79:** 4050–4054.

7. WALL, J., J. HAINFELD, R. H. HASCHEMEYER & M. W. MOSESSON. 1983. Analysis of human fibrinogen by scanning transmission electron microscopy. Ann. N.Y. Acad. Sci. **408:** 164–179.

8. STEVEN, A. C., J. F. HAINFELD, B. L. TRUS, J. S. WALL & P. M. STEINERT. 1983. The distribution of mass in heteropolymer intermediate filaments assembled *in vitro*: STEM analysis of vimentin/desmin and bovine epidermal keratin. J. Biol. Chem. **258:** 8323–8329.

9. STEVEN, A. C., J. F. HAINFELD, B. L. TRUS, J. S. WALL & P. M. STEINERT. 1983. Epidermal keratin filaments assembled *in vitro* have masses-per-unit-length that scale according to average subunit mass: Structural basis for homologous packing of subunits in intermediate filaments. J. Cell Biol. **97:** 1939–1944.

10. STEVEN, A. C., B. L. TRUS, J. F. HAINFELD, J. S. WALL & P. M. STEINERT. 1985. Conformity and diversity in the structures of intermediate filaments. Ann. N.Y. Acad. Sci. **455:** 371–380.

11. YANG, H., J. F. HAINFELD, J. S. WALL & P. A. FREY. 1985. Quaternary structure of pyruvate dehydrogenase complex from *E. coli*. J. Biol. Chem. **260:** 16049–16051.

12. JOHNSON, K. A. & J. S. WALL. 1983. Structure and molecular weight of the dynein ATPase. J. Cell Biol. **96:** 669–678.

13. THOMAS, D., W. W. NEWCOMB, J. C. BROWN, J. S. WALL, J. F. HAINFELD & A. C. STEVEN. 1985. Mass and molecular composition of vesicular stomatitis virus: A STEM analysis. J. Virol. **54:** 598–607.

14. TRACHTENBERG, S., D. J. DEROSIER, S.-I. AIZAWA & R. M. MACNAB. Pairwise perturbation of flagellin subunits—the structural basis for the differences between plain and complex bacterial flagellar filaments. J. Mol. Biol., submitted.

15. HOUGH, P. V. C., I. A. MASTRANGELO, J. S. WALL, J. F. HAINFELD, M. M. SIMON & J. L. MANLEY. 1982. DNA-protein complexes spread on N_2-discharged carbon film and characterized by molecular weight and its projected distribution. J. Mol. Biol. **160:** 375–386.

16. MASTRANGELO, I. A., P. V. C. HOUGH, V. G. WILSON, J. S. WALL, J. F. HAINFELD & P. TEGTMEYER. 1985. Monomers through trimers of large tumor antigen bind in region I and monomers through tetramers bind in region II of simian virus 40 origin of replication DNA as stable structures in solution. Proc. Natl. Acad. Sci. U.S.A. **82:** 3626–3630.

17. STEVEN, A. C., J. F. HAINFELD, B. L. TRUS, P. M. STEINERT & J. S. WALL. 1984. Radial distributions of density within macromolecular complexes determined from dark-field electron micrographs. Proc. Natl. Acad. Sci. U.S.A. **81:** 6363–6367.

18. DUDA, R. L., J. S. WALL, J. F. HAINFELD, R. M. SWEET & F. A. EISERLING. 1985. Mass distribution of a possible tail-length determining protein in bacteriophage T4. Proc. Natl. Acad. Sci. U.S.A. **82:** 5550–5554.

19. STEWART, M. 1975. The location of the troponin binding site on tropomyosin. Proc. R. Soc. London, Ser. B **190:** 257–266.

20. BEER, M., J. W. WIGGINS, R. ALEXANDER, R. SCHETTINO, C. STOCKERT & K. PIEZ. 1979. Electron microscopy of selectively stained collagen. *In* Proc. 37th Annu. Mtg. Electron Microscopy Soc. Amer. 28–29.

21. WALL, J. S., J. F. HAINFELD, P. A. BARTLETT & S. J. SINGER. 1982. Observation of an undecagold cluster compound in the scanning transmission electron microscope. Ultramicroscopy **8:** 397–402.

22. SAFER, D., J. F. HAINFELD, J. S. WALL & J. RIORDAN. 1982. Biospecific labeling with undecagold: Visualization of the biotin binding sites on avidin. Science **218:** 290–291.

23. LIPKA, J. J., J. F. HAINFELD & J. S. WALL. 1983. Undecagold labeling of a glycoprotein: STEM visualization of an undecagoldphosphine cluster labeling the carbohydrate sites of human haptoglobin-hemoglobin complex. J. Ultrastruct. Res. **84:** 120–129.

24. REARDON, J. E. & P. A. FREY. 1984. Synthesis of undecagold cluster molecules as biochemical labeling reagent. 1. Monoacyl and Mono[N-(succinimidooxy)succinyl] undecagold cluster. Biochemistry **23:** 3849–3856.

25. SAFER, D., L. BOLINGER & J. S. LEIGH. 1986. Undecagold cluster for site specific labeling of biological macromolecules; simplified preparation and model applications. J. Inorg. Biochem. **26,** in press.

26. SAFER, D., J. HAINFELD & J. S. WALL. 1985. The localization of Cysteine-374 in F-actin determined by gold cluster labeling and scanning transmission electron microscopy. Biophys. J. **47:** 128a.

27. SCHMID, G., R. PFEIL, R. BOESE, F. BANDERMANN, S. MEYER, G. H. M. CALIS & J. W. A. VAN DER VELDEN. 1981. Au_{55} $[P(C_6H_5)_3]_2$ Cl_6—Ein Goldcluster ungewöhnlicher Größe. Chem. Ber. **114:** 3634–3642.

28. LIPKA, J. J., J. F. HAINFELD & J. S. WALL. 1985. Heavy atom cluster labeling of biological specimens. *In* Proc. 43rd Annu. Mtg. Electron Microscopy Soc. Amer. 718–719.

29. MONSON, K. L., J. S. WALL & J. F. HAINFELD. Visibility and stability of a 12-tungsten atom complex. *In* Proc. 44th Annu. Mtg. Electron Microscopy Soc. Amer., in press.

30. STEVEN, A. C., T. A. SIMPSON, B. L. TRUS, P. S. FURCINITTI, J. F. HAINFELD & J. S. WALL. 1986. Radial density profiles of macromolecular filaments determined from dark-field scanning transmission electron micrographs. This volume.

DISCUSSION OF THE PAPER

C. COLLIEX (*Université Paris-Sud, Orsay, France*): In your mass mapping of unstained RNA, in one image there are a lot of white dots on the RNA and out of the RNA. What is the origin of these dots and how do they come into account when you do some mass mapping on these molecules?

HAINFELD: In these STEM images the contrast is very high in order to visualize unstained RNA. There's always some extraneous material in any preparation that can then appear white in a high-contrast image. Depending on the severity, this foreign material can alter mass measurements somewhat.

COLLIEX: It is important for the accuracy of your mass measurements—the contribution of all these in an image with a lot of very white spots. When you do mass mapping do you bracket your molecules?

HAINFELD: Yes. A reasonably tight boundary around the molecules is drawn to exclude excessive background. Also a study of the mass versus the diameter of a circle around the molecules is usually done to quantitate the background problem. Mass accuracy depends upon beam noise and film noise (which includes both the variations in the carbon film itself and the "dirt" deposited during specimen preparation). Usually any significant sample-related background can be minimized by running the specimen through an appropriate column.

Radial Density Profiles of Macromolecular Filaments Determined from Dark-Field Scanning Transmission Electron Micrographs

Improvements in Technique and Some Applications

ALASDAIR C. STEVEN,[a] TODD A. SIMPSON,[b]
BENES L. TRUS,[b] PAUL S. FURCINITTI,[c] JAMES F.
HAINFELD,[c] AND JOSEPH S. WALL[c]

[a]Laboratory of Cellular and Developmental Biology
National Institute of Arthritis, Diabetes and Digestive and
Kidney Diseases
National Institutes of Health
Bethesda, Maryland 20892

[b]Computer Systems Laboratory
Division of Computer Research and Technology
National Institutes of Health
Bethesda, Maryland 20892

[c]Biology Department
Brookhaven National Laboratory
Upton, New York 11973

INTRODUCTION

Precise specification of the radial distribution of density within a filamentous or spherical macromolecular complex imposes a stringent constraint on structural models for the complex. Several methods have been developed to characterize radial density distributions, including those based on X ray diffraction analysis. Similar information may also be obtained by negative staining and other forms of conventional electron microscopy. However, both approaches have limitations. In the case of X ray diffraction, problems may arise because the phase information required to convert the pattern into a density map must be modelled, and the measured Fourier amplitude spectrum may be distorted by interference effects arising from interparticle aggregation. With negative staining, internal variations in density are not well visualized, and air drying artifacts and coincidental positive staining may complicate the precise localization of edges. Recently we introduced a method that is not affected by these limitations—computational reconstruction of density profiles from dark-field, scanning transmission electron microscope (STEM) images of unstained frozen-dried specimens.[1] Since the primary data is in image form, it is, of course, fully phased; and because the specimens are not stained or shadowed, internal variations in (dry) density may be visualized directly and quantitatively. The method was tested using tobacco mosaic virus (TMV) as a model system,[1] and has since been applied successfully to other stuctures including the tail-tube of bacteriophage T4,[2] the fimbriae of *Bordetella pertussis*,[3] and tubular polymers of the enzyme glutamine synthetase.[4]

The procedure falls into two phases. First, a transverse trace of projected mass is computed that is optimized with respect to signal-to-noise ratio by averaging over many straight segments of filament, and with respect to sampling rate by "Vernier projection" (FIGURE 1).[1] Although the Brookhaven STEM electron probe is about 0.25 nm in diameter, the specimens are usually sampled sparsely, at intervals of 1.0 or 2.0 nm, to reduce radiation damage. In forming a Vernier projection, the data are re-sorted to give finer sampling without loss of resolution at any desired interval (usually chosen to be of the same order as the beam diameter). Second, this optimized transverse projection is transformed into a radial density profile by means of a noniterative real-space algorithm. Unlike calculations based on Fourier-Bessel inversion, this algorithm is relatively stable to discontinuities in the density function or its derivatives and so is much less subject to fluctuations at the particle edges. Here we give a more detailed account of both phases, including innovations that eliminate several somewhat subjective aspects of the procedure as originally described, and allow a substantially faster analysis. Finally, we review the applications made to date, and consider both the limitations of this procedure and its potentialities for further development, in particular for full three-dimensional reconstructions of helical particles.

PROCEDURE

General Outline

The procedure is implemented for interactive image processing based on exploitation of the capabilities of a video frame buffer system. The digital STEM micrographs are displayed on a video monitor at appropriate conditions of contrast, etc., which may include use of a pseudocolor map, zoom, etc.[8] Using a cursor, the operator designates segments of filament that are straight and free of adsorbed contaminants, and specifies the parameters required for computation of an axially averaged Vernier projection trace.[1] This trace is then centered by means of cross-correlation methods, and displayed on the graphics plane of the video monitor for visual appraisal and for comparison with the current running average (FIGURE 2) before the operator moves on to another segment of filament or another micrograph. The sequence of operations followed is listed in FIGURE 3. Finally, a systematic evaluation of the complete set of individual Vernier projections is performed, based on their centro-symmetry residuals, and a statistical screening for mutual consistency, *i.e.* reproducibility.[9] The data thus selected are averaged, and calculation of the resulting radial density profile is completed according to the sequence of operations listed in FIGURE 4.

Determination of Orientation Angle

The most important parameter in computing a Vernier projection is the orientation angle of the filament axis (θ; FIGURE 1). As indicated in FIGURE 5, even a small error in this angle can cause lateral smearing in the projection with consequent loss of resolution. Initially, our practice was to measure this angle several times from the displayed image using a cursor, and to take the average of these values. We have since introduced a quantitative refinement of the orientation angle. A Vernier projection is first calculated using the value for θ estimated as described above, and this trace is cross-correlated with each of a set of transverse line scans obtained by interpolation. These scans are spaced at equal increments of one pixel along the chosen segment of

1-Dimensional (axial) Vernier Projection

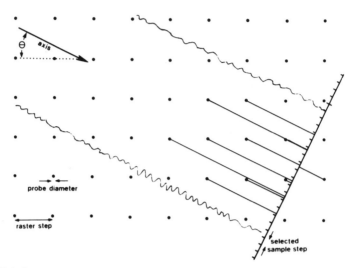

FIGURE 1. Re-sorting of sparsely sampled digital image of a filament to give a fine sampling of the transverse profile in the particle's projected density (Vernier sampling). The particle's axis is oriented at angle θ relative to the original sampling raster (adapted from REFERENCE 1).

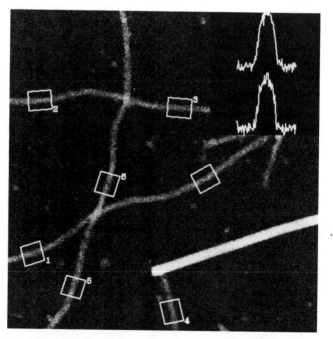

FIGURE 2. Operation of interactive Vernier projection program for filamentous particles (in this case, bovine epidermal keratin intermediate filaments). The micrograph under analysis is displayed on a video frame buffer. The curves displayed (*top right*) represent the Vernier projection of the filament segment currently under examination (*lower curve*) and the average Vernier projection of all segments already analyzed in this experiment (*upper curve*). The *boxes* enclose the straight (and relatively short) segments chosen for analysis. *Bar* = 25 nm.

filament. The coordinates of each cross-correlation maximum are listed, and the straight line fitted by least squares through these points defines our optimized axis direction. The Vernier projection is then recalculated using this refined value of the orientation angle. We have tested the accuracy of this procedure using simulated projections of a cylindrical model under a range of signal-to-noise ratios. The average ratios between the true orientation angles and the values determined are plotted in

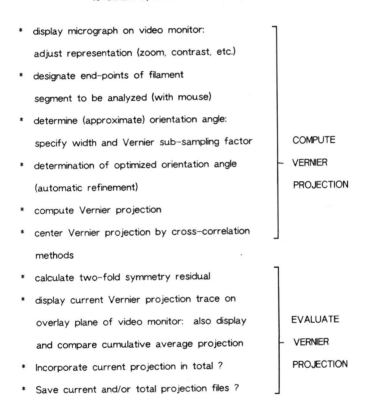

FIGURE 3. Sequence of image processing operations followed in computing optimized one-dimensional Vernier projections.

FIGURE 6. The results show that the method is both stable and precise, and indicate that errors of no worse than 0.1° may generally be encountered in practice, at least for statistically comparable data. (With very noisy data corresponding to extreme low-dose conditions, 20 el/nm², errors as high as 0.3° occurred. In practice, however, the orientation angles of such particles should be determined from subsequent high-dose scans of the same field.)

Sequence of Image Processing Operations to

Compute Radial Density Profiles

* Eliminate spurious data (individual Vernier projections) on basis of centro-symmetry residual values and mutual consistency screening algorithm (OMO).

* Form averaged Vernier projection.

* Centro-symmetrize Vernier projection; (alternatively, proceed separately with left- and right-hand sides to evaluate reproducibility of features present in final reconstructions).

* Smooth center points by least-squares fit of even-order polynomial.

* Smooth projection trace by low-pass Fourier filter.

* Subtract background.

* Set projection values to zero outside designated particle edge.

* Reconstruct radial density profile.

FIGURE 4. Sequence of image processing operations followed in reconstructing radial density profiles from one-dimensional Vernier projections.

Screening of Vernier Projection Data

Empirically, we have observed considerable variability in structural preservation of freeze-dried specimens, as represented in their Vernier projections. A high degree of centro-symmetry in such traces provides a necessary condition (although not a sufficient one) for good preservation of native structure. As a quantitative measure of

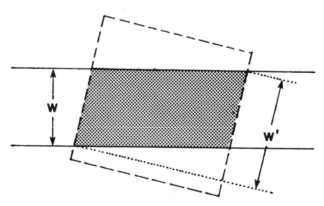

FIGURE 5. Schematic diagram illustrating how errors in determination of the orientation angle of a filament (true width w) can result in anomalously large width values (w') as measured from axial projections.

centro-symmetry, we calculate the residual R, given by

$$R = \frac{\sum_i (x_{c+i} - x_{c-i})^2}{\sum_i (x_{c+i} + x_{c-i})^2}$$

where x_k is the density at point k, and x_c is the center point.

Centro-symmetry essentially reflects bilateral consistency in a given particle. A quantitative measure of mutual consistency also provides a means of discriminating the best (*i.e.* most reproducible) members in a set of projections. An algorithm that was originally devised for screening sets of prealigned two-dimensional images for anomalous members[9] is well suited for this purpose. The algorithm ranks the individual projections according to declining consistency and imposes an acceptability threshold in this ordered list.

Interpretation of STEM Radial Density Profiles

Contrast and Density Scale

With density maps obtained from frozen-dried specimens, unlike those obtained from fully hydrated specimens, the contrast baseline is the density of the vacuum, not that of the solvent. Consequently, the respective contrasts visualized for the same specimen under these two conditions are significantly different. In particular, hydrophobic domains have less water to lose upon drying, and are thus likely to show up in the STEM density profiles as features of relatively high density. An extreme case would be represented by the hydrocarbon chain core of a lipid bilayer, which appears as a low-density feature in maps obtained by X ray diffraction analysis, but should be visualized with strong positive contrast in the STEM profile. Conversely, primarily hydrophilic domains should lose water in substantial amounts upon freeze-drying, and so should appear as relatively low-density features in the reconstructed profiles.

Edge Detection

One major advantage of these profiles is that they allow precise definition of edges, in terms of the outermost extent to which mass is visible above background in averaged Vernier projections. By contrast, with negative staining, diffuse peripheral density tends to be visualized poorly, if at all, on account of ineffective stain exclusion, spatial disordering, and the possibility of coincidental positive staining. Initially, to demarcate edges in interpolated (*i.e.* non-Vernier) projections, we used an operational definition for edges as those points on either side at which the signal dropped to 25% of maximum.[10] Widths were scaled relative to the corresponding spacing for TMV as a reference particle of known width. Others have followed a similar procedure employing a "50% of maximum" criterion.[11] However, this procedure is liable to systematic errors unless the particle's radial density distribution closely matches that of TMV. In particular, it will substantially underestimate the widths of particles that have diffuse peripheral density, TMV having rather sharp edges[12–14,1] (cf. FIGURE 7). A more reliable estimate of full width is to be had either from the Vernier profile directly,[12] or as that radius within which 95% of the total mass, as determined by polar integration of the calculated radial density profile, appears to be contained. (This figure, 95%, is chosen to compensate for the slight amount of lateral smearing that may persist even after optimal Vernier sampling.)

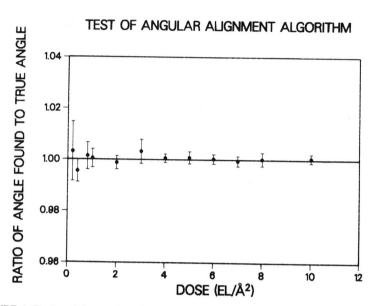

FIGURE 6. Testing of the angular alignment algorithm in the "Vernier sampling" program was performed by computer generation of "TMV particles" modelled as hollow cylinders of inner radius 1.0 nm and outer radius 8.8 nm. Particles were scaled and a background value added corresponding to a carbon film about 2.5 nm in thickness. The simulated images were sampled with a 0.25-nm diameter probe on a 1.0 × 1.0-nm sampling grid at specified angles. The simulation program modelled the STEM dark-field signal from particle plus background, taking into account the detector geometry and collection efficiency as well as Poisson "shot noise" as a function of electron dose. For each dose tested, 8 segments, each 50 nm in length, were analyzed. In each case, the average value is plotted for the ratio between the orientation angle determined by our automatic procedure and the true angle. The *error bars* correspond to one standard deviation.

Flattening

    Biological macromolecules adhering to a carbon film may undergo substantial distortion (flattening) at two stages of specimen preparation: either when adsorbing to the substrate[16] or subsequently through surface tension forces imposed during air drying.[17] Adequately gentle freeze-drying should eliminate the latter effect, but adsorptive flattening remains a potential problem. This effect will distort the calculated profiles in a way that will generally involve widening and diminution of inner densities from their correct values. The most direct way to diagnose flattening of individual particles is by tilting. However, the Brookhaven National Laboratory STEM[5] is not equipped with a goniometer. To assess the resistance of a given specimen to adsorptive flattening, conventional transmission electron microscope (CTEM) images of unidirectionally shadowed frozen-dried specimens may be used. Provided that this test establishes that severe flattening is not widespread, the unflattened particles may then be identified in unstained STEM images of the same specimen, based on the premise that the narrowest particles are least flattened. It is possible to compensate computationally for a limited amount of flattening by a simple modification of the reconstruction algorithm (cf. FIGURE 2 and Equations 1-K of REFERENCE 1) based on concentric isodensity annuli that are elliptical rather than circular (the unflattened case).

Signal-to-Noise Considerations

Like any reconstruction from projection data, radial profiles represent the result of a numerical differentiation operation and, as such, are acutely sensitive to residual noise in the projection data. The degree of smoothing realized in a cumulative Vernier projection trace depends on the total length averaged as well as the electron dose and sampling rate (magnification). It has been our experience that with micrographs recorded with 1 nm sampling and doses of $1-3 \times 10^2$ el/nm^2, the results obtained tend to be quite reproducible in their resolved features (after a final band-limited smoothing—FIGURE 4), provided that at least $0.25 - 0.5$ microns of cumulative length are included. However, residual fluctuations are not uniform as a function of radius. Several factors conspire to make the calculated density values in the immediate vicinity of the axis less stable than elsewhere: (i) Poisson noise[5] in the projection is greatest at the highest signal values, which are usually around the center; (ii) noise at the midpoint of the projection is not further suppressed by centro-symmetrization (FIGURE 4); and (iii) the decreasing areas of the inner annuli mean that the density values solved there are increasingly noise-sensitive. Thus a greater amount of averaging is required to stabilize the near-axial densities than elsewhere.[2,3]

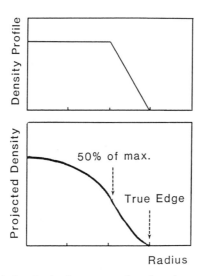

FIGURE 7. Model calculation showing how systematic underestimates of a particle's full width may be given by defining edges as "50% of maximum" points on the transverse mass projection. This model particle has a core of uniform density extending to two-thirds of its full radius, with density tapering off linearly to zero beyond that point. The "50% of maximum" criterion underestimates the full width by 31% in this case.

DISCUSSION

Differential Mass Mapping

In the interpretation of radial density maps, distinctions in hydrophobicity/ hydrophilicity among the various molecules involved may allow some general preliminary inferences (cf. above). However, attribution of density features to specific molecular constituents is accomplished most directly on the basis of quantitative comparisons between the radial profile of a given complex, and those of structural variants related to it in biochemically defined ways. For instance, on comparing the respective profiles of the TMV virion and of RNA-free helical polymers of TMV coat protein, the former was found to contain an additional density peak at a radius of 4 nm.[1] Thus, direct visualization of the virion RNA was achieved, even though the genome accounts for only about 5% of the total virion mass. This finding confirmed an earlier conclusion based on X ray fiber diffraction,[13,14] and has since been further corroborated in an analysis of CTEM micrographs of frozen-hydrated specimens.[7]

A conceptually similar experiment was performed by Duda et al.[2] in comparing tail-tube complexes of bacteriophage T4 prepared with and without guanidine hydrochloride treatment. This denaturant was observed to remove the gp48 protein from the complex, and evaluation of the respective STEM radial density profiles revealed a concomitant reduction in axial density, whereas the profiles were superimposable elsewhere.[2] This study concluded that several (approximately six) molecules of gp48, in a highly extended conformation, constitute the template around which 144 molecules of gp19 polymerize to produce a rodlike tail-tube of precisely determined length. Interestingly, a similar length-determining mechanism has also been attributed, on different experimental grounds, to the assembly pathway of the tail of phage lambda.[18]

These two studies illustrate the principle of differential mass mapping. In general, appropriate structural variants may be produced in several different ways: (i) the binding of specific accessory proteins to a basic polymer (such as the binding of "associated proteins" to cytoskeletal filaments); (ii) the systematic removal of less tightly bound constituents by treatment with reagents such as urea, guanidine hydrochloride, etc; (iii) molecular dissection by limited proteolysis; and (iv) the tagging of specific molecules with electron-dense heavy-atom clusters,[19] either for intact complexes or prior to reconstitution in vitro. In each case, STEM determinations of mass-per-unit-length may be used to calibrate the respective projections prior to calculation of their density profiles, so that no ambiguities arise as to the relative scaling of the curves under comparison.

Future Directions

The applications outlined above indicate the utility of this experimental tool as already implemented. However, further extension of the approach in several directions appears feasible. Improvements in specimen preparation technique may be required to improve the resolution beyond the present level of about 2 nm at which structural details are reproducibly defined. The use of frozen-hydrated specimens as well as frozen-dried ones appears to have considerable potential, as excellent preservation of native structure seems possible by this technique.[7,20] Moreover, the STEM dark-field mode may prove a useful alternative to under-focussed bright-field CTEM imaging as a means of obtaining adequate contrast.

Eventually, it should be possible to perform full three-dimensional reconstructions of helical specimens by this approach. For this purpose, a generalization of the axial Vernier projection to give finely sampled two-dimensional representations of axially repeating helical segments may be calculated as shown in FIGURE 8. Thereafter, reconstruction of the three-dimensional density map may be calculated by conven-

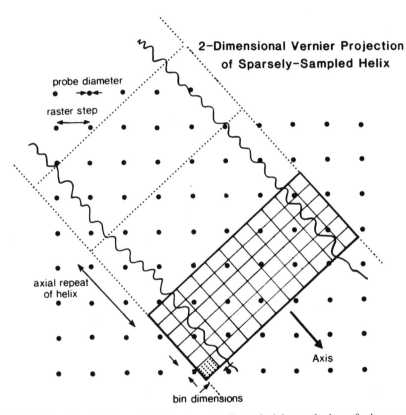

FIGURE 8. Generalization of the Vernier sampling principle to obtain a finely sampled representation of the axially repeating unit of a helical particle that is optimized for signal-to-noise ratio. The two-dimensional array of sampling bins is, in effect, placed serially along the particle at successive steps of the axial repeat. The points thus collected are averaged on a bin-by-bin basis, with density values for "empty" bins assigned by local bilinear interpolation. Optimized orientation angle is determined as in the one-dimensional case (see Procedure section).

tional procedures (*e.g.* REFERENCE 21), although it would appear advisable to constrain the contouring level of the resulting three-dimensional density map to maintain compatibility with the radial profile defined by the axial projection (in effect, the equatorial data in the corresponding Fourier transform). In conclusion, several potentially fertile lines of investigation should allow further development of this novel method of mapping density distributions within macromolecular complexes.

REFERENCES

1. STEVEN, A. C., J. F. HAINFELD, B. L. TRUS, P. M. STEINERT & J. S. WALL. 1984. Radial distributions of density within macromolecular complexes determined from dark-field electron micrographs. Proc. Natl. Acad. Sci. USA **81:** 6363–6367.
2. DUDA R. L., J. S. WALL, J. F. HAINFELD, R. M. SWEET & F. A. EISERLING. 1985. Mass distribution of a probable tail-length determining protein in bacteriophage T4. Proc. Natl. Acad. Sci. USA **82:** 5550–5554.
3. STEVEN, A. C., M. E. BISHER, B. L. TRUS, D. THOMAS, J. -M. ZHANG & J. L. COWELL. 1986. The helical structure of fimbriae of *Bordetella pertussis.* J. Bacteriol., in press.
4. FURCINITTI, P. S., J. F. HAINFELD, J. J. LIPKA & J. S. WALL 1985. Radial mass density analysis of unstained STEM Images. Proc. 43rd Annu. Mtg. Electron Microscopy Soc. Amer. 318–319.
5. WALL, J. S. 1979. Mass measurements with the electron microscope. *In* Introduction to Analytical Electron Microscopy. J. Hren, J. Goldstein & D. Joy, Eds. 333–342. Plenum Publishing Corp. New York, NY.
6. DEROSIER, D. J. & A. KLUG. 1968. Reconstruction of three-dimensional structures from electron micrographs. Nature **217:** 130–134.
7. LEPAULT, J. 1985. Cryo-electron microscopy of helical particles TMV and T4 polyheads. J. Microsc. **140:** 73–80.
8. TRUS, B. L. & A. C. STEVEN. 1981. Digital image processing of electron micrographs—the PIC system. Ultramicroscopy **6:** 383–386.
9. UNSER, M., A. C. STEVEN & B. L. TRUS. 1986. Odd Men Out: a quantitative objective procedure for identifying anomalous members of a set of noisy images of ostensibly identical specimens. Ultramicroscopy, in press.
10. STEVEN, A. C., J. WALL, J. HAINFELD & P. M. STEINERT. 1982. Structure of fibroblastic intermediate filaments: analysis by scanning transmission electron microscopy. Proc. Natl. Acad. Sci. USA **79:** 3101–3105.
11. ENGEL, A., R. EICHNER & U. AEBI. 1985. Polymorphism of reconstituted epidermal keratin filaments: determination of their mass-per-unit-length and width by scanning transmission electron microscopy. J. Ultrastruct. Res. **90:** 323–335.
12. STEVEN, A. C., B. L. TRUS, J. F. HAINFELD, J. S. WALL & P. M. STEINERT. 1985. Conformity and diversity in the structures of intermediate filaments. Ann. N.Y. Acad. Sci. **455:** 371–380.
13. CASPAR, D. L. D. 1956. Radial density distribution in the tobacco mosaic virus particle. Nature **177:** 928.
14. FRANKLIN, R. M. 1956. Location of the ribonucleic acid in the tobacco mosaic virus particle. Nature **177:** 928–930.
15. STUBBS, G., S. WARREN & K. C. HOLMES. 1977. Structure of RNA and RNA-binding site in tobacco mosaic virus from 4-Å map calculated from X-ray fiber diagrams. Nature **267:** 216–221.
16. KISTLER, J. & E. KELLENBERGER. 1977. Collapse phenomena and freeze-drying. J. Ultrastruct. Res. **59:** 70–75.
17. ANDERSON, T. F. 1954. Some fundamental limitations to the preservation of three-dimensional specimens for the electron microscope. Ann. N.Y. Acad. Sci. **16:** 242–249.
18. KATSURA, I. & R. W. HENDRIX. 1984. Length determination in bacteriophage lambda tails. Cell **39:** 691–698.
19. LIPKA, J. J., J. F. HAINFELD & J. S. WALL. 1985. Heavy atom cluster labelling of biological specimens. Proc. 43rd. Annu. Mtg. Electron Microscopy Soc. Amer. 718–719.
20. ADRIAN, M., J. DUBOCHET, J. LEPAULT & A. W. MCDOWALL. 1984. Cryo-electron microscopy of viruses. Nature **308:** 32–36.

DISCUSSION OF THE PAPER

U. AEBI (*Johns Hopkins University, Baltimore, Maryland*): Have you tested your model to date, through mass density profile, on various types of intermediate filaments, to see whether or not this would hold? You have all rod domains in a core and the end domains radiating out. I can see sterical problems when you have multilayered filaments, unless you pack them in a hollow tube. If you have several layers of cores within the filaments, they can't all radiate out, there must be also mass inside.

STEVEN: Let me answer your first question first. I just showed you the data on native vimentin. We have also done reconstituted vimentin density Class-I and density Class-II. We've done them with about four different kinds of specimen preparation for STEM, in terms of different washing as control experiments. We've also done bovine epidermal keratin filaments reconstituted *in vitro*. We were worried about the possibility of flattening, so we also prepared them not supported on a carbon film, but strung over empty holes. The results of all these experiments were consistent.

In terms of whether there has to be successive layers of molecules with accretion of molecules on the outside, I'm not quite sure what sort of evidence you're referring to for such a model.

AEBI: You say that all the rod domains would form a core and the end domains would radiate out. Now if you think in terms of protofilaments and protofibrils and if you have more than one layer, and all the data seem to point to that, it's really a highly coiled structure. So you should have internal contributions from the end domains.

STEVEN: Whoa! I don't think there's any direct evidence that any molecules in IF are buried under other molecules. For instance, in the desmin system, Kaufman *et al.* (J. Mol. Biol. 185: 733–742, 1985) have shown that the carboxy-terminal 30 or so residues can be removed quantitatively by proteolysis from all molecules, implying that none of them are buried or otherwise inaccessible. They took that, among other things, to rule out the "core-plus-ring" model that had been proposed by Frazer and co-workers, at least as applied to desmin IF.

I think that at the periphery of the IF core, say, at the 4.5-nm radius that we're discussing there could be plenty of space to stick more rod domains on, not necessarily with end domains underneath them. The low *average* density of the periphery does not necessarily mean low continuous density, but could also arise from a relatively sparse distribution of domains with normal local density.

J. B. PAWLEY (*University of Wisconsin, Madison, Wisconsin*): I'm a little confused by something else, this multiple binning or whatever you call it here. We have a beam that's 2.5 Å across and we're sampling every once in a while. We're saying this is all right, because the average dose is one electron or 2, or 3, or 4, electrons per square angstrom, and, it seems to me, where the beam actually is, the dose is a good deal more. Now, because of the delocalization of the inelastic collision, it will spread out a little bit more, but do you have some measurement by which you can assure me that these numbers you have are effective average doses?

STEVEN: It's difficult to compare dose rates directly between situations where you've an expanded beam illuminating all parts of the specimen simultaneously and where you have a highly focused probe sampling sparsely, except when you are working at very high magnification and putting your samples close together. In the parts of the specimen immediately sampled by the probe, obviously the primary damage is going to be greater than during conventional EM illumination. However, what one is guarding against in these specimens is secondary damage by diffusion of radiation products generated in the first scattering events, and that's not entirely a theoretical argument. When we did the experiment of looking at intermediate filaments that were strung across empty holes, it was extraordinary how well we could see these spider webs at low magnification. Then at increased magnification, putting the samples closer together, they would just tear until Dr. Wall managed to reduce the illumination to a point where we had images with dose rates about a tenth of the ones that were used to study TMV and intermediate filaments supported by carbon films. So, yes, the primary damage is greater than the corresponding number given for a CTEM image, but it's not a straightforward issue. Dr. Wall, do you have anything to add to that?

J. S. WALL (*Brookhaven National Laboratory, Upton, New York*): We're working on some measurements to test the hypothesis.

F. P. OTTENSMEYER (*University of Toronto, Toronto, Canada*): For someone who has worked in dark-field for ages, I am pleased to see some STEM people trying eventually to aim towards high-resolution work. It pleases me no end also that you are still trying to reduce the dose, even in the STEM. But to ask you to go one step further, are you intending to use this data to reconstruct in three-dimensions?

STEVEN: Yes, that will be the next step. We'd like to go to a vernier-sampled axial repeat and then do a three-dimensional reconstruction of that. I don't think we're getting anywhere close to the sort of resolutions that you have been claiming for your work. In terms of reproducibility of edge detection in TMV, if you do it once or twice, you get the edge in the same place to within 4–5 Å. However, in terms of band-limited resolution, at which particular features reproducibly show up in the profile, I don't think we're doing much better than 20-Å resolution. There are probably two major reasons for that. First, we're probably still working at too high a dose. I hope that with modifications that are under way in Brookhaven, their STEM will be able to work practically at lower doses. Second, there may be a limitation in the freeze-drying technique, as we presently do it, in that substructure is not well preserved. A promising direction is to get frozen-hydrated particles onto the cold stage at the Brookhaven STEM and use a STEM dark-field signal as an alternative to defocused bright-field CTEM imaging and for contrast generation.

D. F. PARSONS (*New York State Department of Health, Albany, New York*): I get the impression that you think that a model for the vimentin intermediate filaments is imminent. I can't really understand how this can be considering the very variable polypeptide composition of several types of intermediate filaments in normal cells and embryonic cells.

STEVEN: Your question is how can they have the same structures when they have such different proteins?

PARSONS: Yes.

STEVEN: They all have common rod domains of about 310 residues. There's not much hard evidence as to molecular structure of intermediate filaments, but it was noticed a few years ago by two groups, Milam and Erickson, and Henderson *et al.*, that if you prepare intermediate filaments by the technique of making up a suspension in glycerol, nebulizing it onto a mica surface, and low-angle rotary shadowing with platinum, an intrinsic repeat of about 22 nm is somehow contrasted by this specimen preparation technique. This has been seen in every type of intermediate filament, including members of all the five main classes, glial filaments, desmin, vimentin, neurofilaments, and many kinds of keratin. Also, of the high-angle X ray patterns that have been recorded for several kinds of intermediate filaments we know that characteristic equatorial and radial reflections are present in all of them. There's also evidence from systematic comparisons of mass per unit length that we have made at Brookhaven. The mass per unit length of an intermediate filament seems to be proportional to the average mass of the subunits in it. That situation is very readily explained if you have a universal backbone building block about 450 Å in length in all cases. Those are basically the main lines of evidence at this time, that there are close similarities between the structures.

Our proposal, which of course needs to be rigorously tested, is that end domains are just sticking around the outside. I think Dr. Reese went over this point very quickly when he showed his freeze-etched picture of cytoskeleton. The cross bridge that he was indicating was protruding from one neurofilament to a neighboring filament. Similar data have also been published by Hirokawa and co-workers (1984. J. Cell Biol. 98:1523–1536). Most people would probably say it was representative of the large carboxy-terminal domains of a large neurofilament subunit.

PARSONS: I have a request for assistance, which I think your system and your microscopy can give. There is the confusion that exists about antibody labeling, fluorescent or whatever, of intermediate filaments in cells, as it is a highly chancy business. For example, if you use monoclonal antibodies, particular monoclonal antibodies against bovine epidermal keratin, which turns out to be one of the best for human squamous carcinoma cells. It would be helpful perhaps if we could have a model system in which some of these gold tag antibodies are checked against some of the principal types of keratin, because at the moment we're reduced to trial and error of a very laborious sort in order to do this kind of cytochemistry.

STEVEN: I expect you're aware of the Pruss monoclonal antibody that will react against every intermediate filament protein as yet characterized, and its epitope seems to have been localized to a sequence up near the carboxy-terminal end of the rod domain, where there is an extraordinarily conservative run of about 20 amino acid residues that is conserved almost absolutely by all intermediate filament proteins. So in terms of immunofluorescence and tissue-specific typing of different intermediate filaments that is obviously a tremendous tool. Henry Sun has also generated specific monoclonal antibodies against Type I and Type II keratins.

Z-Contrast in Biology

A Comparison with Other Imaging Modes[a]

E. KELLENBERGER, E. CARLEMALM, W. VILLIGER,
M. WURTZ, C. MORY,[b] AND CH. COLLIEX[b]

Department of Microbiology
Biozentrum, University of Basel
CH-4056 Basel, Switzerland
and
[b]*Université de Paris-Sud*
Laboratoire de Physique des Solides
F-91405 Orsay, France

INTRODUCTION

Electron microscopy is needed to solve biological problems more often for cases where today's procedures of image processing are not yet ready or applicable and thus not very helpful. Of particular importance are observations *in situ,* in the cell, because here electron microscopy is about the only tool available. Freeze-fracturing and sectioning are the preparation methods to employ. With these procedures we are still orders of magnitude away from being able to exploit the instrumental resolving power. Improvements are possible from the side of specimen preparation but also from that of observation. In the present paper we will consider the latter; for the former we refer the reader to other publications.[1,2]

Too much noise and lack of contrast are the two basic limitations for biological material that we will discuss here. In direct imaging, without exploiting redundancies, the noise cannot simply be removed by the current averaging procedures. One has therefore to evaluate critically the different imaging systems with respect to contrast and signal/noise.

Biological material has to be stained with heavy metals in order to present sufficient contrast for conventional bright-field imaging. We will deal particularly with what is called the positive stain, as is involved with the observation of thin sectioned material. As a reminder of the problem of staining, we just mention the "double track" image of the biological membrane, which shows no trace of the biologically most important integral transmembrane proteins. They become visualized by freeze-fracturing but still not routinely by sectioning, although sectioning should give the only relevant functional view of these proteins that transport molecules and signals across the membrane. Only in rare cases is the arrangement in the plane of the membrane, as it is depicted in freeze-fracture and negative stain, of large importance. Even three-dimensional reconstructions from planar images of membranes suffer from the well-known fact that the precision is the least good in the direction that is also functionally the most important one.

[a]Supported in Basel by the Kanton Basel-Stadt and the Swiss National Science Foundation and in Orsay by the Centre National de la Recherche Scientifique.

Some of us have investigated positive staining[3] that occurs in two steps: by treating the cells or tissues with OsO_4 and uranyl acetate, between 10–20% of Os and U is deposited. This still does not provide very much contrast in conventional bright-field, phase contrast imaging. The on-section staining with lead and uranyl salts increases the contrast enormously, so that we estimate a heavy metal deposit that is about ten times larger than that produced in the first step. In the end we have about equal weights of heavy metal and biological material. If the heavy metal were concentrated in a sphere, its diameter would only be about three times smaller than that, for instance, of a protein. While the latter is invisible, the metal deposit is clearly seen by virtue of its contrast. One interprets it implicitly as being the biological material, although one has not the slightest idea of how and where the metal is deposited, for instance on, or in, or around a protein molecule. In order to study this severe limitation we have to be able to observe embedded material without stain. A comparable situation occurred some 30 years ago in light microscopy: biological material is in general colorless and transparent and thus more or less invisible in normal light microscopy. It had therefore to be fixed and stained with dyes. For these stains to be retained, frequently an acid fixation was useful, because it produced coarse aggregates out of the cytosol. Later, with the phase contrast microscope, unstained material became observable.

Part of our contribution will be devoted to demonstrate that the bright-field phase contrast imaging, as usual in electron microscopy and as can be simulated in the light microscope by chosing a small aperture for the illuminating beam, is not necessarily the best imaging mode for introducing contrast.

For the light microscope we will show that the image obtained with a small illuminating aperture is much less sharp than that obtained with the commercial annular phase contrast. We thus confirm previous work[4] and will discuss possible explanations for this difference.

For the electron microscope we will compare the bright-field phase contrast with dark-field and ratio-contrast imaging. The scanning transmission electron microscope (STEM) has proved advantageous for dark-field imaging, because the efficiency of collection of scattered electrons and the recording is one to two magnitudes better than with a conventional transmission electron microscope (CTEM). Ratio-contrast is based on simultaneous collection and processing of two signals and thus easily possible with a STEM. With the STEM, the electrons scattered by passing through the specimen are usually separated into those scattered into a wide angle and collected by the annular dark-field detector (ADF) and those remaining within a small angle of 10^{-2} to 10^{-3} rad. The ADF detects electrons that are scattered elastically at least once. The unscattered ones and those scattered only into a small angle are separated by a spectrometer according to energy losses. The inelastically scattered electrons with low energy losses (lel) form a relatively large peak near to that of unscattered electrons. By electronic or digital means a ratio can be formed of the lel signal divided by the ADF signal. For thin objects this ratio-contrast depends essentially on the atom number Z, and Crewe therefore called it Z-contrast.[5]

It has been shown that ratio-contrast imaging is also suitable for high-contrast imaging of biological thin sections obtained without any heavy metal treatment.[6-7] The contrast is as good as with the dark-field mode, the image "sharpness," however, is better with ratio-contrast, because this type of imaging is much less sensitive to thickness variations than is the dark-field mode.[7] It is well known that the section surfaces are not evenly flat, but represent a specimen-dependent but strongly distorted image of the structure.[8,9] Ratio-contrast, as opposed to dark-field, therefore, images the biological material within the section in preference to the surface relief.

NOISE AND CONTRAST

For biological specimens it is well known that the visibility of significant details is much less a matter of the instrumental resolution than of contrast and noise. In quantitative terms the physicist expresses this problem as depending on "the signal-to-noise ratio." However, for the practice of electron microscopy this ratio is highly approximative. Images have first to be observed with our eyes. According to the situation of the above variables we will "discover" a new structure or not. To our advantage, our eye has a limited resolution; by varying the distance of observation we are able to smooth out a noise or a raster. That is why the impression of images has become possible. The experienced electron microscopist will vary the magnification or the distance of observation of a micrograph, so that the noise disappears and biologically significant details suddenly emerge from the noise. He then increases the contrast during photographic printing, so that these details appear best. Since only very few people are trained in these techniques today, one might use practical tricks (FIGURE 1) or the computer[10] to achieve this smoothing. The practical tricks are based on a limited, artificial blurring and are achieved, for instance, by placing an adequate frosted glass plate on top of the photographic paper during enlargement. A very fine fabric, like a silk stocking, can achieve the same result. This procedure was used for FIGURE 1.

From the results of such experiments it becomes clear that the size of the noise is of prime importance. It might be defined as the average distance between peaks. If the size of the detail is of the same order as that of the noise then only signal/noise is relevant, as illustrated in FIGURE 2. As soon as the detail is an order of magnitude larger than the noise, then the noise can be smoothed out and, practically speaking, only the contrast will determine whether or not a detail will be visible. The difference of the optical densities of the detail and of the environment has to be above the physiological contrast threshold of about 0.1 optical density units, varying somewhat according to the environment.

Several sources of noise can be distinguished: (i) quantum noise, which increases with lower electron doses; (ii) destruction noise, "shot-out" by some of the electrons during beam-induced destruction; (iii) shadowing noise, represented by the granularity of heavy metal deposited by evaporation; (iv) staining noise, represented, as above, by the granularity of the heavy metal; (v) holographic noise, due to high-coherence illumination, as a corollary to "phase contrast;" and last, but not least, (vi) the deformations of the specimen that are not equal in each copy of the structural detail, which is the most difficult to seize by experimental methods. Such variations of conformation could occur in the native state or only as a consequence of specimen preparation.

The sizes of the first five of the above-mentioned noises cover the range up to about 3 nm. The visibility of details of a size within this range of 1–3 nm thus has to be treated in terms of signal-to-noise ratios.

For larger details, as we are mostly concerned with in direct imaging of biological material, the contrast becomes increasingly important. Contrast is usually described in the literature by the ratio $\Delta S/S$, where S is the number of electrons (intensity) averaged between that of the detail and that of the environment and ΔS the difference between these two numbers.

This definition is in most cases adequate, although one has to be aware that in photographic procedures we never represent S as such on the final print. We also have to take into account that the silver bromide emulsion responds differently towards electrons and light. Valentine[11] has shown that the optical density is linear with the electron dose, while for light it is proportional to the logarithm of the energy received.

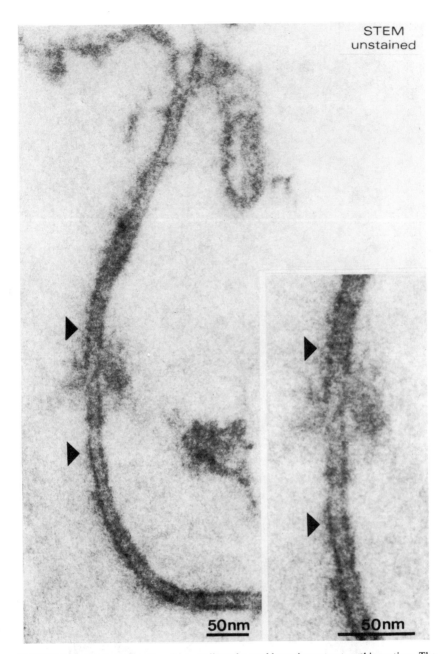

FIGURE 1. Isolated gap junctions from rat liver observed by ratio-contrast on thin sections. The isolated junctions (courtesy Drs. L. Landmann and P. Meier) were fixed with glutaraldehyde and embedded in Lowicryl HM20 by the method of progressive lowering of temperature.[44] The high magnification *inset* was high-frequency filtered with a fabric to remove the quantum noise. Note that the sample has in no instance been treated with heavy metals. The *arrows* indicate regions where a periodic transmembrane structure is revealed. The sections were observed at EMBL in Heidelberg in a STEM VG HB5 equipped with a cryostage. We are very grateful to Dr. W. Tichelaar for having optimized the imaging conditions of this STEM. We strongly believe that the observation at the temperature of liquid N_2 is essential for seeing details in resin-embedded sections.

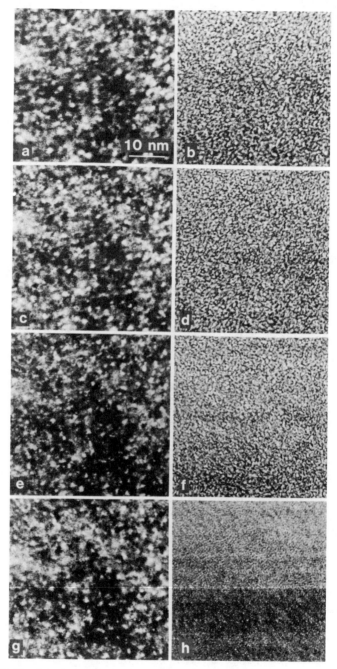

FIGURE 2. Uranium clusters prepared on a carbon film viewed in STEM, dark-field (**a,c,e,g**) and phase contrast (**b,d,f,h**). Note that all the micrographs concern the same region of the specimen.

The possibilities of manipulating a signal in STEM either by analogue or digital computing are similar to those for the experienced photographer, including the reversal of contrast. In both cases experience has to be gained by practical applications. Nevertheless, even approximating theories help enormously in this task, if they are accepted as approximative.

From these qualitative considerations one might have learned that according to the size of the details to be uncovered, noise and contrast have to be considered. The biologist has done that without asking the physicist by steadily increasing the contrast by using more and more heavy metal staining. With most methods associated with the direct observation of micrographs, he reached a biologically significant resolution of some 10 nm, but only very rarely below that. The situation is obviously much better when considering repeat structures and image processing. The latter procedures have been discussed in this symposium, so that we can concentrate on the possibilities offered for reducing components of noise and for improving contrast considered from the point of view of direct imaging. We shall have the final goal in mind of being able to observe unstained biological material and thus of understanding more about the location of stain deposits in respect to the biological macromolecules that constitute cellular structures.

POSSIBILITIES AND LIMITATIONS OF PHASE CONTRAST

In light microscopy, some decades ago, it was known to every operator that unstained material can become visible, although unsharp, when the angular aperture of the illuminating light beam is reduced. When he had to work with living and thus unstained material, he manipulated the condenser aperture adequately.

The practical materialization of the phase contrast according to Zernike[12] was initially very disappointing in its adaptation to light microscopy. By competition between the firms it was further developed into a system that is now rather different from the initial idea. It is, however, successful in producing contrast with transparent (nonabsorbing) objects without losing sharpness, as we do when closing the condenser aperture. Nevertheless, phenomena of refraction disturb good imaging and high resolutions can be achieved only when these phenomena are reduced to a minimum. Bacterial nucleoids, for instance, can be observed only when the refractive index of the suspension medium is adjusted to nearly that of the cell.[13] By this procedure the bacterium is prevented from behaving as a cylindrical lens.

It is interesting to note also that modern, high-magnification objectives no longer have a phase ring for a $\lambda/4$ shift, but only a ring of evaporation-deposited metal providing some 80–95% of absorption. It should further be remembered that modern microscopes use illumination according to Köhler, by which the incandescent filament of the light bulb (arranged in a zig-zag fashion) is projected into the condenser lens. With an open condenser aperture a minimum of coherence is thus achieved. It can be increased by closing the condenser aperture, thus cutting out a very small segment of the image of the filament. The annular aperture of the "phase contrast," however, does not increase the coherence to any comparable degree.

We have compared the annular phase contrast with the narrow-beam phase contrast when applied on a specimen that is essentially a scatterer. Empty bacterial envelopes are thinner than 50 nm, as is cytoplasmic material still adhering to it. As dark-field light microscopy confirms, such empty bacterial envelopes are excellent light scatterers. In FIGURE 3a, we see an electron micrograph of such a preparation, in FIGURE 3b an annular phase contrast image, and in FIGURE 3c a narrow-beam phase contrast image. In all cases an objective of 40X was used. The angular aperture of the illumination in FIGURE 3c was 0.013 radians.

A word should be said here about our definition of a "scattering" object for light; it means that the phase shift is related to a structure that is not resolved by the microscope. The cytoplasmic ribosomes (25 nm) are not visible, but they are excellent scatterers. Such scattering is in general accompanied by a phase shift. An intact bacterium, in a normal medium, is also a strong phase object; its transparent body has a higher refractive index than the surrounding medium. This difference of refractive index per se leads to a phase shift, which can be measured by an interference microscope. However, the refractility of the bacterial body is also the reason for its acting like a cylindrical lens, as discussed above.

From these empty "bacterial ghosts" we have then produced through-focus series, which we analyzed in the optic diffractometer for studying the contrast transfer function. As expected from our experience with electron microscopy, we found nicely defined rings with the narrow-beam phase contrast (FIGURE 4). The through-focus series with the annular phase contrast, however, did not show any rings (FIGURE 5). In some experiments we could find, as expected, a diffuse diffraction corresponding approximately to the size of the empty envelopes.

The images of the two through-focus series are also very different despite the same objective apertures (nA = 0.65) in both cases. Only in the annular phase contrast can we see a significant contrast reversal that is unambiguously object related. In the other case the images of the envelopes very rapidly disappear in a "holographic" noise.

The narrow-beam phase contrast of FIGURE 4 is comparable to the phase contrast produced in electron microscopy (FIGURE 2), although in the latter case we have to take into account an additional influence of the spherical aberration of the lens.[14] The micrographs of FIGURE 2b,d,f,h are phase contrast images of uranium oxide clusters deposited on a thin carbon film. The clusters disappear completely in the holographic noise, but they can be detected in dark-field images (FIGURE 2a,c,e,g).

Two further observations with phase contrast bright-field electron imaging are not immediately explained on a conventional basis.[15,16] Frozen suspension of particles is observed in the bright-field mode of the conventional electron microscope[17] when they are still hydrated by placing them into a cryostage cooled with liquid nitrogen. In order to have contrast, underfocus of several μm has to be used.[18] The contrast, which is related to the structure of the specimen, increases, while the noise apparently stays very low (Dubochet, personal communication). We have applied this possibility of producing contrast to sections of unstained, resin-embedded material. To our surprise, we did not find a notable increase of contrast with underfocus, but only a steady decrease of definition; in other words, the content of biologically significant information decreased.

Interestingly, an observation by Westphal & Froesch[19] points in an opposite direction. They embedded the retina into Nanoplast, a new water-miscible melamine resin,[20] and made through-focus series in phase contrast bright-field electron microscopy, as shown in FIGURE 6. Here we find for the first time a real specimen-related reversal of contrast on a composite structure that is substantially larger (25 nm) than the subunit of a biological crystal (10 nm) where a reversal was also observed.[21] It should be noted that with the retina the structure stays unaltered, independently of the reversal and this also in parts of the specimen that are not periodic. If similar experiments are done with the same material embedded in other resins, nothing like this occurs, not even at larger defocus. This phenomenon can therefore only be due to the persistence of lipids in the Nanoplast embedding, which in all other resins are extracted.[22]

Two aspects emerge from these observations that have to be discussed. The first is that the nature of illumination might be accompanied by properties that are not entirely describable by the coherence, and the second, that ice, lipids, and Nanoplast might act on the phase of electrons in ways that might have been considered theoretically but as yet not applied to the right specimens.

The strong influence of the illuminating system on the quality of the image was observed very early (see textbooks, as for instance Heidenreich[23]). It is well known that the Fresnel fringes on edges increase with narrower beams of illumination. This was also simulated in optical microscopy.[24] Although we did not yet find specific references

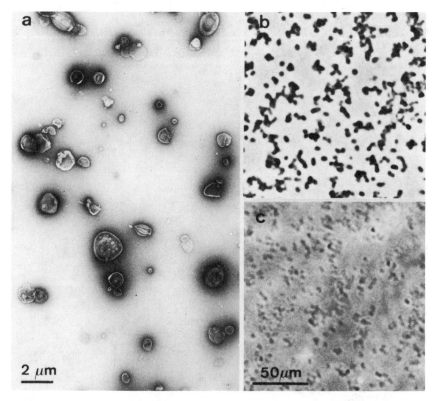

FIGURE 3. Fragments of empty envelopes ("bacterial ghosts") of *E. coli.* Observed (**a**) on a negatively stained preparation in CTEM, bright-field; (**b**) by annular phase contrast in the light microscope; and (**c**) in narrow-beam (0.013 radians) phase contrast.

for it,[25] we think that the Fresnel fringes should be explainable by the contrast transfer theories.[14–16] On the other hand, we feel that the understanding of Fresnel fringes as a result of interference between the electrons just after passage through the specimen should broaden the physical understanding of the transfer functions. Any discontinuity of the specimen, as slight as it might be, would lead to a system of fringes, which, by their superposition would produce the focus-dependent "holographic noise" situated just below the specimen. The carbon film would then not be a simple homogeneous scatterer, randomly "emitting" all space frequencies. It is well known that carbon

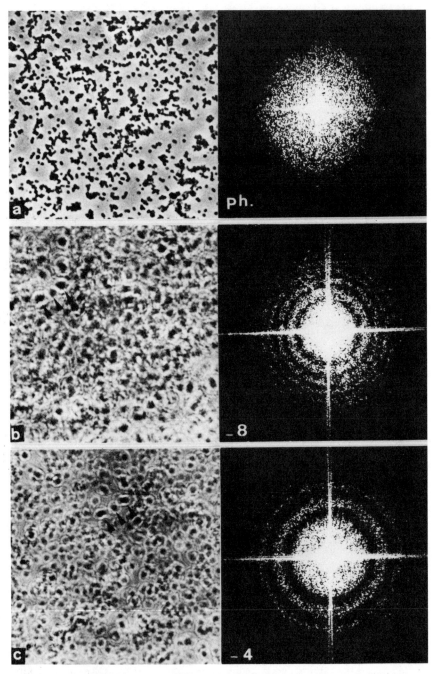

FIGURE 4. Bacterial ghosts (as described in FIGURE 3) observed in (a) with *annular phase contrast*, in focus (as in FIGURE 3b). In (b) to (f) we show a through-focus series (−8 to +8 μm) obtained with *narrow-beam phase contrast*. The diffractograms were obtained with an optical diffractometer. All micrographs were obtained with a Wild microscope and 40X objective (achromat, nA = 0.65). The aperture of the narrow beam of 0.013 radians was obtained by inserting an electron microscopy aperture of 300 μm into the condenser. Illumination according to Köhler.

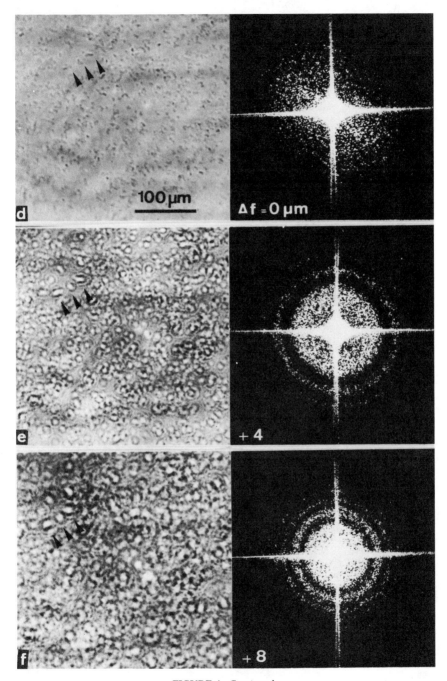

FIGURE 4. *Continued*

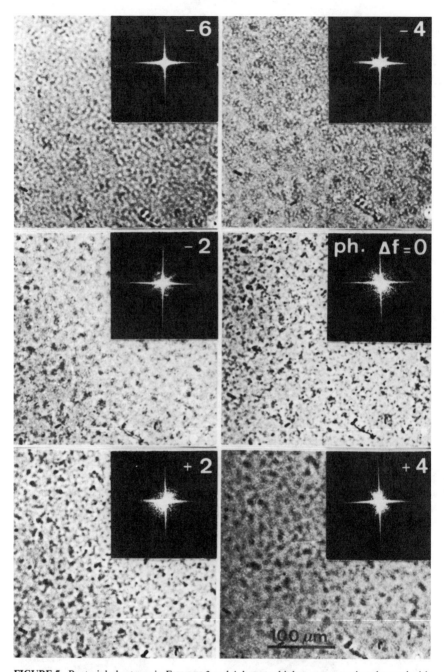

FIGURE 5. Bacterial ghosts, as in FIGURES 3 and 4, but at a higher concentration observed with annular phase contrast with the Wild 40X Ph-achromat. Through-focus series (from −6 to +4). The diffractograms are shown as *insets*. Despite the higher concentration of ghosts no "Thon"-rings appear. The central diffuse area is less well defined than in FIGURE 4a and shows some asymmetry, which we attribute to an irregular light distribution in the annular aperture of the condenser.

films, obtained by evaporation, show diffuse rings in diffraction, which might be due to a microcrystalline structure of the film.[23] Other materials, like vitreous ice, might indeed produce less noise, despite also showing a diffuse band in diffraction.[26] Other

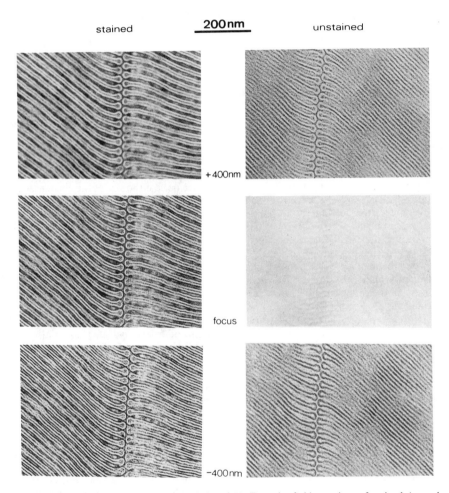

FIGURE 6. Micrographs by C. Westphal and H. Froesch of thin sections of stained (uranyl acetate, block stain) and unstained retina of frog fixed by aldehydes and embedded in Nanoplast.[20] The micrographs are taken in focus and at 0.4 μm over- and underfocus. The *unstained* sections show contrast reversal, explainable by the possible maintenance of lipids. The *stained* section does not show any contrast reversal, because now we deal with (unresolved) scattering particles of heavy metal. (Courtesy of Dr. Froesch *et al.*, University of Ulm).

materials should be reinvestigated when observed with cryostages and in very high vacuum. It was indeed proposed by many (oral communications) that the diffracto-grams observed on various films might often have been the contamination layer instead of that of the film investigated!

The unsharp micrographs in the narrow-beam phase contrast of the light microscope might be partly explainable by Fresnel fringes surrounding the bacterial ghosts. However, these fringes should disappear when in correct focus, which, according to our experience and that of others[4] is not the case. The ghosts stay unsharp, always less defined than those obtained with the annular phase contrast. That this is not simply due to a bad optic is ruled out by the fact that the same objective lens can be used for both imaging modes, with results of the narrow-beam imaging that are indistinct from those with a "normal" objective lens (without phase ring). Possibly one should account for the fact that any beam limitation, therefore, even that caused by the condenser aperture, introduces "contaminating" or "parasitic" waves, which disturb the (approximately) planar waves of the main part of the beam. These parasitic waves must obviously also disturb the ideal imaging with the planar waves.

When comparing light with electron microscopy we also have to be aware of a profound difference in the illumination.[27] While in light microscopy the whole system is designed for illumination according to Köhler, i.e. by imaging the light source into the condenser lens, the electron microscope images the electron source into or just above the specimen. In some types of microscopes this is done precisely, while in others the image is "messed up" by an "incomplete" imaging by the Wehnelt cylinder. These differences are relevant when discussing the influence both of the size of the condenser aperture and of the electron source[28] on the coherence.

Whatever the outcome of ad hoc experiments designed for resolving the above-outlined problems will be, it is certainly reasonable to come back to the idea of considering a phase and intensity distribution of the electrons in the "layer" of space after the specimen, as proposed clearly by Lenz.[16] The contrast transfer function then provides the analytic description of this space in those cases where the coherence is sufficiently high. The space after the specimen we propose to define as a "Fresnel transformation," suggesting by this that any eventually formulated theory should also account for the Fresnel fringes and their increase with a narrower beam.

The new phase contrast observations on Nanoplast sections of biological membranes with their lipid probably still present,[19] and of frozen-hydrated material[17,18] should be recalculated with the inner potentials accounted for. In this context it is important to realize that lipids are the main component of the only nonwatery phases existing in cells. Proteins are either dissolved in water or in lipids, according to their composition. No proper protein phase exists. Even relatively compact protein structures, like virus capsids, contain some water in between subunits. The protein molecules themselves, however, do not contain water within their bulk structure. Therefore, they represent a minute phase-compartment only. Water soluble proteins are solvated with the help of a hydration shell. Hydrophobic proteins dissolved in a lipid bilayer integrate completely into the lipid phase without distinct interface. Nucleic acids and polysaccharides are always hydrophilic and behave like hydrophilic proteins. It is easily imaginable that all the hydrophilic molecules behave similarly when the water is replaced by a resin: they fall out of solution and form aggregates. We try to keep these as small as possible by adequate fixation or by applying low temperatures during processing.

If we accept the idea that narrow-angle illumination boosts the formation of Fresnel fringes, then it also becomes completely clear why bright-field phase contrast electron imaging has given such excellent results with periodic structures. In this case the fringes amplify the periodic structure instead of producing noise!

RATIO-CONTRAST, DARK-FIELD, AND SOME OTHER IMAGING MODES

In order to avoid the strong interference phenomena associated with the narrow-beam, highly coherent illumination, it was proposed to use an annular aperture in the condenser, or to rotate the beam so as to describe a hollow cone with the summit in the specimen.[30,31] The results published[32] confirm what we said above by showing a very much decreased holographic noise.

Dark-field imaging was considered since time immemorial as alternative to bright-field.[33] In CTEM the dose needed for recording usable dark-field images is one to two orders of magnitude higher than for bright-field. For biological material this meant that we had to accept a very "burnt" specimen. By using this "microincineration" interesting results were obtained on very small molecules.[34] With STEM and the annular detector (ADF) the efficiency of electron collection is such that the dose requirements are now comparable to those of CTEM bright-field. Even frozen-hydrated specimens can be observed with this mode (Kellenberger, Carlemalm, Tichelaar (EMBL), to be published).

Dark-field shows no holographic noise, and we believe that what we see represents the specimen rather faithfully. In FIGURE 2 we see that uranium clusters are perfectly imaged with ADF but disappear in the holographic noise of the bright-field phase contrast image.[35]

As we have shown elsewhere[7] both dark-field and bright-field imaging of unstained biological material depicts mainly thickness variations, and the influence of density differences between embedding material and biological structures is negligibly small. This situation is reversed obviously with heavy metal stained material.

Ratio-contrast produced by the ratio of the signals from the low energy loss electrons (lel or plasmon) and the signal from the annular detector,[5] however, is much more sensitive to the properties of matter than to thickness variations.[7] For unstained biological material the variable hydrogen content is the main parameter defining contrast.[7]

These theoretical predictions are experimentally demonstrated by the disappearance of irregularly thick contamination layers from the image of uranium clusters (FIGURE 7) and by the disappearance of knife marks from biological thin sections (FIGURE 8).

These two examples show influences of an esthetic nature only. However, the "relief independence" has much more important consequences: it was shown very early that the surface of thin sections shows a strong, object-related relief.[8] It has since become clear that the cleaved part of the embedded material and the surrounding surface of the resin are strongly distorted as a consequence of differences in plastic flow and the tension of rupture (Acetarin et al., submitted).

Imaging of the surfaces, as with dark-field, thus provides much less good information than does the ratio-contrast that images the differences of atomic composition of the matter inside of the section.[7,9]

The micrographs of the sort represented in FIGURE 7 were submitted to cross correlation of two interlaced micrographs of the same region. Interlaced means that each line is scanned twice and one each is used for the respective micrographs. From the cross correlation one can determine the resolution, d, and the S/N. The data are given in TABLE 1. The results are summarized as follows: (i) The resolution is significantly better with ratio-contrast than with the ADF. One has apparently eliminated organic contaminants that surround the uranium aggregates. (ii) By the division with the inelastic image one has also filtered away low frequency components, acting also in the direction of a higher resolution of the aggregates. (iii) The S/N, however, has slightly suffered from the division.

These results, obtained on uranium clusters, cannot be extrapolated to thicker specimens as are thin sections (400–600 Å). The method of cross correlation on interlaced micrographs is, however, quite adequate and will be applied to sections also.

In FIGURE 9 we show a series of three micrographs of the same unstained thin section of septate junctions, first by ratio-contrast, then by dark-field, and finally again by ratio-contrast. It is obvious that the dark-field image is blurred when compared to the two ratio images. We show here the series of three micrographs because it was frequently proposed that good definition and contrast of the ratio image is due only to beam-induced destruction. As we see here, the subsequent dark-field image is still

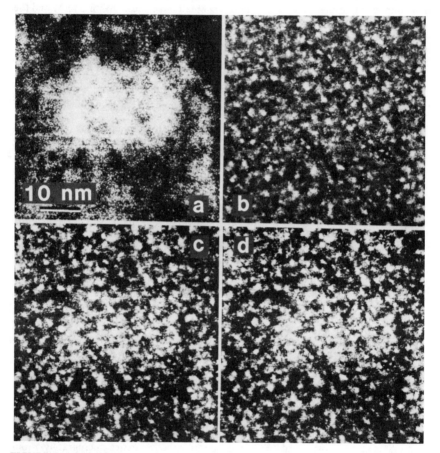

FIGURE 7. Uranium clusters, as in FIGURE 2; the specimen was, however, contaminated by a previous observation on a restricted central area. The micrographs represent the following signals: (a) inelastic, lel dark-field; (c) and (d) elastic dark-field (ADF); and (b) the ratio of lel and ADF.

blurred, independently of the alleged improvement that would be due to beam-induced burning of the first micrograph!

By ratio-contrast imaging it became possible to compare stained with unstained material (FIGURE 10). From these results we conclude tentatively[36] that the hydrophobic surface of proteins is not stained by the usual procedures, and this explains why

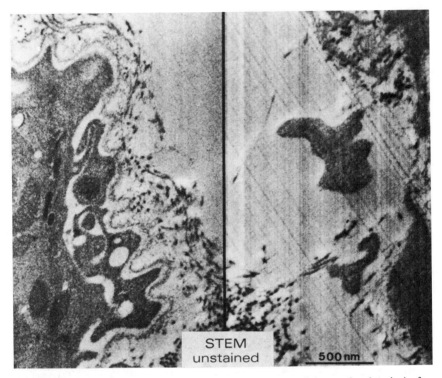

FIGURE 8. Thin section of epithelial cells of the rat intestine (jejunum, basal region) after aldehyde fixation and embedded in a tin-containing resin. During recording the mode of imaging was switched from dark-field (*right*) to ratio-contrast (*left*). The knife marks, quite visible in the *right panel*, continue into the *left panel* with very much reduced intensity.

transmembrane proteins cannot really be observed in thin sections by conventional procedures.

Another case of a transmembrane protein is the gap junction; we show the first results obtained with ratio-contrast in FIGURE 1. Some details are visible on the processed micrograph and encourage further studies.

As will be discussed further below, a sort of negative contrast can also be achieved with thin-sectioned unstained biological material by incorporating heavy metal into the resin.[37,38] In CTEM, BF the 3 atom percent of tin in our resin produces some contrast that is usable for cytological low-magnification micrographs. With the ratio mode the

TABLE 1. Quantitative Evaluation of Resolution (**d**) and Signal-to-Noise (**S/N**) in Two Images of the Same Type as FIGURE 7

	Image 1		Image 2	
	$d_{(\text{Å})}$	S/N	$d_{(\text{Å})}$	S/N
ADF	6.7	1.34	6.1	2.63
Inelastic	20.2	0.85	25.6	1.49
Ratio-Contrast	5.5	1.20	5.2	1.81

STEM
unstained

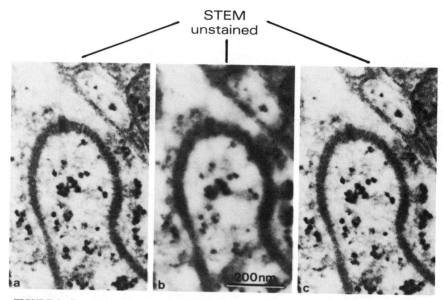

FIGURE 9. Septate junctions in the testis of *Drosophila melanogaster*. Aldehyde fixation and embedding in Lowicryl HM20. No heavy metal treatment. The first ratio-contrast image is shown in (**a**). The *subsequent micrograph* (**b**) was taken at the same focus in the dark-field mode. Note the blurred appearance of all details. From the same area a *third micrograph* (**c**) was taken again with ratio-contrast for checking the focus and for possible beam-induced alterations that could have occurred after taking (a). This series demonstrates that the image blurring of (b) is not the consequence of "burning" the specimen, as is sometimes claimed. The series precludes the idea that the fine details of the ratio-contrast relative to dark-field is only the consequence of beam-induced alterations.

contrast is very strong, as is shown in FIGURE 11, which presents an aldehyde fixed, part of a blood vessel, and an endothelial cell.

Finally, in FIGURE 12 we compare again the results of dark-field with those of ratio-contrast observed on thin sections of cells of the rat skin. Besides the desmosome—a cellular junction—we see the cytoskeleton in the ratio-contrast, but not in the other.

RATIO-CONTRAST AS A CONSEQUENCE OF ATOMIC COMPOSITION AND ITS USE FOR DETERMINING CONCENTRATIONS

The scattering cross sections of atoms depend essentially on Z. The ratio of inelastic to elastic electrons also essentially represents Z. For molecules it is, however, not correct to calculate simply an average Z. One first has to determine average cross sections and from these one can calculate a Z-equivalent. We have plotted a curve representing the ratio signal for various biological substances as a function of an average Z (FIGURE 13a). Comparing with the cross sections as we have done elsewhere[7] shows that instead of Z as variable it is reasonable to use the content of hydrogen atoms, as is shown in FIGURE 13b. Carbon and nitrogen have very similar cross sections, while those of hydrogen (and also of helium) are very different. Fortunately, lipids, hydrocarbons, and proteins have quite different hydrogen contents and correspondingly produce contrast if the embedding material is adequately chosen. Ice is a very adequate embedding medium because it has the highest hydrogen content, followed by the resins Lowicryl HM20[39] and HM23,[40] which were designed for that and other purposes (low temperature embedding). Epoxy resins, currently used in cytology, have a hydrogen content similar to proteins and thus are inadequate for ratio-contrast imaging, because the embedded proteins are "matched out" by the resin. The deliberate use of this property might lead to very useful applications: sulfur and phosphorus have nearly the same cross sections and are also components of biological macromolecules. The sulfur content of proteins (in methionine, cysteine) is, however, negligibly small. Nucleic acids contain large amounts of phosphorus, which put them out of any of the curves of FIGURES 13a and b. The high ratio-contrast of nucleic acids

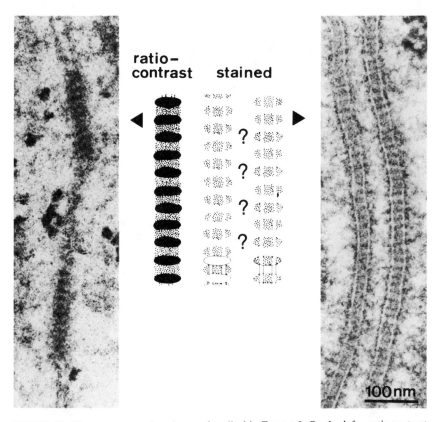

ratio-contrast   stained

FIGURE 10. The same septate junctions as described in FIGURE 9. On the *left* a ratio-contrast image of completely unstained junctions showing the two membranes connected by integral membrane proteins. On the *right*, sections of the same block were stained with uranyl acetate and observed in bright-field CTEM. From the models *in between*, we see the interpretation of the results: hydrophobic portions of the transmembrane protein are not stained. The location of the stain deposits remains unsettled: either in the space between the proteins or in or on the protein itself.

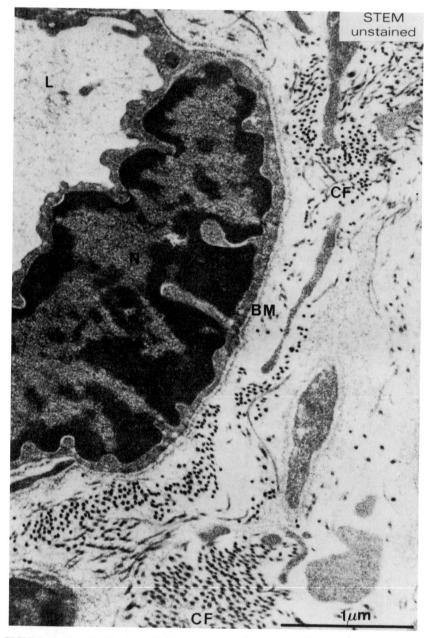

FIGURE 11. Part of a blood vessel of the rat intestine (jejunum). The tissue was aldehyde-fixed and embedded in a tin-containing resin. This micrograph is taken at low magnification by ratio-contrast and shows a remarkably high contrast on this unstained biological structure. **N** = nucleus from an endothelial cell, **L** = lumen, **CF** = collagenous fibers, **BM** = basal lamina material.

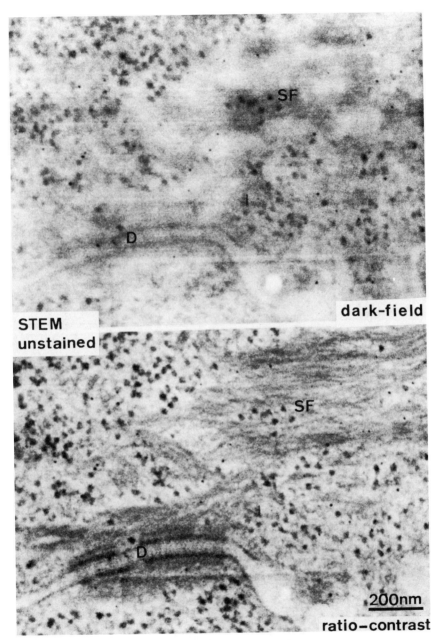

FIGURE 12. STEM micrographs of thin sections of rat skin without any heavy metal treatment. The desmosome (**D**) joins two cells together. The cytoskeleton is attached to the desmosomes. The stress fibers (**SF**) are only visible with ratio-contrast but not in the ADF mode. The tissue was aldehyde-fixed and embedded in Lowicryl K4M by the method of progressive lowering of temperature.[44] The sections were observed at EMBL in Heidelberg in a STEM HB5 equipped with a cryostage. We are grateful to Dr. W. Tichelaar for having optimized the conditions of this STEM. We strongly believe that the observation at the temperature of liquid N_2 is essential for seeing the details of resin-embedded material.

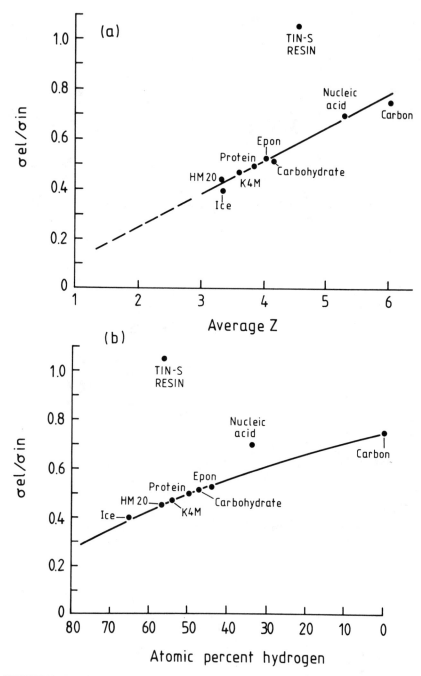

FIGURE 13. The ratio signal $r = \sigma_{el}/\sigma_{in}$ as (**a**) a function of the *average Z* and (**b**) of the hydrogen content of organic matter. On the right scale, in (**b**) we indicate the elastic and inelastic scattering cross sections (σ_{el} and σ_{in}).

could be used in combination with a matching resin, composed such that the other biological substances (particularly proteins) produce no or very little contrast. In this manner the location of nucleic acids could be detected. Unfortunately, RNA and DNA will not be different without an additional differentiation, for example, by a specific DNA stain.[41]

Instead of using a resin with a high hydrogen content and which thus produces a low signal with ratio-contrast, we approached the problem by constructing a resin that produces a very high signal. This was achieved by attaching tin to the aromatic ring of a styrene.[38] As far as we understand, it should be possible to prepare a resin that matches biological material simply by reducing the tin content. This can be done relatively easily by mixing the tin-containing resin with its counterpart without tin. Only incidental observations are as yet available; systematic studies have still to be made.

Since the ratio signal is a consequence of the average atom composition, it is now possible to determine the relative amounts of a biological material in ice or in a resin. The resin content corresponds, to a very fair approximation, to the initial water content of the living cell. By such calculations it should become possible to determine the water content of cellular compartments and of substructures.[42] We already have evidence that mitochondria appear as extremely compact structures, confirming indirect predictions.[43] It would also be possible to follow the condensation of chromatins during the cellular division cycles and many other biologically relevant problems.

Finally, we should not forget to mention that the formation of a ratio is not only possible with the lel or plasmon peak, but also for specific energy losses. The resulting independence from the relief might also turn out to be a real asset here, particularly if the use of cryostages substantially decreases the beam-induced compositional alterations. It should be remembered here that the electrons with specific energy losses show only intensities that are more than 100 times smaller than the lels.

CONCLUDING REMARKS

While on a light micrograph it is possible to detect biologically significant details that are two to three times the theoretical resolving power of the instrument, and this without any image processing, we face a factor of more than ten with electron micrographs. When investigating biological material, we approach the resolving power of the instrument only when applying image processing. It seems to us that the time has come in which efforts must be made to intensify the studies of both the biological specimen with its preparation for observation and the processes of its imaging.

The predominant school of electron optics considers illumination as of little importance; the approximation of representing the incoming electrons as a planar wave is taken for granted. The importance of the illumination, particularly of the angular aperture of the electron beam, is illustrated in the present paper by its simulation in the optical microscope. We show that a contrast transfer function, as definable by the diffraction rings obtained from a through-focus series of micrographs, is clearly establishable with narrow-beam imaging, but not with an annular aperture in the condenser, as is used for the light-optical phase contrast.

We then see that bright-field phase contrast electron imaging is comprehensibly described as a Fresnel transformation (*i.e.* as a superposition of Fresnel fringes) because of its strong dependence on the illumination aperture. The superposition of the Fresnel-fringe system, caused by object inhomogeneities, produces the specific ("holographic") noise. With nonregular objects, this noise is strongly disturbing and leads to the complete disappearance of small objects. With periodic crystalline objects, however, it is likely that the fringes help to amplify the periodicity. For a better understanding of phase contrast it would be most welcome if the influence of narrow-beam illumination would be treated theoretically and experimentally. The results should then be compared with hollow-cone illumination[31,32] and with the dark-field and ratio-contrast obtained with STEM.

The interesting phase behavior of biological matter in ice or in Nanoplast should be investigated. One should try to understand why underfocus of ice-embedded biological material produces so much more contrast than the same material embedded in resins. Similarly one should try to produce a satisfactory theory for explaining high contrast and its reversal with biological material embedded in Nanoplast, which occurs neatly with structures that are two to three times larger than the lattice constant in biological crystals and at very small values of defocusing (approximately 0.5 μm). One should also take into account that the same biological structure, when embedded in a different resin, does not show the reversal at defocus that still allows for sharp, clearly defined micrographs.

The possibility offered by STEM to observe samples in the dark-field and the ratio mode, at conditions of minimal dose, opens new fields of investigation. It becomes possible to observe (i) unstained material and (ii) frozen-hydrated material both with high contrast and without holographic noise. When added to the above-mentioned features of phase contrast, one might learn more about the interrelations between properties of matter and imaging.

Ratio-contrast in particular will allow investigation of the location of the heavy metal deposits relative to the biological macromolecules that results from positive staining.

Over dark-field, the ratio-contrast has the advantage of giving much more weight to the differences of the atomic composition than to differences of specimen thickness. This feature is particularly useful when observing unstained sections, where the surface is strongly deformed by the cleavage process involved in sectioning.[7,9]

The study of these different modes of imaging will be further helped by the new developments associated with the filter lens[45,46] in fixed beam electron microscopy (FBEM) and with improved spectrometers in STEM (M. Haider, A. Jones, H. Rose, personal communication). The dispute about the influence of multiple scatter in the imaging of the "normal" specimen of some 50–70 nm thickness will hopefully be settled, as will that over the contribution of the inelastically scattered electrons to image blurring.

ACKNOWLEDGMENTS

We are particularly indebted to EMBL (Director General, L. Phillipson) for the possibility offered to one of us (E.C.) to use the VG-STEM with cryostage that is under the responsibility of W. Tichelaar. It is the only STEM in Europe with a well functioning "homemade" cryostage; the vital importance of it is mentioned in the text. Regula Niederhauser is thanked for having patiently typed numerous versions associated with so many authors.

REFERENCES

1. KELLENBERGER, E. & J. KISTLER. 1979. The physics of specimen preparation. *In* Unconventional Electron Microscopy for Molecular Structure Determination. W. Hoppe & R. Mason, Eds. 49–79. Fred. Vieweg & Sohn. Braunschweig.

2. KELLENBERGER, E., E. CARLEMALM & W. VILLIGER. 1986. Physics of the preparation and observation of specimens that involve cryo-procedures. *In* The Science of Biological Specimen Preparation. A. Boyde, R. Becker, M. Müller & J. Wolosewick, Eds. Proc. 4th Annu. Pfefferkorn Conf.

3. CARLEMALM, E., C. HOHL, W. BASCHONG, C. BASCHONG, C. KELLENBERGER, H. SEILER & E. KELLENBERGER. Uptake of osmium and uranyl by biological specimens during heavy metal treatment, in preparation

4. BENNET, A. H., H. JUPNIK, H. OSTERBERG & O. W. RICHARD. 1970. Quoted *in* Principles of Optics. M. Born & E. Wolf, Eds. Fig. 8.35, p. 428. Pergamon Press. Oxford.

5. CREWE, A. V., J. P. LANGMORE & M. S. ISAACSON. 1975. Resolution and contrast in the scanning transmission microscope. *In* Physical Aspects of Electron Microscopy and Microbeam Analysis. B.M. Siegel & D.R. Beaman, Eds. 47–62. John Wiley & Sons. New York, NY.

6. CARLEMALM, E. & E. KELLENBERGER. 1982. The reproducible observation of unstained embedded cellular material in thin sections: visualization of an integral membrane protein by a new mode of imaging for STEM. The EMBO Journal **1**: 63–67.

7. CARLEMALM, E., CH. COLLIEX & E. KELLENBERGER. 1985. Contrast formation in electron microscopy of biological material. *In* Advances in Electronics and Electron Physics. P.W. Hawkes, Ed. Vol. **63**: 269–334. Academic Press. New York, NY.

8. WILLIAMS, R. C. & F. KALLMAN. 1955. Interpretations of electron micrographs of single and serial sections. J. Biophys. Biochem. Cytol. **1**: 301–314.

9. KELLENBERGER, E., W. VILLIGER & E. CARLEMALM. 1986. The influence of the relief on thin sections of embedded, unstained biological material on the image quality. Micron Microsc. Acta, anticipated to appear in Vol. 17, No. 4.

10. TENCE, M., N. BOUNEL, C. JEANGUILLAUME, P. TREBBIA & C. COLLIEX. 1985. A digital acquisition and processing unit for STEM images. J. Microsc. Spectrosc. Electron **10**: 65–90.

11. VALENTINE, R. C. 1966. The response of photographic emulsions to electrons. *In* Advances in Optical and Electron Microscopy. R. Barer & V. E. Cosslett, Eds. 180–203. Academic Press. New York, NY.

12. ZERNIKE, F. 1935. Das Phasenkontrastverfahren bei der mikroskopischen Beobachtung. Phys.Z. **36**: 848–851.

13. MASON, D. J. & D. M. POWELSON. 1956. Nuclear division as observed in live bacteria by a new technique. J. Bacteriol. **71**: 474–479.

14. THON, F. 1966. Zur Defokussierungsabhängigkeit des Phasenkontrastes bei der elektronenmikroskopischen Abbildung. Z. Naturforsch. Teil A **21**: 476–478.

15. HANSZEN, K. J. 1971. The optical transfer theory of the electron microscope: fundamental principles and applications. *In* Advances in Optical and Electron Microscopy. R. Barer & V. E. Cosslett, Eds. Vol. **4**: 1–84. Academic Press. New York, NY.

16. LENZ, F. & W. SCHEFFELS. 1958. Das Zusammenwirken von Phasen- und Amplitudenkontrast in der elektronenmikroskopischen Abbildung. Z. Naturforsch. Teil A **13**: 226–230.

17. DUBOCHET J., M. ADRIAN, J. LEPAULT & A. W. MCDOWALL. 1985. Cryo-electron microscopy of vitrified biological specimens. TIBS **10**: 143–146.

18. ADRIAN, M., J. DUBOCHET, J. LEPAULT & A. W. MCDOWALL. 1984. Cryo-electron microscopy of viruses. Nature **308**: 32–36.

19. WESTPHAL, CH. & D. FROESCH. 1984. Electron-phase-contrast imaging of unstained biological materials, embedded in a water-soluble melamine resin. J. Ultrastruct. Res. **88**: 282–286.

20. BACHHUBER, K. & D. FROESCH. 1983. Melamine resins, a new class of water-soluble embedding media for electron microscopy. J. Microsc. **130**: 1–9.

21. BOULESTEIX, C., C. CESARI, C. COLLIEX, A. ELHILI, Z. FAKHFAKH, C. MORY & G. NIHOUL-BOUTANG. 1978. Quantitative analysis of the contrast of stained biological sections. J. Microsc. Spectrosc. Electron **3**: 619–632.

22. WEIBULL, C., W. VILLIGER & E. CARLEMALM. 1984. Extraction of lipids during freezesubstitution of *Acholeplasma laidlawii* cells for electron microscopy. J. Microsc. **134**: 213–216.

23. HEIDENREICH, R. D. 1964. Fundamentals of Transmission Electron Microscopy. Interscience Publishers. New York, NY.

24. KINDER, E. & A. RECKNAGEL. 1947. Ueber Fresnelsche Beugung beim Licht- und Elektronenmikroskop. Optik **12:** 346–363.
25. FUKUSHIMA, K., H. KAWAKATSU & A. FUKAMI. 1974. Fresnel fringes in electron microscope images. J. Phys. D **7:** 257–266.
26. DUBOCHET, J., M. ADRIAN & R. H. VOGEL. 1983. Amorphous solid water obtained by vapour condensation or by liquid cooling: A comparison in the electron microscope. Cryo Lett. **4:** 233–240.
27. MORY, C. & C. COLLIEX. 1981. Model and experimental study of the double condenser illumination system in CTEM. Optik **59:** 311–334.
28. HIBI, T. & S. TAKAHASHI. 1971. Relation between coherence of electron beam and contrast of electron image of biological substances. J. Electron Microsc. **20:** 17–22.
29. LENZ, F. 1965. Kann man Biprisma-Interferenzstreifen zur Messung des elektronenmikroskopischen Auflösungsvermögen verwenden? Optik **22:** 270–288.
30. HANSSEN, K. -J. & L. TREPTE. 1971. Die Kontrastübertragung im Elektronenmikroskop bei partiell kohärenter Beleuchtung. Teil A: Ringkondensor. Optik **33:** 166–181.
31. HANSSEN, K. -J., R. LAUER & G. ADE. 1985. Beiträge zur nachverarbeitung elektronenmikroskopischer Aufnahmen und zur Elektronenholographie. Dreiländertagung Konstanz. PTB-Bericht A Ph-25. Physikalisch-Technische Bundesanstalt. Braunschweig.
32. KUNATH, W., F. ZEMLIN & K. WEISS. 1985. Apodization in phase-contrast electron microscopy realized with hollow-cone illumination. Ultramicroscopy **16:** 123–138.
33. DUBOCHET, J. 1973. High resolution dark field electron microscopy. *In* Principles & Techniques of EM. Biological Applications. M. A. Hayat, Ed. Vol. 3: 115–151. Van Nostrand Reinhold. New York, NY.
34. OTTENSMEYER, F. P., D. P. BAZETT-JONES, R. M. HENKELMAN, A. P. KORN & R. F. WHITING. 1978–79. The imaging of atoms: its application to the structure determination of biological macromolecules. Chem. Scr. **14:** 1–5.
35. MORY, C. & CH. COLLIEX. 1985. Experimental study of the resolution, contrast and signal/noise ratio in different STEM imaging modes. J. Microsc. Spectrosc. Electron **10:** 389–394.
36. GARAVITO, R. M., E. CARLEMALM, C. COLLIEX & W. VILLIGER. 1982. Septate junction ultrastructure as visualized in unstained and stained preparations. J. Ultrastruct. Res. **80:** 344–353.
37. CARLEMALM, E., J. -D. ACETARIN, W. VILLIGER, CH. COLLIEX & E. KELLENBERGER. 1982. Heavy metal-containing surroundings provide much more "negative" contrast by Z-imaging in STEM than with conventional modes. J. Ultrastruct. Res. **80:** 339–343.
38. ACETARIN J. -D., W. VILLIGER & E. CARLEMALM. A new heavy-metal containing resin for low-temperature embedding and imaging of unstained sections of biological material. J. Electron Microsc. Tech., in press.
39. CARLEMALM, E., R. M. GARAVITO & W. VILLIGER. 1982. Resin development for electron microscopy and an analysis of embedding at low temperatures. J. Microsc. **126:** 123–143.
40. ACETARIN, J. -D., E. CARLEMALM & W. VILLIGER. 1986. Development of new Lowicryl sections for embedding biological specimens at even lower temperatures. J. Microsc. **143:** 81–88.
41. GAUTIER, A. 1976. Ultrastructural locations of DNA in ultrathin tissue sections. Int. Rev. Cytol. **44:** 113–185.
42. REICHELT, R., E. CARLEMALM, W. VILLIGER & A. ENGEL. 1985. Concentration determination of embedded biological matter by scanning transmission electron microscopy. Ultramicroscopy **16:** 69–80.
43. SRERE, P. A. 1981. Protein crystals as a model for mitochondrial matrix proteins. TIBS **6:** 4–6.
44. CARLEMALM, E., W. VILLIGER, J. A. HOBOT, J. -D. ACETARIN & E. KELLENBERGER. 1985. Low temperature embedding with Lowicryl resins: Two new formulations and some applications. J. Microsc. **140:** 55–63.
45. CASTAING, R. & L. HENRY. 1962. Filtrage magnetique de vitesse en microscopie électronique. C. R. Acad. Sc. Paris **B255:** 76–78.
46. HENKELMAN, R. M. & F. P. OTTENSMEYER. 1973. An energy filter for biological electron microscopy. J. Microsc. **102:** 79–94.

DISCUSSION OF THE PAPER

D. F. PARSONS: (*New York State Department of Health, Albany, New York*): What is your opinion on a slightly different use of STEM? One frequently sees it proposed as a penetration instrument, and with the penetration of thick specimens, people often expect to get good stereo photographs, often forgetting about the top-to-bottom effect and the highly convergent beam.

KELLENBERGER: I can only say how I feel, and what I feel is exactly like you, that one should do more. At a symposium in Paris a long time ago, thick specimens were observed by STEM and were compared with high voltage, and there was no very big difference. Therefore, everybody thought it was very good. But it was not on very selective or very suitable specimens, and I think it should be done again. But physicists don't want to do it, because they don't want to have thick specimens in which there is much multiple scattering, and you cannot calculate it anymore. It's very difficult. I hope we will eventually come to that point where we can look at a series of sections of variable thicknesses. Then maybe we can answer your question.

A. P. SOMLYO: (*University of Pennsylvania School of Medicine, Philadelphia, Pennsylvania*): You related the elastic-inelastic ratio in unstained organic material to composition. Shouldn't we consider a second-order correction that, since this is going to be clearly Z-dependent, we might have to worry about what the local composition is after radiation damage? Egerton showed some years ago with EELS that, at least from nonbiological polymers, the sequence of loss was hydrogen, oxygen, nitrogen, and carbon.

KELLENBERGER: You made a very important point. We have discussed that a lot, because in our laboratory we had somebody who was so absolutely strict that he did not want to work with more than one electron per angstrom square. It's clear that we have beam damage and we have to live with some, even in these cases that I have shown.

SOMLYO: Where you didn't see contrast reversal, is it possible that you used a very small aperture, and the image may have been dominated by amplitude contrast? That would have been the reason why the phase reversal was not observed.

KELLENBERGER: In normal cases you don't see a contrast reversal. Only in ice and with this new embedding material, Nanoplast.

J. B. PAWLEY (*University of Wisconsin, Madison, Wisconsin*): You mentioned that you'd like to be able to measure the inelastic scattering and the elastic scattering and make a ratio at every point in the image. You did mention briefly that you could now do this with the Zeiss-902. Have you used this in this way to make ratio images?

KELLENBERGER: We are still convinced that on the same image point you should collect two signals at once as we do it in STEM, rather than successively. I should have elaborated on that more, because what we call the elastic signal is the angular dark-field detector signal and what we call the inelastic signal is the low energy-loss or plasma-loss electrons exclusively. It could obviously also be done by the specific losses, but these are the two H sources that we use. There is also multiple scattering in both cases, which one may eliminate with new spectrometers or with the filter lens. We have tried, but Peter Ottensmeyer (Toronto, Canada) has done much more work on the dividing of two separately collected images. Superposition of the two images is awfully difficult. For the moment he has not used completely unstained material, but osmium stained, and this also makes a difference.

COLLIEX: I want to make three comments. The first one is that there are some physicists who are willing to put biological specimens in the microscope. The second one is that I have tried even to put thick specimens in the microscope, that is to answer one question there. But what happened is that, as you have explained very clearly, the contrast is so high in dark-field that as soon as the signals become very important we do not recognize anything. So this method should be applied to very well defined specimens. This is a problem whenever you have a thick specimen, thicker than 5,000 Å, and you have too much contrast and too much information.

The third point is, that to answer the other questions about the comparison of STEM and CTEM, STEM is definitely better in terms of acquiring the same information for a given dose. So, because you have all the signals simultaneously, you have a better use of the signal; and this CTEM cannot do.

Electron-Probe X Ray Microanalysis of *In Situ* Calcium and Other Ion Movements in Muscle and Liver

A.V. SOMLYO, M. BOND, H. SHUMAN,

AND A.P. SOMLYO

The mechanism of excitation-contraction coupling, whereby an agonist or depolarization of the cell membrane releases Ca from the sarcoplasmic reticulum (SR) to activate contraction, is one of the major remaining problems to be solved in muscle physiology. The reuptake of Ca by the Ca ATPase of the SR has been extensively studied in isolated SR vesicles. Although Ca release can be induced by a variety of conditions, physiological release is difficult to produce and to identify in these isolated SR fractions. Redistribution of ions occurs during isolation procedures, and the composition of the intracellular milieu in a live working muscle fiber cannot be precisely mimicked. Therefore, it was of considerable interest to obtain *in situ* measurements of the elemental composition of the SR at rest and during activation and recovery, and to directly measure the movements of Ca and the accompanying counter and co-ions. The method used was electron-probe X ray microanalysis (EPMA) of dry cryosections prepared from rapidly frozen tissues. These studies have led to some new insights into excitation-contraction coupling as well as to the ruling out of certain models. Our methods have been described elsewhere[1-3] and shown to preserve the ultrastructure of the tissues as well as the *in vivo* distribution of diffusible monovalent ions across plasma membranes, intracellular vacuoles, and organelles. These techniques have been used to study relaxed, contracted, and fatigued skeletal,[1,4,5,10] cardiac,[6] and smooth muscle,[7-9] Ca and Mg binding to the A- and I-band of skinned skeletal muscle fibers,[11] as well as light- and dark-adapted retinal rods,[12] sickled red blood cells,[13] liver,[14] and brain.[15]

RAPID FREEZING AND ELECTRON-PROBE ANALYSIS: THE APPROACH

Rapid freezing is the essential preparatory method required for retaining Ca and other ions measured with EPMA in the SR and has the potential for freeze-trapping physiological events on a millisecond time scale.[16] Ultrathin sectioning of frozen tissue at ambient temperatures of $-130°$ C, followed by freeze-drying *in vacuo* has been described in detail.[1-3] Given the preservation of the SR and other structures in muscle by suitable rapid freezing methods, it is possible to identify the triad and its components as well as mitochondria, nuclei, I- and A-bands, etc., in dry, ultrathin cryosections suitable for quantitative EPMA. In such cryosections, the contrast is achieved mainly by elastic electron scattering due to different cellular regions having different mass densities. Regions that are more hydrated *in vivo,* such as the I-bands, are relatively more electron lucent, and those that are less hydrated, such as mitochondria, are more electron opaque in the dry cryosections (FIGURE 1).

The electron beam of an electron mieroscope can be focussed to a probe of desired diameter that just covers the organelle or area of interest. The fast electrons in the focussed probe can eject inner shell electrons from atoms in the specimen. Through this process of inelastic scattering, incident electrons lose energy and can generate X rays. Electron-probe X ray microanalysis (EPMA) is based on the fact that the energy of the X rays emitted, as the result of core shell ionization, is characteristic of the specific core shell ionized and, more specifically, of the transition of the outer shell electron filling the vacancy created by ionization. For example, the ejection of a K-shell electron from a Ca atom and the transition of an L-shell electron into the resulting vacancy has a certain probability (fluorescence yield) of producing a Ca Kα X ray photon with an energy of 3.69 KeV. Therefore, the appearance of a peak at 3.69 KeV in the X ray spectrum is indicative of the presence of Ca. In addition to characteristic X

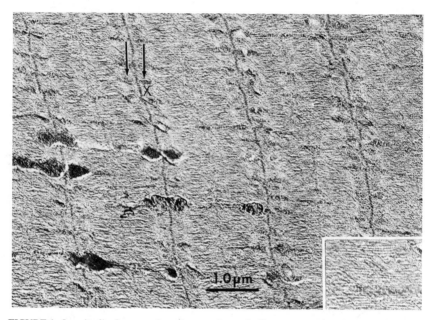

FIGURE 1. Longitudinal cryosection of a control muscle. Sarcomeres, Z- and M-lines, as well as SR, are apparent. Typical paired TC (*arrows*) such as those used for analysis and the adjacent cytoplasm in the I-band (*X*) are indicated. The right hand side of the image is toward the outer aspect of the bundle and has less ice-crystal damage than the left hand aspect. Note that myosin filaments can be imaged at the outer aspect (see *inset*). Mitochondria and cristae can be seen. 14,000X. (From A.V. Somlyo *et al.*[5] Reprinted by permission from the *Journal of Cell Biology*.)

rays that have well defined energies, X rays in a broad energy band are also emitted under electron irradiation. This broad band, called continuum or bremsstrahlung radiation, is the result of inelastic electrons scattering by atomic nuclei. The number of continuum X rays generated is proportional to the mass of *all* the atoms in the microvolume irradiated. Therefore, in ultrathin specimens, the ratio of characteristic X rays to continuum X ray radiation is directly proportional to the concentration of the element giving rise to the characteristic radiation.[17-20] The use of this principle, together with appropriate standards for calibration, has been the most successful approach for quantitative biological EPMA, and has already produced a considerable amount of information, not accessible by other methods, about cellular and intracellular transport. The spatial resolution of EPMA at present is better than 10 nm, and the practical limit of sensitivity for detecting Ca is approximately 300 μmol/kg dry wt.[9] However, the attainment of this level of sensitivity requires long collection times and/or a large number of spectra, because of the inherent limits on peak-to-background ratios and geometric efficiency of detectors (even the present-day energy-dispersive X ray detectors can detect only about 1% of the X rays emitted from the specimen). The X ray counting process follows Poisson statistics and, therefore, improvement in sensitivity follows a square law, i.e., twofold improvement in sensitivity requires four times longer counting times (or higher currents). In addition, in muscle and in most other eukaryotic cells, Ca occurs in the presence of high concentrations of K, and the K $K\beta$ peak overlaps the Ca $K\alpha$ peak. The overlap of the two peaks can introduce significant errors in Ca quantitation, unless the calibration (centroid position and resolution) is scrupulously maintained (to within 1 eV) at the same values as were used for the collection of the reference standards. Such errors (ordinarily about 1–2 mmol/kg)[20] can be reduced to insignificant levels through the use of a multiple least squares fitting technique that includes the first and second derivative of the K X ray peaks.[21] Comparison of EPMA quantitation of thin film Ca-EGTA/albumin standards with the results of atomic absorption spectrophotometry on the same samples shows a linear correlation through the origin with r = 0.99. The Ca content in a thin film Ca standard (containing 500 mmol/kg K) was 1 ± 0.2 S.E.M. by EPMA, indistinguishable from the 0.9 ± 0.01 S.E.M. measured by atomic absorption spectrophotometry of the same sample.

A scanning electron beam can be used in conjunction with EPMA, instead of the more conventional fixed beam mode, to produce X ray maps reflecting the two-dimensional distribution of the elements that give rise to characteristic X rays.[22] In the simplest application of this method, an energy window that includes only the characteristic energy of the X rays generated by the element of interest is used to modulate the brightness of the oscilloscope screen. Since the spatial resolution of the method cannot be better than the probe size, X ray maps are best obtained with microscopes equipped with a field emission gun to provide sufficient current into small probes. Even with a field emission gun, the acquisition of such maps is time-consuming. Quantitation, as well as the acquisition of X ray maps of elements occurring at relatively low concentrations, can be achieved through longer collection times (dwell time/pixel) by computer-fitting the spectrum obtained from each picture point (pixel), in the same manner as would be done for a conventional, "spot mode" EPMA spectrum.[22] In spite of the limitation of the long scanning times, it is expected that this approach to X ray mapping will yield highly reliable results in determining the spatial distribution of Ca and other elements within subcellular regions such as the triad in skeletal muscle.

SUBCELLULAR COMPOSITION OF RESTING AND CONTRACTING MUSCLE: THE RESULTS

The composition of the terminal cisternae (TC) of the SR in cryosections of rapidly frozen, small bundles of frog semitendinosus muscles (FIGURE 1) was determined by focussing the beam to 50 nm over the TC, followed by the collection of a paired spectrum, using the same parameters, with the probe position over the adjacent I-band region. Comparison of resting muscles with paired muscles frozen at 1.2 seconds following the onset of tetanic stimulation showed that approximately 60–70% of the total fiber Ca was localized in the TC in resting frog muscle, and that 60% (69 mmol/kg dry TC) of this Ca was released during tetanus. This quantity is sufficient to raise the Ca concentration by nearly 1 mmol/l-fiber H_2O, or equivalent to the concentration of high-affinity Ca binding sites on troponin and on parvalbumin.[5,23] Ca release was associated with a significant uptake of Mg and K into the TC, but the amount of Ca released exceeded the total measured cation accumulation by 62 mEq/kg dry wt. The resultant "*apparent* charge deficit" is so large that, if real, it would lead to an unrealistic (13 V) electrical potential across the SR membrane. Therefore, we suggested that charge compensation is achieved by a movement of protons into the SR which is not detectable by energy-dispersive EPMA.[5] In view of the high cytoplasmic K^+ concentration and the high K^+ permeability of isolated SR fractions, it was surprising that K^+ was not the sole or even major cation providing charge neutralization for the Ca release. This relatively small K^+ movement argued against the presence of an *in vivo* K-conductance as high as attributed to the voltage-gated K channels of the SR membrane.[24,25] A submaximal K conductance during activation was confirmed by finding that valinomycin, a K ionophore, increased, by 50%, the amount of K movement into the TC and abolished the apparent charge deficit, without having a major effect on the amount of Ca released.[26] These results suggest that Ca^{2+} release from activated muscle is an electrogenic process and that the K^+ conductance of the SR in untreated frog muscles is insufficient to allow charge neutralization of the Ca^{2+} current during release solely by K^+-fluxes. That the K^+ permeability is lower *in situ*[27] than in fragmented SR,[24,28] has also been confirmed by K^+ flux studies of skinned muscle fibers.[29] The EPMA results also argue against an excitation-contraction coupling mechanism based on the direct invasion of the TC by ionic current from the T-tubules.[30]

The uptake of Mg into the SR remained highly significant even in the presence of valinomycin. Again, this is an example of EPMA providing new information about the Mg permeability of the *in situ* SR, not anticipated on the basis of studies of fragmented SR that suggested that the Mg permeability of the SR is low.[31] Mg uptake into the TC was also observed during caffeine or quinine contractures.[32,33] Quinine blocks Ca pumping by the Ca ATPase, and Ca^{2+} is released to the contractile regulatory proteins due to the uncompensated (by Ca pump) leak through the SR membrane. Therefore, this finding suggests that Mg uptake into the SR is more clearly linked to Ca release than to Ca pumping by the Ca ATPase, and probably occurs through the Ca^{2+} (SR) channel.

The Na and Cl content of the TC is similar to that of the adjacent cytoplasm, indicating that the SR is *not* in ionic communication with the extracellular space. These EPMA results were also confirmed by more recent ion flux studies on single fibers.[34] The unchanged distribution of Cl during tetanus argued against the existence of a large and sustained change in trans-SR potential, since Cl^- would be expected to be redistributed by such a potential change. Furthermore, valinomycin has no effect on the K content of the TC in resting muscles, further supporting the conclusion that there is neither a large electrical potential nor a chemical gradient of K across the SR membrane in resting muscle.

Ca was selectively localized to the TC of the SR, where the low-affinity Ca-binding protein, calsequestrin, is also localized.[35,36] The concentration of Ca in the longitudinal SR was low, supporting the conclusion, based on studies of fragmented reticulum,[37] that the *free* Ca^{2+} in the lumen of the SR is in the low millimolar range. During tetanus, the measured Ca released from the TC was reflected in the increased concentrations measured in the cytoplasm.

In another series of experiments, we determined the time course of reversal of the tetanus-induced changes by freezing muscles at, respectively, 0.6, 1.3, 2.2, and 4.6

seconds and five minutes after a 1.2-second tetanus.[27] At 600 milliseconds following a 1.2-second tetanus at room temperature, the force had relaxed to baseline, and 300 μmol of Ca^{2+}/l of cytoplasmic H_2O had been pumped by the SR (FIGURE 2), indicating that the in situ rate of Ca^{2+} pumping by the SR Ca ATPase is sufficiently high to account for the removal of Ca from the Ca^{2+}-specific sites of troponin (180 μmol of Ca^{2+}-specific sites/l of cytoplasmic H_2O) and for the rate of relaxation from a tetanus at room temperature. The halftime of the return of the total 1.0 mmol of Ca^{2+}/l cytoplasmic H_2O released during a tetanus was 1.1 seconds, comparable to the slow $K_{off\text{-}rate}$ of Ca^{2+} from (carp) parvalbumin ($1\ s^{-1}$) and consistent with the hypothesis that the return of this Ca^{2+} to the TC is rate-limited by the Ca^{2+} off-rate from parvalbumin (FIGURE 2). The somewhat lower Ca^{2+} off-rate in situ and in vitro could reflect a slight difference in the Ca^{2+} affinity of, respectively, fish and frog parvalbumins or a difference in the rate of removal of Ca^{2+} from parvalbumin in vitro and in situ. It is worth noting that during a tetanus, unlike a twitch, the parvalbumin sites would be fully saturated with Ca^{2+} and, therefore, would not contribute to the relaxation rate of the tetanus. The return of the Mg^{2+} taken up by the TC during a tetanus to resting level was significantly slower than the time course of the Ca^{2+} movements, suggesting that the Mg^{2+} permeability of the SR in situ is low and may be transiently increased during tetanic stimulation. This observation strengthens our suggestion that Mg^{2+} moves through the Ca^{2+} channels of the SR membrane.

MITOCHONDRIAL Ca^{2+} IN SITU

The Ca content of mitochondria in situ is another biological problem that is uniquely and directly approachable by EPMA.[14] Considerable controversy has existed over the years as to whether mitochondria contain sufficient Ca to effectively regulate cytoplasmic free Ca^{2+} via the mitochondrial Ca efflux pathway[38] or whether mitochondial Ca is very low and compatible with modulation of Ca-sensitive mitochondrial enzymes through small [micromolar] fluctuations in mitochondrial matrix free [Ca^{2+}]. The source of this controversy, to a large extent, was the wide range of Ca concentrations measured in isolated mitochondria and the uncertainty of the effects of isolation on Ca content. In view of the importance of the question whether hepatic mitochondrial matrix Ca^{2+} regulates metabolism, much of the controversy centered on the Ca^{2+} content of mitochondria in the liver. In order to avoid the trauma and anoxia associated with removal of the liver and subsequent freezing, we developed a solid freon popsicle clamp which was used to snap-freeze a lobe of liver partially protruding from a small incision in the abdominal wall of an anesthetized rat suspended over an environmental chamber held at 37° with high humidity.[14] Cryosections revealed minimal ice crystal damage, and intracellular organelles including stacks of rough endoplasmic reticulum could be easily identified within the hepatocytes (FIGURE 3). Analysis of the red blood cells within the hepatic sinusoids showed that their Ca content was below the limit of detectability, 0.1 ± 0.2 mmol/kg dry wt ($\pm$ S.E.M.; n = 34), consistent with the μmolar total Ca content of red blood cells,[39] and provided an internal estimate of the very small errors of the EPMA measurements.

The concentrations of elements in the mitochondria and stacks of rough endoplasmic reticulum as well as the nuclei and total cell are shown in TABLE 1. The mitochondrial Ca concentration was 0.8 ± 0.1 ($\pm$ S.E.M.) mmol/kg mitochondria (n = 97; 29 cells, 4 rats: 2 fasted, 2 fed), equivalent to 0.8 nmol/mg mitochondrial protein measured by the biuret method with albumin standards.[40] The ratio of free

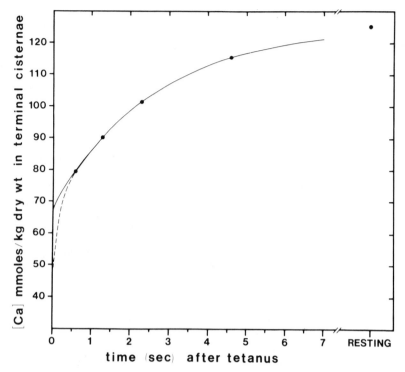

FIGURE 2. The reuptake of Ca^{2+} by the sarcoplasmic reticulum during the first seconds of the recovery period following a tetanus appears to follow an exponential time course. Using weighted regression analysis and standard least squares criteria, the early points can be fit by the single exponential equation:

$$Y = 57.5 (1 - e - 0.38t) + 68.1$$

where Y = concentration of Ca in the terminal cisternae in mmol/kg, dry wt, and t = time in seconds. The curve described by this equation is illustrated by the *solid line.* This curve fits the points very closely with the regression accounting for greater than 99.99% of the observed variation in the data. The intercept when t = 0 is 68.1 mmol of Ca/kg, dry wt, which is greater than previous measurements[5] of 48 mmol of Ca/kg, dry wt, made at 1.2 seconds from the onset of tetanus. The *dashed line* represents the fit when this tetanus value is included and is described by the equation

$$Y = 76.4 (1 - 0.25 e - 7.0t - 0.75 e - 0.38t) + 8.0.$$

(From A.V. Somlyo *et al.*[27] Reprinted by permission from the *Journal of Biological Chemistry.*)

matrix Ca^{2+} to total mitochondrial Ca, in mitochondria isolated from liver, has been estimated as 7×10^{-4} M.[41] Therefore, based on 64% mitochondrial H_2O, the free Ca^{2+} concentration in the mitochondrial matrix of rat liver is about 0.4 μM (95% confidence limit 0.3–0.5 μmol), well within the range (0–2 μM) in which mitochondrial dehydrogenases are regulated.[38,42] Note that if the total mitochondrial Ca content was significantly greater than 2–3 mmol/kg dry wt, the Ca^{2+} regulatory sites on the

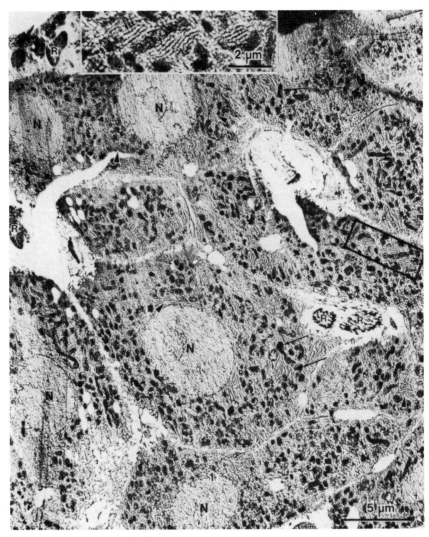

FIGURE 3. Electron micrograph of an ultrathin, freeze-dried cryosection obtained from rat liver rapidly clamp-frozen *in situ* with frozen Freon 22. Nuclei (*N*) and numerous mitochondria (*arrows*) are indicated, as well as stacks of rough ER (*arrowheads*). The *boxed region* is enlarged (*inset*) to show the stacks of rough ER; some regions show periodic bumps (*small arrows*) along the membrane which are thought to be ribosomes. The EPMA measurements of the Ca content of ER were obtained from such stacked tubules. Red blood cells (*R*) are also evident in the liver sinusoids. Rats were anesthetized intraperitoneally with Nembutal (50 mg/kg), an incision made in the abdominal wall over the liver and the rat placed on its abdomen in a cloth hammock above an environmental chamber maintained at 37° C with high humidity. The liver was exposed gently, without touching its surface, through a slit in the cloth hammock and suspended in the atmosphere of the environmental chamber. The freezing clamp with two opposing aluminum cups filled with solidified Freon 22 was removed from the nitrogen, and when the surface of the solid Freon had melted the door of the environmental chamber was opened and the lobe of the liver clamped rapidly. The temperature of the melted Freon surface is at the melting point of Freon 22 and moulds to the irregular tissue surface, promoting a good contact between the coolant and the tissue. The liver was sectioned at −130° C in a cryoultramicrotome, then freeze-dried. (From A.P. Somlyo *et al.*[14] Reprinted by permission from *Nature*.)

TABLE 1. Subcellular Distribution of Calcium and Other Elements in Rat Liver Parenchymal Cells (Four Animals) (mmol/kg dry wt)[a]

	No. of Spectra	Na	Mg	P	S	Cl	K	Ca	% Cell Volume	% Total Cell Ca
Mitochondria	97	54 ± 3	43 ± 1	360 ± 7	316 ± 4	75 ± 3	314 ± 5	0.8 ± 0.1	20	5
RER	24	80 ± 8	76 ± 2	920 ± 21	224 ± 8	94 ± 5	579 ± 17	5.0 ± 0.4	9.4	14–23[b]
Nuclei	21	77 ± 6	53 ± 2	557 ± 16	245 ± 8	117 ± 6	625 ± 22	0.8 ± 0.4	6	1
Other regions[c]	29	75 ± 6	53 ± 2	653 ± 20	241 ± 9	103 ± 5	465 ± 15	4.0 ± 0.4		
Total cell	20	53 ± 4	46 ± 2	564 ± 12	230 ± 9	106 ± 4	479 ± 20	3.4 ± 0.4		

[a]The instrumentation and methods used for EPMA, including calibration, have been published elsewhere. To obtain the precision required for these measurements, individual spectra were collected for 400–600 seconds, until the measurement error of a single spectrum was 1.1 mmol Ca/kg dry wt. Spectra were collected from mitochondria with 75–100-nm diameter probes; from tubules of ER with 100-nm or with 1,000-nm diameter probes over stacks of ER tubules that included the intervening cytoplasm and, therefore, give an underestimate of the Ca concentrations in the ER; from nuclei using probes of diameter 3–10 μm and for total cell measurements, using defocused probes covering whole cell profiles. Values are the mean ± S.E.M. (From A.P. Somlyo et al.[14] Reprinted by permission from *Nature*.)

[b]The lower value is based on the Ca concentration ([Ca]) of the stacks of rough ER, the higher value on the [Ca] measured with 10–20-nm diameter probes over the ER lumen.

[c]Regions in the cell exclusive of nuclei, mitochondria, and stacks of rough ER, which, based on known liver structure, would include the smooth ER tubules that were not imaged in the cryosections.

enzymes would be saturated with Ca^{2+}. The mitochondrial Ca content measured in these experiments is too low for significant activation of the mitochondrial Ca efflux pathway (apparent K_m = 9.7 μM in isolated liver mitochondria), which is required for the effective regulation of cytoplasmic free Ca^{2+}.

The concentration of Ca in the stacks of rough endoplasmic reticulum was 5.0 ± 0.4 mmol/kg dry wt, which is very much higher than that measured in the mitochondria or nuclei. As the analyses were done by focusing the beam over stacks of ER tubules rather than within the lumen of individual tubules, this Ca concentration is probably an underestimate by at least a factor of 2. Therefore, the true Ca content of the rough ER is about 8–10 mmol/kg dry wt, which is comparable to the steady state Ca content of 10 nmol Ca/mg protein measured in microsomes equilibrated with 70 nM free Ca^{2+}.[43] The high Ca content of the ER in liver and in retinal rods[12] is consistent with the view that the ER is the main organelle regulating the cytoplasmic Ca^{2+} concentration in nonmuscle cells and that this is the source of Ca released by inositol triphosphate.[44-48]

CONCLUSION

The *in situ* composition of organelles and of their surrounding cytoplasmic matrix can be directly measured with EPMA at high spatial resolution and sensitivity, in cells frozen in resting or activated states. Utilizing this technology, studies of excitation-contraction coupling in skeletal muscle and of Ca distribution in liver have led to the following conclusions:

- The sarcoplasmic reticulum is *not* in ionic communication with the extracellular space, either at rest or during tetanus.
- The electrical potential across the SR is near zero at rest, and undergoes little or no prolonged change during tetanus.
- Ca release is *not* electrically silent, but is accompanied by the (partial) electrophoretic exchange of K^+ + Mg^{2+} for Ca^{2+}; proton movement may compensate for the remaining *apparent* charge deficit.
- The tetanic uptake of Mg^{2+} into the SR is through passive channels, not by active (Ca ATPase) transport. These passive channels may be the Ca^{2+} release channels.
- The Ca^{2+} released (0.6–0.8 mmol/l cell H_2O) during tetanus is in the cytoplasm, presumably bound to troponin (0.18 mmol Ca^{2+} specific sites/l cell H_2O) and to parvalbumin (0.7 mmol sites/l cell H_2O).
- The posttetanic return of Ca^{2+} to the SR consists of two processes, one fast and one slow. The fast process returns the amount of Ca^{2+} that can be bound to the regulatory sites on troponin, and its rate indicates that the SR Ca^{2+} pump can account for relaxation; the rate of the slower process is comparable to the off-rate of Ca^{2+} from parvalbumin.
- Quantitation of the subcellular distribution of *total* Ca in liver cells frozen in a physiological condition *in vivo* has resolved the long-standing controversy regarding the role of mitochondrial Ca, and supports the view that mitochondria do not play a major role in the physiological regulation of cytoplasmic Ca^{2+}, but that mitochondrial enzymes may be regulated by matrix free Ca^{2+}. The high Ca^{2+} content of the endoplasmic reticulum is consistent with a major Ca^{2+}-regulatory function of this organelle.

REFERENCES

1. SOMLYO, A. V., H. SHUMAN & A. P. SOMLYO. 1977. Elemental distribution in striated muscle and effects of hypertonicity: electron probe analysis of cryosections. J. Cell Biol. **74:** 828–857.
2. SOMLYO, A. V. & J. SILCOX. 1979. Cryoultramicrotomy for electron probe analysis. *In* Microbeam Analysis in Biology. C. Lechene & R. Warner, Eds. 535–555. Academic Press. New York, NY.
3. KARP, R., J. C. SILCOX & A. V. SOMLYO. 1982. Evidence against melting and the use of a low temperature cement for specimen orientation. J. Microsc. (Oxford) **125:** 157–165.
4. SOMLYO, A. V., H. SHUMAN & A. P. SOMLYO. 1977. Composition of the sarcoplasmic reticulum *in situ* by electron probe X-ray microanalysis of cryosections. Nature (London) **258:** 556–558.
5. SOMLYO, A. V., H. GONZALEZ-SERRATOS, H. SHUMAN, G. MCCLELLAN & A. P. SOMLYO. 1981. Calcium release and ionic changes in the sarcoplasmic reticulum of tetanized muscle: an electron probe study. J. Cell Biol. **90:** 577–594.
6. CHIESI, M., M. M. HO, G. INESI, A. V. SOMLYO & A. P. SOMLYO. 1981. Primary role of sarcoplasmic reticulum in phasic contractile activation of cardiac myocytes with shunted myolemma. J. Cell Biol. **91:** 728–742.
7. SOMLYO, A. P., A. V. SOMLYO & H. SHUMAN. 1979. Electron probe analysis of vascular smooth muscle: composition of mitochondria, nuclei and cytoplasm. J. Cell Biol. **81:** 316–335.
8. BOND, M., T. KITAZAWA, A. P. SOMLYO & A. V. SOMLYO. 1984. Release and recycling of calcium by the sarcoplasmic reticulum in guinea pig portal vein smooth muscle. J. Physiol. (London) **355:** 677–695.
9. BOND, M., H. SHUMAN, A. P. SOMLYO & A. V. SOMLYO. 1984. Total cytoplasmic calcium in relaxed and maximally contracted rabbit portal vein smooth muscle. J. Physiol. (London) **357:** 185–201.
10. GONZALEZ-SERRATOS, H., A. V. SOMLYO, G. MCCLELLAN, H. SHUMAN, L. M. BORRERO & A. P. SOMLYO. 1978. The composition of vacuoles and sarcoplasmic reticulum in fatigued muscle: electron probe analysis. Proc. Natl. Acad. Sci. USA **75:** 1329–1333.
11. KITAZAWA, T., H. SHUMAN & A. P. SOMLYO. 1982. Calcium and magnesium binding to thin and thick filaments in skinned muscle fibres: electron probe analysis. J. Muscle Res. Cell Motil. **3:** 437–454.
12. SOMLYO, A. P. & B. WALZ. 1985. Elemental distribution in *Rana pipiens* retinal rods: quantitative electron probe analysis. J. Physiol. (London) **83:** 183–195.
13. LEW, V. L., A. HOCKADAY, M.-I. SEPULVEDA, A. P. SOMLYO, A. V. SOMLYO, O. E. ORTIZ & R. M. BOOKCHIN. 1985. Compartmentalization of sickle-cell calcium in endocytic inside-out vesicles. Nature **315:** 586–589.
14. SOMLYO, A. P., M. BOND & A. V. SOMLYO. 1985. The calcium content of mitochondria and endoplasmic reticulum in liver rapidly frozen *in situ*. Nature **314:** 622–625.
15. SOMLYO, A. P., R. URBANICS, G. VADASZ, A. G. B. KOVACH & A. V. SOMLYO. 1985. Mitochondrial calcium and cellular electrolytes in brain cortex frozen in situ: Electron probe analysis. Biochem. Biophys. Res. Commun. **132:** 1071–1078.
16. JONES, G. J. 1983. On estimating freezing times during tissue rapid freezing. J. Microsc. **136:** 349–360.
17. HALL, T. A. 1971. The microprobe assay of chemical elements. *In* Physical Techniques in Biological Research. G. Oster, Ed.: Vol. 1A: 157–275. Academic Press. New York, NY.
18. HALL, T. & B. L. GUPTA. 1983. The localization and assay of chemical elements by microprobe methods. Q. Rev. Biophys. **16:** 279–339.
19. SHUMAN, H., A. V. SOMLYO & A. P. SOMLYO. 1976. Quantitative electron probe microanalysis of biological thin sections: methods and validity. Ultramicroscopy **1:** 317–339.
20. SHUMAN, H., A. V. SOMLYO & A. P. SOMLYO. 1977. Theoretical and practical limits of ED X-ray analysis of biological thin sections. *In* Scanning Electron Microscopy. O. Johari, Ed. Vol. 1: 663–672. ITT Research Institute. Chicago, IL.

21. KITAZAWA, T., H. SHUMAN & A. P. SOMLYO. 1983. Quantitative electron probe analysis: problems and solutions. Ultramicroscopy **11**: 251–262.
22. SOMLYO, A. P. 1985. Compositional mapping in biology: X-rays and electrons. J. Ultrastruct. Res. **83**: 135–142.
23. BAYLOR, S. M., W. K. CHANDLER & M. W. MARSHALL. 1983. Sarcoplasmic reticulum calcium release in frog skeletal muscle fibres estimated from arsenazo III calcium transients. J. Physiol. (London) **344**: 625–666.
24. MILLER, C. 1978. Voltage-gated cation conductance channel from fragmented sarcoplasmic reticulum: steady-state electrical properties. J. Membr. Biol. **40**: 1–23.
25. HASSELBACH, W. & H. OETLIKER. 1983. Energetics and electrogenicity of the sarcoplasmic reticulum calcium pump. Annu. Rev. Physiol. **45**: 325–339.
26. KITAZAWA, T., A. P. SOMLYO & A. V. SOMLYO. 1984. The effects of valinomycin on ion movements across the sarcoplasmic reticulum in frog muscle. J. Physiol. (London) **350**: 253–268.
27. SOMLYO, A. V., G. MCCLELLAN, H. GONZALEZ-SERRATOS & A. P. SOMLYO. 1985. Electron probe X-ray microanalysis of post tetanic Ca and Mg movements across the sarcoplasmic reticulum in situ. J. Biol. Chem. **260**: 6801–6807.
28. GARCIA, A. M. & C. MILLER. 1984. Channel-mediated monovalent cation fluxes in isolated sarcoplasmic reticulum vesicles. J. Gen. Physiol. **83**: 819–839.
29. BEST, P. M. & C. W. ABRAMCHECK. 1985. Potassium efflux from single skinned skeletal muscle fibers. Biophys. J. **48**: 907–913.
30. MATHIAS, R. T., R. A. LEVIS & R. S. EISENBERG. 1980. Electrical models of excitation-contraction coupling and charge movement in skeletal muscle. J. Gen. Physiol. **76**: 1–31.
31. NAGASAKI, K. & M. KASAI. 1979. Magnesium permeability of sarcoplasmic reticulum vesicles monitored in terms of chlortetracycline fluorescence. J. Biochem. (Tokyo) **87**: 709–716.
32. YOSHIOKA, T. & A. P. SOMLYO. 1984. The calcium and magnesium contents and volume of the terminal cisternae in caffeine-treated skeletal muscle. J. Cell Biol. **99**: 558–568.
33. YOSHIOKA, T. & A. P. SOMLYO. 1984. Changes in Ca and Mg during quinine contracture in the sarcoplasmic reticulum (SR) of frog muscle. Proc. 3rd Int. Congr. Cell Biol. Tokyo: 514a.
34. NEVILLE, M. C. 1979. The extracellular compartments of frog skeletal muscle. J. Physiol. (London) **288**: 45–70.
35. JORGENSEN, A. O., V. KALNINS & D. H. MACLENNAN. 1979. Localization of sarcoplasmic reticulum proteins in rat skeletal muscle by immunofluorescence. J. Cell Biol. **80**: 372–384.
36. JORGENSEN, A. O., A. C.-Y. SHEN, K. P. CAMPBELL & D. H. MACLENNAN. 1984. Ultrastructural localization of calsequestrin in rat skeletal muscle by immunoferritin labeling of ultrathin frozen sections. J. Cell Biol. **97**: 1573–1581.
37. WEBER, A. 1971. Regulatory mechanisms of the calcium transport system of fragmented rabbit sarcoplasmic reticulum. I. The effect of accumulated calcium on transport and adenosine triphosphate hydrolysis. J. Gen. Physiol. **57**: 64–70.
38. HANSFORD, R. G. 1985. Relation between mitochondrial calcium transport and control of energy metabolism. Rev. Physiol. Biochem. Pharmacol. **102**: 1–72.
39. BOOKCHIN, R. M. & V. L. LEW. 1980. Progressive inhibition of the Ca pump and Ca: Ca exchange in sickle red cell. Nature **284**: 561–563.
40. BEAVIS, A. D., R. D. BRANNAN & K. D. GARLID. 1985. Swelling and contraction of the mitochondrial matrix. I. A structural interpretation of the relationship between light scattering and matrix volume. J. Biol. Chem. **260**: 13424–13433.
41. COLL, K. E., S. K. JOSEPH, B. E. CORKEY & J. R. WILLIAMSON. 1982. Determination of the matrix free Ca^{2+} concentration and kinetics of Ca^{2+} efflux in liver and heart mitochondria. J. Biol. Chem. **257**: 8696–8704.
42. DENTON, R. M. & J. G. MCCORMACK. 1980. On the role of the calcium transport cycle in heart and other mammalian mitochondria. FEBS Lett. **119**: 1–8.
43. DAWSON, A. P. 1982. Kinetic properties of the Ca^{2+}-accumulation system of a rat liver microsomal fraction. Biochem. J. **206**: 73–79.
44. BERRIDGE, M. J. 1983. Rapid accumulation of inositol triphosphate reveals that agonists

hydrolyse polyphosphoinositides instead of phosphatidylinositol. Biochem. J. **212:** 849–858.

45. BURGESS, G. M., P. P. GODFREY, J. S. MCKINNEY, M. J. BERRIDGE, R. F. IRVINE & J. W. PUTNEY. 1984. The second messenger linking receptor activation to internal Ca release in liver. Nature **309:** 6366.

46. DAWSON, A. P. & R. F. IRVINE. 1984. Inositol(1,4,5)-trisphosphate-promoted Ca^{2+} release from microsomal fractions of rat liver. Biochem. Biophys. Res. Commun. **120:** 858–864.

47. JOSEPH, S. K., A. P. THOMAS, R. J. WILLIAMS, R. F. IRVINE & J. R. WILLIAMSON. 1984. MYO-inositol-1,4,5-trisphosphate: A second messenger for the hormonal mobilization of intracellular Ca^{2+} in liver. J. Biol. Chem. **259:** 3077–3081.

48. SOMLYO, A. P. 1984. Cellular site of calcium regulation. Nature **308:** 516–517.

DISCUSSION OF THE PAPER

C. LECHENE (*Harvard Medical School, Boston, Massachusetts*): I'm not sure I understand your conclusion of TABLE 1. You have 20% calcium in the mitochondria, which means that 80% is in the cytoplasm or in the endoplasmic reticulum. Do you say that the action of hormone on the calcium cycle is due to movement at the ER level and not at the plasma membrane close to the cytoplasm?

A. V. SOMLYO: There have been several studies in the literature showing that in permeabilized liver, calcium can be released with inositol trisphosphate. This occurs in the presence of mitochondrial blockers, suggesting that the reticulum is the source. TABLE 1 shows that mitochondria occupy 20% of cell volume and contain only 5% of total cell Ca. Based on a rough endoplasmic reticulum volume of 10% and the measured Ca content, this reticulum contains approximately 20% of the total cell Ca.

In addition, we know from conventional electronmicrographs that other regions throughout the cell that do not contain mitochondria or stacks of rough ER, contain a fairly extensive smooth reticulum, and this probably also represents a calcium sink. Nuclear calcium is low, as in other cell types that we have measured. A significant proportion of other calcium is probably in the smooth endoplasmic reticulum, and the proportion of release from, respectively, the smooth and the rough reticulum has not been resolved. Meredith Bond, with G. Vadasz, in our lab find that vasopressin injection reduces the calcium content of rough endoplasmic reticulum in the intact liver.

A. P. SOMLYO: In the, by now, rather extensive literature on inositol 1,4-5 trisphosphate there is no evidence suggesting that it can release calcium from any site other than the endoplasmic reticulum. A few, earlier reports that claimed to have shown release from mitochondria have now been withdrawn, because the authors have recognized that these effects were due to endoplasmic reticulum that was coisolated with the mitochondrial fractions.

Analytical Electron Microscopy in the Study of Biological Systems

DALE E. JOHNSON

Center for Bioengineering
University of Washington
Seattle, Washington 98195

INTRODUCTION

The use of Analytical Electron Microscopy (AEM) in the study of biological systems typically involves energy dispersive X ray analysis (EDS) of freeze-dried cryosections in a conventional transmission electron microscope (CTEM) or a scanning transmission electron microscope (STEM). In fact, the AEM is unique in being able to answer questions regarding elemental distributions at the subcellular level required. Current technology (including specimen preparation) permits elemental concentration determinations down to about 10^{-5} by weight, in regions as small as tens of nanometers in diameter. This technique is already being successfully applied in a number of basic physiological studies in several laboratories (most notably the Somlyos'—see REFERENCES 1, 2, and 6), and has potential application in a wide variety of areas. Since the movement and segregation of ions play such a crucial role in the function of biological systems, it seems clear that the AEM can contribute to the understanding of biological function in a comparable way to the role of the transmission electron microscope (TEM) in the study of biological ultrastructure.

The purpose of this presentation will be to discuss briefly: 1) the limitations to the sensitivity of the technique, and 2) developments in instrumentation relevant to these limitations.

ANALYTICAL TECHNIQUE (EDS)

As successful as the use of EDS has been in the subcellular microanalysis of biological systems, two fundamental limitations to the sensitivity of the technique are obvious. First, the combination of low collection efficiency (about 0.3% for Na) and low fluorescence yield for low atomic number elements (about 2.5% for Na) means that in the analysis of Na, for example, about 10^4 Na atoms must be ionized for each detected X ray. Second, although the emitted characteristic X rays have natural energy widths of a few electron volts, these peaks are broadened to $>10^2$ eV by the EDS detection system, resulting in decreased peak-to-background ratios and, in the detection of Ca, a serious overlap of the Ca, $K\alpha$ peak with the K, $K\beta$ peak.

Because of these limitations of EDS, considerable effort has been invested in the application of electron energy-loss spectrometry (ELS) to biological microanalysis. ELS typically should have collection efficiencies and energy resolution 1 to 2 orders of magnitude higher than EDS, and this is true experimentally. There are two other aspects of ELS, however, that serve to limit its general applicability to biological microanalysis. First, the background intensity due to collision processes in ELS is much higher (by about 2 orders of magnitude) than the background due to radiative processes (bremsstrahlung) in EDS. Second, both the ELS peak and background intensities are very dependent on specimen thickness, principally since any collision process other than the inelastic excitation of interest will reduce the energy-loss peak of interest. This limits usable specimen thickness to $\simeq$ the mean free path for all collision processes (about 0.1 μm) but equally important, means that small variations in projected mass thickness will produce significant variations in a given energy-loss signal.

In its most efficient form (*i.e.* with parallel detection)[3,4] and with careful analysis, it is clear that ELS can result in higher sensitivity (see Shuman *et al.*, these proceedings).[7] What is less clear, however, is the role ELS will play in the routine analysis of multiple elements (Na-Ca) in trace concentrations (1% to .001% by weight) in specimens with large spatial variations in mass thickness (10%–50%).

The effect of specimen mass thickness variations is much less in EDS analysis, since the main effect is through variations in X ray absorption with a typical mean free path of about 1 μm in carbonlike materials.

ELECTRON OPTICS

Given the basic sensitivity of the analytical technique and a certain elemental concentration to be detected, the spatial resolution obtainable on sufficiently thin specimens will be determined by the minimum diameter beam into which the required number of electrons can be focussed. This in turn will be determined by the electron source and electron optics used. In a hot-filament CTEM/STEM this is typically about 500 Å and could be reduced to at least 50 Å in a field emission STEM. The field emission STEM also provides a clean, ultrahigh vacuum specimen environment reducing any effects of contamination or etching and also allowing analysis at low temperatures without effects of residual gas condensation onto the specimen.

The advantages of intermediate high voltages (200–400 KeV, either with or without field emission sources) for biological microanalysis are not as clear. The increased voltage certainly reduces any effect of beam spread in a specimen of given thickness, and also reduces multiple scattering limitations in energy-loss spectrometry. However, the increased voltage also allows atomic displacement damage for the elements of interest Na-Ca; a damage mechanism possible at 100 KeV only for C, N, and O. The extent and other characteristics of such damage in biological specimens has not been experimentally determined, although the cross sections for such damage are comparable to the small cross sections for ionization *followed by* fluorescence and detection of characteristic X rays.

SPECIMEN PREPARATION

Many approaches to the cryofixation of biological tissue have been used successfully, ranging from the simple (hand plunging into cryogenic fluids) to the sophisticated ("slamming" into liquid helium cooled copper blocks). The limitations imposed in this crucial step relate mainly to the depth and reproducibility of adequate cryofixation: *i.e.* structural changes and elemental movements due to ice formation small compared to the spatial resolution required in the microanalysis. While the more advanced techniques may provide somewhat greater depth of freeze, most techniques now in use can provide a region of cryofixation typically frozen to a better spatial resolution than can be analyzed. The thin cryosectioning necessary for high spatial resolution is also now routinely accomplished, followed by freeze-drying within a separate vacuum chamber or within the specimen chamber of the electron microscope. Some questions have recently been raised regarding the effects of any rehydration of these freeze-dried sections during transfer between an external freeze-drier and the microscope. Based on the extensive microanalytical results already obtained with external freeze-drying, the limitations imposed by any such rehydration appear to range from minor to negligible.

BEAM-INDUCED SPECIMEN CHANGES

Depending on the magnitude, such changes can clearly result in limitations to biological AEM. The effects can be grouped in two categories: instrumentation-related and fundamental ionization effects. While practical instrumentation-related types of beam-induced changes, such as contamination and etching, can in principle be completely eliminated, fundamental, ionization-related changes generally can only be slowed through specimen cooling. These include partial loss of some specimen elements (typically C, N, and O) and corresponding ultrastructural changes.[5]

Because of the nature of the vacuum systems of many analytical electron microscopes, and such additional factors as the temperature of the anticontaminator relative to the specimen, it is likely that in many applications of AEM, significant beam-induced changes (*e.g.* etching) take place in addition to fundamental ionization damage.

At spatial resolutions currently attainable ($\simeq 50$ nm) ionization-related structural changes may not be significant, but with the higher resolutions possible with field emission STEMs, such damage may ultimately limit spatial resolution and should be carefully assessed for a given type of specimen using minimum dose imaging. Fortunately elemental changes associated with ionization damage appear to be limited to matrix elements (C, N, and O) and can be measured and corrected for. The one exception is S, for which the loss at room temperature can be major (about 80%), but also for which specimen cooling reduces not only the rate but the total amount lost.

SUMMARY

The AEM is a powerful tool in biological research, capable of providing information simply not available by other means. The use of a field emission STEM for this application can lead to a significant improvement in spatial resolution in most cases now allowed by the quality of the specimen preparation but perhaps ultimately limited by the effects of radiation damage. Increased elemental sensitivity is at least possible in selected cases with electron energy-loss spectrometry, but fundamental aspects of ELS will probably confine its role to that of a limited complement to EDS.

The considerable margin for improvement in sensitivity of the basic analytical technique means that the search for technological improvement will continue. Fortunately, however, current technology can also continue to answer important biological questions.

REFERENCES

1. SOMLYO, A. V., H. GONZALEZ-SERRATOS, H. SHUMAN, G. McCLELLAN & A. P. SOMLYO. 1981. Calcium release and ionic changes in the sarcoplasmic reticulum of tetanized muscle: An electron-probe study. J. Cell Biol. **90:** 577–594.

2. JOHNSON, D. E. & M. CANTINO. High resolution analytical electron microscopy of biological systems. *In* Advanced Techniques in Biological Research. J. K. Koehler, Ed. Vol. III. Springer-Verlag. New York, NY, in press.
3. SHUMAN, H. 1981. Parallel recording of electron loss spectra. Ultramicroscopy **6:** 163–168.
4. JOHNSON, D. E., K. L. MONSON, S. CSILLAG & E. A. STERN. 1981. An approach to parallel-detection electron energy loss spectrometry. *In* Analytical Electron Microscopy. R. H. Geiss, Ed. 205–209. San Francisco Press. San Francisco, CA.
5. CANTINO, M., M. GODDARD, L. WILKINSON & D. JOHNSON. Mass loss in the microanalysis of thin cryosections. J. Microsc., in press.
6. SOMLYO, A. V., M. BOND, H. SHUMAN & A. P. SOMLYO. 1986. Electron-probe X ray microanalysis of *in situ* calcium and other ion movements in muscle and liver. This volume.
7. SHUMAN, H., C.-F. CHANG, E. L. BUHLE, JR. & A. P. SOMLYO. 1986. Electron energy-loss spectroscopy: quantitation and imaging. This volume.

DISCUSSION OF THE PAPER

A. P. SOMLYO: (*University of Pennsylvania School of Medicine, Philadelphia, Pennsylvania*): Would you agree that even in terms of radiation damage, one could benefit greatly from improving the microscope vacuum?

JOHNSON: Definitely.

J. B. PAWLEY (*University of Wisconsin, Madison, Wisconsin*): In keeping with the multidisciplinary approach of this meeting, for measuring calcium, how would you compare the electron probe techniques with things like Fura, the fluorescent dye that changes its excitation spectrum in the presence or absence of calcium and can be used in live cells to measure calcium in the physiological range?

JOHNSON: I would say these methods are complementary in giving different kinds of information. Dr. Somlyo may also like to comment on this.

SOMLYO: I agree with Dr. Johnson. We have used Fura and Quin. These dyes measure free calcium in the range of say 50 nM to 5 μM or so. They do not allow you to measure total calcium, or calcium bound at regulatory sites that is functionally very important.

Dyes measure free calcium at relatively low spatial resolution; electron probe analysis measures total calcium at high resolution. The techniques are complementary.

Electron-Probe X Ray Microanalysis of Transepithelial Ion Transport[a]

ROGER RICK,[b] ADOLF DÖRGE,[c] FRANZ X. BECK,[c] AND
KLAUS THURAU[c]

[b]Department of Physiology and Biophysics
Nephrology Research and Training Center
University of Alabama at Birmingham
Birmingham, Alabama 35294
and
[c]Department of Physiology
University of Munich
D-8000 Munich 2, Federal Republic of Germany

INTRODUCTION

Active transepithelial ion transport can be envisioned as a two-barrier process, uptake and extrusion of the ion across the two limiting cell membranes of the transporting epithelial cell. In such a two-barrier or three-compartment model the intracellular space of the epithelial cell can be considered as a transport compartment. Intracellular electrical potential and intracellular ion concentrations in this space define the electrochemical gradients across each of the transport barriers, and, vice versa, the ion fluxes across the two limiting cell membranes determine the membrane potential, intracellular ion concentrations, as well as the volume of an epithelial cell.

It is, therefore, immediately evident that the knowledge of the intracellular ion concentrations in the transport compartment is essential for elucidation of the individual transport mechanisms. Unfortunately, chemical or radiochemical analyses, which often have been employed to measure the intracellular ion concentrations in epithelia, generally provide inconsistent and conflicting data. In part, this can be attributed to the fact that these techniques measure only mean values for all epithelial cells, comprising cells with possibly different transport functions. Also, it has been shown that chemical analyses, at least in the intact skin, yield too high intracellular Na concentrations and too low K concentrations, mainly because of an underestimation of the extracellular space.[1] Some of the difficulties involved in accurately estimating the extracellular space can be overcome by using isolated epithelia in which the size of the extracellular space is drastically reduced.

During recent years two new methods have been increasingly applied for measuring intracellular ion concentrations in transporting epithelia: ion selective microelectrodes and electron microprobe analysis. While impalement with ion selective electrodes provides intracellular ion activities, electron microprobe analysis measures the total chemical concentration of an ion, irrespective of binding or activity state. The two methods have different limitations, advantages as well as disadvantages and, therefore, should be considered as complementary, rather than alternative techniques. Ion selective electrodes allow the recording of temporal changes in the ion concentration, whereas electron microprobe analysis yields concentration "frozen" at only one

[a]Supported by grants from the Deutsche Forschungsgemeinschaft.

point in time. On the other hand, when performed on sufficiently thin sections in a scanning transmission or conventional transmission electron microscope, electron microprobe analysis provides a much better spatial resolution, even down to subcellular structures. This is of particular importance in histologically complex epithelia that are composed of different cell layers or cell types with possibly different transport functions.

The present paper illustrates the feasibility of electron microprobe analysis for transepithelial transport studies. The scope will be limited to data obtained from the Na-transporting frog skin and the Cl-secreting frog cornea. In addition, our group has applied the same technique to a variety of other epithelial tissues including the toad skin,[2] toad urinary bladder,[3] rat kidney cortex,[4] and rat renal papilla.[5] Further studies have been performed using various pathophysiological models.[6,7] While no attempt will be made to review the available literature in this field, some important publications by others should be noted here.[8–13]

METHODS

Energy dispersive X ray microanalysis was performed on freeze-dried cryosections in a scanning electron microscope (Cambridge S150) to which an Si-Li solid state X ray detecting system (LINK) was attached. The cryosections were obtained from specimens snap frozen in a liquid propane/isopentane mixture at $-196°$ C. 1-μm thick sections were cut at $-90°$ C in a modified Reichert cryoultramicrotome (OmU 3) and freeze-dried at $-80°$ C and 10^{-6} Torr in a custom made freeze-dryer based on a turbomolecular pumping unit (Balzers). The measuring conditions were 20 kV acceleration voltage and between 0.2 and 0.5 nA beam current (determined in a Faraday cup). During analysis small areas of between 0.1 and 5 μm^2 of the section were scanned for 100 sec. The emitted X rays were analysed in the energy range from 0 to 20 keV comprising the K-alpha lines of the biologically relevant light elements Na, Mg, P, S, Cl, K, and Ca. Quantification of the cellular element concentrations and cellular dry weight was achieved by a comparison with an internal standard. The standard was obtained by covering the specimen immediately prior to freezing with a thin layer of an albumin solution. The albumin standard solution contained an extracellular composition of electrolytes and 20g/100g solution bovine serum albumin (Behringwerke, Marburg). A detailed description of the methods was given previously.[14–16]

Frog Skin

The experiments were performed on isolated abdominal skins of *Rana temporaria* and *Rana esculenta* (Stein, Lauingen). Several pieces (up to four) obtained from one skin were incubated in Ussing-type chambers in which short-circuit current, transepithelial potential difference, and skin conductance were continuously monitored. The normal bathing solution was NaCl Ringer containing 2.5 mM bicarbonate. After appropriate incubation under various experimental conditions, the skin pieces were quickly removed from the chambers and snap frozen by plunging into the liquid coolant.

Frog Cornea

The experiments were performed on the isolated cornea of the American bullfrog, *Rana catesbeiana* (West Jersey Biological Supply Farm, Wenonah, NJ). After careful dissection, the corneas were mounted in Ussing-type incubation chambers in which short-circuit current, transepithelial potential difference, and skin conductance were continuously monitored. The natural bulging of the cornea was maintained by applying a 1-cm H_2O hydrostatic pressure difference, anterior chamber-side positive. The normal incubation solution was Conway's solution (25 mM bicarbonate). Usually one of the corneas remained untreated serving as a control. At the end of the experiments the rings were quickly removed from the incubation chambers for freezing.

RESULTS

Transepithelial Na Transport in the Frog Skin

According to the model proposed by Koefoed-Johnsen and Ussing[17] transepithelial Na transport involves two different transport mechanisms. Na influx into the epithelial cell across the apical membrane is thought to be passive, driven by an inwardly directed electrochemical gradient, whereas Na extrusion across the basolateral membrane is assumed to be active, directly coupled to an energy-providing metabolic reaction. The model predicts that the Na concentration of the cells participating in transepithelial transport, usually referred to as the Na transport compartment, should vary according to the activity of the individual transport steps. Surprisingly, chemical or radiochemical analysis of the intracellular Na concentration in transporting epithelia such as frog skin or toad urinary bladder often fails to show the expected behavior, lending some support to the view that Na may take a noncellular transport route.[18] Ussing himself entertained the possibilities that actually only a small fraction of the epithelial cell is engaged in transepithelial transport,[19] or that Na transport proceeds along specialized intracytoplasmatic structures.[20]

To answer the question which of the different epithelial cell types in a histologically complex epithelium such as the frog skin are actually engaged in transepithelial Na transport, we measured the intracellular electrolyte concentration using electron microprobe analysis.[21] As shown in FIGURE 1, under control conditions in all layers of the multilayered epithelium, except in the outer cornified layer, a typical cellular distribution of Na and K was observed. The data are displayed as mean intracellular concentrations, since for small cations such as Na and K no significant differences between the nuclear and cytoplasmic concentration values were detectable.[20] After inhibition of the Na-K pump by the cardiac glycoside ouabain the Na concentrations in all living cell layers were increased by about 100 mmol/kg w.w., while the K concentrations showed an equivalent drop (FIGURE 1). Since such a Na/K exchange may also occur in cells not participating in transepithelial transport, we further tested whether the Na originates mainly from the outer bathing medium, as is expected only for transporting cells. For this purpose we abolished the apical Na influx either by using a Na-free solution or by adding the diuretic amiloride, which renders the apical membrane impermeable to Na. As shown in FIGURE 1, in both cases the Na increase after ouabain was completely abolished, suggesting that all epithelial layers participate in transepithelial Na transport. No significant changes in the cellular Cl and P concentrations and the dry weight were observed, indicating that despite drastic changes in the transepithelial Na transport activity the epithelial cell volume remains essentially unchanged.

The behavior of the Na concentration confirms the two-barrier concept of Koefoed-Johnsen and Ussing.[17] Moreover, the results shown in FIGURE 1 provide a direct proof for a syncytial organization of the epithelium. Since the paracellular shunt pathway between the cells is sealed at the level of the stratum granulosum,[22] only this cell layer can directly communicate with the outer medium. The apparent Na influx

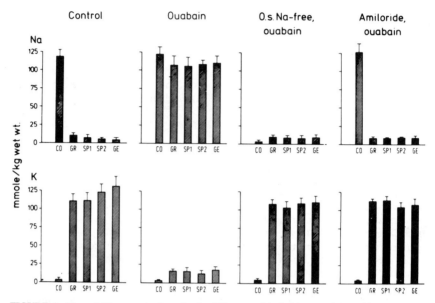

FIGURE 1. Na and K concentrations in the different epithelial layers of frog skin in *control*, after *ouabain* (100 μM, 90 min) and when, simultaneous with the application of ouabain, either the outside was incubated with a *Na-free* choline Ringer's solution or *amiloride* (100 μM) was added. The different epithelial layers are: stratum corneum (**CO**), granulosum (**GR**), superficial and deeper spinosum (**SP1, SP2**), and germinativum (**GE**). Mean ± 2 S.E.M. (From Rick *et al.*[21] Reprinted by permission from *Microscopica Acta.*)

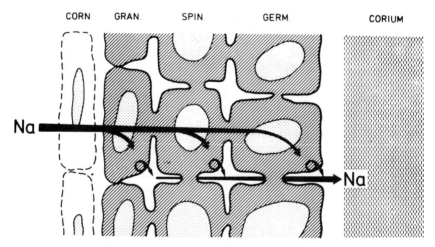

FIGURE 2. Pathway of transepithelial Na transport in frog skin.

from the outer bath into deeper lying epithelial layers, therefore, can only be explained if the different epithelial cell layers are interconnected. Thus, the pathway for transepithelial Na transport across the frog skin can be described as schematically depicted in FIGURE 2. Na enters the epithelium, virtually unimpeded by the stratum corneum, across the outer-facing membranes of the outermost living cell layer, the stratum granulosum. From there, it can readily diffuse via low-resistance junctions to all deeper epithelial layers, and is finally extruded across the basolateral membranes.

However, the syncytial Na transport compartment does not comprise all cells of the epithelium. The stratum corneum obviously represents a layer of dead cells, exchanging their electrolytes freely with the outer bathing medium (see FIGURE 1). As shown in FIGURE 3, a further exception is the mitochondria-rich cell. After ouabain the Na

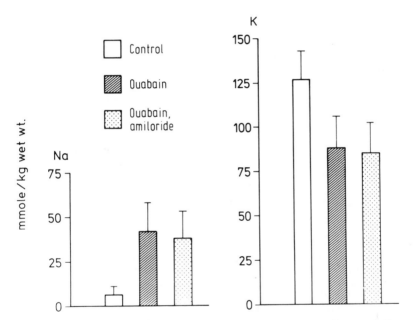

FIGURE 3. Na and K concentrations in mitochondria-rich cells (same experiment as in FIGURE 1). Mean ± 2 S.E.M. (From Rick *et al.*[40] Reprinted by permission from Academic Press.)

concentration increases to a much lesser extent than in adjacent granular or spiny cells. More significantly, the Na increase cannot be prevented by amiloride, which, at least in frog skin, completely abolishes transepithelial Na transport. A further significant difference between mitochondria-rich cells and syncytium cells is the intracellular Cl concentration (TABLE 1). Despite some variability from skin to skin, the Cl concentration in the mitochondria-rich cells was always found to be lower than in syncytium cells. Moreover it is dependent on the availability of Cl in either bathing solution, whereas in the syncytium the intracellular Cl seems to exchange only with the inner bath. This finding is consistent with the idea that the mitochondria-rich cells constitute a Cl shunt pathway.[23] Furthermore, the intracellular Cl concentration showed a differential response to removal of Na from the inner bath.[24] In syncytium cells this

TABLE 1. Intracellular Cl Concentrations in Syncytium Cells and
Mitochondria-Rich Cells of the Frog Skin Epithelium[a]

	Syncytium Cells		Mitochondria-Rich Cells	
	Number of Measurements	mmol/kg w.w.	Number of Measurements	mmol/kg w.w.
Control	(13)	42.2 ± 5.3	(41)	22.3 ± 7.2
O.s. Cl-free	(7)	44.7 ± 4.8	(47)	9.2 ± 0.9
I.s. Cl-free	(11)	6.5 ± 1.9	(36)	10.1 ± 4.7

[a]Intracellular Cl concentrations are given in control and after removal of Cl from the outer
(*O.s. Cl-free*) or inner bathing solution (*I.s. Cl-free*). Experiments were performed on skins of
Rana temporaria; incubation time was 60 minutes. Chloride was replaced with sulfate, and the
osmolarity was adjusted by addition of sucrose. Means ± SD.

procedure led to a marked reduction of the Cl concentration, while in mitochondria-rich cells the Cl concentration remained largely unchanged.

Further cells not involved in transepithelial Na transport are the cells of the mucous and seromucous glands, which seem to be responsible for the Cl secretion elicited by beta-agonists.[25] Similar studies performed in the toad urinary bladder have shown that only the granular cells, which in this epithelium constitute the majority of cells, are involved in transepithelial Na transport.[3]

Electron microprobe analysis of intracellular electrolytes in transporting epithelia can also provide information regarding the mechanism of action of modulators of transepithelial transport. FIGURE 4 shows the effect on the intracellular concentrations of Na, K, and Cl when the Na transport, as measured by the short-circuit current, was stimulated by arginine vasopressin (AVP) to about 250% of the control value. At the same time, with the exception of mitochondria-rich cells, in all epithelial cells a marked increase in the Na concentration and an equivalent drop in the K concentration is observed. The vasopressin-induced changes could be completely abolished by additional application of amiloride. The Na concentration under this condition was even lower than in the control. Under all three experimental conditions no significant changes in the Cl and P concentrations or dry weight were detectable.

The increase in the Na concentration suggests that vasopressin stimulates transepithelial Na transport primarily by increasing the apical Na influx. In fact, if combined with measurements of the apical membrane potential obtained in the same preparation,[26] a threefold to fivefold increase in the apical Na permeability can be calculated. The data, however, do not rule out the possibility of an additional effect on the active transport step. In a recent study on the effect of vasopressin[24] we observed that at identical rates of transepithelial Na transport the vasopressin-stimulated skin generally showed significantly lower Na concentrations, consistent with an additional stimulation of the Na-K pump. Identical rates of transport were achieved either by full inhibition of Na transport with supramaximal doses of amiloride or by reducing the short-circuit current of the stimulated skin piece to the level observed in the control using low amiloride concentrations.

The parallel changes in the Na and K concentrations are, again, consistent with a syncytial cooperation between the different cell layers. The coupling between the epithelial cell layers does not necessarily imply that all layers are actually engaged in active transepithelial Na transport. It is conceivable that Na is extruded only from the outermost living cell layer, while the other layers merely exchange their electrolytes with this layer. However, the finding of an almost continuous Na concentration gradient between the various cell layers argues against such a possibility. After AVP and to a lesser extent under control conditions (see FIGURES 1 and 4), the Na concentrations in the outer epithelial cell layers are slightly higher than in the deeper lying layers. If the cells are coupled, such a concentration gradient should result in an inwardly directed net flux of Na which can only be maintained if the deeper epithelial layers are capable of extruding Na. Consistent with the assumption of an appreciable net Na flux between the outer and inner epithelial cell layers is the fact that, under conditions where the transepithelial Na transport was stimulated as with AVP, the Na gradient was significantly enhanced, whereas, when the Na transport was abolished, as with amiloride or ouabain,[3] virtually no Na gradient could be detected.

Cl Secretion in the Frog Cornea

Cl secretion in the isolated frog cornea, as in many other secretory epithelia, is electrogenic and depends on the presence of Na in the inner bath, although under short-circuited conditions Na is not transported transepithelially.[27] In the model of Cl secretion originally proposed by Silva *et al.*[28] the Na dependence is explained by assuming that Cl uptake across the basolateral membrane is a secondary active process, driven by the extra-/intracellular Na gradient. Cl efflux across the apical membrane is thought to be passive, most likely through a channel.[29] Two key predictions of this hypothesis were recently verified for the corneal epithelium of the

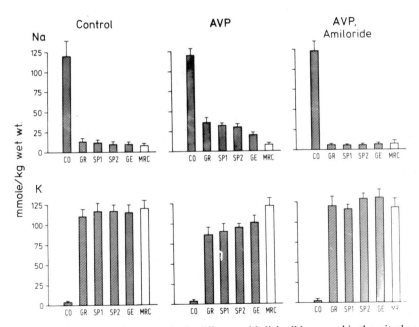

FIGURE 4. Na and K concentrations in the different epithelial cell layers and in the mitochondria-rich cells (**MRC**) of the frog skin epithelium in *control, after AVP* (150 mU/ml, 45 min), and after subsequent additional application of *amiloride* (100 μM). Mean $\pm$ 2 S.E.M. (From Rick *et al.*[40] Reprinted by permission from Academic Press.)

bullfrog. Reuss et al.,[30] using ion selective microelectrodes, demonstrated that the intracellular Cl activity is above electrochemical equilibrium and that the apical membrane exhibits a high Cl conductance.

In two recent studies we measured the intracellular electrolyte concentrations in the cornea epithelium during changes in the pH and ionic composition of the bathing medium[31] and application of inhibitors and stimulators of Cl secretion.[32] FIGURE 5 shows the transepithelial concentration profile of the major electrolytes Na, Cl, and K and of P under control conditions. The values are strikingly similar to those observed in the frog skin under comparable conditions, except for the significantly lower Cl concentration. It may be speculated that the only major difference between the two epithelia is the absence or presence of an additional Cl leak in the apical membrane. Using published values for the membrane potential,[33] the Cl concentration nevertheless is still significantly above the expected equilibrium value. Thus, the Cl efflux across the apical membrane may be an entirely passive process.

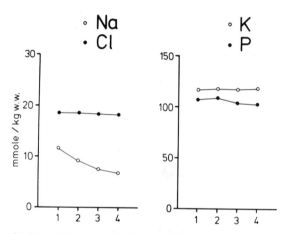

FIGURE 5. Na, Cl, K, and P concentrations in the different epithelial cell layers of the frog cornea. Cell layer 1 is the outermost (surface cells), 4 the innermost layer of the epithelium (basal cells). Mean of 23 corneas incubated with normal Conway solution. (From Rick et al.[31] Reprinted by permission from the Journal of Membrane Biology.)

Surprisingly, as in the Na-transporting frog skin epithelium, an inwardly directed Na concentration gradient was observed, although in the frog cornea the rate of Na transport is negligibly small.[34] The gradient was found to be decreased when Cl secretion was inhibited (see FIGURE 6) and greatly enhanced when Cl secretion was stimulated (see FIGURE 7), suggesting that it is somehow linked to the Cl transport activity. In fact, the Na gradient can be explained by assuming that the Cl uptake across the basolateral membranes is coupled to Na.[31] If the rate of NaCl cotransport were the same in all epithelial layers and the outer layers had a lesser capacity to extrude Na, such a gradient should exist. Moreover, a higher NaCl influx into the outer epithelial layer could account for the gradient as well. Like the frog skin epithelium, the cornea can be considered a functional syncytium. Evidence for this is the fact that removal of Cl from the outer, tear-side bathing solution resulted in a drop of the Cl concentration in all epithelial layers and not only in the outermost cell layer,

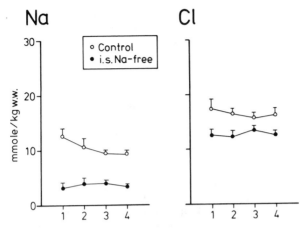

FIGURE 6. Na and Cl concentrations in the different epithelial cell layers of the frog cornea in control and after removal of Na from the inner bath (substitution with choline). Mean ± S.E.M. (Modified from Rick *et al.*, REFERENCE 31.)

which is the only one directly communicating with the outer bath. In contrast, removal of Na or K had no detectable effect on the intracellular concentrations.

The effect of changing the ionic composition of the inner, anterior chamber-side bathing medium is shown in FIGURE 6 and TABLE 2. A marked reduction in the intracellular Na and Cl concentrations can be observed in all epithelial layers after removal of Na. At the same time, the Cl secretion, as measured by the short-circuit current (SCC), was reduced by about 70%. A similar drop in the Na and Cl concentrations was observed when Cl was removed from the inner bath. (TABLE 2). As to be expected, the SCC under this condition was completely abolished. TABLE 2 also

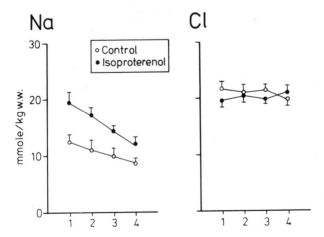

FIGURE 7. Na and Cl concentrations in the different epithelial cell layers of the frog cornea in control and stimulation with isoproterenol (30 min, 10^{-6} M). Mean ± S.E.M. (Modified from Rick *et al.*, REFERENCE 32.)

shows the effect of removal of K from the inner bath. Again, the Cl concentration and the SCC are reduced. The increased Na concentration during K removal can be explained by an inhibition of the Na-K pump.

The results suggest that the basolateral uptake of Cl is mediated by a Na, K, 2Cl cotransport system as previously described for red blood cells,[35] Ehrlich ascites cells,[36] or the ascending limb of the loop of Henle.[4] The parallel operation of Na/H and Cl/HCO_3 exchangers as suggested for the small intestine[38] can be ruled out, since neither amiloride, nor SITS or DIDS had any significant effect on the Cl secretion and intracellular electrolytes when added in sufficiently high concentrations for inhibition of these transport systems (own unpublished data). The presence of a Na, K, 2Cl transporter was confirmed by the finding that the intracellular Na and Cl concentrations and Cl secretion were sensitive to bumetanide,[31] a highly specific inhibitor of this transport system.

Cl secretion in the cornea, as in many other secretory epithelia, can be stimulated by a variety of procedures that finally lead to an increase in the intracellular Ca activity or cAMP concentration. To determine whether the increased transport rate is accomplished by an effect on the apical or basolateral membrane, we stimulated Cl

TABLE 2. Mean Intracellular Na, Cl, and K Concentrations and Cellular Dry Weight of the Cornea Epithelium of *Rana catesbeiana*[a]

	Number of Measurements	Na	Cl	K	Dry Weight g/100 g
		mmol/kg w.w.			
Control	(78)	12.3 ± 4.6	14.9 ± 3.8	117.1 ± 15.0	22.0 ± 1.9
I.s. Na-free	(73)	6.5 ± 3.8	11.6 ± 3.1	124.6 ± 15.7	23.3 ± 2.4
Control	(81)	14.2 ± 5.8	14.8 ± 3.7	116.3 ± 17.1	21.7 ± 3.2
I.s. Cl-free	(89)	7.0 ± 3.8	5.5 ± 2.2	118.7 ± 19.7	22.6 ± 4.0
Control	(27)	7.1 ± 5.8	16.4 ± 3.1	112.6 ± 17.0	18.6 ± 2.1
I.s. K-free	(49)	19.3 ± 8.8	13.0 ± 2.8	110.6 ± 16.1	20.0 ± 2.4

[a]Na, Cl, and K concentrations and cellular dry weight are given in control and after removal of Na (*I.s. Na-free*), Cl (*I.s. Cl-free*) or K (*I.s.K-free*) from the inner bath. Na, Cl, and K were replaced with choline, SO_4, and Na, respectively. Osmolarity was adjusted by sucrose. Means ± SD.

secretion by isoproterenol (FIGURE 7). To our surprise, in none of the seven paired experiments performed could we detect any significant change in the intracellular Cl concentration, suggesting that the hormone equally stimulates both transport steps. Alternatively, it is possible that the hormone directly affects only one transport process and that the activity of the other process is secondarily adjusted by some cellular control mechanism. Signals that conceivably might be exchanged between the two membranes are the electrical potential, intracellular pH or pCa, and the epithelial cell volume. In this respect it is of interest to note that the cell volume as reflected in the dry weight or P concentration was not detectably changed after isoproterenol stimulation. Similar results were obtained after stimulation with epinephrine, Ca ionophore A23187, or adenosine.[32]

DISCUSSION

In models of transepithelial transport the epithelium is generally considered as functionally homogeneous and often represented by a single black box. This is largely due to the fact that most of the techniques used in studying transepithelial ion transport mechanisms such as recording of electrical parameters, flux measurements, and chemical analyses of the intracellular ion concentrations are applicable only to the epithelium as a whole. However, most transporting epithelia such as renal tubules, airway epithelia, intestine, frog skin, or toad urinary bladder are histologically complex tissues, consisting of different epithelial layers or cell types. This raises the question whether the various transport functions of an epithelium are shared by all epithelial cells, or whether a particular ion transport process might reside in a particular cell type. To open the lid of the black box, methods are needed that will allow us to analyze epithelial transport processes at a cellular level.

In principle, electron microprobe analysis provides such a method. In fact, the attainable spatial resolution is much higher than required for analysis of ion concentrations on a cellular scale. In the studies reviewed in this paper the measurements were actually performed separately in the nucleus and in an organelle-free region of the cytoplasm and are given as mean values for the total intracellular space. Justification for this is that under all experimental conditions employed in these studies the concentrations of diffusable ions in the two cellular compartments were virtually identical, and no significant ion concentration gradients were detectable within one cell. Therefore, with regard to transepithelial ion transport, the intracellular space can be considered as only one distributional space. The lack of a detectable concentration gradient within an epithelial cell supports the notion that the intracellular resistance to flow of small monovalent ions is negligibly small.

The capability of analyzing electrolytes at the level of an individual cell in an intact epithelium permits us to study aspects of epithelial ion transport that otherwise can be approached only indirectly: the localization of a transport compartment and transport cooperation between different epithelial cells. Identification of cells involved in a particular transport process, however, requires that we be able to make reasonable predictions about the behavior of the intracellular ion concentrations in the transporting epithelial cells. An increase in the intracellular Na concentration after ouabain is certainly not sufficient evidence for identification of a Na transport compartment. Such a Na increase may also occur in nonpolar cells or in cells involved in other transport processes. It remains to be shown that the Na accumulating during ouabain is in a transcellular transport pathway. In the case of the frog skin, this was proved by the fact that the Na increase in the syncytium cells could be abolished by preventing the Na influx from the outer bath. However, even this result can be misleading. In mitochondria-rich cells the Na increase after ouabain was partially inhibited by using a Na-free outer bathing solution suggesting that this cell type also constitutes a possible transepithelial Na transport pathway.[2] It is the lack of response to amiloride, a highly selective inhibitor of transepithelial Na transport in the frog skin, that rules out the possibility that this cell type constitutes the Na transport compartment.

The behavior of the intracellular Cl concentrations in the mitochondria-rich cells suggests that this cell type may play a role as a transcellular Cl shunt. However, the evidence for this is not very convincing, since a priori other pathways, such as a paracellular route or the epithelial glands, cannot be excluded.

All cells of the cornea epithelium and, with the exception of several specialized cells, the majority of the cells in the frog skin epithelium can be considered as part of a functional syncytium. This conclusion is based on the observation that the ion concentrations in the various epithelial layers were almost identical and showed parallel changes under all experimental conditions tested. The demonstration of the syncytial organization was aided by the fact that both epithelia are tight and have a shunt pathway that is sealed at the level of the outermost living cell layer. Therefore, addition of amiloride in the case of the frog skin or removal of Cl from the outer bath in the case of the cornea should directly affect only the intracellular concentrations of the outermost epithelial layer, unless there is direct cell-to-cell communication between the layers. For a multilayered epithelium the advantages of the syncytial organization are obvious: without cell-to-cell communication the burden of transepithelial transport would have to be carried by only one layer. In most cases, because of the location of the tight junction, this cell layer would be the outermost, oldest, and least viable cell layer of the epithelium. In fact, for the frog skin it has been shown that this cell layer shows much less ouabain binding than all deeper lying cell layers.[39] For an epithelium composed of a single layer, the functional advantage of the syncytial organization is of lesser importance. Nevertheless, the exchange of ions between neighboring epithelial cells may lead to a better adjustment between the transport activities of the two limiting membranes and may prevent cell death in cells with partially impaired energy metabolism or imbalance between leak and pump.

The syncytial organization of the epithelium does not necessarily imply that all epithelial layers share equally in the transport activity. Quantification of the extent to which the different epithelial layers exchange their electrolytes would require a direct measurement of the permeability of the cell-to-cell pathway. However, the existence of a Na gradient between the layers of the epithelium and the variation of this gradient with the rate of transepithelial transport strongly suggest that all epithelial layers participate significantly in the transport activity. If this is so, then variations in the coupling between the cells can be expected to alter the rate of transepithelial transport. In fact, for the frog skin we have shown that removal of Na from the inner bath leads to uncoupling between the outermost and second epithelial layer, which was found to be correlated with an inhibition of the short-circuit current.[24] Thus, at least for a multilayered epithelium, the coupling between the cell layers can be considered a third barrier to transepithelial transport.

Quantification of the ion concentrations in the transport compartment can provide further insights into the site and mode of action of modulators of transepithelial transport. The drop in the Na concentration after amiloride and the rise in the Na concentration after ouabain is consistent with the generally accepted tenet that these inhibitors of transepithelial Na transport act on the passive and active transport steps, respectively. The increased Na concentration after vasopressin supports the view that the natriferic effect of the hormone is due to stimulation of the passive Na influx rather than an effect on the pump. The increased pump rate can be explained by the higher intracellular Na levels, although a slight direct stimulation of the pump cannot be ruled out completely.[24] Using the same line of reasoning, we suggest that the stimulation of Cl secretion in the frog cornea by isoproterenol can be explained by an about equal effect on each of the two transport steps. However, it is entirely possible, if not likely, that isoproterenol primarily affects only one of the two transport processes and that the other one is secondarily adjusted via a cellular control mechanism. Since Cl is a major determinant of the epithelial cell volume, it is likely that a volume regulatory mechanism is involved.[31]

The concentration values obtained by electron microprobe analysis can be used to calculate the chemical gradient for the transported ions across each of the limiting cell membranes. For this purpose, the wet weight concentrations obtained by the internal standard method may be recalculated in concentration per liter cell water on the basis of the cellular dry weight values. This procedure, however, presumes that the intracellular ions are essentially free and not compartmentalized. For small monovalent ions such as Na, K, and Cl this assumption may, indeed, be correct. In the frog skin, the intracellular concentration of Na was reduced practically to zero when the outer medium was replaced with distilled water.[2] Also, using Cl-free bathing solutions, one can observe very low intracellular Cl values (see TABLE 1). Furthermore, after ouabain an almost complete Na/K exchange can be detected, suggesting that K as well is largely unbound. Comparison of the ion concentrations obtained in the frog cornea with ion activities reported for the same preparation[30] reveals an excellent agreement. The calculated mean intracellular activity coefficient for Na and K is 0.75, almost identical with the activity coefficient of an electrolyte solution of similar salt composition.

REFERENCES

1. MACKNIGHT, A. D. C. & A. LEAF. 1978. The sodium transport pool. Am. J. Physiol. **234:** F1–F9.
2. RICK, R., A. DÖRGE, R. BAUER, F. X. BECK, J. MASON, CH. ROLOFF & K. THURAU. 1980. Quantitative determination of electrolyte concentrations in epithelial tissues by electron microprobe analysis. Curr. Top. Membr. Transp. **13:** 107–120.
3. RICK, R., A. DÖRGE, E. VON ARNIM & K. THURAU. 1978b. Electron microprobe analysis of frog skin epithelium: Evidence for a syncytial sodium transport compartment. J. Membr. Biol. **39:** 313–331.
4. BECK, F. X., R. BAUER, U. BAUER, J. MASON, A. DÖRGE, R. RICK & K. THURAU. 1980. Electron microprobe analysis of intracellular elements in the rat kidney. Kidney Int. **17:** 756–763.
5. BECK, F. X., A. DÖRGE, R. RICK & K. THURAU. 1985. Osmoregulation of renal papillary cells. Pfluegers Arch. **405:** 528–532.
6. MASON, J., F. X. BECK, A. DÖRGE, R. RICK & K. THURAU. 1981. Intracellular electrolyte composition following renal ischaemia. Kidney Int. **20:** 61–70.
7. BECK F. X., G. BIANCHI, A. DÖRGE, R. RICK, M. SCHRAMM & K. THURAU. 1983. Sodium and potassium concentrations of renal cortical cells in two animal models of primary arterial hypertension. J. Hypertension **1** (supp. 2): 38–39.
8. ANDREWS, S. B., J. E. MAZURKIEWICZ & G. R. KIRK. 1983. The distribution of intracellular ions in the avian salt gland. J. Cell Biol. **96:** 1389–1399.
9. BULGER, R. E., R. BEEUWKES III & A. J. SAUBERMANN. 1981. Application of scanning electron microscopy to x-ray analysis of frozen-hydrated sections. III. Elemental content of cells in the rat renal papillary tip. J. Cell Biol. **88:** 274–280.
10. CIVAN, M. M., T. A. HALL & B. L. GUPTA. 1980. Microprobe study of toad urinary bladder in absence of serosal K^+. J. Membr. Biol. **55:** 187–202.
11. GUPTA, B. L., T. A. HALL, S. H. P. MADRELL & R. B. MORETON. 1976. Distribution of ions in a fluid-transporting epithelium determined by electron-probe X-ray microanalysis. Nature. **264:** 284–287.
12. GUPTA, B. L., M. J. BERRIDGE, T. A. HALL & R. B. MORETON. 1978. Electron microprobe and ion-selective microelectrode studies of fluid secretion in the salivary glands of Calliphora. J. Exp. Biol. **72:** 261–284.
13. GUPTA, B. L. & T. A. HALL. 1979. Quantitative electron probe X-ray microanalysis of electrolyte elements within epithelial tissue compartments. Fed. Proc. Fed. Am. Soc. Exp. **38:** 144–153.
14. DÖRGE, A., R. RICK, K. GEHRING & K. THURAU. 1978. Preparation of freeze-dried cryosections for quantitative X-ray microanalysis of electrolytes in biological soft tissues. Pfluegers Arch. **373:** 85–97.
15. BAUER, R. & R. RICK. 1978. Computer analysis of X-ray spectra (EDS) from thin biological specimens. X-Ray Spectrom. **7:** 63–69.
16. RICK, R., A. DÖRGE & K. THURAU. 1982. Quantitative analysis of electrolytes in frozen dried sections. J. Microsc. (Oxford) **125:** 239–247.
17. KOEFOED-JOHNSEN, V. & H. H. USSING. 1958. The nature of the frog skin potential. Acta Physiol. Scand. **42:** 298–308.
18. CEREIJIDO, M. & C. A. ROTUNNO. 1968. Fluxes and distribution of sodium in frog skin. A new model. J. Gen. Physiol. **51:** 280–289.
19. VOUTE, C. L. & H. H. USSING. 1968. Some morphological aspects of active sodium transport. J. Cell Biol. **36:** 625–638.
20. VOUTE, C. L., K. MOLLGARD & H. H. USSING. 1975. Quantitative relationship between

active sodium transport, expansion of endoplasmic reticulum and specialized vacuoles ("scalloped sacs") in the outermost living cell layer of the frog skin epithelium (*Rana temporaria*). J. Membr. Biol. **21**: 273–289.

21. RICK, R., A. DÖRGE & E. VON ARNIM. 1978a. X-ray microanalysis of frog skin epithelium: Evidence for a syncytial Na transport compartment. Microsc. Acta Suppl. **2**: 156–165.
22. ERLIJ, D. 1971. Salt transport across isolated frog skin. Philos. Trans. R. Soc. London, Ser. B **262**: 153–161.
23. VOUTE, C. L. & W. MEIER. 1978. The mitochondria-rich cell of frog skin as hormone sensitive 'shunt path.' J. Membr. Biol. **405**: 141–165.
24. RICK, R., CH. ROLOFF, A. DÖRGE, F. X. BECK & K. THURAU. 1984. Intracellular electrolyte concentrations in the frog skin epithelium: Effect of vasopressin and dependence on the Na concentration in the bathing media. J. Membr. Biol. **78**: 129–145.
25. MILLS, J. W., K. THURAU, A. DÖRGE & R. RICK. 1985. Electron microprobe analysis of intracellular electrolytes in resting and isoproterenol-stimulated glands of frog skin. J. Membr. Biol. **86**: 211–220.
26 NAGEL, W. 1978. Effects of antidiuretic hormone upon electrical potential and resistance of apical and basolateral membranes of frog skin. J. Membr. Biol. **42**: 99–122.
27. FRIZZELL, R. A., M. FIELD & S. G. SCHULTZ. 1979. Sodium coupled-chloride transport by epithelial tissues. Am. J. Physiol. **236**: F1–F8.
28. SILVA, P., J. STOFF, M. FIELD, L. FINE, J. N. FORREST & F. H. EPSTEIN. 1977. Mechanism of active chloride secretion by shark rectal gland: Role of Na-K-ATPase in chloride transport. Am. J. Physiol. **233**: F298–F306.
29. SHOEMAKER, R. L., G. RECHKEMMER & R. A. FRIZZELL. 1986. Chloride channel regulation by cAMP or Ca^{++} in cultured human tracheal cells. Biophys. J. **49**: 158a.
30. REUSS, L., P. REINACH, S. A. WEINMAN & T. P. GRADY. 1983. Intracellular ion activities and Cl^- transport mechanisms in bullfrog corneal epithelium. Am. J. Physiol. **244**: C336–C347.
31. RICK, R., F. X. BECK, A. DÖRGE & K. THURAU. 1985. Cl transport in the frog cornea: an electron microprobe analysis. J. Membr. Biol. **83**: 235–250.
32. RICK, R., A. DÖRGE & K. THURAU. 1986. Electrolyte concentration in the frog cornea during stimulation and inhibition of Cl secretion. J. Membr. Biol., submitted.
33. NAGEL, W. & P. REINACH. 1980. Mechanism of stimulation by epinephrine of active transepithelial Cl transport in isolated frog cornea. J. Membr. Biol. **56**: 73–79.
34. CANDIA, O. A. & W. A. ASKEW. 1968. Active sodium transport in the isolated bullfrog cornea. Biochim. Biophys. Acta **162**: 262–265.
35. DUNHAM, P. B., G. W. STEWART & J. C. ELLORY. 1980. Chloride-activated passive potassium transport in human erythrocytes. Proc. Natl. Acad. Sci. USA **77**: 1711–1715.
36. BAKKER-GRUNWALD, T. 1978. Effect of anions on potassium self-exchange in ascites tumor cells. Biochim. Biophys. Acta **513**: 292–295.
37. GREGER, R. & E. SCHLATTER. 1983. Properties of the basolateral membrane of the cortical thick ascending limb of Henle's loop of rabbit kidney. Pfluegers Arch. **396**: 325–334.
38. TURNBERG, L. A., F. A. BIEBERDORF, S. G. MORAWSKI & J. S. FORDTRAN. 1970. Interrelationships of chloride bicarbonate, sodium and hydrogen transport in the human ileum. J. Clin. Invest. **49**: 557–567.
39. MILLS, J. W., S. A. ERNST & D. R. DiBONA. 1977. Localization of Na^+-pump sites in frog skin. J. Cell Biol. **73**: 88–110.
40. RICK, R., A. DÖRGE, C. ROLOFF & K. THURAU. 1981. Electron microprobe analysis of Na transporting epithelia. *In* Microprobe Analysis of Biological Systems. T. E. Hutchinson & A. P. Somlyo, Eds. 47–64. Academic Press. New York, NY.

DISCUSSION OF THE PAPER

D. F. PARSONS (*New York State Department of Health, Albany, New York*):
What were the tear-shaped cells in the frog skin?

RICK: They are mitochondria-rich cells. They are just packed with mitochondria
and appear to be involved in secretion of organic acids. Also, mitochondria-rich cells
are known to accumulate certain dyes. In addition, they may constitute a shunt
pathway for chloride as suggested by our Cl measurements (unpublished data).
Normally, *i.e.* under nonshort-circuited conditions, Na reabsorption will be accompanied by a reabsorption of Cl.

Resolution of Sarcolemma and Sarcoplasmic Reticulum in Electron-Probe X Ray Microanalysis of Cardiac Muscle[a]

JOHN McD. TORMEY AND ELLYN S. WHEELER-CLARK

Department of Physiology
Center for Health Sciences
University of California at Los Angeles School of Medicine
Los Angeles, California 90024

INTRODUCTION

Most microprobe studies on cardiac muscle (*e.g.* REFERENCES 1,2) have measured only the larger subcellular compartments, namely, myofibrils and mitochondria. A few reports, emanating from Tubingen[3-6] and more recently from our laboratory,[7] have attempted in addition to measure the sarcolemma (SL) and the sarcoplasmic reticulum (SR). Microanalysis of these smaller structures is of crucial importance for understanding the mechanisms of excitation-contraction coupling in cardiac muscle. However, although the terminal cistern of skeletal muscle SR has been successfully studied by microprobe analysis,[8-10] the comparable structure in cardiac muscle (junctional SR) is much more difficult to work with because of its order of magnitude smaller size.

This paper addresses the practical ability of electron-probe analysis to quantify such small structures as the SR and SL. The SL, for instance, is less than 10 nm wide and occurs at the interface of the extracellular fluid and the cytosol, which have radically different compositions. This gives rise to at least two issues: (1) An elevated concentration of Ca measured near the cell surface does not necessarily indicate Ca accumulation and hence can be very misleading. When concentrations are reported in mmol/kg dry wt,[11,12] an elevated concentration can simply reflect the fact that dry mass is much lower in extracellular fluid than in cells. How can changes in Ca concentration that might have functional significance be distinguished from those due only to a local variation in dry mass? (2) Electron-probe diameters typically used for biological microanalysis are an order of magnitude larger than the width of the SL. Is it possible, using an electron gun of the type available in most laboratories, to resolve such small structures and to distinguish their composition from that of adjacent structures?

Both questions are dealt with by the approach to be described. It is based (1) on a knowledge of the spatial distribution of electrons in the probing beam and (2) on sets of analyses that include separate measurements of the two bulk phases that flank the primary structue to be analyzed. These permit a spatial deconvolution in which the X rays that arise from the structure of interest can be distinguished from those that arise

[a]Supported by National Heart, Lung, and Blood Institute Grant HL 31249 and by grants from the American Heart Association and the Muscular Dystrophy Association. This work was done during the tenure by ESW of an Advanced Fellowship of the American Heart Association, Greater Los Angeles Affiliate.

from adjoining phases. This approach will be applied here to measurements of concentration changes at the SL and JSR of cardiac muscle.

METHODS

Biological Specimens

Papillary muscles are taken from the right ventricles of anesthetized adult rabbits. After isolation they are superperfused at 28° C and electrically stimulated at 0.3 Hz while their force development is monitored. The oxygenated perfusing solution includes 1 mM Ca, 5 mM K, 5% w/v dextran (MW 83,000) and HEPES buffer at pH 7.4. The dextran is added to provide an extracellular matrix that retains extracellular ions in place, e.g. without piling up against membranes, during cryosectioning and freeze-drying. Although higher concentrations of dextran have been reported to affect papillary muscle performance,[6] dextran at this concentration did not affect the strength or kinetics of muscle contraction.

After an initial period of equilibration during which performance is ascertained, the muscles are bathed either in a *control* solution that includes 139 mM NaCl or in a *low-Na* solution that contains only 36 mM NaCl with sucrose added to maintain osmolarity. After 30 minutes in either of these solutions, the muscles are frozen against a liquid-helium cooled copper block. Cryosections ca. 150 nm thick are cut at $-120°$ C, stored under liquid nitrogen on folding grids, and finally transferred in the frozen-hydrated state into a JEOL 100CX electron microscope. Freeze-drying is carried out in the microscope at -110 to $-90°$ C, followed by microanalysis at $-110°$ C.

Electron-Probe X Ray Microanalysis (EPMA)

Electron-probe microanalysis is carried out using the JEOL 100CX operating at 100 kV in the scanning transmission electron microscope (STEM) mode. The probe current normally employed is 3 nA, obtained using a conventional tungsten hairpin filament and a 200-um condenser aperture. Instrumentation includes a Gatan cryotransfer stage and a Tracor Northern NS 800 energy dispersive X ray spectrometer. Electron dosages approach 10^5 coulomb/cm^2 in our smaller rasters. These huge dosages produce no visible local effects. This is attributable in part to the 3×10^{-8} Torr vacuum routinely obtained in the specimen chamber region, and in part to the fact that the sections have already experienced maximum radiation damage by the time local analyses are carried out.

Initial quantitative data is obtained from X ray spectra using the Hall peak-to-continuum normalization method[11] as implemented by Shuman et al.[12] Deconvolution of the K-Ca overlap is performed as described previously.[2,13] Additional calculations are performed as described under Results and Discussion.

The focused electron probe cannot be directly imaged by the 100CX in the STEM mode. Therefore an aluminum knife edge specimen, kindly supplied by JEOL USA, is used for measuring the distribution of electrons in the 3 nA electron beam. The production of Al X rays is measured as the function of static probe position with respect to the knife edge. The resulting cumulative X ray intensity distribution corresponds to the cumulative distribution of electrons perpendicular to the knife edge. This distribution is analyzed[14,15] to obtain values for the standard deviation and full width at half maximum (FWHM) of the distribution of electron intensities with

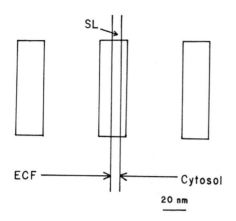

FIGURE 1. Illustration of the strategy used for placing rasters for microprobe analysis of the sarcolemma (SL). The *rectangular SL raster* (22 nm wide) is centered over the extracellular edge of the *SL* itself. After an X ray spectrum is collected using this raster, additional spectra are collected from adjacent rasters centered over the bulk *ECF* and *cytosol*. Drawn to indicated scale.

respect to the center of the beam. These parameters are calculated for both the two-dimensional distribution of electrons in the specimen plane and the one-dimensional projection of this distribution perpendicular to the knife edge.

The latter (one-dimensional) distribution is the appropriate one for the analysis presented in this paper, because the cardiac SL and junctional SR are probed as linear structures and we are concerned about only those electrons that are distributed perpendicularly into the adjacent phases (cf. FIGURES 1 and 5).

The values for (radial) standard deviation and FWHM are much larger in the two-dimensional case. For example, the one-dimensional distribution of electrons for the 3-nA probe has a standard deviation of 14.8 nm, while the corresponding probe diameter that includes 95% of the electrons is 72 nm.

RESULTS AND DISCUSSION

FIGURE 1 illustrates the approach we use to analyze sarcolemma (SL). Scanning rasters 22 nm wide are positioned along a cell edge. The SL is not visualized as such, but its exterior surface is indicated by an abrupt density change between cytosol and extracellular fluid (ECF). While X ray spectra are being acquired, the position of this transition is continuously visualized in the raster, and adjustments are made for any movement of the sample with respect to the probing electrons. The X ray intensities generated from this so-called SL raster are compared with those sequentially obtained from adjacent rasters over pure ECF and cytosol.

Such triads of measurements can be used to obtain better estimates of elemental accumulation associated with SL proper. We begin with the assumption that the ECF and cytosol phases are homogenous. Although the comparison of concentrations in mmol/kg dry weight among these three regions can be misleading (cf. Introduction), the electron beam intensity and sample thickness in each region remain essentially constant during the analysis of each of the three regions while the triad of measurements (FIGURE 1) is being performed. Therefore, the X ray intensity of one element in a single region is proportional to its concentration per unit volume in that region. These intensities from all three regions can be directly compared, and furthermore they can be added as shown in the following equation.

$$I^{sl} = F^{ecf} I^{ecf} + F^{csl} I^{csl} + F^{sl'} I^{sl'}, \tag{1}$$

where F is fraction of electrons impinging on a phase, and I is either continuum or characteristic X ray intensity, and superscripts refer to structural phases (*e.g.*, csl = cytosol), of which only "true" SL (sl′) cannot be directly measured. The interpretation of the equation is as follows. When the SL is probed, many electrons do not excite the SL proper. A fraction of them, F^{ecf}, excites the extracellular phase, another fraction, F^{csl}, excites the cytosol, and only the remaining fraction, $F^{sl'}$, excites the true SL. Therefore, any intensity measured using the SL raster is the sum of the weighted intensities excited from the ECF, the cytosol, and the true SL. If these weighting factors, *i.e.* the fractions of electrons impinging on each phase, are known, the X ray intensities attributable to the true SL can be calculated. Rearranging Equation 1, gives

$$I^{sl'} = (I^{sl} - F^{ecf} I^{ecf} - F^{csl} I^{csl}) \, / \, F^{sl'} \qquad (1a)$$

I^{sl}, I^{ecf}, and I^{csl} are measured directly in each triad of rasters. The weighting factors $F^{sl'}$, F^{sl}, F^{ecf}, and F^{csl} can be estimated, if the spatial distribution of electron intensities with respect to the SL raster is known.

The first step towards estimating the distribution of electrons in a raster is to measure their distribution in a static, unrastered probe. To do this, the 3 nA probe used in this study is positioned at precise distances perpendicular to a knife edge specimen, and the production of X rays as a function of distance is measured. The standard

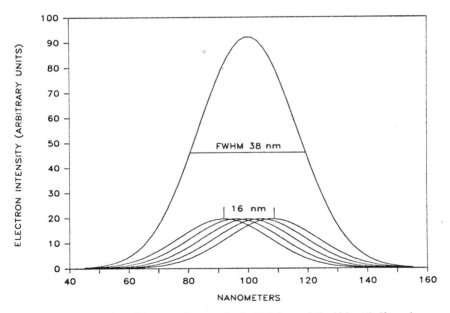

FIGURE 2. Summation of five gaussians, each of which has a full width at half maximum (FWHM) of 36 nm and is distributed at 4-nm intervals over a distance of 16 nm. Their sum is approximately guassian and has a FWHM of 38 nm. This shows how the distribution of electrons around the individual line scans that comprise a rectangular raster can be added to give the distribution of electrons around the center of that raster. It also illustrates that the spatial distribution of electrons around a raster need not be much broader than that around a static probe.

deviation of 14.8 nm and a FWHM of 34.9 nm are calculated[13,14] from the resulting cumulative intensity distribution (see Methods). These parameters describe the gaussian distribution of electrons projected perpendicular to any straight line in the specimen plane over which the static probe might be centered.

If the same probe is scanned along a straight line, these same parameters describe the distribution of electrons perpendicular to the line. Since the rasters we use are essentially a parallel array of such line scans, the electron distribution at right angles to this parallel array is merely the sum of a set of gaussians that are distributed across the raster width. This is illustrated in FIGURE 2, where each small gaussian has a FWHM of 36 nm and represents the distribution of electrons at right angles to a single line scan. Since these gaussians are distributed over a distance of 16 nm, their sum

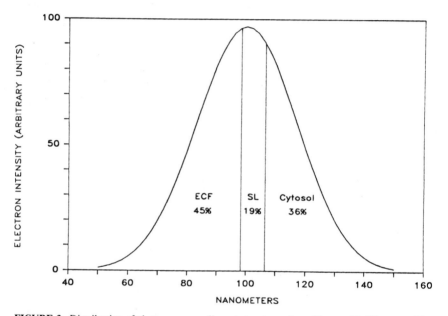

FIGURE 3. Distribution of electrons over adjacent structures for a 22-nm wide SL raster. The intensity distribution of electrons (centered at the 100 nm mark) is integrated to show the percentages of electrons exciting X rays from the *SL* proper and from the adjacent *ECF* and *cytosol*.

represents the distribution of electrons at right angles to a 16-nm wide raster. This sum is close to a gaussian and has a FWHM only 2 nm greater than each of the component gaussians. Therefore, in our situation, rastering gives controlled probe placement without significant deterioration of spatial resolution.

FIGURE 3 presents the distribution of electrons around the center of the 22 nm rectangular raster used for SL analysis. Knowing the width of the SL and how it is centered in the raster, one can readily integrate this distribution to estimate the fraction of electrons falling on the SL and on the two adjoining phases. F^{sl} is 0.19, F^{ecf} is 0.45, and F^{csl} is 0.36. When these fractions are used in Equation 1a, the contributions of X rays from the adjoining ECF and cytosol can be factored out to leave a corrected true SL value.

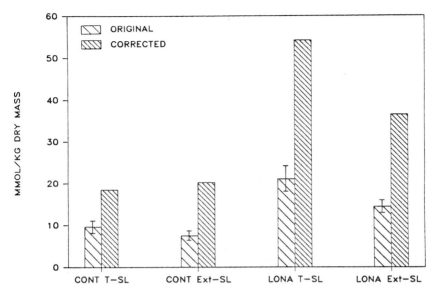

FIGURE 4. Sarcolemmal (SL) Ca concentrations measured originally using SL rasters (*left hand bars*) compared with values corrected using Equation 1a as described in text. Control (*CONT*) Ca values originally obtained over the SL in T tubules (*T-SL*) and at the exterior cell surface (*Ext-SL*) are similar to each other, but differ significantly from those obtained over the same structures in low-Na experiments (*LONA*). The correction for X rays coming from adjacent phases causes all values to double but preserves the original observed differences.

FIGURE 4 shows the effect of such a correction to Ca concentrations measured from either T tubule (T) SL or external (E) SL of heart muscle perfused with either control or low-Na solution. Before correction, a highly significant doubling of Ca was measured in both locations in the presence of low Na. After correction, all SL concentrations increased markedly without changing the overall result. The corrected

FIGURE 5. Illustration of strategy used for placing rasters for microprobe analysis of the junctional sarcoplasmic reticulum (JSR). The 18-nm wide *JSR raster* is centered within the *JSR complex* (cf. text). After an X ray spectrum is collected using this raster, additional spectra are collected from adjacent rasters centered over the bulk *ECF* and *cytosol*. Drawn to indicated scale.

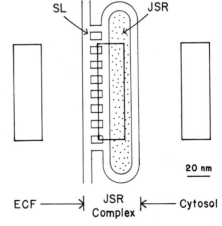

control values agree well with those measured by Philipson *et al.*[16] in sarcolemmal vesicles isolated from rabbit heart, namely, approximately 70 mmole Ca/kg *protein* in the presence of 1 mM Ca in a medium with physiological ionic strength.

We have applied a similar approach to analysis of junctional sarcoplasmic reticulum (JSR). The JSR is visualized as an elongate, dense thickening on the cytosolic side of the cell edge, and an 18 nm-wide rectangular raster is centered over it. As indicated in FIGURE 5, this raster over the JSR excites a complex of structures that includes the JSR proper along with the adjoining SL and the junctional feet that connect them. The raster is presumably centered within the complex as such. As with analyses of the SL, adjacent rasters are made sequentially over the bulk ECF and cytosol phases. A separate measurement of SL from a region of the same T tubule where JSR is not apparent is also obtained where possible.

FIGURE 6 presents the distribution of electrons around the center of the 18-nm wide raster used for the JSR complex. Using the width of the entire complex and of the SL, the electron distribution has been integrated to obtain estimates of the fraction of electron falling on SL, JSR, ECF, and cytosol. An equation similar to Equation 1a but with an added term for SL is then used to factor out contributions of adjoining phases and to arrive at a corrected "true" JSR value.

Control JSR measurements of seven elements are presented in FIGURE 7. The left-hand bars are uncorrected, while the right-hand bars have been corrected using the fractions indicated in FIGURE 6. The corrections are relatively minor. The concentration pattern is similar to that of cytosol and also resembles that reported by the Somlyos for terminal cisternae of skeletal muscles, save for the lower Ca concentration.

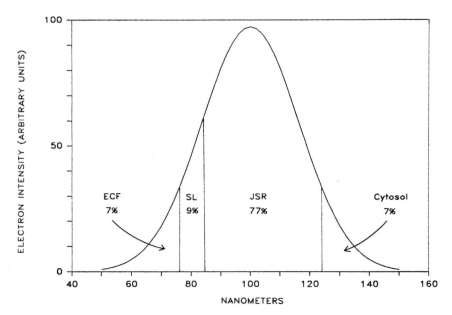

FIGURE 6. Distribution of electrons over adjacent structures for a 18-nm wide JSR raster. The intensity distribution of electrons is integrated to show the percentages of electrons exciting X rays from the *JSR* and from *adjacent structures*. In this case, the JSR is defined as the JSR complex minus the SL.

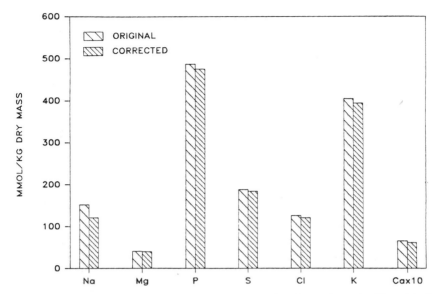

FIGURE 7. Elemental concentrations measured from the JSR of control muscles. The original uncorrected data (*left hand bars*) is compared with values corrected using Equation 1a. X rays coming from adjacent phases have only a small effect on these JSR measurements.

An additional interesting result is that the JSR Ca concentration doubles in low-Na muscles (p < 0.001). Application of the same type of correction to these low-Na muscles shows that this doubling is virtually independent of the doubling of Ca concentration that occurs simultaneously in the SL.

CONCLUSION

Electron-probe microanalysis of sarcolemma and sarcoplasmic reticulum in cardiac muscle is important for understanding cardiac contractility, but it has been necessary to demonstrate that satisfactory analysis of such small structures is not prevented by limitations due to electron beam spread. In order to minimize potential errors related to this limitation, we propose the following:

- Addition of low concentrations of dextran to the perfusate in order to help retain extracellular material *in situ, i.e.* to prevent "pile up" of extracellular ions on the structures in question.
- Analysis in groups, *e.g.*, the structure in question along with the surrounding bulk phases.
- Knowledge of the distribution of electrons around the raster probe in order to provide a rational basis for estimating the relative contributions of X rays from adjacent phases.

Additional refinement of this approach is possible. For instance, the use of specific phase markers is currently under investigation in our laboratory.

Other types of electron source might provide better spatial resolution, even at the relatively high probe currents used here for analyzing low levels of Ca. However, the more fundamental limitations are likely to be related to the thickness and low inherent contrast of cryosections.

We conclude that the size of the electron probe we use does not prevent one from obtaining useful information from these narrow structures at the cell surface. With proper care, spatial deconvolutions can be carried out to assist in the interpretation of data, and specifically to evaluate the likely contributions of adjoining phases to concentrations measured at the SL and JSR.

ACKNOWLEDGMENTS

The authors wish to thank Dr. A.P. Somlyo for helpful discussions.

REFERENCES

1. BUJA, L. M., H. K. HAGLER, D. PARSONS, K. CHIEN, R. C. REYNOLDS & J. T. WILLERSON. 1985. Alterations of ultrastructure and elemental composition in cultured neonatal rat cardiac myocytes after metabolic inhibition with iodoacetic acid. Lab. Invest. 53: 397–412.

2. TORMEY, J. McD. 1984. Accuracy and precision of biological x-ray microanalysis: Cardiac muscle. *In* Microbeam Analysis-1984. A.D. Romig, Jr. & J.I. Goldstein, Eds. 272–276. San Francisco Press. San Francisco, CA.

3. WENDT-GALLITELLI, M. F. & R. JACOB. 1982. Rhythm-dependent role of different calcium stores in cardiac muscle: X-ray microanalysis. J. Mol. Cell. Cardiol. 14: 487–492.

4. WENDT-GALLITELLI, M. F. & R. JACOB. 1982. Intracellular membranes as boundaries for ionic distribution. In situ elemental distribution in guinea pig heart muscle in different defined electro-mechanical coupling states. Z. Naturforsch. Teil C 37: 712–720.

5. WENDT-GALLITELLI, M. F. & R. JACOB. 1984. *In* Cardiac Glycoside Receptors and Positive Inotropy. E. Erdmann, Ed. 79–86. Steinkopff. Darmstadt.

6. WENDT-GALLITELLI, M. F. & H. WOLBURG. 1984. Rapid freezing, cryosectioning, and x-ray microanalysis on cardiac muscle preparations in defined functional states. J. Electron Microsc. Tech. 1: 151–174.

7. WHEELER-CLARK, E. S. & J. McD. TORMEY. 1985. Redistribution of subcellular electrolytes accompanying the increased myocardial contractility produced by low Na. *In* Microbeam Analysis-1985. J. Armstrong, Ed. 116–118. San Francisco Press. San Francisco, CA.

8. SOMLYO, A. V., H. SHUMAN & A. P. SOMLYO. 1977. Elemental distribution in striated muscle and the effects of hypertonicity. Electron probe analysis of cryo sections. J. Cell Biol. 74: 828–857.

9. SOMLYO, A. V., H. GONZALEZ-SERRATOS, H. SHUMAN, G. McCLELLAN & A. P. SOMLYO. 1981. Calcium release and ionic changes in the sarcoplasmic reticulum of tentanized muscle: An electron-probe study. J. Cell Biol. 90: 577–594.

10. SOMLYO, A. V., G. McCLELLAN, H. GONZALEZ-SERRATOS & A. P. SOMLYO. 1985. Electron probe x-ray microanalysis of post-tetanic CA^{2+} and Mg^{2+} movements across the sarcoplasmic reticulum in situ. J. Biol. Chem. 260: 6801–6807.

11. HALL, T. 1971. The microprobe assay of chemical elements. *In* Physical Techniques in Biological Research, 2nd Ed. G. Oster, Ed. Vol. 1A: 157–265. Academic Press. New York, NY.

12. SHUMAN, H., A. V. SOMLYO & A. P. SOMLYO. 1976. Quantitative electron probe microanalysis of biological thin sections: Methods and validity. Ultramicroscopy 1: 317–339.

13. TORMEY, J.McD. 1983. Improved methods for x-ray microanalysis of cardiac muscle. *In* Microbeam Analysis-1983. R. Gooley, Ed. 221–228. San Francisco Press. San Francisco, CA.

14. JOY, D. C. 1974. SEM parameters and their measurement. *In* Scanning Electron Microscopy/1974. Vol. 1: 327–334. IIT Research Institute. Chicago, IL.
15. VAUGHAN, W. H. The direct determination of SEM beam parameters. *In* Scanning Electron Microscopy/1976. Vol. 1: 745–752. IIT Research Institute. Chicago, IL.
16. PHILIPSON, K. D., D. M. BERS, A. Y. NISHIMOTO & G. A. LANGER. 1980. Binding of Ca^{2+} and Na^+ to sarcolemmal membranes: relation to control of myocardial contractility. Am. J. Physiol. **238:** H373–H378.

DISCUSSION OF THE PAPER

A. V. SOMLYO (*University of Pennsylvania School of Medicine, Philadelphia, Pennsylvania*): I wondered whether the junctional SR in the cardiac muscle would extend through the entire thickness of a 1,000-Å section?

TORMEY: I don't believe it will extend through the entire thickness of *every* 1,000-Å section. We do try to select regions where it is well defined, and suspect that it usually does in our case. There is always the possibility that the SR does not extend through the entire section and this could "dilute" our results.

T. A. HALL (*University of Cambridge, Cambridge, England*): A somewhat related question, do you think your determination of the fractions of volume excited might be affected by electron scattering? Or is this negligible?

TORMEY: We assume that it's negligible, though we are dealing with relatively dense structures. What is your opinion?

HALL: My guess is that it would make a slight correction to the fractions you already have.

TORMEY: I think any slight correction would not be warranted, since we're already basing our corrections on a number of assumptions.

A. P. SOMLYO (*University of Pennsylvania School of Medicine, Philadelphia, Pennsylvania*): Did you try to make the sarcolemmal measurements under nominally calcium-free conditions?

TORMEY: Are you referring to our measurements of elevated membrane-associated calcium in the presence of low sodium?

SOMLYO: I am not referring only to the low sodium, but also to the calcium present in your solutions. Did you try to make the same measurements while perfusing your preparation with a solution that contains no extracellular calcium?

TORMEY: No, we have not tried to measure this in near-zero calcium containing media. Secondly, we are measuring the total concentration of several forms of calcium at or very near the cell surface. In the case of the low-sodium results, much of the increased calcium is not physically bound to the membrane. Due to the lower ionic strength of the solution, fixed negative charges on the membrane exert a stronger effect on nearby ions in solution. We're picking up a cloud, as it were, of calcium ions accumulated in a local electrostatic field.

Electron-Probe Analysis of Cultured Cells

C. LECHENE

Harvard Medical School
and
Brigham & Women's Hospital
Boston, Massachusetts 02115

INTRODUCTION

Electron-probe analysis of cultured cells presents unique advantages to the study of transport phenomenon in cellular physiology. A very small number of cells, only a few hundred, is sufficient for measuring intracellular K, Na, and Cl content and their net fluxes. Thus, a broad variety of cells may be studied. Experiments are not limited to established cell lines or to primary cultures of long duration, as when it is necessary to harvest a quantity of cells large enough for usual methods of analysis. The cells can be obtained by culturing precisely identified, microdissected microsegments of an organ. Thus, both the histological origin of the cells is easy to assert, and the transport properties of cell types difficult or impossible to isolate in bulk can be studied. Because only a small number of cells is needed for analysis, the cells can be cultured for no more than a few days; culture of short duration may be important for the cells to keep their terminally differentiated functions as they were expressed in the organ from which they originated. Measuring intracellular K, Na, and Cl content on the same cells provides for a direct evaluation of the stoichiometry between their net fluxes, of the interplay between effective permeability pathways and Na-K pump activity, and of cell volume variations during experimental challenges. Because cells are analyzed individually, functional differences among cells believed to be from the same histological origin, plated simultaneously, and growing together on the same supports, may be sorted out. An important adjunct to electron-probe analysis of cultured cells is the use on individual cells of microfluorescence analysis of dyes sensitive to pH. It allows us, using a small number of cells, to study the relation of intracellular pH with K, Na, and Cl permeability pathways and with Na-K pump activity.

METHODOLOGY

The methods for electron-probe analysis of cultured cells and for isolating and culturing renal proximal tubule epithelial cells have been described.[1-4] We shall only mention some salient methodological aspects.

After isolation, the cells are plated onto small silicon pieces (2 × 2 mm) that have been extensively washed. The silicon pieces (approximately one hundred) are contained in one 35-mm Petri dish, and the cells are cultured for a period of from 2 to 4 days. The cells are observed during growth in the culture medium on the silicon pieces using incident light microscopy. During an experiment, cells on one silicon piece are successively treated for a given time in the appropriate experimental medium, washed, and frozen in liquid nitrogen. At the end of an experiment, the cells attached to the silicon pieces are freeze-dried. All the silicon pieces are mounted onto a single support so that they can be analyzed under the same electron-probe setting. In this manner, except for the experimental maneuver, all the cells are cultured and analyzed in the

270

same condition. This gives the analytical results great reliability. The cells are not stripped from the silicon piece onto which they have grown; this may avoid disturbing some membrane receptors.

Washing off the culture medium is essential in order to avoid leaving on the preparation a crust of dry salts that may decrease by absorption or increase by contamination characteristic X rays emitted by the cells. For electron-probe analysis of a small number of cells attached to silicon pieces, washing with distilled water at 4° C for a few seconds is better than washing with isotonic media using either nonionic (sucrose) or heteroionic ($Mg(NO_3)_2$, $MgSO_4$, LiCl, $LiNO_3$) washing solution.[1] After brief washing with distilled water and freeze-drying, control cells keep a normal high intracellular K and low intracellular Na content; the correlation coefficient between K and P or C, an index of the mass of the analyzed cell, is high; the intercept of the regression between P and K intensity counts is not different from zero; there is no contamination of the silicon surface between the cells with Na, Cl, or K—an indication of cellular or incubating medium elements. Microscopic observation shows well delineated cells and a clean surface of the silicon support between the cells. There is no visible swelling or cell detachment from the silicon support. It should be noted that in contrast to red blood cells, all the cells types we have observed using interference contrast microscopy after washing with distilled water, fibroblasts of different origins, MDCK and LLCPK renal cell lines, rat renal proximal tubular cells (RPTC), explants from microdissected mouse renal thick ascending limbs, rat renal papillary cells, and rat gastric surface cell can withstand the osmotic shock of distilled water for at least one minute without bursting. Moreover, the cell membrane permeability does not appear to increase, at least with respect to the cells' organic content, as verified by the equivalence of P or C cellular characteristic X ray counts between control conditions and up to one minute washing in distilled water.

We compared Na and K content in RPTC using a wash in distilled water, in isotonic ammonium acetate, and (for K content only) in isotonic sodium chloride. The results were similar to results we had observed with human diploid fibroblasts.[1] K content was slightly higher when cells were washed with distilled water than with ammonium acetate. Na content may be lower when washed with distilled water than with ammonium acetate. P content was equivalent (S. Rane and C. Lechene, unpublished data, TABLE 1).

PROTOCOLS

Two general protocols are used. With one protocol the net ionic permeability pathways are studied. With the other protocol the Na-K pump activity is evaluated.

Permeability pathways are studied after inhibition of the Na-K pump either with ouabain or by incubation in a medium without any K. Intracellular content of Na, K, and Cl are measured as a function of time until a new steady state is achieved. The best fit to the changes in intracellular Na and K content with time is computed. It is, in general, described by a single exponential. The rate coefficients of the exponential fit to Na influx and to K efflux give an estimate of their effective permeability. The effective permeability, obviously, encompasses conductive, symport and antiport pathways. The value of the first derivative at time $\emptyset$ of the exponential fit is the value of the initial rate of flux. Initial rates of flux may also be calculated from measurements taken at few seconds' intervals over a period of a few minutes. Equivalent values of initial rates were obtained from experiments in RPTC by fitting the data either to a single exponential using the data obtained from a three-hour-long experiment or to a linear

TABLE 1. Comparison of the Effect of Washing Media on Ionic Content in Renal Proximal Tubule Cells

Wash Solution[a]	P (fEq)	K/P (M/M cell P)	Na/P (M/M cell P)	Cl/P (M/M cell P)
First Experiment				
$H_2O(30)^b$	122 ± 8	1.46 ± 0.019	0.279 ± 0.049	0.413 ± 0.038
CH_3COONH_4				
$(30)^b$	114 ± 7	1.31 ± 0.039[c]	0.240 ± 0.080	0.435 ± 0.064
$NaCl(20)^b$		1.17 ± 0.104[c]		
Second Experiment				
$H_2O(30)^b$	129 ± 7	1.49 ± 0.037	0.266 ± 0.054	0.359 ± 0.034
CH_3COONH_4				
$(30)^b$	119 ± 6	1.25 ± 0.080[c]	0.471 ± 0.060[c]	0.621 ± 0.052[c]
$NaCl(28)^b$		1.31 ± 0.074[c]		

[a]Results of two experiments. Solutions of CH_3COONH_4 and NaCl were approximately 300 mOsm. P is in femtoequivalents/100 μ^2 dryed cell.
[b]( ): number of cells—values are mean ± SE.
[c]p < 0.05 with respect to H_2O.

regression by using multiple early time points obtained within the first 3 min of the experiment. (0.037 mM Na/mM P/min from linear fitting over 3 min compared to 0.038 mM Na/mM P/min calculated from the derivative at time $\emptyset$ of monoexponential curve fitting of experimental points obtained over a period of 3 hours). Initial rates are different among cell lines or among experimental conditions, so that the times during which they are approximately linear is variable. Thus, we found it preferable to calculate initial rates from the first derivative of the fit to experimental points measured during a long experiment than from the linear regression fit to early time points. Comparison of rate coefficients or initial rates of fluxes in the presence or absence of "specific" inhibitors of ionic movement pathways may uncover the presence of these pathways and their magnitude.[5] Changes in K, Na, and Cl intracellular content can be measured over a short period of time (seconds) and in the presence of an active Na-K pump. For example, in cultured RPTC, intracellular Na fall within a few minutes in the presence of amiloride or increase within a few seconds after intracellular acidification.[5]

Na-K pump activity can be measured in two experimental conditions. The first condition is when the Na-K pump is close to Vmax activity. This is obtained by loading the cells with Na using preincubation in a K-free medium. After a return in a 5-mM K-containing medium, the ouabain-sensitive initial rates of Na efflux and of K influx may be taken as a good estimate of the Na-K pump activity.[1,5] The second condition is when the Na-K pump functions in a control, steady state situation with the cells containing a normal, low intracellular Na content. Cells are preloaded with Rb by incubation in a K-free medium containing 5 mM ultrapure RbCl; in this situation the cells exchange all the intracellular K for Rb. After the cells are returned in a medium containing 5 mM KCl, Rb-free, the ouabain-sensitive rate of K influx may be taken as the rate of Na-K pump activity for K.[5] Simultaneous measurements of Rb efflux and K influx allow us to compare leak and pump rate.[5] Na-K pump rate for Rb can be measured from the ouabain-sensitive rate of Rb influx and can be compared to the K pumping rate. Similarly, the K and Rb leaks may be compared.

In RPTC, the majority of K influx measured under the conditions of Rb preloading is mediated by the Na-K pump because:[5] 1: Ouabain inhibits at least 87% of net K influx; 2) in the absence of ouabain, Rb efflux and K influx are equivalent; 3) K efflux from normal high K cells exposed to ouabain is equivalent to K efflux from normal high K cells exposed to a medium to which Rb was substituted for external K (0.025 ± 0.004 mM K/mM P/min vs, 0.029 ± 0.005 mM K/mM P/min; (n = 3); and 4) K efflux from normal high K cells exposed to ouabain is not significantly different from Rb efflux from Rb-loaded cells in which K was substituted for external Rb (0.033 ± 0.001 mM K/mM P/min (n = 4) vs. 0.042 ± 0.007 mM Rb/mM P/min (n = 6)(P > 0.05)).

Because intracellular K has been shown to decrease the affinity of Na for the internal binding site of the Na-K pump,[6] it is possible that Rb has similar effects. Compared to Na-loaded cells the slower rate of K influx measured in cells with low Na (Rb loaded) may reflect in part competition of intracellular Rb with Na binding to the pump.

RENAL PROXIMAL TUBULAR CELLS (RPTC)

Na Effective Permeability, Na-H Exchange, Na-K Pump Activity, pH Regulation, Volume Regulation, and Chloride Permeability in RPTC

Na effective permeability of mature rat RPTC is the highest that we have observed among any of the cell types that we have studied, polymorphonuclear neutrophiles, fibroblasts, and epithelial cell lines (MDCK, LLCPK) and is equivalent to the high transport rate of Na estimated from micropuncture in the rat proximal tubule.[4] Na permeability pathway is at least 80% through a Na-H exchanger.[5] The magnitude of the Na-H exchange and the magnitude of its expression in the absence of serum is a unique characteristic of these cells. Na-K pump activity can increase three times between low intracellular Na RPTC and Na-loaded RPTC (FIGURE 1). After RPTC are acid-loaded, there is an activation of the Na-H exchanger, able to correct the acid load in 5 minutes. The activation of the Na-H exchanger increases the intracellular Na concentration in approximately 40 seconds. Increased intracellular Na concentration leads in turn to an increase in the rate of Na-K pump activity (FIGURE 2), resulting in a secondary return of intracellular Na to control values. This may be a general mechanism for intracellular pH homeostasis.[5]

A small component of the increase in Na content after intracellular acidification

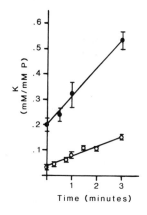

FIGURE 1. Comparison of rates of K influx into RPTC with high intracellular Na (1.06 mM Na/mM P) after intracellular K depletion by preincubation in DMEM + 0.2 mM K (*filled circles*), and K influx into RPTC with low, basal intracellular Na (0.121 mM Na/mM P) after K replacement by preincubation in DMEM-containing RbCl (*open circles*).

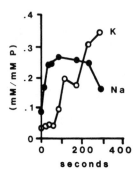

FIGURE 2. Simultaneous time course of increase in Na (*filled circles*) and K (*open circles*) content after intracellular acidification by NH_4Cl exposure and removal. RPTC were preincubated 180 min in DMEM + 5 mM Rb, then incubated with 15 mM NH_4Cl, and then returned at time $\varnothing$ to DMEM without NH_4Cl and RbCl, and with 5 mM KCl.

may be due to an inhibition of the Na-K pump. In RPTC, however, our estimate of Na-K pump activity measured by ouabain-sensitive K influx into Rb-loaded cell is 0.032 mM K/mM P/min. Assuming a 3–2 stoichiometry for the Na-K pump, the maximum Na that would accumulate by one minute if the Na-K pump was completely inhibited would be 0.048 mM Na/mM P. In the acid-loaded cells, the Na content increased by 0.215 mM Na/mM P within one minute. Therefore, almost 80% of the increase may be attributed to an increase in Na influx. Moreover, Na influx was inhibited by 0.5 mM amiloride, a concentration that has negligible effects on the RPTC Na-K pump. Thus, the amount of Na influx that could have been caused by at most a complete inhibition of the Na-K pump could not represent more than 20% of the change in intracellular Na content following cytoplasmic acidification.

Preparation and analysis of cells on impermeable silicon support does not allow us to differentiate between apical and basolateral membrane transport pathways. Na influx pathways that are confined to either apical or basolateral membranes *in vivo* may contribute to measurement of net Na influx in the present preparation of RPTC. The large component of amiloride-inhibitable Na influx may, however, represent predominantly an apical pathway, since, unlike the amphibian proximal tubule, the Na-H exchanger is confined to the apical membrane in the mammalian proximal tubule.[7]

It was surprising to find that short-term primary cultures of RPTC growing on nonpermeable supports not only displayed all the properties found in the luminal and the basolateral side of proximal tubules but did so with a magnitude equivalent to that measured in proximal tubules *in vivo* from micropuncture data. The luminal properties found in RPTC are Na-glucose, Na-amino-acids, Na-phosphate cotransports, parathyroid hormone inhibition of Na influx, and Na-H exchange. The basolateral properties are ouabain-inhibitable Na-K pumping activity and barium-inhibitable K influx. One may wonder if RPTC have kept a polarized organization of these different transmembrane pathways or if they are all exposed on the surface of the cells facing the medium. Judging from the observations and discussion of Cereijido *et al* on MDCK cell line,[8] it is possible that small colonies of RPTC maintain a polar organization of membrane pathways, and that the basolateral side of the cells is immediately accessible, for example to ouabain, through "tight" junctions that remain permeable.

RPTC transferred in hypotonic media regulate their volume. Volume regulatory decrease is through an increase in K and Cl leaks. RPTC returned from hypotonic to isotonic media regulate their volume by increasing Na entry and Na-K pump activity. Cells preloaded with Na demonstrate volume regulatory decrease. Thus RPTC volume regulation is not predicated upon the presence of a high intracellular K content.[9] This may be explained by the gelled state of the cytoplasm and by the physical chemistry of gels.[10]

The effective permeability of RPTC to Cl is much higher than to K. After replacement of Cl by gluconate in the incubating medium, RPTC lose Cl with a half-time of 20 sec. Cl permeability pathway is in part through a Cl-HCO$_3$ exchanger.[11] Despite the high permeability of RPTC to Cl, due to the fact that intracellular Cl concentration is higher than Cl electrochemical equilibrium, there cannot be massive net conductive transcellular Cl movement. This means that the bulk of proximal tubule Cl reabsorption ought to be through paracellular pathway.

Coupling Ratio of the NA/K Pump

The coupling ratio Na/K of the Na-K pump may be evaluated both in cells containing a high Na concentration[1] and in cells in normal situation with a low intracellular Na. In this last condition, Na/K pumping ratio may be estimated from the initial rates of K and Na leaks after Na-K ATPase inhibition with ouabain. Such a calculation makes the obvious assumption that in steady state Na efflux and K influx pump rates are equal to Na influx and K efflux leak rates. A question may arise, however, with respect to a possible instantaneous modification of the initial rates of leak after ouabain addition. Such is unlikely in RPTC for the following reasons. As discussed previously, in RPTC, initial rate of K efflux, J_k, after Na-K pump inhibition with ouabain is equivalent to J_k measured in steady state condition in Rb-preloaded cells with a functional Na-K pump. Thus, ouabain does not affect the initial rate of K efflux compared to steady state. By analogy, it is also unlikely that the initial rate of Na influx is modified after pump inhibition with ouabain.

Na-K pump coupling ratios in two conditions of pump activity—at low intracellular Na, normal situation, and a high intracellular Na, with a pump activated[5]—are displayed in TABLE 2. The coupling ratio is not significantly different from 1 in low Na cells and not significantly different from 1.5 in high Na cells. The variability of the coupling ratio with pump activity has been reported in frog skin by Cox and Helman.[12] Using ^{22}Na and ^{42}K, Cereijido et al.[8] have reported that in monolayers of MDCK cells at physiological Na content, the pump transports 1 Na$^+$ for 1 K$^+$, but that a high Na content the pump becomes electrogenic.

Ontogeny

In RPTC cultured for two days, the effective Na and K effective permeabilities are twice as high in cells from adult rat kidneys terminally differentiated as in cells from young rat kidneys not terminally differentiated.[4]

TABLE 2. Coupling Ratio of the Na-K Pump in RPTC[a]

	K/Na	Δ from 1	Δ from 1.5
Low Na (N = 15)[b]	1.08 ± 0.04 SE	t = 1.8	t = 11.3[c]
High Na (N = 5)[b]	1.44 ± 0.12	t = 3.51[c]	t = 0.48

[a]R. C. Harris, S. Larsson, and C. Lechene, unpublished data.
[b]N = number of experiments.
[c]p < .05.

When RPTC originating from kidneys of young rats 10 to 16 days old during the span of age of terminal differentiation are cultured for two days, there is an increase in K and Na effective permeability in function of the age of the rat (FIGURE 3).[4] RPTC from immature rat kidneys do not undergo a change in Na and K permeability with the duration of culture from 2 to 5 days. On the contrary, Na and K effective permeability of RPTC from adult rat kidneys terminally differentiated decrease by a factor of two between 2 and 5 days of culture (FIGURE 4). After 5 days of culture, they reach the lower level of permeability observed in cells from young rat kidneys not terminally differentiated.[4] Thus, terminally differentiated RPTC after only 4 to 5 days of culture regress to the permeability levels observed in cells that are not terminally differentiated, indicating a dedifferentiation of the cells after a very short time in culture.

Parallel to the decrease in Na and K effective permeability observed in adult rat RPTC between 2 and 5 days of culture, there is a decrease in Na-K pump activity.[13] During terminal differentiation we found an increase in the expression of Na-H exchange and a decrease in amiloride-insensitive Na influx pathways.[14]

Growth

In a colony of RPTC, the most peripheral cells, located at the edge of the colony, are the only cells synthesizing DNA, as demonstrated by 3H thymidine incorporation. The K/Na ratio in the cells located at the edge of the colony is significantly higher than in the cells more centrally located in the colony. The intracellular pH of these cells is alkaline compared to the centrally located cells. Thus, in a colony of RPTC, growth occurs at the edge, where the K/Na ratio is higher and where the cells are more

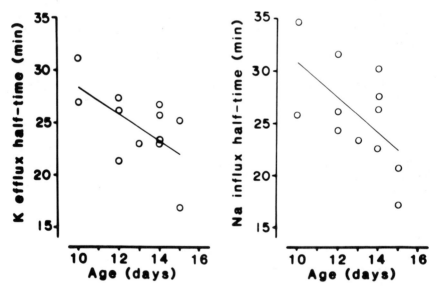

FIGURE 3. Effective K and Na permeability in young rat RPTC in relation to age of animal. Cells were cultured for 3 days. Effective permeability was measured as half-time of K efflux and Na influx after exposure of cells to ouabain for variable periods of time. Cells were continuously grown in a medium containing 10% FBS. A significant correlation was found between age of animal and half-time for K efflux ($r = -0.61$, $n = 12$) or Na influx ($r = -0.60$, $n = 12$).

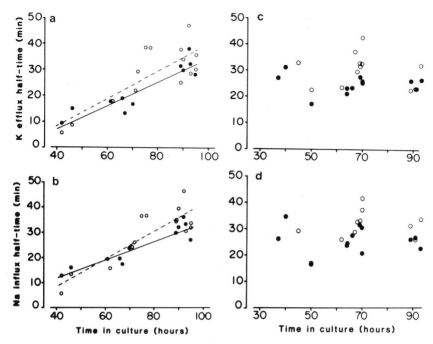

FIGURE 4. (a) Half-time of K efflux in relation to culture time in adult rat RPTC. Adult rat RPTC were cultured for 2–4 days, and K efflux was measured after exposing cells to 1 mM ouabain. Half-time for K exit was used as an estimate of effective permeability. Studies were performed on RPTC continuously grown in DMEM + 10% FBS (*filled circles*) and on RPTC that had been serum-deprived for 24 hrs before and during experiment (*open circles*). Half-times were significantly correlated to time in culture both during serum conditions (*continuous line;* r = 0.90, n = 11) and during serum-deprived conditions (*broken line;* r = 0.80, n = 14). Slopes during serum and serum-deprived conditions were not significantly different. (b) Half-time of Na influx in relation to culture time in adult rat RPTC. Experimental conditions and symbols as in (a). Half-times were significantly correlated to time in culture both during serum conditions (r = 0.90, n = 11) and during serum-deprived conditions (r = 0.80, n = 86). Slopes during serum and serum-deprived conditions were not significantly different. (c) Half-time for K efflux in relation to culture time of young rat RPTC. Experimental conditions and symbols as in (a). Half-times were not correlated to time in culture either during serum conditions (r = − 0.5, n = 12) or during serum-deprived conditions (r = 0.13, n = 12). (d) Half-time for Na influx in relation to culture time of young rat RPTC. Experimental conditions and symbols as in (a). Half-times were not correlated to time in culture either during serum conditions (r = − 0.14, n = 12) or during serum-deprived conditions (r = − 0.32, n = 12).

alkaline (FIGURE 5).[15] This observation is in line with the studies relating DNA synthesis and cellular division with alkalinization of intracellular pH.[16,17]

INTERCELLULAR COMMUNICATION[18]

In a coculture of a wild type of Madin Darby kidney cells, ouabain-sensitive (W-MDCK) with an ouabain-resistant mutant (R-MDCK), the wild type of cells were protected from the effect of ouabain up to concentrations as high as 100 μM. Ouabain

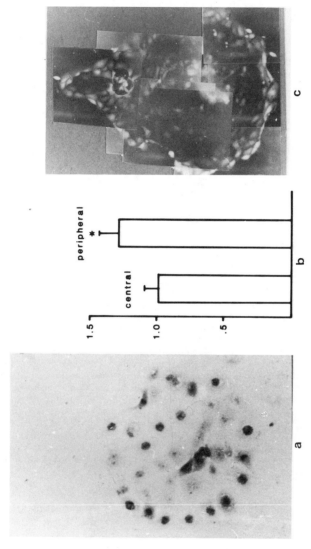

FIGURE 5. DNA incorporation, ionic content, and intracellular pH in small colonies of RPTC during growth. (**a**) Autoradiography of an RPTC colony after incubation with tritiated thymidine. Labelled nuclei are in the remnant of a proximal tubule segment, the origin of the cells' outgrowth. (**b**) Intracellular K/Na ratio. The value of the ratio is 12.9 ± 1.5 SE in the *peripheral* cells and 9.95 ± 1.0 in the *centrally located* cells (n = 5, p < 0.05). (**c**) Fluorescence of dimethylcarboxyfluorescein in a colony loaded with the pH-sensitive dye. The cells *at the edge* of the colony are more intensely fluorescent than the *more centrally located* cells. The *central* fluorescent mass is the remnant of a proximal tubule segment. Ratio of fluorescent intensity measurement at two excitation wavelengths, 490 and 460 nm, confirmed that the cells *at the periphery* of the colony were more alkaline than the cells *more centrally located*.

revealed two types of binding sites in R-MDCK cells, one with high and the other with low affinity. Only the high affinity site was present in W-MDCK cells. Electron-probe analysis of individual rescued cells revealed that they kept a high K and a low Na intracellular content, similar to the control cells. Histograms of K/Na ratios in W- and R-MDCK cells cocultured and treated with ouabain were unimodal, indicating that the rescued W-MDCK and the R-MDCK behaved as a single population with respect to their K and Na content. By microinjection of Lucifer Yellow or electrophysiology techniques it was estimated that 13% at most of the R and W cells may be connected at a given time through cell-to-cell junctions. Therefore, cell-to-cell communication of a permanent type did not seem to play an essential role in the rescue. W-MDCK cells in coculture with R-MDCK cells and subsequently separated were not rescued. Thus, rescue did not seem to depend on the transfer from R-MDCK to W-MDCK cells either of ouabain-resistant Na-K pumps, or of information to synthesize them. A possible explanation for these findings is that intercellular communications were sporadic events, so that cells may have become intermittently connected and rescued.

TABLE 3. K Influx and Na Efflux in Ouabain Resistance-Induced Oua^{r6} Cells^a

	JK	JNa
	(M/M cell P/min)	
Control	0.039 ± 0.004	0.038 ± 0.001
+ 1 mM ouabain	0.042 ± 0.010	0.041 ± 0.003
+ 0.5 mM amiloride	0.012 ± 0.003	0.014 ± 0.005

^{a}Data from Epstein and Lechene.[23]

BRAIN LESION AND KCl COTRANSPORT

K-Cl Efflux in 3T3 Fibroblasts

Using electron-probe analysis of cultured 3T3 fibroblasts, we found the following. 1) Bumetanide blocks a KCl efflux pathway but does not affect net Na influx. 2) Total net K efflux is not modified in the presence of low external Cl concentration, although the bumetanide-sensitive fraction is decreased. 3) Net K efflux is decreased by replacement of external Na with tetramethylammonium (TMA), although the fraction of bumetanide-sensitive K efflux appears unchanged. Since TMA is impermeable to the cell membrane, a reduced K efflux coupled electrically or chemically to Na influx is suggested 4) The contribution of the various pathways of K efflux to net K movement are variable under differing conditions, although net K flux may remain unchanged.[19] This may indicate that cells may maintain a net ionic flux constant by using different pathways at different rates.

K-Cl Influx: Relation to Brain Edema in Anoxia

In other series of experiments fibroblasts were loaded with sodium by preincubation in a K-deprived medium. They were then returned in a K-containing medium with 10^{-3} M ouabain and different external K concentration, maintaining constant the sum K + Na. When external K concentration reached a threshold of approximately 30 mM K, there was a massive entry of KCl leading to important cellular swelling. The net KCl entry was bumetamide-inhibitable (P. Cheng-Sing and C. Lechene, unpublished observations). The expression of such a KCl cotransport influx may be of importance to explain cellular damage in some situations of brain anoxia. A focal lesion could lead to potassium leak from intracellular store, and increase potassium concentration in the intercellular fluid, which is only in small volume in the brain. Increased K in the interstitial cerebral fluid could in turn lead to astrocytes swelling through activation of a KCl cotransport, leading itself to increased anoxia by capillary compression. Thus, it would create a vicious circle by increasing the extent of the original lesion.

TRANSPORT CHARACTERISTICS OF TRANSFORMED CELLS

Relations between cellular ionic events and cellular division have aroused an increasing interest in the past years. It is thought that transformed cells do not maintain exactly the same intracellular K and Na concentration as normal cells. The steady state cellular composition, however, does not provide any information on the rate of leaks and pump, although greatly different leak and pump rates may lead to equivalent steady states of intracellular K and Na concentrations.

With Dr. R. Bassin, National Cancer Institute, National Institutes of Health, we are studying clones of 3T3 fibroblasts, of Kirsten sarcoma virus transformed cells, and of flat nontransformed variants (revertants) isolated from populations of retrovirus-transformed cells. Two striking observations have been made. Compared to their parent lines and to their flat revertant, the cells transformed by the retrovirus: 1) undergo dramatic changes in morphology when the Na pump is inhibited, and 2) are more permeable to K by a factor of at least 1.5. It is likely that the later observation will be found to be part of the phenotype of transformed cells. Compared to their nontransformed parents, transformed cells grow more actively, and thus are likely to have to maintain a higher rate of Na and K leak rates and of Na-K pump activity in order to perform the metabolic work of growth, which spends the potential energy of the Na and K transmembrane gradients in the cotransport of metabolic substrates.

INDUCIBLE OUABAIN RESISTANCE

Transfection of CV-1 green monkey kidney cells with a 6.5-Kb ouabain resistance gene by Levenson et al.[20] has resulted in a cell line oua' 6, which is not a ouabain-resistant mutant, but to which ouabain resistance may be induced by culture with 10 μM ouabain for at least 24 hours. The genetic expression of the resistance was a 1.2 Kb mRNA.[21] We found that CV-1 cells and uninduced oua' 6 cells Na-loaded by preincubation in a medium without K and returned in a medium containing 5 mM K expressed a Na-K pumping activity, totally inhibitable with 10^{-3} M ouabain. Induced oua' 6 cells also expressed a Na-K pumping activity, but with two differences compared to the uninduced cells: total ouabain resistance and striking sensitivity to amiloride (TABLE 3). Although some inhibitory effect of amiloride on Na-K ATPase has been reported,[22] it is much less than that observed in the induced oua' 6 cells. Thus, the functional expression of the ouabain resistance gene is a Na-K pump that is resistant to 1 mM ouabain but is inhibited by 0.5 mM amiloride.[23] What the relations are between the genetic expression of the ouabain resistance gene, the induction with ouabain of a Na-K pump that is ouabain resistant, and the high sensitivity of this Na-K pump to amiloride is not yet known. It is indicative, however, of some homology between the structure of Na channel or Na-H exchanger and regulatory molecules of Na-K ATPase activity.

CONCLUSION

With electron-probe analysis of cultured cells, a few hundred cells is sufficent to measure simultaneously Na, K, and Cl movements. Many more parameters in a much smaller number of cells can be measured in comparison to usual methods of analysis. Important consequences follow. Regulation of leak pathways and of Na-K pump activity can be extensively studied in cell types in which they have been difficult or impossible to measure. Interplay between leak pathways and Na-K pump activity is also easier to study in cell lines, although the culture yield is sufficient to use other methods of analysis. However, simultaneous measurement of the kinetic of intracellular K, Na, and Cl movements, allowed with electron-probe analysis, is a unique advantage. Cell terminally differentiated functions that may disappear after only a few days of culture can be maintained and studied, because the number of cells sufficient for analysis is provided after 2 or 3 days of culture. Intercellular communication or heterogeneity of functions in a growing colony of cells may be observed, because cells are anlayzed individually. We have attempted to illustrate the power of electron-probe analysis of cultured cells.

REFERENCES

1. ABRAHAM, E. H., J. L. BREWSLOW, J. EPSTEIN, P. CHANG-SING & C. LECHENE. 1985. Preparation of individual human diploid fibroblasts and study of ion transport. Am. J. Physiol. (Cell Physiol.) 17:C154–C164.
2. LARSSON, L., A. APERIA & C. LECHENE. 1986. Ionic transport in individual renal epithelial cells from adult and young rats. Acta Physiol. Scand. 126:321–332.
3. LECHENE, C. & R. C. HARRIS. Electron probe analysis of cultured renal cells. In Contemporary Issues in Nephrology: Modern Techniques of Ion Transport. B. M. Brenner & J. H. Stein, Eds. Churchill Livingstone. New York, N.Y. In press.
4. LARSSON, S., A. APERIA & C. LECHENE. 1986. Studies on final differentiation of rat renal proximal tubular cells in culture. Am. J. Physiol. (Cell Physiol. 20) 251:C455–C464.
5. HARRIS, R., J. L. SEIFTER & C. P. LECHENE. Coupling of Na/H exchange and Na-K pump activity in cultured rat proximal tubule cells. Am. J. Physiol. (Cell Physiol), in press.
6. SOLTOFF, S. P. & L. J. MANDEL. 1984. Active ion transport in the renal proximal tubule. II. Ionic dependence of the Na pump. J. Gen. Physiol. 84:623–642.
7. IVES, H. E., V. H. YEE & D. G. WARNOCK. 1983. Asymmetric distribution of the Na^+/H^+ antiporter in the renal proximal tubule epithelial cell. J. Biol. Chem. 258:13513–13516.
8. CEREIJIDO, M., J. EHRENFELD, S. FERNANDEX-CASTELO & J. MEZA. 1981. Fluxes, junctions, and blisters in cultured monolayers of epithelioid cells (MDCK). Ann. N.Y. Acad. Sci. 372:422–441.
9. SAVIN, V. J., R. C. HARRIS, J. L. SEIFTER, B. M. BRENNER & C. LECHENE. 1986. Hypotonicity causes net efflux of K and Cl in normal rat renal proximal tubular cells (RPTC) and net efflux of Na and Cl in Na loaded RPTC. Kidney Int. 29:406.
10. LECHENE, C. 1985. Cellular volume and cytoplasmic gel. Biol. Cell 55:177–180.
11. LIPMAN, R. D., R. C. HARRIS & C. LECHENE. 1986. High chloride permeability of rat proximal tubule cells (RPTC) in primary cultures. Proc. 19th Annu. Mtg. Am. Soc. Nephrol.
12. COX, T. C. & S. I. HELMAN. 1986. Na^+ and K^+ transport at basolateral membranes of epithelial cells. II. K^+ efflux and stoichiometry of the Na, K-ATPase. J. Gen. Physiol. 87:485–502.
13. LARSSON, S., A. APERIA & C. LECHENE. 1986. Ontogeny of Na-K pump activity in rat renal proximal tubular cells (RPTC). Proc. 19th Annu. Mtg. Am. Soc. Nephrol.
14. RANE, S., A. APERIA & C. LECHENE. 1986. Serum stimulated, amiloride inhibitable Na influx is expressed at terminal differentiation of renal proximal tubular cells (RPTC). Proc. 19th Annu. Mtg. Am. Soc. Nephrol.

15. LIPMAN, R. D., R. C. HARRIS, J. L. SEIFTER, B. M. BRENNER & C. LECHENE. 1986. Growth of rat proximal tubular cells (RPTC) occurs at the periphery of a colony, where cells are more alkaline and have a higher K/Na ratio. Kidney Int. **29:**402.
16. AERTS, R. J., A. J. DURSTON & W. H. MOOLENAAR. 1985. Cytoplasmic pH and the regulation of the dictyostelium cell cycle. Cell **43:**653–657.
17. POUYSSEGUR, J., A. FRANCHI, G. LALLEMAIN & S. PARIS. 1985. Cytoplasmic pH, a key determinant of growth factor-induced DNA synthesis in quiescent fibroblasts. FEBS Lett. **190:**115–119.
18. BOLIVAR, J. J., A. LARARO, S. FERNANDEZ, E. STEFANI, V. PENA-CRUZ, C. LECHENE & M. CEREIJIDO. Metabolic cooperation between a ouabain-resistant mutant of MDCK cell and the wild type in culture. Am. J. Physiol. (Cell Physiol.), in press.
19. BOSEK, G. & C. LECHENE. 1985. K-Cl contransport in 3T3 fibroblasts. Fed. Proc. **44(6):** 3663a.
20. LEVENSON, R., V. RACANIELLO, L. ALBRITTON & D. HOUSMAN. 1984. Molecular cloning of the mouse ouabain-resistance gene. Proc. Natl. Acad. Sci. USA **81:** 1489–1493.
21. ENGLISH, L. H., J. EPSTEIN, L. CANTLEY, D. HOUSMAN & R. LEVENSON. 1985. Expression of an ouabain resistance gene in transfected cells. **260(2):** 1114–1119.
22. SOLTOFF, S. P. & L. J. MANDEL. 1983. Amiloride directly inhibits the Na-K-ATPase activity of rabbit kidney proximal tubules. Science **220:** 957–959.
23. EPSTEIN, J. A. & C. LECHENE. 1986. Ouabain induces an amiloride sensitive, ouabain resistant Na-K pump in Oua[6] cells. Proc. 19th Annu. Mtg. Am. Soc. Nephrol.

DISCUSSION OF THE PAPER

D. F. PARSONS (*New York State Department of Health, Albany, New York*): We have found with tissue cultured cells that there is a new specimen preparation problem. In working with what are usually considered to be highly viable cell lines, mammary ascites carcinoma cells, we decided to look at ^{31}P spectrum. We were shocked to find that there are only special conditions in which the full complement of energy-rich phosphate compounds are present in these cells. We had to change our cell media, we had to add nucleoside, we had to oxygenate and work at lower temperature in order to get anything like a reasonable distribution of energy-rich compounds. We didn't know about this, and wanting to monitor them by electron microscopy, we thought the cells were intact and "hardy." We found in the literature that others made the same observations. It is, in fact, very difficult to take tissue cultured cells and manipulate them, take them off one substrate, change the medium, do other things, and keep them in the intact state. If, on the other hand, you do the ^{31}P NMR spectroscopy on the live animal, you'll find that ATP-ATP ratios are very, very constant. Have you found anything like this in your work?

LECHENE: These cells that I have shown here exhibit every characteristic function of proximal tubular cells: for example, they are sensitive to parathyroid hormone, which inhibits sodium flux and have a rate of Na flux very close, within a factor of 1.5, to what one finds by micropuncture. I think the cells are in good shape, but again I emphasize that we are culturing them for no more than 3 days.

A. V. SOMLYO (*University of Pennsylvania School of Medicine, Philadelphia, Pennsylvania*): I have a question concerning the use of expressing the concentrations on the basis of phosphorus. In terms of the nucleotide concentrations just mentioned, they would only be a really small fraction of the total cell phosphorus; but I wondered whether, in the developing cells, there would be large changes in the amounts of membranes and ribosomes containing high concentrations of phosphorus.

LECHENE: This is discussed in the paper in the *American Journal of Physiology*.[4] It doesn't seem to make much difference, as much as we know about the morphometry of these cells. One could also use the same measurements normalized to carbon, although there would be some changes in carbon content during development.

Intracellular Structure and Elemental Analysis in Rapid-Frozen Neurons

S. BRIAN ANDREWS[a] AND THOMAS S. REESE[b]

[a,b]*Laboratory of Neurobiology,*
National Institute of Neurological and
Communicative Disorders and Stroke
National Institutes of Health,
Bethesda, Maryland 20892
and
[b]*Marine Biological Laboratory,*
Woods Hole, Massachusetts 02543

INTRODUCTION

It is now recognized that cryofixation is an excellent approach to preserving both the structural[1] and diffusible[2] components of unfixed tissue. Fine detail at the level of macromolecular organization can be demonstrated by electron microscopy of aperiodic, cellular structures that have been rapidly frozen and processed by newer techniques, *e.g.,* freeze-fracture and shallow-etch replication, which maximize structural information.[3,4] At the same time, sustained progress has led to impressive capabilities in analytical electron microscopy. In favorable preparations electron energy-loss analysis can now detect a few atoms at a resolution of 1–3 nm, while X ray methods are sensitive to a few hundred atoms; in realistic biological specimens, X ray microanalysis can routinely measure important elements such as calcium and potassium at submillimolar concentrations within small subcellular organelles.[5] However, the spatial resolution of analytical microscopy in cells and tissues is still about an order of magnitude less than that of structural microscopy, mainly because of limitations on the dried, frozen sections of unfixed tissue that must be used for analytical experiments.

This paper discusses, from the structural as well as the analytical viewpoint, the status of techniques for preparing directly frozen tissues for electron microscopy; it considers why, in practice, structural preparations yield more information than analytical specimens, and speculates on the prospects for improvement of structure in analytical preparations. These points are discussed in the context of recent work on the structure and composition of synapses between parallel fiber axons and Purkinje cell dendritic spines from the mammalian cerebellar cortex (FIGURE 1).[4,6] These synapses are particularly interesting because the presynaptic terminals are well-suited for studying the changes in calcium distribution that accompany synaptic transmitter release,[7] while the structure and calcium-handling ability of the postsynaptic dendritic spines has been the subject of considerable speculation with regard to the regulation of synaptic potency and plasticity.[8,9]

STRUCTURE IN DIRECTLY FROZEN NEURONS

Direct freezing of tissues has become a starting point for several new methods of analyzing the structure of cells at the macromolecular level of organization. The

284

success of these methods has largely depended on the development of rapid freezing techniques, such as cryofixation with a cold metal block.[1,10] Freezing rates measured when tissue is pressed against a copper block cooled by liquid helium are rapid enough to suppress ice crystal formation to the point where, in the most superficial 10 μm of the tissue, rearrangements of tissue components are limited to the nanometer scale. Thus, a method that relies on freeze-substitution of concentrated ferritin solutions for measuring displacements caused by ice failed to detect any ice crystals in the 2–3-μm-thick layer nearest to the cold copper block.[11] This suggested that the outermost layer may actually have been frozen vitreously, that is, without ordering of

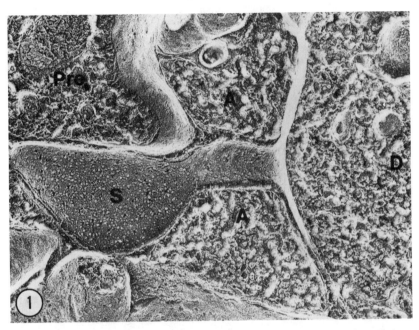

FIGURE 1. Synapse between a cross-fractured parallel fiber synapse (**Pre**) and a membrane-fractured Purkinje cell dendritic spine (**S**), from the superficial cortex of the mouse cerebellum, prepared by rapid freezing, freeze-fracturing, shallow-etching, and rotary shadowing.[3,4] Slices of cerebellum that were rapidly frozen within 20–30 seconds of decapitating the mouse[4] were frozen on their pial (uncut) surfaces. Cross-fractured astrocytic processes (**A**) enclose the base of the spine. The cross-fractured views show the fine details of cytoplasmic structure revealed by these techniques. 78,000X.

water molecules—a suggestion that has since been confirmed by Chang *et al.*,[12] as well as in our laboratories, using electron diffraction of hydrated thin cryosections.

Even with the best available freezing methods, however, the nature of ice crystal gradients requires that important structural features of a preparation be near a natural surface, or that the tissue be sufficiently hardy to permit exposure of a useful surface by dissection. For studies of neurons the choices are quite limited. One useful tissue is the mouse cerebellum, which has a sheath less than one μm thick[13]; in addition, the cerebellum of young mice (with a thin calvarium) can be removed and frozen in less than 30 seconds.[4] The turtle brain exemplifies the alternative approach, in that it survives more readily *in vitro* than the mammalian brain.[14] The turtle optic nerve also has a very thin sheath and, for studies of axons, this nerve can be dissected and maintained *in vitro* until it is frozen.

Several methods have been used to visualize the interiors of cells once they have been directly frozen. If tissues are simply broken open by freeze-fracture and then freeze-dried ("deep-etched"), the tissue collapses into a mat of salts, aggregated proteins, and filamentous cytoskeletal elements that is difficult to interpret.[15] One of the first ways to avoid this artifact was to extract and lightly fix tissue in the presence of a detergent to remove salts and soluble proteins.[15] A modification of this approach was to extract the tissue briefly in distilled water just before freezing it, thereby presumably diluting the soluble components.[16] Unfortunately, both of these approaches extract some cytoplasmic components and, at best, distort relationships of the cytoskeletal filaments and membranous organelles to the matrix of finer, shorter filaments and associated globular proteins that comprise the cytoplasm.[3,17] A combina-

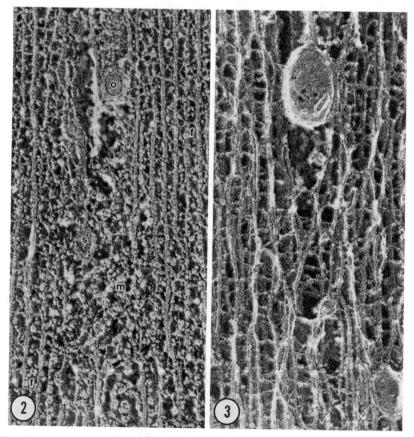

FIGURES 2 and 3. Cytoplasmic structure in a rapidly frozen, freeze-fractured, and shallow-etched turtle optic nerve (**FIGURE 2**), compared with cytoplasmic structure in a similar but more deeply etched axon (**FIGURE 3**).[3] FIGURE 3 shows a reticular pattern characteristic of ice crystal damage. Organelles (**o**), smooth endoplasmic reticulum (**s**), neurofilaments (**f**), and granular axoplasm (**m**) are indicated in FIGURE 2. 92,000X.

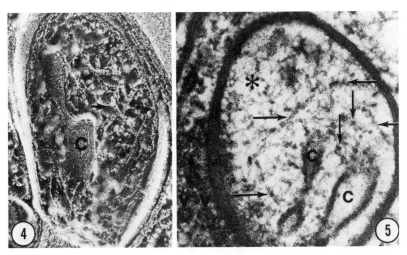

FIGURES 4 and 5. Comparison of rapid-frozen dendritic spines from mouse cerebellar cortex prepared by shallow-etching (FIGURE 4) and freeze-substitution (FIGURE 5). The shallow-etched and cross-fractured spine reveals several classes of cytoplasmic filaments, some as small as 4–6 nm in diameter, within a granular matrix. A fractured membrane cistern (C) is also present. FIGURE 5 is a similar view of a spine prepared by freeze-substitution in osmium tetroxide dissolved in acetone. Two cisterns (C) and 9–10-nm microfilaments (*arrows*) are evident within the spine, as well as some smaller filaments near the *asterisk*. However, the granular matrix is not so evident as in the freeze-etch view. A few synaptic vesicles (*under the* V's) characterize the presynaptic component of the synapse. FIGURE 4—136,000X. FIGURE 5—104,000X.

tion of small ice crystals and deep-etching yields a picture in which most of the finer details of cytoplasmic organization are missing (compare FIGURES 2 and 3).

One solution to the problems arising from freeze-drying or deep-etching is to limit the depth of etching. A quartz crystal monitor head cooled with liquid nitrogen is positioned directly over the freshly fractured specimen to measure and control how much water comes off the fractured surface, so that a layer of water considerably less than 100 nm thick is removed (FIGURES 1, 2, and 4).[3,4] In such preparations the ice table below the filaments and granules on the surface is still seen through numerous open spaces in the cytoplasm (FIGURE 2). A simple calculation shows that salts in the amounts found in tissue are not concentrated enough (provided that their concentration is not increased by excessive drying) to account for very many of the globular or fine filamentous structures revealed by shallow-etching. Thus the matrices of fine filaments and granules are presumed to represent the myriad cytoplasmic proteins that surround the intermediate filaments, actin filaments, and microtubules.

Another approach to achieving realistic structural views of cytoplasm is to dissolve the ice from tissue water with an organic solvent, a process known as "freeze-substitution" (FIGURE 5).[10] Various fixatives and stains can be used in conjunction with or after the substitution,[17] but even with the best available stains the finest details of cytoplasmic structure are barely visible when the tissue is embedded in epoxy resin and sectioned (compare FIGURES 4 and 5). However, details comparable to those seen with shallow-etching become apparent if the plastic embedding medium is avoided by freeze-substituting whole mounts of cells that subsequently can be critical-point dried and examined directly.[17]

It is apparent that structural details even finer than the smallest, membrane-limited organelles are preserved by current direct freezing methods. To the degree that spatial resolution in analytical microscopy is any less detailed, it must be attributed to the particular techniques used to prepare the tissue for analysis or to inherent limitations in analytical methods.

STRUCTURE AND ANALYSIS IN CRYOSECTIONED NEURONS

As a result of the pioneering work of Hall and colleagues[18] and the subsequent efforts of many other investigators (reviewed in REFERENCES 2 and 19), biological electron probe X ray microanalysis (EPMA) is now a mature, practical technique. The measurement of millimolar amounts of elements within subcellular organelles by the

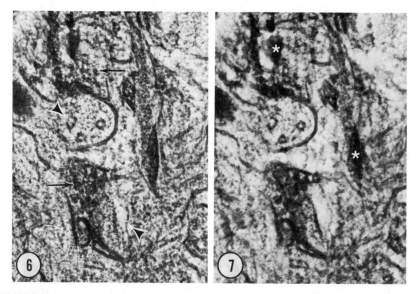

FIGURES 6 and 7. Thin, freeze-dried cryosection from the molecular layer of a directly frozen slice of cerebellum. **FIGURE 6** was photographed in the transmission mode at a nominal temperature of $-145°$ C immediately after freeze-drying in the electron microscope at approximately $-110°$ C with a Gatan Model 626 cold transfer device; the accumulated electron dose was less than 100 e/A^2. These sections, approximately 100 nm thick, were cut at $-140°$ C (gas temperature) in an FC-4-equipped Reichert Ultracut E; other procedures were essentially as previously described.[21] Synapses and other elements of the neuropil are recognized in this field. Small membrane-bound organelles, such as endoplasmic reticulum in both cross and longitudinal sections of spines (*arrowheads*) and synaptic vesicles in presynaptic parallel fiber terminals (*arrows*) are readily identified. These organelles are embedded in a granular cytoplasmic material which, however, is not resolved into the discrete components seen with the shallow-etch technique (compare with FIGURE 4). **FIGURE 7** shows the same section after warming the cold stage to room temperature overnight, recooling to $-145°$ C, and rephotographing the same field. Some cutting and drying artifacts are evident, including surface chatter (*asterisks*) and collapse and aggregation of cytomatrix. However, the collapse is now much more marked, presumably due to beam damage in conjunction with the additional water loss when the tissue was warmed to room temperature. 35,000X.

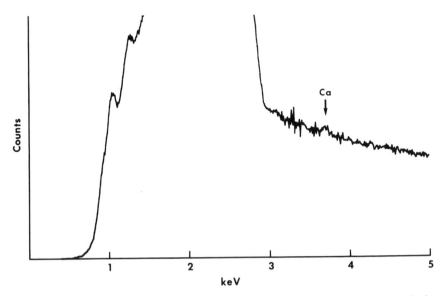

FIGURE 8. Average X ray spectrum from synaptic vesicle clusters within presynaptic terminals of cerebellar synapses. Spectra were fit and quantitative results extracted using the multiple-least-squares method.[22,23] The calculated potassium K peaks have been stripped in order to show the size and peak-to-continuum ratio of the calcium peak, which represents 400 μmol per liter wet tissue.

analysis of electron-induced X ray emission, which is the basis of EPMA, is now routine and can even be carried out using commercial hardware and software. Moreover, it is apparent that biologically useful sensitivity and resolution are not limited by the technical aspects of EPMA. X ray maps of the distribution of molybdenum in stained catalase crystals, obtained by using an analytical electron microscope equipped with a field-emission gun, demonstrate a resolution of better than 10 nm.[20] Even in tissues EPMA detects as few as 100 calcium atoms localized within a mitochondrion.[5] In analytical experiments on brain synapses, we have been able to measure 400 μM calcium in clusters of 1–4 small (45-nm diameter) synaptic vesicles within irradiated microvolumes of approximately 5×10^{-4} μm³; this represents the detection of about 100 atoms of calcium using a conventional tungsten filament (FIGURE 8).

Structure in Cryosections

The resolution needed for many analytical applications in biology is, however, limited by the preparative methods. The spectrum shown in FIGURE 8 was obtained from brain slices subjected to three essential preparative steps: 1) the slices were rapidly frozen; 2) cryosections (80–120 nm thick in the hydrated state) were cut from the well frozen face of the tissue block; and 3) the frozen sections were slowly freeze-dried (in the electron microscope, as illustrated in FIGURES 6 and 7). The primary cryofixation step is very important for EPMA, since its aim is to preserve the distribution of even highly diffusible cell constituents such as sodium and potassium.

However, rapid freezing appears likely to be satisfactory for analytical purposes in view of the fine details—presumed to represent cytoplasmic proteins—that are seen after shallow-etching. In contrast to freezing, the subsequent cryosectioning and freeze-drying steps introduce significant artifacts into the final preparation—artifacts that up to now have had to be tolerated, because the specimen must be both *thin* and *dry* for probing with an electron beam.

Thin specimens are necessary in order to avoid multiple electron and X ray scattering, which otherwise results in a loss both of analytical spatial resolution and of definition in transmission images. In addition, thin specimens conform to the assumptions about the mechanisms of continuum X ray production that are implicit in the Hall quantitation scheme.[18] To this time cryosectioning is the only direct approach for obtaining thin, unfixed, and unextracted samples from bulk tissues that have retained the natural distribution of cell components. There are freeze-substitution schemes which have a limited application to the *qualitative* analysis of certain elements in specific situations,[24,25] but freeze-substitution methods extract and translocate diffusible elements to an unknown extent, and therefore cannot serve as a general alternative to cryosectioning.

Unfortunately, cryosections, particularly if they are thinner than 100 nm, are difficult to prepare. Recent studies have documented the important requirements for and major problems with cryosectioning unfixed material.[12,26] Our experience is generally consistent with these reports; nevertheless, some of the critical or controversial findings are worth reviewing. In order to be sectioned, frozen tissue must be embedded in amorphous ("vitreous") ice and cut below the devitrification temperature. Even then, ice is not a good embedding medium; serious compression has been unavoidable so far, and may be the reason that it is difficult to cut sections thinner than 80 nm.[12] Furthermore, substantial and irregular surface relief is present on the "chip" side of the section while knife marks are evident on the "block-face" side (the side of the section contacting the knife edge). In heterogeneous tissues the surface defects ("chatter") appear to be dominated by regional variations in compressibility, and may well be compounded by microtome instabilities. Stereo views and replicas of frozen-hydrated cryosections show that these surface artifacts typically extend 10–15 nm into a 100-nm thick section. In view of the low inherent contrast of unstained sections, this is enough to make a significant and confusing contribution to the overall contrast of a transmission electron microscopic image of a cryosection. The manifestations of several of these cutting artifacts, as seen in *freeze-dried* (and therefore somewhat shrunken) cryosections, are illustrated in FIGURE 7.

The requirement for dry sections, in contrast to that for thin ones, is based wholly on practical considerations. In fact, it should ultimately be possible to carry out high-resolution EPMA as well as electron energy-loss analyses on frozen-hydrated sections. At present, however, this is precluded by two major considerations: 1) cells are typically 75–80% water, so that analytical sensitivity in dried sections is four to five times better than in hydrated sections; 2) the contrast in unstained hydrated sections is very low, so that observation of structure at a resolution of 100 nm or better is, although not impossible,[27] very difficult. In addition, there must be concern about the ability of frozen-hydrated sections to withstand the high electron doses used in analytical electron microscopy. These problems can be overcome by freeze-drying the sections, but additional and serious structural rearrangements are introduced in the process. The structural integrity of hydrated tissues, especially those not stabilized by fixatives or protectants, is mainly dependent on hydration forces. The removal of these forces promotes changes in the interactions of macromolecules—for example, in protein components of the cytoskeleton—which result in shrinkage and collapse of the cytoplasm. This collapse occurs even at the lowest temperatures (around −110° C) at which water sublimes from tissue, and further structural degradation becomes apparent when the at least partially dried tissue is warmed to room temperature (and then cooled again before being examined; compare FIGURES 6 and 7).

Freeze-drying can be viewed as complete deep-etching; thus, it is pertinent to recall the differences between shallow-etched (FIGURE 2) and deep-etched tissue (FIGURE 3). Furthermore, partially-etched preparations are stabilized against collapse by firm attachment to a substratum (often the unetched layer of ice), whereas cryosections may be only loosely attached to their substrate. In view of the forces operating during sectioning and particularly drying, it is not surprising that *unfixed, freeze-dried* cryosections yield only limited interpretable fine structure (FIGURE 6).

Electron Probe Microanalysis of Cryosections

Although this discussion has so far focused on structural rearrangements and distortions as limitations to the resolution of EPMA, it is not intended to be overly pessimistic, especially in view of the advances that have already been made in recent years. Cryosections are now sufficiently good that even small, membrane-limited organelles, like synaptic vesicles and smooth endoplasmic reticulum in complex tissues such as the neuropil of the cerebellar cortex, can be readily identified and analyzed (FIGURE 6). Moreover, it is clear that those organelles are contained within a cytoplasmic matrix, although this matrix cannot be resolved into the filamentous network that is known to be present in cytoplasm of these synapses (compare FIGURES 4 and 6).[4] Elemental analysis demonstrates that differences in the distribution of diffusible ions between small organelles can be determined. For example, EPMA is sensitive enough to measure submillimolar levels of total calcium in the smooth endoplasmic reticulum and in synaptic vesicles of cerebellar synapses, and to demonstrate a fivefold increase of calcium in the smooth endoplasmic reticulum of depolarized pre-and postsynaptic endings, even though the content of synaptic vesicles remains unchanged (FIGURE 9); these measurements represent the detection of as few as 400 calcium atoms.

An important limitation on the spatial resolution of elemental analysis appears when the smallest organelles are analyzed. The smooth-membrane cisterns of dendritic spines are typically 90 nm in diameter; in instances where the cisterns pass entirely through a 100-nm thick section, a small probe beam can be placed on the cistern, yielding reasonable values for its composition. By contrast, a synaptic vesicle 45 nm in diameter, roughly half the thickness of the section, occupies only about 25% of the volume of a concentrically placed 50-nm probe. If only the core volume of the vesicle is of interest, and if the probe is not centered exactly on a single vesicle, then the internal contents will contribute only a small fraction of the X rays from a cluster of vesicles. This does not mean that concentration changes cannot be attributed to the vesicles, but it does demand caution in any such interpretation.

FUTURE PROSPECTS

How can the resolution limits for analytical microscopy be overcome? There are certain improvements that, once realized, will yield immediate improvements in resolution. The foremost among these is to use thinner frozen sections, because thinner sections would reduce overlap of small structures, and because freeze-drying thinner sections, which is analogous to shallow-etching of fractured specimens, should lessen shrinkage and collapse artifacts. Additionally, thinner sections would open up the possibility of exploiting, in cells and tissues, the higher resolution and sensitivity of electron energy-loss spectroscopy.

A second possibility for improvement depends on continued progress in the analysis of hydrated sections. The main advantage for high-resolution studies would be the avoidance of structural artifacts induced by freeze-drying. The beam sensitivity and contrast problems mentioned above may yield to newer, stabler, and contamination-free cold stages, and to new video and digital image processing techniques. Given that the hydrated state is the natural condition of biological materials, there should be

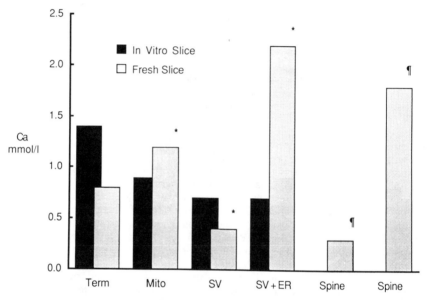

FIGURE 9. Comparison of the calcium concentrations (mmol per liter wet tissue) in different subcellular regions of fresh, depolarized cerebellar synapses compared to those from a preparation that was resting *in vitro*. Abbreviations: **Term**—whole presynaptic terminals; **Mito**—intraterminal mitochondria; **SV**—synaptic vesicle clusters; **SV+ER**—synaptic vesicle clusters that include ER elements; **Spine**—ER cisterns within dendritic spines. The group of cisterns within spines is divided into two groups, because the distribution of means was bimodal. SEMs were ±0.2 for all compartments except terminals, for which the SEMs were ±0.5. Statistically different paired populations (95% confidence) are indicated by * and ¶.

considerable advantage in using frozen-hydrated tissues for studying both composition and structure.

A third strategy would use mild fixation to stabilize the cytoskeleton against collapse upon drying. This may be quite effective for some problems; for example, it might be sufficiently nonextracting for high-resolution compositional studies of specific fractions of bound elements, such as phosphorus and calcium. This approach can also be viewed as an alternative to prepare tissues for structural studies. Like whole-mounts of cells,[17] mildly fixed or freeze-substituted cryosections might be critical-point dried to further reduce cytoskeletal collapse,[28] so that cryosectioning could also become the basis for improved *structural* analysis of solid tissues.

Among the cryotechniques applied to subcellular studies, cryosectioning has proved to be resistant to rapid development, though past perseverance is beginning to pay off. For extracting purely structural information from rapidly frozen tissues, cryosectioning is not up to the standards of freeze-fracture and freeze-etch, but there are good prospects for closing the resolution gap. Advances in this area may, at the same time, lead the way to mapping the elemental composition of even the smallest organelles.

ACKNOWLEDGMENTS

We thank the many collaborators and associates who have contributed to these ongoing studies—R.D. Leapman, D.M.D. Landis, B.J. Schnapp, P.C. Bridgman, R.L. Ornberg, and M.F. O'Connell. We also thank John Murphy for excellent photographic assistance.

REFERENCES

1. HEUSER, J. E., T. S. REESE, M. J. DENNIS, Y. JAN, L. JAN & L. EVANS. 1979. J. Cell Biol. **81:** 275–300.
2. SOMLYO, A. P. & H. SHUMAN. 1982. Ultramicroscopy **8:** 219–234.
3. SCHNAPP, B. J. & T. S. REESE. 1982. J. Cell Biol. **94:** 667–679.
4. LANDIS, D. M. D. & T. S. REESE. 1983. J. Cell Biol. **97:** 1169–1178.
5. SOMLYO, A. P., M. BOND & A. V. SOMLYO. 1985. Nature **314:** 622–625.
6. ANDREWS, S. B., R. D. LEAPMAN, D. M. D. LANDIS & T. S. REESE. 1985. Proc. Soc. Neurosci. **11:** 644.
7. BLAUSTEIN, M. P., C. F. MCGRAW, A. V. SOMLYO & E. S. SCHWEITZER. 1980. J. Physiol. (Paris) **76:** 459–470.
8. RALL, W. 1974. *In* Cellular Mechanisms Subserving Changes in Neuronal Activity. C. Woody, K. Brown, T. Crow & J. Knipsel, Eds. Brain Information Service. University of California, Los Angeles, CA.
9. CRICK, F. 1982. Trends Neurosci. **5:** 44–46.
10. VAN HARREVELD, A. & J. CROWELL. 1964. Anat. Rec. **149:** 381–386.
11. ORNBERG, R. L. & T. S. REESE. 1979. *In* Freeze-Fracture: Methods, Artifacts and Interpretations. J. E. Rash & C. S. Hudson, Eds. 89–97. Raven Press. New York, NY.
12. CHANG, J.-J., A. W. MCDOWALL, J. LEPAULT, R. FREEMAN, C. A. WALTER & J. DUBOCHET. 1983. J. Microsc. **132:** 109–123.
13. BRIGHTMAN, M. W. & T. S. REESE. 1969. J. Cell Biol. **40:** 648–677.
14. MORI, K., M. C. NOWYCKY & G. M. SHEPHERD. 1981. J. Physiol. (London) **314:** 281–294.
15. HEUSER, J. E. & M. W. KIRSCHNER. 1980. J. Cell Biol. **86:** 212–234.
16. GULLEY, R. L. & T. S. REESE. 1981. J. Cell Biol. **91:** 298–302.
17. BRIDGMAN, P. C. & T. S. REESE. 1984. J. Cell Biol. **99:** 1655–1668.
18. HALL, T. A. 1971. *In* Physical Techniques in Biological Research. G. Oster, Ed. 2nd ed. Vol. 1A: 157–275. Academic Press. New York, NY.
19. HALL, T. A. & B. L. GUPTA. 1983. Q. Rev. Biophys. **16:** 279–330.
20. SOMLYO, A. P. 1984. J. Ultrastruct. Res. **88:** 135–142.
21. ANDREWS, S. B., J. E. MAZURKIEWICZ & R. G. KIRK. 1983. J. Cell Biol. **96:** 1389–1399.
22. SHUMAN, H., A. V. SOMLYO, & A. P. SOMLYO. 1976. Ultramicroscopy **1:** 317–339.
23. KITAZAWA, T., H. SHUMAN & A. P. SOMLYO. 1983. Ultramicroscopy **11:** 251–262.
24. MCGRAW, C. F., A. V. SOMLYO & M. P. BLAUSTEIN. 1980. J. Cell Biol. **85:** 228–241.

25. ORNBERG, R. L. & T. S. REESE. 1980. Fed. Proc. **39:** 2802–2808.
26. ZIEROLD, K. 1984. Ultramicroscopy **14:** 201–210.
27. McDOWALL, A. W., J.-J. CHANG, R. FREEMAN, J. LEPAULT, C. A. WALTER & J. DUBOCHET. 1983. J. Microsc. **131:** 1–9.
28. RIS, H. 1981. J. Cell Biol. **91:** 305a.

DISCUSSION OF THE PAPER

D. F. PARSONS (*New York State Department of Health, Albany, New York*): FIGURE 2 shows such close packing that it raises serious questions about the diffusion coefficients—rotational and translational—for water and small molecules in this structure. I wonder if you have any ideas as to what the impact of such a closely packed structure is on rotation or translation diffusion of water or small molecules?

REESE: You'd be a bit happier if you could see them in stereo. One has to remember that we're looking through an ordinary thin section, 500 to 1,000 Å thick, so that one tends to flatten it out and see it as much more closely packed than it is. We haven't attempted to measure this, but I would have thought that the open space in here was 30, 40, 50% at least.

U. AEBI (*Johns Hopkins University, Baltimore, Maryland*): Have you tried shadow fractions of the neurofilaments with the side arms? The distribution of side arms looks relatively irregular (FIGURE 2) Do you think that this is a specimen preparation artifact, or is the distribution of side arms along the intermediate filaments irregular? Is it due to a distortion during the specimen preparation?

REESE: The filament you saw had been fractured right on top, and that removed a subset of side arms.

AEBI: I realize that. However, if it were cleaved close, you would still see some sort of regular structure, as in muscle.

REESE: I would be concerned that there could be some shrinkage in this material, which could tip the positions and make them appear less regular. I've no way of knowing really.

J. R. SOMMERS (*Duke University, Durham, North Carolina*): You mentioned the possible contamination by salts; what about by soluble proteins?

REESE: I wouldn't consider that contamination. That was the question we were trying to ask. It may be that they could collapse and aggregate.

SOMMERS: So you would not trust your geometry?

REESE: I would not necessarily trust that particular aspect of it.

R. N. McBURNEY (*Medical Research Center, Newcastle-upon-Tyne, England*): You have identified (FIGURE 9) a subpopulation of ER or synaptic vesicles that accumulates calcium but is not as well characterized as other organelles. Do you think that this could represent a population that might contain a second transmitter substance?

ANDREWS: That is possible. The identity of the transmitter in this particular terminal is a matter of some contention, and it's not clear that there is only one.

Electron Energy-Loss Spectroscopy: Quantitation and Imaging[a]

H. SHUMAN, C.-F. CHANG, E.L. BUHLE, JR., AND A.P. SOMLYO

Pennsylvania Muscle Institute
University of Pennsylvania School of Medicine
Philadelphia, Pennsylvania 19104

INTRODUCTION

Electron energy-loss spectroscopy (EELS) derives information about the atomic and molecular characteristics of a specimen from the energy lost, through inelastic scattering, by an incident electron.[1] At present, the greatest potential for providing useful biological information from EELS is through the detection of the characteristic energy losses arising from the ionization of core-shell electrons. These energy losses depend on the element and the specific core shell ionized. For example, electrons that ionize the K shells of carbon, nitrogen, and oxygen produce characteristic "edges" in the electron energy-loss spectrum at, respectively, 284, 401, and 532 electron volts (FIGURE 1), while the L-ionization edges for phosphorus, sulfur, and calcium are at, respectively, 130, 160, and 346 eV. Implicit in the foregoing statement are two advantages of EELS: first, the feasibility of detecting elements having low atomic numbers (*e.g.*, B, Li, C, N, and O) that are not detectable by conventional X ray detectors, due to the absorption of the characteristic X rays by the beryllium window, and second, the use of L-shell excitations, which have higher probabilities (hence better signal-to-noise) than the K-shell excitation detected by electron probe microanalysis (EPMA). Because of these physical characteristics, and because electron spectrometers have higher collection efficiencies than X ray detectors,[2,3] EELS can extend the range and sensitivity of analytical electron microscopy as a complementary technique to EPMA. However, the advantages of EELS are accompanied by difficulties related to the generally large background and the more stringent requirements for ultrathin specimens in EELS than in electron-probe X ray microanalysis. Therefore, EELS is best suited for the detection, at high resolution, of elements occurring at high local concentrations. Nevertheless, we shall demonstrate here that EELS can exceed the sensitivity of EPMA, and is also suitable for the reliable measurement of near-trace concentrations of some elements, such as calcium.

Theoretically, EELS can also provide information about the positions of neighboring atoms that scatter secondary electrons and so modulate the extended electron energy-loss fine strucutre (EXELFS) of the core edges.[4,5] This is analagous to the extended X ray absorption fine structure (EXAFS), usually determined with synchrotron radiation.[6,7] However, the EXELFS is radiation-sensitive and, in organic materials, readily destroyed by the currents required for analysis at high spatial resolution.

---

[a]Supported by National Institutes of Health Grants HL15835 and HL07499 to the Pennsylvania Muscle Institute.

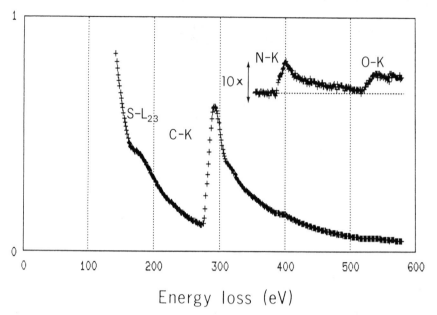

Energy loss (eV)

FIGURE 1. Carbon (284 eV), nitrogen (401 eV), and oxygen (531 eV) K edges and S-L$_{2,3}$ (160 eV) edge acquired from a solvent-deposited thin film of methionine.

The low energy-loss region of the spectrum contains energy-loss peaks equivalent to the visible and UV absorption spectra of the materials irradiated. The UV absorption peaks of amino acids,[8] adenine nucleotides,[9] cholesterol,[10] and visible peaks of β-carotene (FIGURE 2) and the blue color of F centers in radiation-damaged KCl[11] have been detected in this manner. These features of the spectra are also radiation-sensitive, and the detection of organic molecules with low loss spectra, at high spatial resolution, will only be possible through the reduction of radiation damage by cryogenic stages, and the use of image averaging methods comparable to those used in conventional, high-resolution "low-dose" electron microscopy.[12–14]

INSTRUMENTATION

EELS can be used in two ways: the first is to produce electron energy-loss line spectra with the probe focussed to a point in the specimen, and the second is to produce energy-filtered images representative of elemental distribution. Regardless of subsequent processing or purpose, the first requirement for EELS is the separation of electrons that have different energies determined by the inelastic scattering in the specimen. Most electron spectrometers use a magnetic field to disperse electrons having different velocities.[15,16] Modern spectrometers are corrected for second-order aberrations,[17] and we anticipate the eventual appearance of spectrometers corrected to third-order.[18] Following dispersion by the spectrometer, the energy-loss spectrum of a thin specimen can be recorded on photographic film that is subsequently densitometered or, in a serial detection mode, with a scintillator-photomultiplier, while electrons having different energies are ramped across an energy-selecting slit. The best approach, particularly for biological specimens, is parallel detection of a relatively wide energy range of the energy-loss spectrum, in a manner suitable for subsequent computer processing.[19]

The characteristic edges in EELS spectra can be used, coupled with a scanning transmission electron microscope (STEM), to map local concentrations of elements[20-23] at a spatial resolution determined by the electron-probe diameter (1 nm or less) and delocalization of the inelastic scattering.[24] We shall illustrate the power of this method by describing the quantitation of calcium in an organic matrix with EELS at a sensitivity equivalent to three Ca^{2+} atoms in a 1.2×10^3 nm^3 microvolume.[25]

In energy-filtered conventional transmission electron microscopy (EFEM), a transmission image is also generated by using only electrons having experienced losses due to the characteristic excitation of the elements of interest. The energy-selected electrons pass through a slit that excludes electrons of different energies, and are imaged directly on the microscope screen and photographed, or recorded with a television or other two-dimensional array detector and stored in computer memory. The first and most common mode of filtered CTEM imaging is with a Castaing-Henry

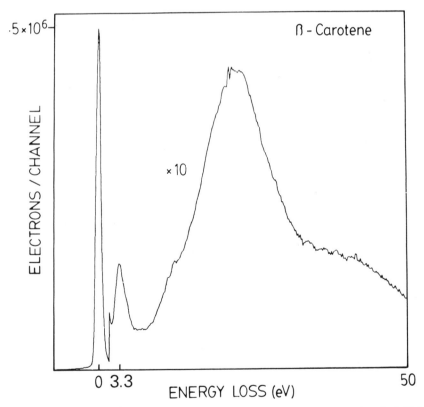

FIGURE 2. The zero and low-loss region of a spectrum acquired from a solvent-deposited thick film of β-carotene. The *large peak* at 3.3 eV is the electron energy-loss equivalent of the orange color of carrots.

spectrometer,[26,27] and a commercial version of the system is now available.[28] Alternatively, it is also possible to produce filtered CTEM images with a magnetic sector spectrometer mounted outside the electron microscope,[13,29–31] albeit with somewhat greater technical difficulty. We shall illustrate EFEM images that show elemental distribution at 3.4-nm resolution.[13]

Energy-filtered images, whether obtained with EFEM or scanning transmission energy-filtered electron microscopy (STEFEM), are based on the principle that images generated by only those electrons that ionized atomic core shells represent the two-dimensional distribution of the scattering element. The importance of accurate background subtraction for reliable imaging of elemental composition cannot be overemphasized. This requires the removal of background electrons that are due to other inelastic scattering processes, but have energies identical to those due to the characteristic scattering. Besides radiation damage, the need for accurate background

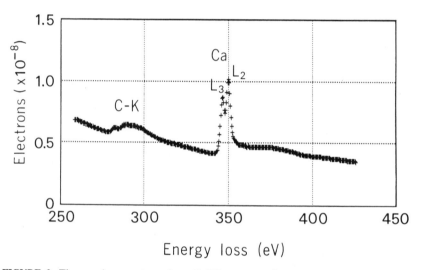

FIGURE 3. The core-loss spectrum from $CaCO_3$ evaporated onto carbon film. Some of the $CaCO_3$ has been converted to CaO by radiation damage. The Ca-$L_{2,3}$ edge is at 346 eV. (FIGURES 3–6 from Shuman and Somlyo.[25] Reprinted by permission from *Ultramicroscopy*.)

subtraction is the most restrictive aspect of any of the EELS methods, and the criteria for background subtraction become increasingly stringent with increasing specimen thickness and at low elemental concentrations.

QUANTITATIVE EELS ANALYSIS: LOW CONCENTRATIONS OF CALCIUM

The characteristic $L_{2,3}$ edge obtained from a thin, evaporated film of $CaCO_3$, but probably containing CaO due to radiation damage, is shown in FIGURE 3. The abrupt shape of this sharp $L_{2,3}$ transition is easier to detect than, for example, the more slowly rising phosphorus or sulfur edges (FIGURE 1).

Quantitation with EELS is based on the measurement of the number of characteristic loss electrons resulting from core-shell ionizations. The accuracy of this measurement is very much dependent on accurate background subtraction. Elemental concentrations are determined by relating the number of characteristic loss electrons to atomic concentrations, either through theoretical models or by comparison with elemental standards. In STEFEM imaging of elements occurring at very high local concentrations in ultrathin specimens, background subtraction from the large characteristic edge peak can be readily performed with a simple algorithm based on determining the "pre-edge" background, and assuming that it varies as BE^{-r} where E is the energy loss and r is some empirically determinable value (usually between 2.5 and 4) dependent on the energy and scattering angle.[32-34] The extrapolated BE^{-r} background is subtracted from the region containing the characteristic edge, as illustrated in FIGURE 1.

In the case of small signal-to-background, such as the Ca edge in spectra obtained from biological (soft tissue) specimens, it is advantageous to enhance the signal-to-background ratio by electronic processing that accentuates sharp features and reduces the effects of slowly varying or DC components. These procedures are analogous to the digital filtering methods used to remove the background from X ray spectra.[35,36] In the case of EELS, first difference and second difference processing of spectra have been used to accentuate the characteristic $L_{2,3}$ edge.[3,37,38] The results illustrated here were obtained with the first difference method. FIGURE 4a-c shows the "raw" spectrum, before and after background subtraction, and the first difference spectrum of a specimen containing 100 mmol Ca/kg dry weight. The enhancement of the Ca-$L_{2,3}$ edge by first difference processing is obvious, although it results in biphasic, positive negative features, both of which represent characteristic loss electrons (fitting shown in FIGURE 5 for 6 mmol/kg). For those unaccustomed to viewing values in biphasic signals, the absolute values of the data can be equally readily displayed. In order to relate the number of characteristic loss electrons to concentration, we have used standards of Ca EGTA dissolved in PVP,[39] and determined the ratio of Ca-L loss to C-K loss electrons at various Ca concentrations. Furthermore, to reduce the effects of plural electron scattering, only the first 8 volts of the carbon K edge were used in the fit.

The concentration of Ca in an organic matrix was quantitated by EELS using the ratio of the first difference Ca-$L_{2,3}$ signal and the partial integral of the raw carbon K absorption edge signal in polyvinylpyrrolidone films containing various concentrations of calcium. The EELS measurements were compared to the results of simultaneous EPMA, as this is the only independently validated method that is suitable for this type of microanalysis and is not subject to errors due to microheterogeneity of the specimen. The results (FIGURE 6) show excellent correlation between the two methods, but the measurement errors indicated by the error bars are considerably greater with EPMA, indicating the greater sensitivity of EELS. The maximum systematic error of the EELS measurement, determined from the intercept of FIGURE 6, is 0.28 ± 0.49 S.D.mmol/kg dry wt.[25] The experimentally determined limit for 100 sec measurements with 4.6 namp beam current, for $t/\lambda = 0.25$ where t is the specimen thickness and λ is the inelastic mean free path, is 0.7 mmol/kg dry wt. Given a 10-nm diameter stationary probe analyzing a volume of 1,200 nm^3, the mass analyzed would be 2.4 × 10^{-21} kg of the standard, and the experimentally measured standard deviation (0.7 mmol/kg dry wt) is equivalent to 1 atom of calcium. Therefore, the minimal detectable mass in a biological matrix with the above parameters at a conservative (99%) confidence level is three calcium atoms.

Lest the potential for single atom detection with EELS generate an untempered level of enthusiasm, we would take into account the fact that the detection of three

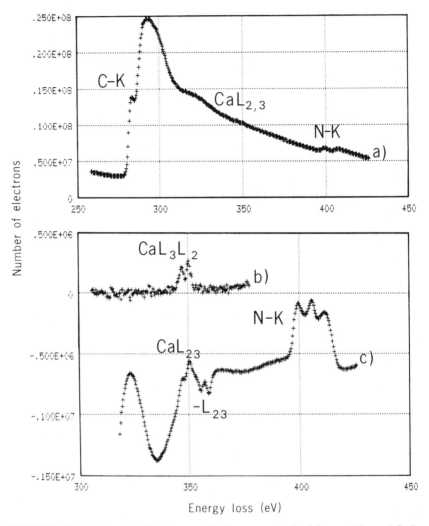

FIGURE 4. (a) Dark-current and gain-normalized spectrum obtained from a 100-mmol Ca/kg PVP specimen, acquired in 50 sec with an incident current of 4.6 nA. (b) The same spectrum as (a) with the carbon K edge background removed by subtracting the spectrum from an otherwise identical film of pure PVP. (c) First difference spectrum from the same specimen, acquired in 100 sec.

calcium atoms in a 1.2×10^3 nm^3 volume corresponds to a concentration of 2.1 mmol/kg, and also note two significant limitations of elemental analysis with EELS. The first, and best known, of these is that due to variations in sample thickness. Due to plural electron scattering, detection with EELS becomes increasingly difficult with increase in sample thickness. The inelastic mean free path for 100 kV electrons in carbon is about 60 nm, hence the $t/\lambda = 0.25$ corresponds to a cryosection with thickness equivalent to a 15-nm carbon film. The second, and in some instances equally serious, limitation is that due to fluctuations in energy calibration. The measurements cited[25] were obtained with the spectrometer aligned with a peak stabilization system to within 0.2 eV energy throughout calibration and microanalysis.[40] An intentional energy-loss shift of 1.0 eV produced a systematic error of 2.0 mmol/kg. A shift of this magnitude or greater could easily occur with commercial spectrometers lacking peak stabilization, and may lead to significant errors in the quantitation of Ca. The source of this error, in organic specimens, is the extended energy-loss fine structure of the carbon K edge. The error in quantitation could be reduced by improving peak stabilization further or compensating for it by fitting to the first derivative of the carbon standard, in a manner previously employed in EPMA.[39] Such errors will be further reduced by obtaining the most accurate representation of the background under the calcium $L_{2,3}$ edge, including the extended energy-loss fine structure of the carbon K and potassium $L_{2,3}$ edges.

SCANNING TRANSMISSION ENERGY-FILTERED ELECTRON MICROSCOPY (STEFEM): PHOSPHORUS IN BACTERIOPHAGE T4

STEFEM has the distinct advantage of efficient signal collection.[20] At each pixel in the scanned image, an entire EELS spectrum can be obtained. This parallel collection reduces the possibility of image artifacts due to changes in instrument stability. In EFEM (see below), three images are needed to get a reasonably accurate estimate of the background variations in the EELS spectra. Any variations in the microscope stability will have a drastic influence on the outcome of the resulting images.

To demonstrate the potential of STEFEM to map the distribution of phosphorus in a biological specimen, we used as a test object T4 bacteriophage that is known to contain approximately 1.5 mol/kg of phosphorus in the form of DNA in the head

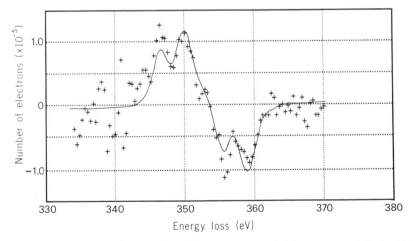

FIGURE 5. Background-subtracted first difference spectrum $(+)$ from a 6-mmol Ca/kg PVP specimen with the best fit Ca-L edge superimposed $(-)$. The spectrum was acquired in 1,000 sec with an incident beam current of 4.6 nA.

region. The tail is composed mostly of protein, although it has the potential to contain some T4 DNA. We previously examined the T4 bacteriophage using EFEM.[41]

A spectrum acquired in 0.66 sec, with a 0.5 namp, 4.6-nm probe from the head region of T4 phage, with E^{-r} background subtraction is shown in FIGURE 7. The three prominent features are the phosphorus $L_{2,3}$, chlorine $L_{2,3}$, and carbon K edges at 130, 200 and 284 eV, respectively.

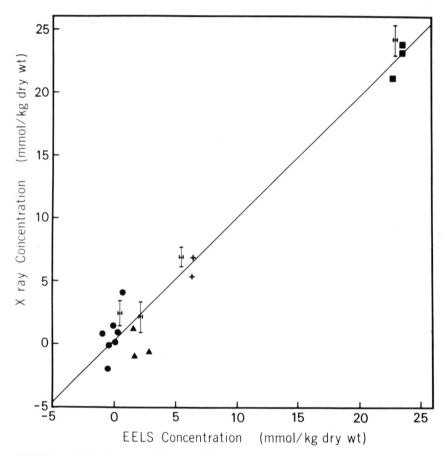

FIGURE 6. Calibration curve of X ray and EELS Ca concentration measurements for specimens containing 0, 2, 6, and 20 mmol Ca/kg PVP. The best fit linear regression line shown is with slope = .99 and intercept = 0.28 mmol Ca/kg PVP. The *error bars* shown are the standard deviations of ten measurements of the same specimen area.

The compositional images of a representative T4 bacteriophage are shown in FIGURE 8. Protein is composed largely of carbon, nitrogen, oxygen, and hydrogen, while DNA contains, in addition, a substantial amount of phosphorus. The carbon image (FIGURE 8b) shows the basic profile of the head and tail of the phage. Phosphorus (FIGURE 8c) appears localized in the head of the bacteriophage. FIGURE 8a shows the background image and is proportional to the mass-thickness of the specimen.

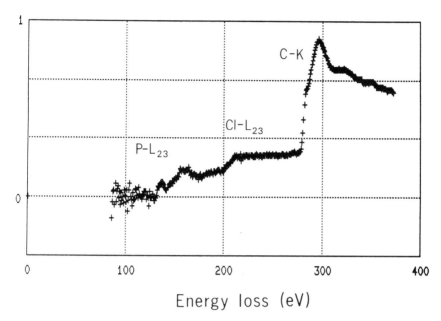

Energy loss (eV)

FIGURE 7. Background-subtracted spectrum acquired in 0.66 sec with a 4.5-nm diameter, 0.5-nA electron beam from a portion of the head of a T4 bacteriophage. The three features clearly visible are the phosphorus (130 eV) and chlorine (200 eV) $L_{2,3}$ edges and the carbon K edge (284 eV). A dilute buffered solution of T4 bacteriophage was placed on a glow-discharged carbon-coated copper grid and permitted to air dry, with crystallization of NaCl accounting for the appearance of the chlorine edge.

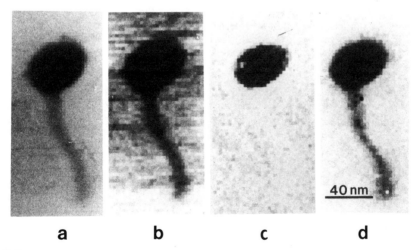

a   b   c   d

FIGURE 8. A set of 32 × 64 pixels STEFEM images of T4 bacteriophage. At each pixel a spectrum as in FIGURE 7 was collected and analyzed for partially integrated background counts and background-subtracted edge counts. (**a**) The background image of phage with an integration window of 100–125 eV. The element edge images were converted to approximate concentration maps by dividing by the background image. The three concentration maps shown are for: (**b**) carbon, (**c**) phosphorus, and (**d**) chlorine.

More quantitative results at higher sensitivity will be required to determine whether the tail does or does not contain a strand of DNA.

The practical difficulties of STEFEM are radiation damage and specimen drift. The amount of radiation used to obtain compositional images of nonperiodic specimens with a signal-to-noise ratio of greater than 1, by either EFEM or STEFEM, is usually many orders of magnitude greater than the electron doses used for high-resolution electron microscopy of biological specimens. The T4 data presented here were obtained using a dose of approximately 1,000,000 e/A^2. Transmission EFEM or STEFEM images taken at these radiation doses may eventually provide information at better than 2.0-nm resolution, but the effect of radiation damage on such information will have to be evaluated.

Specimen drift in STEFEM is manifested by the wavy appearance of the tail of the bacteriophage in FIGURE 8. In order to reduce the radiation damage, the specimen was analyzed in a cold stage, operated at $-100°$ C. The drift is due to instability of the cold stage. The spectrum for each pixel was acquired in 0.66 secs, so that the full 32×64 pixel image was obtained in 1,350 secs. Correction for drift during the long collection times required in STEFEM will necessitate cross-correlation against bright- or dark-field images and raster repositioning during acquisition.

ENERGY-FILTERED "CONVENTIONAL" TRANSMISSION ELECTRON MICROSCOPY (EFEM)

The energy-loss spectrum obtained from a large area of a uranyl acetate stained catalase crystal is shown in FIGURE 9. The two prominent features in the spectrum are the characteristic edges of uranium ($O_{4,5}$, E = 114 eV) and carbon (K, E = 284 eV). Since the average concentrations of these two elements are high, the edges are readily visible above the background, making this specimen particularly suitable for high-resolution energy-filtered elemental mapping.[13]

The EFEM images were electronically recorded with a slow scan silicon intensified target vidicon from five small areas (10×10 unit cells) of a single thin crystal of catalase stained with uranyl acetate. A minimum contrast elastic image, and five images, at the five energy losses indicated in FIGURE 9, were obtained. After the SIT tube artifacts (dark current and gain nonuniformities) were removed, each image was Fourier transformed and the diffraction intensities and phases averaged, giving reconstructed unit cell images that had a high signal-to-noise ratio for each energy loss. These reconstructed images were then further processed to give concentration maps of carbon and uranium in catalase. The highest order reflections observed were the (2,3) in the elastic image, corresponding to a resolution of 2.9 nm, and the (2,1), for both the U-O edge and C-K edge images, corresponding to 3.4-nm resolution.

In the absence of background signal, due primarily to the ionization of the C, N, O, and U valence electrons, it would be a simple matter to use the images with energy losses equal to the edge maximum and to form a distribution map of the chosen element. This is not the case, as the background at the U edge is about 50% of the total signal. The next simplest processing step is to assume that the background contribution is proportional to the number of valence electrons under the beam. An image at an energy loss lower than the ionization edge is then acquired (for the U edge this would be E = 89 eV), and subtracted from the edge image.[27] The implicit assumption with this method is that the shape of the energy-loss spectrum in the region near an edge is independent of thickness and elemental composition at a single picture point, although its magnitude can change. Background images at two different energy losses (E = 64 eV and 89 eV), both lower than the U-O edge, are shown in FIGURE 10. Since they are clearly different, the shape of the background spectrum is indeed not a constant. In order to improve the image processing, a model for the dependence of the background intensity B(E), on energy loss, specimen thickness, and composition is required. A simple but useful model for the background shape, as verified by Egerton,[33] is that the

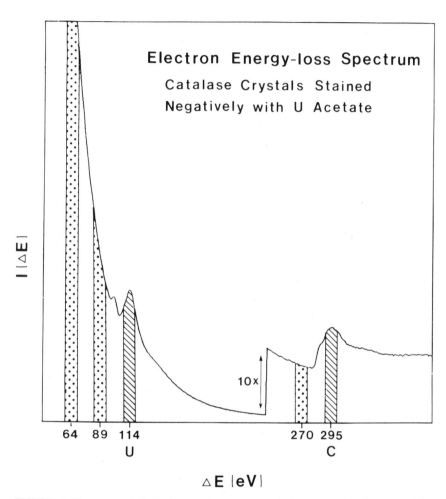

FIGURE 9. The averaged electron energy-loss spectrum of negatively stained catalase. The *shaded area* represents the selected electrons in the specific energy windows that contributed to the EFEM images. The width of the energy window is about 10 eV. The peaks of the uranium $O_{4,5}$ and carbon K edges are at 114 and 295 eV respectively. (FIGURES 9–11 from Shuman *et al.*[13] Reprinted by permission from *Ultramicroscopy.*)

intensity varies as power of the energy loss, $B(E) = B \cdot E^{-r}$, where B and r depend on specimen and instrument parameters. The two pre-edge images are used to determine the background shape, the parameters B and r, at each picture point, and the *model* is used to compute an artificial background image at the energy loss corresponding to the

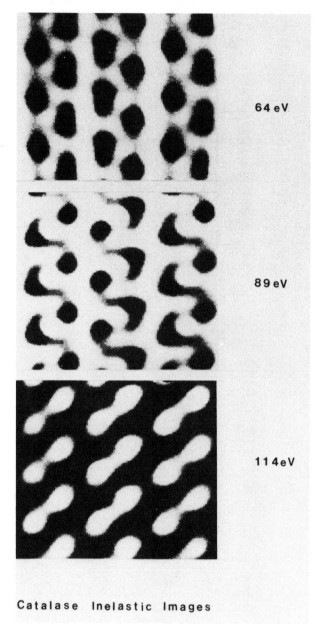

64 eV

89 eV

114eV

Catalase   Inelastic   Images

FIGURE 10. Three reconstructed images used to determine the uranium concentration distribution. (**a**) E = 64 eV and (**b**) E = 89 eV are the pre-edge images. (**c**) E = 114 eV is the U-$O_{4,5}$ edge image.

edge. This procedure has also been used to remove the thickness-dependent artifact found in STEFEM imaging of nonuniform carbon films.[32]

With the extrapolated background and edge images, maps of the distributions of atoms or their concentrations can be created. To obtain the atom distributions, the background is subtracted from the edge image to form a "net" edge image. However, an additional artifact, due to plural elastic and inelastic scattering events, must be eliminated. All the inelastic images are modulated by the elastic amplitude contrast. In other words, some of the energy-loss electrons are also scattered outside the objective aperture and not detected. A minimum contrast (in focus) elastic image is primarily due to large-angle scattering by the uranyl stain. If the net edge image is divided by the amplitude contrast, then a faithful atomic distribution should be obtained. This is

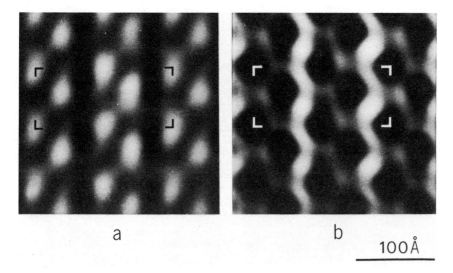

a   b

100Å

FIGURE 11. (a) Uranium map and (b) carbon map, free of elastic amplitude contrast. The *unit cell* of catalase is marked, and the *white regions* correspond to high concentration of the relevant element. The two images are complementary, consistent with the mechanism of negative staining.

particularly important if high atomic number elements are present, such as in negatively stained or osmium-fixed specimens. To form the concentration map, the net edge image is divided by the background (or valence electron) image. As pointed out originally by Johnson,[42] this procedure also corrects for the amplitude contrast artifact. Concentration maps of uranium and carbon are shown in FIGURE 11. Up to a resolution of 3.4 nm, the two maps are nearly complementary, as expected from the mechanism of negative staining. However, the carbon map appears to be sharper (also indicated in the Fourier coefficients of the reconstruction) than the uranium map. The potential sources for this difference are: chromatic aberration, delocalized scattering, and radiation damage. The first two account for resolution limits of 0.6 and 0.2 nm for U and C respectively,[13] indicating that the stain may be more radiation sensitive than the underlying organic crystal. This result suggests that further improvement in resolution may be possible through the reduction of radiation damage, by averaging images of larger areas.

REFERENCES

1. JOY, D. C. 1979. *In* Introduction to Analytical Electron Microscopy. J. J. Hren, J. I. Goldstein & D. C. Joy, Eds. 223–244. Plenum Press. New York, NY.
2. ISAACSON, M. & D. E. JOHNSON. 1975. Ultramicroscopy **1**: 33–52.
3. SHUMAN, H. & P. KRUIT. 1985. Rev. Sci. Instrum. **56**: 231–239.
4. KAMBE, K., D. KRAHL & K. -H. HERRMANN. 1981. Ultramicroscopy **6**: 157–162.
5. CSILLIG, S., D. E. JOHNSON & E. A. STERN. 1983. *In* EXAFS Spectroscopy. B. K. Teo & D. C. Joy, Eds. 241–254. Plenum Press. New York, NY.
6. SAYRES, D. E., F. W. LYTLE & E. A. STERN. 1970. *In* Advances in X-Ray Analysis. Vol. 13: 248. Plenum Press. New York, NY.
7. CHANCE, B. C., C. KUMAR, L. POWERS & Y. -C. CHING. 1983. Biophys. Soc. **44**: 353–363.
8. JOHNSON, D. E. 1972. Radiat. Res. **49**: 63–84.
9. ISAACSON, M. S. 1972. J. Chem. Phys. **56**: 1803–1812.
10. HAINFELD, J. & M. S. ISAACSON. 1978. Ultramicroscopy **3**: 87–95.
11. SHUMAN, H., A. P. SOMLYO & A. V. SOMLYO. 1981. *In* Microprobe Analysis of Biological Systems. T. Hutchinson & A. P. Somlyo, Eds. 273–288. Academic Press. New York, NY.
12. UNWIN, P. N. T. & HENDERSON, R. 1975. J. Mol. Biol. **94**: 425–429.
13. SHUMAN, H., C. -F. CHANG & A. P. SOMLYO. 1986. Ultramicroscopy **19**: 121–133.
14. JENG, T. W. & W. CHIU. 1983. J. Mol. Biol. **164**: 329–346.
15. ENGE, H. A. 1967. *In* Focusing of Charged Particles. A. Septier, Ed. 203–264. Academic Press. New York, NY.
16. CREWE, A. V., M. ISAACSON & D. JOHNSON. 1971. Rev. Sci. Instrum. **42**: 411–415.
17. SHUMAN, H. 1980. Ultramicroscopy **5**: 45–54.
18. SHEINFEIN, M. & M. ISAACSON. 1984. *In* Scanning Electron Microscopy/1984. O. Johari, Ed. 1681. SEM. AMF O'Hare, IL.
19. SHUMAN, H. 1981. Ultramicroscopy **6**: 163–168.
20. ISAACSON, M. S. & A. V. CREWE. 1975. Annu. Rev. Biophys. Bioeng. **4**: 165–184.
21. SOMLYO, A. P. 1985. J. Ultrastruct. Res. **83**: 135–142.
22. SHEINFEIN, M. & M. S. ISAACSON. 1986. J. Vac. Sci. Technol. **B4(1)**: 326.
23. COLLIEX, C. 1985. Ultramicroscopy **18**: 131–150.
24. BOURDILLON, A. J., P. G. SELF & W. M. STOBBS. 1981. Philos. Mag., Ser. A **44**: 1335–1339.
25. SHUMAN, H. S. & A. P. SOMLYO. Ultramicroscopy, in press.
26. CASTAING, R., J. F. HENEQUIN, L. HENRY & G. SLODZIAN. 1967. *In* Focusing of Charged Particles. A. Septier, Ed. 265–293. Academic Press. New York, NY.
27. OTTENSMEYER, F. P. & J. W. ANDREWS. 1980. J. Ultrastruct. Res. **72**: 336–348.
28. ENGLE, W., D. KURZ & A. RIK. 1984. Am. Lab.
29. SHUMAN, H. & A. P. SOMLYO. 1982. Proc. Natl. Acad. Sci. USA **79**: 106–107.
30. SHUMAN, H., A. V. SOMLYO, D. SAFER, T. FREY & A. P. SOMLYO. 1983. *In* Scanning Electron Microscopy/1983. O. Johari, Ed. Vol. 2: 737–743. SEM. AMF O'Hare, IL.
31. KRAHL, D., K. HERMANN & E. ZEITLER. 1981. *In* Proc. 39th Annu. Mtg. Electron Microscopy Soc. Amer. 336–367.
32. LEAPMAN, R. D. 1986. Scanning transmission electron microscope (STEM) elemental mapping by electron energy-loss spectroscopy. This volume.
33. EGERTON, R. F. 1975. Philos. Mag. **31**: 199–206.
34. Colliex, C. 1986. Electron energy-loss spectroscopy analysis and imaging of biological specimens. This volume.
35. SCHAMBER, F. H. 1973. *In* Proc. 8th Annu. Conf. Microbeam Analysis Society. 85A–85D.
36. SHUMAN, H., A. V. SOMLYO & A. P. SOMLYO. 1976. Ultramicroscopy **1**: 317–339.
37. SHUMAN, H. 1986. *In* Principles of Analytical Electron Microscopy. D. C. Joy, Ed. 393–411. Plenum Press. New York, NY.

38. SHUMAN, H., P. KRUIT & A. P. SOMLYO. 1983. *In* Microbeam Analysis Society of America. R. Gooley, Ed. 247–251. San Francisco Press. San Francisco, CA.
39. KITAZAWA, T., H. SHUMAN & A. P. SOMLYO. 1983. Ultramicroscopy **11:** 251–262.
40. KRUIT, P. & H. SHUMAN. 1985. J. Electron Microsc. Tech. **2:** 167–189.
41. BUHLE, E. L., JR., H. SHUMAN & A. P. SOMLYO. 1985. *In* Proc. 43rd Annu. Mtg. Electron Microscopy Soc. Amer. 314–315.
42. JOHNSON, D. E. 1979. *In* Introduction to Analytical Electron Microscopy. J. J. Hren, J. I. Goldstein & D. C. Joy, Eds. 245–258. Plenum Press. New York, NY.

DISCUSSION OF THE PAPER

R. D. LEAPMAN (*National Institutes of Health, Bethesda, Maryland*): Could you make some comment about how the thickness variations would affect that detection limit for calcium? What type of thickness range could one look at?

SHUMAN: When I did the measurements on the calcium at a half mean free path, the error came out to be almost identical to the quarter mean free path. So the quarter mean free path measurements are not statistically limited. You can predict very easily from theoretical grounds that if you only look at the first 8 volts of the calcium $L_{2,3}$ edge, the signal is going to go down as $(t/\lambda)e^{-t/\lambda}$. So it reaches a maximum signal at $t = \lambda$ and begins to go down after that. Unless you work at a thickness of less than one mean free path you will lose signal.

But, in the cases where EELS is useful, when you are interested in high-resolution information, you have to work with thin specimens to begin with or you don't get high-resolution information. So that problem basically resolves itself.

C. COLLIEX (*Université Paris-Sud, Orsay, France*): I have one comment and one question. The question is about your set of four images of the phage (FIGURE 8). You have chloride and sulphur. I want to know which edge you used and how you processed the image.

SHUMAN: $L_{2,3}$.

COLLIEX: And you took several images? Because there must be some overlap between the chloride and the sulphur.

SHUMAN: They're fairly well separated—30 volts apart. We saw a serious problem earlier when, instead of using sodium chloride in solution, we used some boron salt; I don't remember what it was, but it was contaminated with silicon. The silicon peak was in front of the phosphorus $L_{2,3}$ and had lots of structure in it as well. So when we tried to do the analysis, not realizing that there was silicon there, we got large negative phosphorus concentrations. We realized that there was obviously a problem, so we went back and looked at the spectra directly and saw silicon.

COLLIEX: About the spatial resolution that you discussed in your last image (FIGURE 11), your catalase specimen and your Fourier transformed elastic images. We did the same experiments, but not with the same data processing. We have clusters of a few uranium atoms on carbon, and it has the same two edges, uranium and carbon. We did the measurements by cross-correlation of two images, instead of Fourier filtering. By this technique the numbers that I have in my head are much better than yours, because we have about one nm for uranium and a little less for carbon. This is normal, because carbon K losses are at higher energy.

SHUMAN: We know from earlier measurements that the uranium on carbon doesn't move around, but it stays pretty well localized. Within a nanometer. And you can observe it with elastic scattering.

COLLIEX: But on the inelastic characteristic uranium signal we found that the localization was about one nm for uranium, and for the carbon the localization was between 6 and 7 Å. Which is what we got recently.

SHUMAN: That's what we calculated.

F. P. OTTENSMEYER (*University of Toronto, Toronto, Canada*): First of all, congratulations for detecting a single calcium atom. That's something that has been aimed for, and I'm glad it's finally done.

Second, you said that you are stepping across the virus at 5 nm so that obviously you are not going to see the coiling of the DNA inside. You have to do a little better than that, but you can't.

Then let me ask a biological question. You had a hole in the T4 phage head. Is that real?

SHUMAN: One thing to remember is that these pictures are taken at doses where there's no possible reason to believe the structure is real.

OTTENSMEYER: You said that, but that was a pretty big hole. To have been lost, due to mass loss or what? Can you comment on that?

SHUMAN: I don't know that for this particular measurement, but phosphorus, when it's damaged, turns completely into something that looks like an inorganic phosphate that is not going to evaporate. I don't expect the phosphorus itself to go, but it may move 5 nm or it may move 1 nm. That I don't know.

Electron Energy-Loss Spectroscopy Analysis and Imaging of Biological Specimens[a]

CHRISTIAN COLLIEX[b]

Laboratoire de Physique des Solides
Université Paris Sud
94105 Orsay, France

INTRODUCTION

A new dimension has been introduced in electron microscopy (EM) by the use of energy analyzing and selecting devices operating on the transmitted electron beam. The associated electron energy-loss spectroscopy (EELS) is nowadays recognized as a useful complementary technique to X ray microanalysis in many materials-science problems (see REFERENCE 1 for a review of the present state of EELS in the electron microscopic environment). However, EELS has remained a relatively virgin technique when applied to biological material. Why is this so? What are the present capabilities and limits of EELS imaging in biology and medicine? What can the foreseeable future be for this technique? This paper intends to provide a subjective, illustrated answer to such questions; it relies on our past experience in Orsay obtained through stimulating collaboration with many colleagues in biology.

The potential field of application of EELS to biological material has already been discussed by several authors and by ourselves in recent years.[2-4] Specific characteristics of such samples can be enumerated as follows:

- Specimen preparation consists of a succession of complex and lengthy procedures (dehydration, fixation, embedding, sectioning, staining, etc.) so that the relationship between the original product and the final one within the EM column can be questioned.
- Biological specimens are very beam-sensitive and for increasing primary doses exhibit molecular bond rupture, order destruction, preferential desorption of given species, and, finally, complete evaporation as deduced from energy-loss measurements.[5]
- The concentrations of many elements of interest are extremely low within biological structures. For instance, a mass fraction of 0.5%, which is at the lower level of detectability in materials science, is approximately equivalent to the highest local concentration of Ca normally attained in soft tissues.[6] Such a technique can only be useful for microscopically "trace" concentrations if aggregation occurs on a local scale.[7]
- The information embodied in an EELS spectrum is simultaneously very rich—Johnson[2] points out dielectric properties, chemical bonding studies, and molecu-

[a]The Orsay Scanning Transmission Electron Microscope (STEM) is operated as Unite de Service 041: "Microscopie Electronique Analytique et Quantitative" with CNRS support.
[b]Present address: IBM Almaden Research Center, K31/802, 650 Harry Road, San Jose, CA 95120–6099, USA.

lar and elemental analysis within a single EELS spectrum on cytosine—and very frustrating, because the highly interesting information is contained within faint structures (which can moreover vanish under the beam) while the major part of the inelastic spectrum is not truly characteristic. However, some direct use of the total inelastic cross section has recently provided an access to rather subtle modes of contrast matching between organic substances and embedding materials (See REFERENCES 8–10).

In the present paper, our discussion is restricted to core-level signals, which appear in EELS spectra as edges superposed over a noncharacteristic, decreasing background. Their physical origin is understood within a simple atomic model. A high-energy incident electron promotes an inner-shell electron to an empty bound or vacuum state. It is the excitation process detectable in the energy-loss spectrum of the scattered electron. This is a very short lifetime excitation, because both radiative (with X ray emission) and nonradiative (with Auger electron emission) processes tend to fill the primary hole on the core level. EELS deals only with one aspect: the excitation process. But de-excitation processes can also be used to carry the same chemical information. The three different spectroscopies were employed by J. Wroblewski on the same specimen, crystalline inclusions in the rabbit eye lens (FIGURE 1). X ray microanalysis showed the presence of P and Ca (and the common artifact Cu peak of the supporting grid); with the use of Auger probe, O could also be detected. With EELS, it is rather difficult to obtain the P signal (generally visible as a change of slope on a rapidly decreasing background), while Ca and O can easily be seen on top of a prominent C edge. This example illustrates only a qualitative point of view, because the spectroscopies were performed on distinct instruments, with variable primary energies, recording times, probe sizes, and intensities. More particularly, very different specimen volumes have been analyzed (a surface layer for Auger analysis, as opposed to the bulk of the thin section for energy dispersive X ray analysis (EDX) and EELS, the latter being at least one or two orders of magnitude more spatially localized).

METHODS AND MATERIALS FOR EELS IMAGING IN BIOLOGY

Electron Microscopy Instrumentation

An ideal situation would consist in performing all analyses on the same specimen area within one single instrument and, if possible, under the same primary irradiation condition. This is the goal of the latest developments in analytical EM instruments: the general layout of the future optimized machine is shown in FIGURE 2. For the present time, only partial channels of spectroscopic information are simultaneously available (EELS serial acquisition, EDX). The general description of the Orsay scanning transmission electron microscope (STEM) VG-HB501 coupled with a Gatan spectrometer has been given in REFERENCE 11.

The scanning mode consists in focusing the incident electrons into a probe on the specimen and in recording all characteristic signals generated from this volume of material (roughly defined by the probe section and the specimen thickness). In a "point analysis mode," one fixes the incident probe over a given specimen feature and records spectra, among them the EELS one. In the "image selection mode," one selects one or several channels in the spectra and records the associated images by scanning the probe over the specimen, generally following a digitally controlled raster. This mode is used for the acquisition of elemental distribution maps with any signal, but we shall only consider here the use of energy-loss filtered images for chemical mapping.

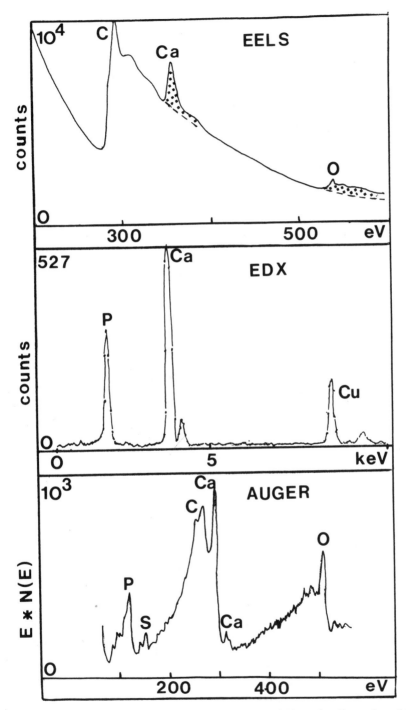

FIGURE 1. Typical examples of analytical signals on a biological thin section. Comparison of the *EELS* core-loss spectrum with the *EDX* and the *Auger* spectrum on a rabbit lens tissue section (courtesy of J. Wroblewski).

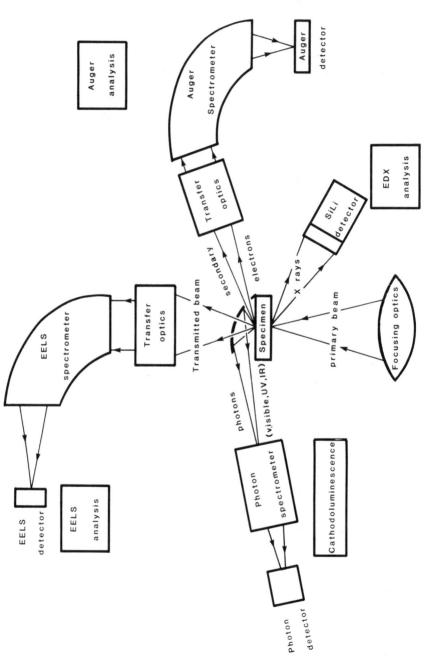

FIGURE 2. The general design of an optimized analytical scanning transmission electron microscope column.

A conventional transmission electron microscope (CTEM) configuration can also be used if a suitable energy filter is introduced between the specimen and the two-dimensional recording medium (either a photographic plate or a direct television camera readout). Following the original design of Castaing and Henry,[12] electron microscopists have built and used several systems, either of the magnetic prism and electrostatic mirror type or of the all magnetic type. This technique of electron spectroscopy imaging (ESI) has been largely applied to biological studies by Ottensmeyer et al.[13]

Computer assistance has become an essential component in all aspects of analytical EM, such as data acquisition, data processing, and display of results. In the EELS domain, it is used to record and process spectra from a given specimen area. Such a spectrum consists of $\simeq 10^3$ channels defining the energy-loss scale. In the mapping mode, the computer governs both the scanning of the probe on the specimen (i.e. the definition of the image pixels) and the spectrometer setting (i.e. the energy-loss position and width for the image). A spectroscopic image is made of $\simeq 10^4$–10^5 pixels. As a consequence of the larger data content, several requirements concern both the memory size and the CPU capacity and access speed. An interactive unit for STEM images has been developed in Orsay together with the necessary software.[14] Another acquisition system designed for similar purposes has been described by Leapman et al.[15]

Materials

A major interest of EELS lies in its higher spatial resolution for microanalysis, i.e. in its inherent capability of localizing with high accuracy elements in their biological matrix. It is a rather new advance in biology, and it raises the challenge of developing preparation techniques that preserve the biochemical and ultrastructural state at the level of interest.

Until now, EELS has been mostly applied to biological specimens, such as isolated proteins on amorphous supports with different concentrations of staining material, or chemically fixed thin tissue sections with no or relatively little staining. This is a satisfactory solution when the elements of interest are securely anchored to a biological structure; it covers all applications of a pathological character in which one is interested by mineralization processes or metal segregation within tissues.[16–17]

For application to cytochemistry, however, cryofixation methods seem to be better suited for stabilizing instantaneously the spatial distribution of diffusive elements. Once the material has been frozen, many cryopreparation methods are then offered, such as cryosubstitution or cryosectioning.[18–20] But more important is the quality of the primary cryofixation step, which determines to a high degree the morphological preservation of the specimen. An objective choice of a proper method from the frozen fresh tissue to the EELS analysis in the microscope is not yet possible. We have begun to investigate with various collaborators the relevance of several of these approaches. With J. and R. Wroblewski (Karolinska Institute, Stockholm) we have worked on unstained sections about 70 nm thick of freeze-substituted tissue embedded in polarbed, and some micrographs have already been published.[9] We have checked that a structure (in this specific case, the nuclear membrane of eukaryotic cells) is more clearly visible when the resin does not contain similar amounts of the considered element as the biological material: the nitrogen image is more interesting than the oxygen one, because the nitrogen concentration in nucleic acids and proteins is not negligible (a few percent) when compared to the embedding resin. The result is less clear for oxygen present both in the specimen and in the resin at a similar level of concentration (around 10%).

More recent observations in collaboration with the same group have shown that it should be possible to carry out the same type of EELS mapping on freeze-dried sections subsequently embedded in Lowicryl resins, in which case the morphological preservation is quite adequate for the identification of objects such as mitochondria, lysosomes, etc. Finally, a future goal should be to apply EELS mapping techniques on frozen-hydrated sections, such as was done by Zierold in EDX microanalysis,[21] in spite of all possible artifacts due to ice crystallization and bubbling under the electron beam.

Data Processing

Spectrum Processing for Point Analysis

A typical spectrum from a biological section is shown in FIGURE 3. It exhibits the major features frequently visible in the core-loss excitation part: *i.e.* mainly a carbon K edge on the decreasing tail of which one sees weaker edges, which, in the present case, are $Ca-L_{2,3}$, N-K, O-K, $Fe-L_{2,3}$, etc. A first straightforward application is to measure the corresponding signals after background subtraction and to obtain a semi-quantitative chemical analysis. One uses the formulae of Egerton[22] and the available partial cross-section calculations. Currently attainable accuracy is of the order of $\pm$ 10%, depending on many parameters related either to the specimen (thickness and multiple scattering, composition and superposed edges, etc.) or to the instrument (efficiency of collection, spectrometer and post specimen lenses aberrations, detection chains, etc.) It must be pointed out that most applications of this type on biological sections have been devoted to the study of radiation damage on molecules and preferential element loss.[23-24]

A more recent field of interest lies in the analysis of core-edge fine structures—see Colliex *et al.* for a review.[25] The most famous example is the change of threshold fine structure on the carbon K edge when the C atoms are involved in molecular bonds with atoms of different nature—see Isaacson[26] for the comparison of carbon K edges in six types of nucleic acid bases. This constitutes a test situation and to our knowledge has not been used for studying molecular composition—a preliminary report by Hainfeld and Isaacson[27] used the fine structures in the energy-loss region less than 15 eV.

Many other cases can be studied and used, such as the detailed behavior of the oxygen K edge in different close-neighbor environments, or the change of white-line ratio between the L_3/L_2 lines of transition elements as a function of the valence state. However, it must be once more recognized that at the present state we have to follow a rather pragmatic approach: we have gathered collections of edge shapes in test compounds, and we rely on the comparison with such standards. There does not exist a satisfactory a priori theoretical prediction of the edge shape as a function of well defined sample parameters.

Image Processing for Elemental Mapping

The characteristic information is contained in a small fraction of the energy-loss spectrum—see FIGURE 4 for an illustration—and the problem associated with chemical mapping is to have an unambiguous access to this specific signal, while satisfying certain considerations of time and dose minimization. It is obvious that one single energy-filtered image is insufficient, because in most cases the ratio of signal to background is $\ll 1$ (see the spectrum of FIGURE 3). A two-image method with a

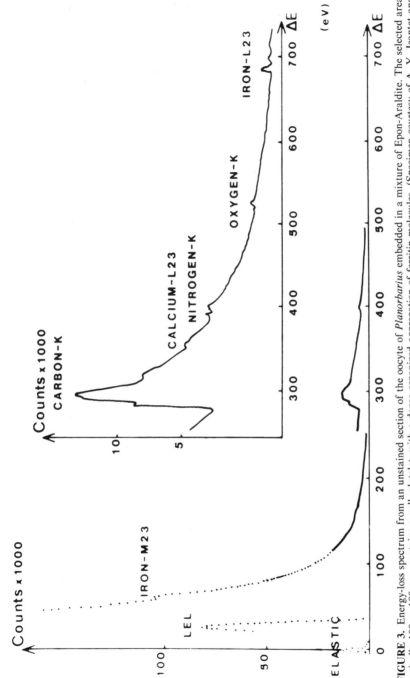

FIGURE 3. Energy-loss spectrum from an unstained section of the oocyte of *Planorbarius* embedded in a mixture of Epon-Araldite. The selected area typically 100 × 100 nm, contains a yolk platelet, with a dense organized aggregation of ferritin molecules. (Specimen courtesy of A. Y. Jeantet and C. Quintana. From Colliex *et al.*[9] Reprinted by permission from the *Journal of Ultrastructure Research.*)

one-parameter fit was used by Ottensmeyer *et al.* But as early as 1978, both Hainfeld and Isaacson[27] and we ourselves[28] suggested that a three-image method with a two-parameter fit is obviously superior and in many cases constitutes a prerequisite to avoid confusion between chemical composition changes and mass-thickness variation. The principle of the method is as follows (see FIGURE 4). From the two images I_1 and I_2 recorded for energy losses E_1 and E_2 below the edge, one estimates for each pixel the couple of best fitting parameters (A, r) for the background model curve:

$$B = A. (\Delta E)^{-r}$$

and then extrapolates it for the third image I_3 recorded at E_3 above the edge. The characteristic signal S is then the difference between I_3 and B_3.

In a one-parameter fit, one assumes a constant average $\bar{r}$ value over the whole image. However, both we (REFERENCES 28 and 33 and unpublished data) and Leapman and his co-workers[29] arrived at the same conclusion, that a one-parameter fit should be generally avoided. It offers a clear advantage: the statistical error is reduced

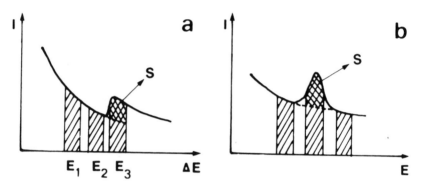

FIGURE 4. Two methods of subtracting the EELS background for elemental mapping purposes.

by as much as a factor of $\simeq 5$, because the shape of the spectrum is forced to be stable over a whole image. On the other hand, it introduces systematic errors in background subtraction. It can be of interest if the counting rates are very low and the statistical errors too high, or if the specimen is highly uniform. To check its validity, it is interesting to look at maps of r values, such as shown in previous works.[9–16] If the standard deviation is larger than anticipated from statistical considerations only, the use of two images should be avoided because of associated mass thickness artifacts. The r image is generally correlated to the annular dark-field image with reversed contrast, that is, the more strongly scattering parts of the specimen also give rise to smaller absolute values of the r parameter; it can also be interpreted as due to an enhanced contribution of multiple scattering. Consequently, in a one-parameter fit, such regions are confused with greater concentrations of the searched-for element. To improve the accuracy in two- (or more) parameters fits, it is important to record several images before and after the edge, so that a least-squares fitting can be achieved for background modelization, and statistical errors can be reduced. Important headway is to be anticipated from parallel-spectrum recording techniques in which all energy-loss channels are stored for the same incident dose.

Correlation techniques have recently been applied to extract in a more quantitative way image characteristics such as contrast, resolution, and signal-to-noise ratio.[30] Jeanguillaume[31] has proposed a multivariate statistical approach to correlate pixel contents between different micrographs, such as energy-filtered ones with distinct energy windows. It offers new possibilities, such as the reconstruction of areas of interest within chemical maps using selection criteria in the multiparameter pattern. It also constitutes an easy way to check slope variations between different energy-filtered images. In conclusion, it is worth pointing out that the enormous amount of data available in a series of energy-filtered images has to be handled in a more efficient way than it now is. It is an obvious field of application for more elaborate statistical methods.

ABOUT A FEW APPLICATIONS

Survey of Useful Edges

In the EELS atlas of Ahn and Krivanek,[32] edges are shown for all elements for $Z = 1$ to $Z = 83$ (plus Th and U). It does not mean that all of them can be of interest for practical EELS investigations on biological tissues. For very low Z, the H and He K edges, which have been recorded on gases, are useless. They lie, in the solid state, in the middle of the intense noncharacteristic low energy-loss region, which consists, for most biological specimens, of a major peak with its maximum at $\simeq$ 20–25 eV and a full width at half maximum (FWHM) of $\simeq$ 10–15 eV. Li may be the first useful element with its K edge at $\simeq$ 60 eV; however, it must be present in rather large concentrations and in very thin specimens to reduce as far as possible the influence of the intense background due to the low energy-loss tail. For practical purposes, K edges can be useful for elements from Be ($Z = 4$) to Ne ($Z = 10$); they have sharp hydrogenic profiles with some fine structures at threshold. Carbon K edge is, of course, the major feature in biological applications, and it is not very sensitive. However, it can be useful as a reference signal representative of the local mass thickness for scaling all other signals. Oxygen and nitrogen, more particularly, are very good candidates for cytochemical studies.[9,33] Beryllium has been searched in pathological sections (lung, kidney, etc.).

For intermediate Z elements (*i.e.* $11 \leq Z \leq 17$)—Na, Mg, Al, Si, P, S, and Cl—the K-loss lies above 1,000 eV, and higher primary voltages such as 1 MV have proved to be of interest.[34] For these elements, the $L_{2,3}$ edge offers an alternative. It lies at lower energy-loss (between 60 and 200 eV) and has more intense cross sections ($\simeq 10^{-20}$ cm² per atom in normal 100 kV operating conditions). But it suffers from two limitations: it is superposed on a very intense and rapidly decreasing background, and its shape is not as sharp; it corresponds to a "delayed edge". In an EELS spectrum, such signals appear as weak modulations over the background. Very good spectra, with a high signal-to-noise ratio over this energy-loss range, are required to detect their presence unambiguously when they are in small concentrations. More refined techniques of edge detection have been elaborated to identify them in spectra.[35–37] However, in this class of elements, there are very important candidates for biological applications. Phosphorus ionization edge at 132 eV has been extensively used by Ottensmeyer and his co-workers to recognize DNA in nucleosomes[38] or RNA in ribonucleoproteins.[39]

For heavier elements ($Z = 19$ for potassium, to $Z = 28$ for nickel), the $L_{2,3}$ edge is a very useful signal, because it displays a characteristic, intense, white line, which is visible as a pair of lines, spin-orbit split, when the spectrometer energy resolution is sufficiently good. Unfortunately, potassium is a very poor element for EELS studies in biology because of its exact edge superposition on top of the carbon K edge (284 eV and 292 eV). On the other hand, calcium can be easily handled; Ca maps on the lens section used in FIGURE 1 have been obtained by extrapolation of the background between two energy-loss windows on each side of the characteristic Ca white line.[40] Another interesting element is iron; the $L_{2,3}$ edge is intense in a domain free from frequent edge superposition problems. It has been used to detect individual ferritin or hemosiderin molecules in sections.[33,41]

For $Z \geq 30$, the most interesting cases are Cs, Ba, and all rare-earth elements (La to Yb). They display a clear and intense pair of white lines at the $M_{4,5}$ edge (the position varies from $\simeq 750$ eV for Cs to $\simeq 1,500$ eV for Yb) and a strong, well identified $N_{4,5}$ edge between $\simeq 100$ and 150 eV. An application will be shown in the next paragraph. One must avoid using sections stained with uranium salts, because U has a clear couple of $O_{4,5}$ peaks at 95 and 105 eV. They can be a source of confusion and artifacts in the study of any characteristic edge in the energy-loss range 80 to 150 eV. On the other hand, this uranium signal has been fruitfully used to study the fundamental detection limits in the EELS technique; the experiments were performed on very small uranium clusters present on the carbon foils that support stained biomolecules.[42]

We must finally mention that a few poorer edges have been used in the literature—Ga-$L_{2,3}$ at 1,100 eV, Cd-$M_{4,5}$ at 400 eV, In-$M_{4,5}$ at 450 eV, Pt-$M_{4,5}$ at 2,300 eV, etc.—to follow the intracellular metal distribution after injection. Analyses were performed with 1 MeV primary beams, but one must be careful about the quality of spectra, because the above edges are not very well recognizable.[33] When EELS analysis fails, it is evident that a combined analytical approach involving EDX on the same specimen area is highly recommended.

Ultrastructural Localization of Alkaline Phosphatase with Cerium as the Capture Metal

The following example illustrates the present capability of the Orsay STEM unit for high spatial resolution localization of elements in cell sections. It has been done in collaboration with members of Harvard Medical School: M. J. Karnovsky, C. Lechene, and J. M. Robinson. Robinson and Karnovsky have shown that Ce ions can be used as a capture agent for phosphatase, more uniform and reproducible than lead in these cytochemical reactions.[43] They also found that the amount of cerium phosphate reaction product is proportional to the amount of enzyme present in a cell-free model system. In conventional EM observations, cerium phosphates appear as fine electron-dense precipitates, but the ultrastructure localization is not unambiguous in sections stained with uranyl acetate and lead citrate. A positive identification of the reaction product in the EM specimen is then required before further studies. As shown from the preliminary results gathered in FIGURE 5, EELS mapping of the Ce-$N_{4,5}$ signal offers an excellent spatial resolution and more accurate localization. These micrographs concern a portion of a guinea pig polymorphonuclear leukocyte incubated for the localization of alkaline phosphatase with Ce, as described in REFERENCE 43. These are unstained sections with a slight OsO_4 postfixation before embedding. Energy-filtered images at 81, 96, and 111 eV below the Ce-$N_{4,5}$ edge and at 126 eV on the peak have been recorded (FIGURE 5a). It allows a satisfactory definition of the background for each pixel, and the processed Ce map is displayed in FIGURE 5b. Cerium is localized on the outer cell surface and on cellular membranes. In the images the spatial resolution is of the order of 5 nm and the signal-to-noise ratio is sufficiently high to achieve unambiguous Ce detection. This experiment shows the feasibility of positive identifica-

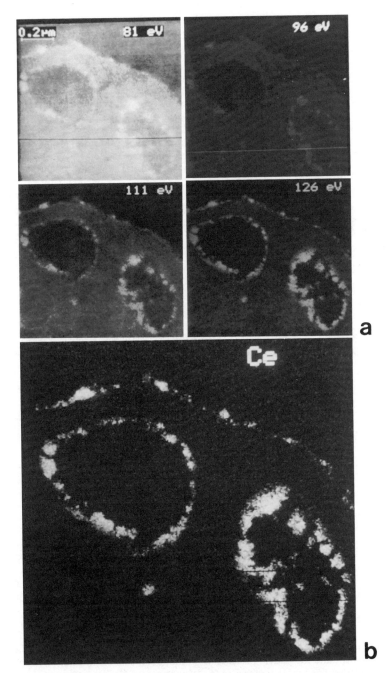

FIGURE 5. Ultrastructural localization of Cerium reaction product within thin cellular sections. (Specimen courtesy of J. Robinson, M. Karnovsky, and C. Lechene.) (a) shows the used family of energy filtered images while (b) displays the high-resolution map of the Cerium distribution after processing. (Micrographs were recorded by C. Jeanguillaume, M. Tence, and P. Trebbia.)

tion of selected reaction products with EELS signals. It could be extended to other types of reaction in which specific molecules are tagged with markers within the cellular environment.

Limits of Detection

In a previous paper,[9] we investigated the sensitivity of EELS analysis in terms of spatial resolution and minimum detectable mass on a test specimen, the inorganic iron-rich core of the ferritin molecule, which contains $\simeq$ 5,000 iron atoms (identified with their Fe-$L_{2,3}$ signal at 702 eV) and $\simeq$ 500 phosphorus atoms (searched for with their P-$L_{2,3}$ signal at 135 eV). Experimental conditions were: primary voltage = 100 kV; beam current = 5.10^{-10} A; probe size $\simeq$ 1.5 nm; energy window width = 10 eV; dwell time per pixel respectively 2 ms for P and 10 ms for Fe. We estimated per pixel: $S/\delta S \simeq$ 1.4 for P and $S/\delta S \simeq$ 4 for Fe, which allowed a clear visibility of iron cores with a 2-nm resolution ($\simeq$ 300 Fe atoms) and a nondetection of P at this resolution level (30

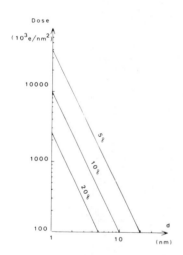

FIGURE 6. Typical doses required for visualizing a level of chemical concentration at a given resolution scale. Calculations are made for the following numbers: $\sigma_B = \sigma_S = 10^{-21}$ cm^2, $\rho = 10^{23}$ cm^{-3}, t = 50 nm. (From Colliex et al.[9] Reprinted by permission from the *Journal of Ultrastructure Research*.)

phosphorus atoms). If we note that an object can be detected with a signal-to-noise ratio $\simeq$ 1 with a resolution of a few pixels in an image because the observer is visually able to integrate information, the level of detection of phosphorus in the present case is rather in the 500–1,000 atoms range, that is, several times greater than the iron one ($\simeq$ 100 atoms), though the cross sections for P detection are about 20 times larger than for iron. This relatively poor performance of EELS for detecting phosphorus is due to the very intense background in this energy domain as compared to that underlying the Fe signal.

The above discussion was intended to provide some feeling concerning the presently accessible limits of detection; the mentioned numbers only carry qualitative information. Minimum concentration and minimum amounts of material depend very heavily on the element and the matrix in which it occurs, but also on instrumental properties. Other contributions to this conference (e.g. Shuman et al.[44]) discuss it more extensively; the advances due to parallel recording of EELS spectra should improve the minimum detectable mass fraction by at least one or two orders of magnitude.[45] We have considered here only statistical arguments, but have neglected the probe size distribution in STEM (which can be made reasonably smaller than 1 nm in a FEG instrument), the chromatic aberration of the objective lens in CTEM, which introduces a degradation of spatial resolution of a few nm when a 10 eV energy window and an angular acceptance of 20 mrad are used to record an image.

Finally we have neglected a most important effect, the radiation damage induced by the primary dose. FIGURE 6 recalls typical doses that are required to visualize a given level of chemical concentration at a given resolution scale. For a 2-nm pixel size and typical concentration of 10 to 20% (which are actually large values when referred to the carbonaceous matrix), it means 10^6 or more e^-/nm^2. These values are far more intense than the critical doses generally measured for radiation damage in organic materials. However, the situation is not as desperate as could be deduced from such simple considerations. In a microanalytical experiment, we are interested in the local concentration of a given element within the selected volume, rather independently of the molecular or structural organization. Consequently, the specimen can remain stable for such a diagnosis as long as no detectable loss or transfer of an element is reached. This critical dose is surely higher than the standard critical dose for high-resolution imaging of biological structures but is yet unknown in most situations.

CONCLUSIONS

Even if it still remains of marginal use in electron microscopic studies of biological specimens—only a few groups have really been involved in such research—EELS's impact has progressed greatly during recent years, as is clearly seen from a comparison of the present state of achievements with the perspectives outlined several years ago. It is the consequence of improvements in spectrometer design and coupling with an electron microscope column, and in computer assistance for rapid and reliable data acquisition and more elaborate data processing. It has also been associated with a more careful consideration of specific requirements for specimen preparation compatible with the newly available high spatial resolution microanalytical techniques.

For the future, one can anticipate a larger use of these techniques and a new decisive progress in terms of dose reduction for the same signal-to-noise ratio, when highly efficient parallel recording devices are available on the market.

ACKNOWLEDGMENTS

At the Orsay STEM unit, the collaboration with C. Jeanguillaume, M. Tencé, P. Trebbia, C. Mory, H. Kohl, and N. Bonnet has been very helpful. Thanks are also due to all our colleagues in biology who came with their specimens and ideas and who were actively involved in the progress of the technique: J. and R. Wroblewski from Stockholm, E. Carlemalm and E. Kellenberger from Basel, C. Lechene from Boston, and C. Quintana, A. Y. Jeantet, and E. Delain from Paris. The IBM Almaden Research Center has provided a very stimulating and efficient environment during the preparation of this manuscript.

REFERENCES

1. COLLIEX, C. 1984. Electron energy loss spectroscopy in the electron microscope. *In* Advances in Optical and Electron Microscopy. E. Barer & V. E. Cosslett, Eds. Vol. 9: 65–177. Academic Press. New York, NY.

2. ISAACSON, M. 1981. *In* Microprobe Analysis of Biological Systems. T. E. Hutchinson & A. P. Somlyo, Eds. 289–307. Academic Press. New York, NY.
   JOHNSON, D. E. 1981. Ibidem. 351–363.
3. EGERTON, R. F. 1982. *In* Proc. 10th Int. Conf. on Electron Microscopy, Hamburg, I: 151–158.
4. COLLIEX, C. & P. TREBBIA. 1979. *In* Microbeam Analysis in Biology. C. P. Lechene & R. R. Warner, Eds. 65–86. Academic Press. New York, NY.
   COLLIEX, C., C. JEANGUILLAUME & P. TREBBIA. 1981. *In* Microprobe Analysis of Biological Systems. T. E. Hutchinson & A. P. Somlyo. Eds. 251–271. Academic Press. New York, NY.
5. GLAESER, R. M. 1975. Radiation damage and biological electron microscopy. *In* Physical Aspects of Electron Microscopy and Microbeam Analysis. B. M. Siegel & D. R. Beaman, Eds. 205–230. Wiley. New York, NY.
   REIMER, L. 1975. A review of the radiation damage problem of organic specimens in electron microscopy. Ibidem. 231–246.
   ISAACSON, M. 1977. Specimen damage in the electron microscope. *In* Principles and Techniques of Electron Microscopy. M. A. Hayat, Ed. Vol. 7: 1–78. Van Nostrand-Reinhold. New York, NY.
6. SOMLYO, A. P. 1984. J. Ultrastruct. Res. **88:** 135–142.
7. LEAPMAN, R. D., C. E. FIORI & K. E. GORLEN. 1986. To be published in Biol. Trace Elem. Res.
8. CARLEMALM, E. & E. KELLENBERGER. 1982. The EMBO Journal **1:** 63–67.
   CARLEMALM, E., C. COLLIEX & E. KELLENBERGER. 1985. Contrast formation in electron microscopy of biological material. *In* Advances in Electronics and Electron Physics. P. Hawkes, Ed. Vol. 63: 269–334. Academic Press. New York, NY.
9. COLLIEX, C., C. JEANGUILLAUME & C. MORY. 1984. J. Ultrastruct. Res. **88:** 177–206.
10. KELLENBERGER, E., E. CARLEMALM, W. VILLIGER, M. WURTZ, C. MORY & CH. COLLIEX. 1986. Z-Contrast in biology: a comparison with other imaging modes. This volume.
11. COLLIEX, C. & P. TREBBIA. 1982. Ultramicroscopy **9:** 259–266.
    COLLIEX, C. & C. MORY. 1984. *In* Quantitative Electron Microscopy. J. Chapman & A. J. Craven, Eds. Vol. 25: 143–216. Scottish Universities Summer School in Physics.
    COLLIEX, C. 1985. J. Microsc. Spectrosc. Electron **10:** 313–332.
12. CASTAING, R. & L. HENRY. 1962. C. R. Acad. Sci. Ser. B **255:** 76–78.
13. OTTENSMEYER, F. P., D. BAZETT-JONES & K. M. ADAMSON-SHARPE. 1981. *In* Microprobe Analysis of Biological Systems. T. E. Hutchinson and A. P. Somlyo, Eds. 309–324. Academic Press. New York, NY.
    ADAMSON-SHARPE, K. M. & F. P. OTTENSMEYER. 1981. J. Microsc. **122:** 309–314.
14. TENCE, M., N. BONNET, C. JEANGUILLAUME, P. TREBBIA & C. COLLIEX. 1985. J. Microsc. Spectrosc. Electron **10:** 65–84.
15. LEAPMAN, R. D., C. E. FIORI, K. E. GORLEN, C. C. GIBSON & C. R. SWYT. 1984. Ultramicroscopy **12:** 281–292.
16. JEANGUILLAUME, C., J. P. BERRY, C. COLLIEX, P. GALLE, M. TENCE & P. TREBBIA. 1984. J. Phys. Suppl. Coll. **45 C2:** 577–580.
17. BERRY, J. P., P. GALLE, Y. KINH, G. ZANCHI, J. SEVELY & B. JOUFFREY. 1984. J. Phys. Suppl. Coll. **45 C2:** 581–584.
    MIGNON-CONTE, M., F. CARENTZ, J. POURRAT, J. J. CONTE, Y. KINH & J. SEVELY. 1984. Ibidem. 603–606.
18. ROOMANS, G. 1981. Scanning Electron Microsc. **2:** 345–356.
19. FREDERIK, P. M., P. H. H. BOMANS, W. M. BUSING & R. ODSELIUS. 1984. J. Phys. Suppl. Coll. **45 C2:** 451–455.
20. KELLENBERGER, E., E. CARLEMALM & W. VILLIGER. 1986. To be published in Proc. 4th Annu. Pfefferkorn Conf.
21. ZIEROLD, K. 1982. Ultramicroscopy **10:** 45–54.
    ZIEROLD, K. 1982. J. Microsc. **125:** 149–156.
22. EGERTON, R. F. 1978. Ultramicroscopy **3:** 243–251.
23. ISAACSON, M., M. L. COLLINS & M. LISTVAN. 1978. Proc. 9th Int. Conf. on Electron Microscopy, Toronto. **3:** 61–69.
24. EGERTON, R. F. 1980. Ultramicroscopy **5:** 521–523.

MISRA, M. & R. F. EGERTON. 1984. Ultramicroscopy **15:** 337–344.
25. COLLIEX, C., T. MANOUBI, M. GASGNIER & L. M. BROWN. 1985. Scanning Electron Microsc. **2:** 489–512.
26. ISAACSON, M. 1979. *In* Microbeam Analysis in Biology. C. P. Lechene & R. R. Warner, Eds. 53–61. Academic Press. New York, NY.
27. HAINFELD, J., & M. ISAACSON. 1978. Ultramicroscopy **3:** 87–95.
28. JEANGUILLAUME, C., C. COLLIEX & P. TREBBIA. 1978. Ultramicroscopy **3:** 137–142.
29. LEAPMAN, R. D., K. E. GORLEN & C. SWYT. 1985. Scanning Electron Microsc. **1:** 1–13.
30. MORY, C. & C. COLLIEX. 1985. J. Microsc. Spectrosc. Electron **10:** 389–394.
31. JEANGUILLAUME, C. 1985. J. Microsc. Spectrosc. Electron **10:** 409–416.
32. AHN, C. C., & O. L. KRIVANEK. 1982. An EELS Atlas. Center for Solid State Science, Arizona State University. Tempe, AZ.
33. JEANGUILLAUME, C., M. TENCE, P. TREBBIA & C. COLLIEX. 1983. Scanning Electron Microsc. **2:** 745–756.
34. SEVELY, J. 1985. Inst. Phys. Conf. Ser. **78:** 155–160.
35. FIORI, C. E., C. R. SWYT & K. E. GORLEN. 1984. *In* Microbeam Analysis. A. D. Romig & J. I. Goldstein, Eds. 179–183. San Francisco Press. San Francisco, CA.
36. HOSOI, J., Y. OIKAWA, M. INOUE & Y. KOKUBO. 1985. J. Electron Microsc. **34:**1–7.
37. ZALUZEC, N. J. 1985. Ultramicroscopy **18:** 185–190.
38. BAZETT-JONES, D. P. & F. P. OTTENSMEYER. 1982. Can. J. Biochem. **60:** 364–370.
39. KORN, A. P., P. SPITNIK-ELSON, D. ELSON & F. P. OTTENSMEYER. 1983. Eur. J. Cell Biol. **31:** 334–340.
40. COLLIEX, C. 1985. Ultramicroscopy. **18:** 131–150.
41. COLLIEX, C. 1982. J. Microsc. Spectrosc. Electron **7:** 525–542.
JEANTET, A. Y. & C. QUINTANA. 1984. Private communication and unpublished work.
42. COLLIEX, C., O. L. KRIVANEK & P. TREBBIA. 1982. Inst. Phys. Conf. Ser. **61:** 183–188.
MORY, C., H. KOHL & M. TENCE. 1986. Private communication and unpublished results.
43. ROBINSON, J. M. & M. J. KARNOVSKY. 1983. J. Histochem. Cytochem. **31:** 1190–1196 and 1197–1208.
ROBINSON J. M. 1985. J. Histochem. Cytochem. **33:** 749–754.
44. SHUMAN, H., C.-F. CHANG, E. L. BUHLE, JR. & A. P. SOMLYO. 1986. Electron energy-loss spectroscopy: quantitation and imaging. This volume.
45. SHUMAN, H. & P. KRUIT. 1985. Rev. Sci. Instrum. **56:** 231–239.

Scanning Transmission Electron Microscope (STEM) Elemental Mapping by Electron Energy-Loss Spectroscopy

R. D. LEAPMAN

Biomedical Engineering and Instrumentation Branch
National Institutes of Health
Bethesda, Maryland 20892

INTRODUCTION

Electron energy-loss spectroscopy (EELS) in the analytical electron microscope provides a unique method for determining the concentration of the low atomic number elements, carbon, nitrogen, and oxygen, that contribute most to the total mass of biological molecules inside cells. In addition, EELS is competitive with energy dispersive X ray spectroscopy (EDXS) for determining phosphorus, sulfur, and calcium concentrations in thin dried cryosections of biological tissue. Recently, it has become possible not only to obtain information from a point in the sample but also to obtain elemental distributions. This is achieved by digitally scanning the electron beam over the sample and acquiring EELS data at a number of pixels so that a chemical image is built up. The elemental map provides an important correlation between morphology and chemical composition. Furthermore, it allows an unbiased sampling of an entire area of the specimen. It is worthwhile at the outset to compare the different techniques of EELS mapping as performed in the scanning transmission electron microscope (STEM) and in the energy filtering transmission electron microscope (EFTEM). We shall then present some examples of EELS mapping applied to thin sections in the STEM mode of the analytical electron microscope. Low atomic number mapping may be useful for determining relative concentrations of different biological molecules such as proteins, nucleic acids, and lipids. However, it is important to consider here the question of preferential mass loss due to radiation damage.

METHODS

Instrumentation and Data Processing

Elemental maps were obtained with a Hitachi H700H transmission electron microscope operated at 100 to 200 keV beam energy and equipped with a magnetic sector electron energy-loss spectrometer and an energy dispersive X ray detector, and a Kevex 7000 multichannel analyzer. A conventional heated tungsten filament source provided a probe of approximately 20-nm diameter with a current of 1 nA using the STEM imaging optics. A Gatan model 626 cryotransfer sample holder allowed cooling of the specimen down to $-150°$ C. The electron microscope has been interfaced to a Digital Equipment Corporation PDP 11-60 computer and a satellite LSI 11-23 processor as described previously.[1] Only a brief outline of the most important features is given here. The satellite controls data acquisition after appropriate software is first downloaded from the host computer. The STEM probe is raster-scanned over a rectangular target area containing up to 512 × 512 pixels. Data are sent from the satellite via a fast direct memory access link to the host computer where they are processed further and stored on disk. In collection of X ray images the satellite controls the voltage applied to the spectrometer electrostatic deflection plates and also controls a 32-bit fast counter that measures the amplified and discriminated photomultiplier pulses. In collection of X ray images the satellite controls a timer and gate that allow the multichannel analyzer (MCA) to accumulate counts at each pixel for an appropriate period; in this case the host computer reads the MCA memory directly. The spectral data (EELS and EDXS) are processed dynamically by the host computer, and the resulting elemental maps are displayed on a DeAnza IP6400 image display system with four 512 × 512 × 8-bit image planes and a color monitor. This system also supports basic image processing capabilities such as addition, subtraction, division, and smooth and contrast enhancement. For EELS image acquisition, the counts are accumulated in several channels below and above an edge (typically 10 in all). An inverse power law is used to extrapolate the pre-edge background by means of a linear least squares fit and this background is subtracted from the counts in the post-edge region to give the net signal. The spectrum and background fit at each pixel are continuously displayed on the Kevex 7000 monitor. This on-line processing provides the operator with useful feedback about the progress of the image acquisition.[2] Dwell times per pixel for recording EELS maps are typically 50 to 100 ms requiring 15 to 30 minutes to record a 128 × 128 pixel image. Drift of the spectrum energy offset is corrected automatically during EELS acquisition by periodically tracking the carbon K edge that is present strongly throughout the sample. For EDXS image acquisition it is often necessary to remove the bremsstrahlung background. A method that eliminates the background by means of a top-hat digital filter is applied to the spectrum at each pixel, and this has been demonstrated even on noisy data.[3] Background-corrected images for up to four elements may be acquired concurrently in both EELS and EDXS mapping. It is also possible to acquire bright-field and dark-field images simultaneously with the elemental maps to help correlate chemical composition with morphology.

Sample Preparation

1. Cultured rat islet of Langerhans beta and alpha cells were directly quick-frozen against a liquid nitrogen cooled copper block[4] 30 minutes after increasing insulin output by exposure to elevated glucose. The cells were freeze-dried, lightly vapor-osmicated, embedded in araldite, and thin sectioned on glycerol to a thickness of about 50 nm. The sections were otherwise left unstained.

2. Lavaged human alveolar macrophages were centrifuged into a pellet and fixed in 2.5% glutaraldehyde. They were then post-fixed with 1% osmium tetroxide, dehydrated with alcohol and propylene oxide, and embedded in epon. Ultrathin sections were cut to a thickness of about 70 nm and were otherwise left unstained.

3. Cultured bovine adrenal chromaffin cells were pelleted and rapidly frozen against a liquid helium cooled copper block.[5] 100-nm to 120-nm thick cryosections were cut and deposited on a thin formvar and carbon coated folding copper grid. The samples were cryotransferred to the microscope specimen chamber by means of the Gatan model 626 holder. The temperature during transfer was maintained below −160° C, after which the sample was slowly warmed to −90° C in order to achieve freeze-drying conditions. Data were recorded either at room temperature or at −80° C to reduce selective mass loss of elements such as sulfur.

COMPARISON OF STEM WITH EFTEM AND THE
PROBLEM OF QUANTITATION

In the STEM a finely focused probe of electrons is raster-scanned across a sample, and spectral data in the neighborhood of a characteristic core edge are obtained from each pixel sequentially. In the EFTEM the image is formed in parallel by adjusting an energy filter situated after the specimen to select one energy loss at a time.[6,7] The spatial resolution in the EFTEM is mainly limited by the performance of the objective lens and by chromatic aberration, whereas in STEM resolution is limited by the minimum useful probe size and hence by gun brightness. This means that a field emission source is generally required for resolution less than 5 nm in the STEM.[8,9] The EFTEM is often operated with about a microampere current on the sample, whereas in the STEM probe currents in the nanoampere range are typically used. EELS images in the EFTEM can therefore often be recorded much more rapidly than in STEM. However, these advantages of the EFTEM may be offset by important features of the STEM.

1. The EELS signal in STEM may be detected as single electron pulses, which can be fed into a digital counter and into a computer-controlled data acquisition system. Several channels can be acquired at each pixel, so it is possible to achieve a precise and accurate background subtraction.[10] Hence relatively low elemental concentrations can be reliably detected. Since the different channels of data from each pixel can be acquired almost at the same time (or perhaps simultaneously) the different pre-edge images are automatically in precise registration, whereas in the EFTEM they are not.

2. In STEM several different signals may be acquired simultaneously such as the EDXS signal, which provides complementary information about the heavier elements and ions. Also the simultaneously acquired annular dark-field elastic signal gives information about the sample morphology.[2]

3. EELS parallel detection[11–13] is possible in the STEM but not in the EFTEM. With this improvement STEM becomes the most efficient technique of recording inelastic scattering and is therefore the method of choice for samples where radiation damage is a limiting factor.

4. In STEM it is straightforward to stop the probe on the sample and record a spectrum from a particular region where the elemental concentration may be too low to be detected by mapping. In the EFTEM it is generally more difficult to obtain spectral information from small regions.

On the other hand the EFTEM can have substantial advantages in terms of recording time and resolution for EELS mapping of large image fields, provided reliable background subtraction is possible. Usually the EELS signal-to-background ratio (S/B) in biological samples is low; for example, the phosphorus $L_{2,3}$ S/B ratio in a typical unstained membrane (porin) has been determined experimentally to be about 5 to 10%. In order to detect the membrane phosphorus in a cell preparation (especially if fixed and embedded) it would be necessary to detect S/B levels that were significantly lower than this figure.

An important question is how best to perform the background estimation. In STEM an inverse power law AE^{-r} has been used to fit the background in the pre-edge region and to extrapolate it to the post-edge region.[14] Here E is the energy loss and A and r are constants. This two-parameter fit, though to some extent empirical, allows for changes in background shape due to variations in mass thickness or composition.[15] Changes in mass thickness determine the amount of plural inelastic scattering that alters the distribution of intensities in the spectrum; changes in elemental composition are reflected in variations of the scattering cross section. Both factors are known to influence spectral shape. The background estimation in the EFTEM is normally

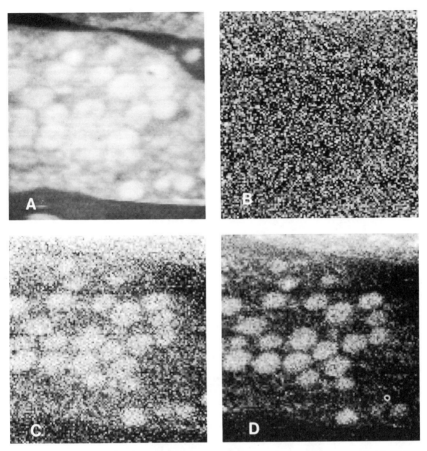

FIGURE 1. (A) Contrast-enhanced inelastic image at energy losses between 95 and 125 eV from a thin (50-nm) section of embedded freeze-dried pancreatic beta cells showing secretory granules with increased scattering and hence higher intensity. Beam energy 100 keV. Full field 2 μm. (B) Phosphorus $L_{2,3}$ image obtained from same area with a two-parameter fit for background (see text). No P is detected. (C) r-image generated from background fit at each pixel in (B). (D) Phosphorus $L_{2,3}$ image obtained from same data as in (B) but assuming a constant value for parameter r (equal to mean r over entire image). Apparent detection of P in granules is an artifact.

performed by subtracting an arbitrarily scaled pre-edge image from the post-edge image, *i.e.,* a one-parameter model with constant background shape is implicitly assumed.[6,16] The statistical noise resulting from the one-parameter fit is always less than for the two-parameter fit, but systematic errors due to actual variations in background shape are likely to be greater.

We have tested the effect of a one-parameter and two-parameter background fit on a thin section of pancreatic beta cells prepared by rapid-freezing, freeze-drying, and embedding. We are interested in determining the phosphorus distribution in the sample. FIGURE 1a is a pre-edge inelastic image showing part of a beta cell containing a number of secretory granules. The image was recorded with 128×128 pixels and five channels from 95 eV to 125 eV (below the phosphorus $L_{2,3}$ edge at 132 eV). Five post-edge channels were selected from 140 eV to 160 eV resulting in a mean total number of counts per pixel of 6,500. The total recording time for the image was 20 minutes. FIGURE 1b is the corresponding phosphorus $L_{2,3}$ image obtained in the STEM mode by subtracting the extrapolated pre-edge background, which has been fitted to

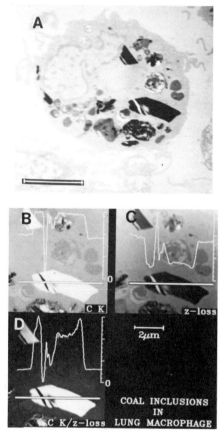

FIGURE 2. (A) Bright-field STEM of thin section of embedded alveolar macrophage prepared from lung lavage of coal miner showing respired particles taken up by cell. (B) Carbon K edge image. (C) Zero-loss image. (D) Carbon distribution obtained by dividing C image by zero-loss image to correct for plural scattering. Amount of C in coal particle relative to surrounding cell is dramatically increased as shown by intensity plots across images.

the two-parameter inverse power law. No phosphorus is detected in the granules, a result that is expected if the composition of the granule core is essentially pure insulin. The same data can be analyzed assuming that the background shape is constant over the entire image, *i.e.,* assuming a one-parameter model for the background (as is usual in analysis of EFTEM data). This result is presented in FIGURE 1c, where the noise is much reduced, and now phosphorus seems to be present in the secretory granules. We can attribute the apparent phosphorus to an artifact due to small systematic variations in background shape, as is evident in the r-image (FIGURE 1d), which reveals directly how the background shape varies across the sample. The mean value of r over the image is 3.62, but the value varies from 3.47 ± 0.01 in the granules to 3.65 ± 0.01 in the cytoplasm. Although such findings indicate possible problems with data analysis in the EFTEM, it must be added that in principle a two-parameter fit for the background could be performed in both instruments, as has been pointed out by Ottensmeyer.[17] More sophisticated analytical methods should be facilitated by electronic detection systems such as the SIT vidicon that are becoming available for the EFTEM.[11-13] Furthermore it is noted that under certain conditions a one-parameter background fit will be valid.[15]

There is another important factor in quantifying EELS mapping. In general the background-subtracted core edge signal does not directly reflect the elemental concentration. In fact, the raw core loss image can give a misleading impression of the elemental distribution. This is so because the measured core edge signal depends on the amount of plural scattering,[18,19] which can vary significantly from region to region, especially in a cryosection. The number of atoms n of an element detected by the core excitation c in a pixel of dimension d is given by

$$n = I_c(\alpha, \Delta) \, d^2 / I_0(\alpha, \Delta) \, \sigma_c(\alpha, \Delta) \qquad (1)$$

where $I_c(\alpha, \Delta)$ is the measured core edge intensity for collection semiangle α and energy-loss integration range Δ above threshold, $I_0(\alpha, \Delta)$ is the measured zero-loss and low-loss intensity and $\sigma_c(\alpha, \Delta)$ is the partial cross section for the parameters α and Δ. In Equation 1 division by $I_0(\alpha, \Delta)$ takes account of the decrease in core loss signal by plural elastic and inelastic scattering. The division of the core loss image by the summed zero-loss and low-loss image is achieved digitally in our system by loading each image into the DeAnza IP6400 display system and utilizing the built-in array processor. FIGURE 2a is a STEM bright-field image of a plastic-embedded and sectioned alveolar macrophage from the lung of a coal miner, showing several lysosomes and respired particles. FIGURE 2b is the carbon K edge image from a region of the cell containing two coal particles. FIGURE 2c is the corresponding summed zero-loss and low-loss image, while the quotient image reflecting the true elemental distribution according to Equation 1 is shown in FIGURE 2d. The line scans across the coal particle demonstrate the large correction factor that can occur due to plural scattering even in thin samples. It is evident that there is four times as much carbon in the coal particle as in the surrounding embedded cytoplasm, whereas the raw carbon K edge image suggests an increase of only 50%.

APPLICATIONS: POTENTIAL AND LIMITATIONS

Embedded Samples

Elemental distributions have been recorded from rapidly frozen, freeze-dried, and embedded preparations of cultured rat islets of Langerhans in order to investigate the composition of insulin storage granules. FIGURE 3a is a bright-field STEM image of an alpha and beta cell showing the presence of numerous granules and mitochondria in each. The nitrogen K edge EELS map recorded from the same area with 128 × 128 pixels, a 60-ms dwell time per pixel, and a 1-nA probe current is shown in FIGURE 3b. The pre-edge window contained five channels ranging from 370 to 395 eV, and the post-edge window contained another five channels from 405 to 420 eV. High

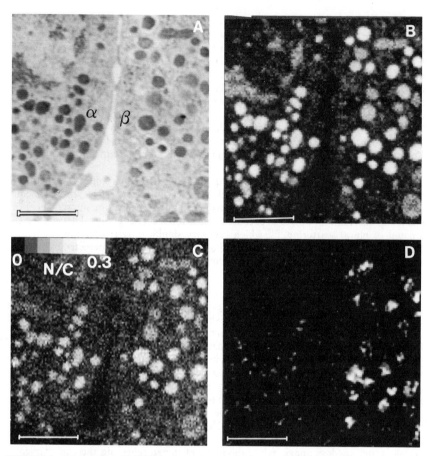

FIGURE 3. (A) Bright-field STEM of rapidly frozen, freeze-dried, and embedded rat islet of Langerhans, showing regions of *alpha* and *beta* cell. *Bar* = 1 μm. **(B)** N-K edge EELS image showing highest signal in secretory granules. **(C)** N/C ratio map with calibrated intensity scale. N/C = 0.23 in secretory granules. **(D)** Smoothed S Kα X ray map showing higher sulfur in beta granules than in alpha granules.

concentrations of carbon and nitrogen are observed in the secretory granules of both cells. Relative amounts of nitrogen and carbon at each pixel could be estimated by dividing the N K by the C K core edge image and multiplying by the inverse ratio of the dwell times and by the inverse ratio of the corresponding partial ionization cross sections, obtained from the hydrogenic model.[20] This procedure nearly eliminates the effects of plural scattering in the ratio map, in the same way as normalization by the summed zero and low loss image mentioned earlier. The N/C ratio map obtained using a two-parameter background fit for each element is shown in FIGURE 3c together with a calibrated intensity scale. Here a factor of 1.2 has been applied to correct for the thin 10-nm formvar support film; this was justifiable since the section thickness was quite uniform. The mean N/C ratio in the beta cell secretory granules was found to be

0.23 ± 0.02, which is close to the value of 0.241 for the protein insulin, known to be a major component of the beta cell granule. It is noted that the sample preparation appeared to leave the granule cores intact, and penetration by the plastic was not evident. The total dose on the sample during EELS image acquisition was 10^4 e Å^{-2}, and the specimen temperature was maintained at $-80°$ C. We conclude that the change in the N/C ratio was not very significant under these experimental conditions. On the other hand there was evidence that loss of oxygen occurred. It was not possible to obtain an EELS sulfur map of the sample due to the high background intensity at the S-$L_{2,3}$ edge. However FIGURE 3d shows the S $K\alpha$ EDXS map recorded from the same sample region with a dwell time of 100 ms per pixel. The map contained a maximum of 10 signal counts per pixel, so the signal-to-noise was improved by performing a 3×3 pixel digital smoothing operation at the cost of degrading spatial resolution to 80 nm. The sulfur is more concentrated in the beta cell granules. This result is explained by the presence of insulin, which has nearly five times the weight fraction of sulfur as glucagon, present in the alpha cell granules. A typical energy-loss spectrum from a beta cell granule is shown in FIGURE 4. Here the background has been removed from the sulfur $L_{2,3}$ edge. Quantitative analysis of such spectra, using hydrogenic cross sections,[20] yielded a sulfur-to-nitrogen ratio of 11.2 ± 1:1, which is close to the value for pure insulin (10.1:1). Spectra were also recorded from pure embedded insulin crystals in order to test the quantitation procedures. These data on the pancreatic beta cell serve to demonstrate that EELS is capable of providing information about the distribution of biological molecules in cellular compartments at

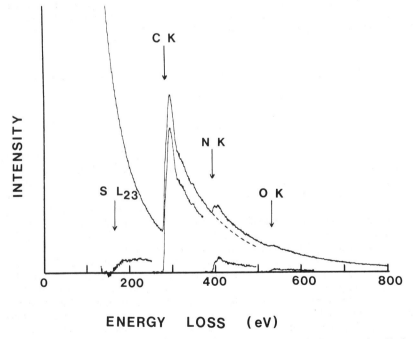

FIGURE 4. Energy-loss spectrum from beta cell granule showing sulfur $L_{2,3}$ edge, carbon K edge, and N-K edge with background subtracted. N/S = 11.2 ± 1.0.

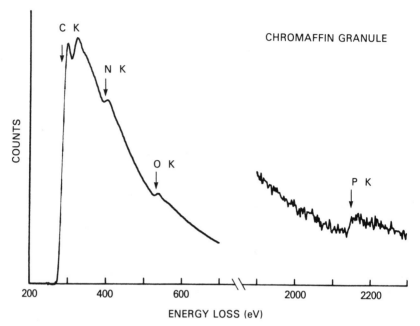

FIGURE 5. Sum of energy-loss spectra recorded at 200 keV beam energy from ten secretory granules of adrenal chromaffin cell. Sample prepared by rapid-freezing, cryosectioning, and freeze-drying. C-, N-, O-, and P-K edges are observed. Estimated value of $N/P = 5.0 \pm 0.5$.

a resolution of 30 nm provided a radiation exposure of 10^4 e Å^{-2} can be tolerated by the sample.

Cryosections

It is clearly advantageous to obtain elemental information from cryosections of rapidly frozen cells.[21,22] Distributions of ions can then be preserved and light element analysis is not frustrated by uncertainties concerning the effects of embedding.[23] However, cryosections are much more difficult to prepare, especially if they are to be thin enough for EELS. Preliminary EELS data from cryosections of another secretory system, the adrenal chromaffin cell, have been obtained in order to investigate the composition of the secretory granules.[24] FIGURE 5 is the summed spectrum recorded at 200 keV beam energy from ten different chromaffin granules. The K edges of carbon, nitrogen, oxygen, and phosphorus are all clearly visible. A quantitative analysis gives a N:P ratio of 5.0 ± 0.5:1. Since the N:P ratio in ATP is 1.67:1, the result suggests that only 33% of the granule nitrogen is accounted for by trinucleotides, which are known to be present. FIGURE 6a is a digitized bright-field STEM micrograph of a chromaffin cell. The carbon and nitrogen EELS images in FIGURES 6b and 6c respectively show the increased nitrogen signal relative to carbon in the secretory granules. The N:C ratio map (not shown) has a mean value of about 0.2 in the granules after a correction is made for the carbon support film. Since N/C for ATP is 0.5, the measured ratio also suggests the presence of nitrogenous compounds other than ATP in the granule core.

These data would be greatly enhanced if concurrent maps of the phosphorus and sulfur distribution could be obtained under conditions where mass loss was not problematic. It might then be possible to deduce the amount of nitrogen tied up with phosphorus in the trinucleotides, the amount linked with sulfur in protein, and that remaining in other important molecules such as epinephrine in the case of the chromaffin granule. We have not yet been able to achieve this result in EELS mapping because of our relatively low sensitivity for the elements, phosphorus and sulfur, which have core edges at high energy losses. However, recently available parallel detection devices may make such measurements feasible in the future. It would be very interesting, for example, to measure directly the relative amounts of nucleotides, proteins, and catecholamines in the chromaffin granules under resting and stimulated conditions.

High-Resolution EELS Mapping

With a field emission source STEM it is, in principle, possible to obtain spectral data and elemental maps at a resolution of about 1 nm, provided the sample is suitably prepared. Unfortunately, despite the fact that the STEM offers the most efficient means of detecting inelastic electron scattering, the problem of radiation damage is still a limiting factor. Thus, in order to detect nitrogen atoms at 1-nm resolution at a concentration of 3% the required dose is estimated to be 10^6 e Å^{-2}. Under these

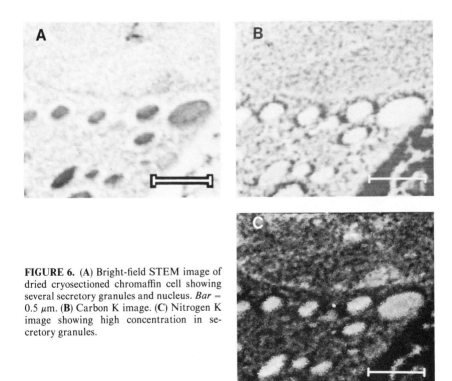

FIGURE 6. (A) Bright-field STEM image of dried cryosectioned chromaffin cell showing several secretory granules and nucleus. *Bar* = 0.5 μm. (B) Carbon K image. (C) Nitrogen K image showing high concentration in secretory granules.

conditions mass loss is inevitable even if the sample is kept at low temperatures. In order to overcome this problem it is necessary to average EELS data over many identical structures and hence achieve relatively low dose conditions (< 100 e Å^{-2}).

For example, estimation of the attainable signal-to-noise suggests that low-dose, high-resolution mapping of regular two-dimensional arrays of reconstituted porin membrane[25] should prove feasible. In such an experiment it is envisaged by Engel *et al.*[26] that low-dose EELS maps of the phosphorus and nitrogen distributions at 1 nm resolution could be averged over some 10^4 unit cells. The resulting map could provide information about the location of phospholipid (P) and protein (N) components of the membrane.

DISCUSSION

EELS mapping can potentially help us visualize the distributions of the major elemental constituents of cells at a resolution of 10 to 50 nm. In the STEM the counting time per pixel is necessarily limited, since data must be acquired sequentially. Thus the dwell time per pixel for a 128×128 pixel image recorded in 30 minutes is only 100 ms. Under these conditions it is difficult to obtain sufficient counting statistics to measure physiological levels of important elements like calcium even with a highly optimized detection system. However, it has been demonstrated by Shuman *et al.*[27] that EELS analysis of calcium from a single point (or line) in a cryosectioned cell is competitive with EDXS analysis in terms of sensitivity. The technique of EELS mapping is probably better suited to determining distributions of abundant atoms like C, N, and O as well as P and S. The latter two can also be imaged by EDXS. In combination the maps should provide some insight into relative distributions of various biological compounds such as proteins (C, N, O, and S), nucleotides (C, N, O, and P), lipids (C, O, and P), and other molecules such as catecholamines (C, N, and O). Secretory cells are one example where light element mapping may be of value; in this case secretory granules often contain one or more compounds in high concentration.

It is not yet clear to what extent radiation damage will limit the application of this technique. Our data showed that with modest cooling down to $-80°$ C and with doses of around 10^4 e Å^{-2}, mass loss did not greatly affect measured ratio of nitrogen and carbon in our samples. There was evidence for loss of oxygen, however. This result must depend on the type of compound being analyzed. For higher resolution studies radiation damage and mass loss undoubtedly become a serious problem. We must then turn to a low-dose approach where image processing is used to average EELS data from many identical structures.

ACKNOWLEDGMENTS

The author is indebted to Dr. A.F. Boyne (Department of Pharmacology and Experimental Therapeutics, University of Maryland) and Dr. R. Dudek (Department of Anatomy, East Carolina University) for the pancreatic beta cell sample, to Dr. R. L. Ornberg (NIDDK-National Institute for Diabetes, Digestive, and Kidney Diseases, National Institutes of Health) for the cryosectioned adrenal chromaffin cells, and to Dr. W. Rom (NHLBI-National Heart, Lung, and Blood Institute, National Institutes of Health) for the alveolar macrophage specimen. The author is also indebted to Kieth Gorlen, Carol Swyt, and Charles Fiori for development of the computer system and software.

REFERENCES

1. GORLEN, K. E., L. K. BARDEN, J. S. DEL PRIORE, C. E. FIORI, C. C. GIBSON & R. D. LEAPMAN. 1984. Rev. Sci. Instrum. **55:** 912–921.
2. LEAPMAN, R. D., C. E. FIORI, K. E. GORLEN, C. C. GIBSON & C. R. SWYT. 1984. Ultramicroscopy **12:** 281–292.
3. FIORI, C. E., C. R. SWYT & K. E. GORLEN. 1984. *In* Microbeam Analysis-1984. A. D. Romig and J. I. Goldstein, Eds. 179–185. San Francisco Press. San Francisco, CA.
4. PHILLIPS, T. E. & A. F. BOYNE. 1984. J. Electron Micros. Tech.**1:** 9–29.
5. ORNBERG, R. L., to be published.
6. OTTENSMEYER, F. P. & J. W. ANDREW. 1980. J. Ultrastruct. Res. **72:** 336–348.
7. SHUMAN, H. & A. P. SOMLYO. 1982. Proc. Natl. Acad. Sci. USA **79:** 106–107.
8. JEANGUILLAUME, C., M. TENCÉ, P. TREBBIA & C. COLLIEX. 1983. Scanning Electron Microsc. **2:** 745–756.
9. ISAACSON, M. & D. E. JOHNSON. 1975. Ultramicroscopy **1:** 33–52.
10. JEANGUILLAUME, C., P. TREBBIA & C. COLLIEX. 1978. Ultramicroscopy **3:** 237–242.
11. SHUMAN, H. 1981. Ultramicroscopy **6:** 385–396.
12. JOHNSON, E. E., K. L. MONSON, S. CSILLAG & E. A. STERN. 1981. *In* Analytical Electron Microscopy. R. H. Geiss, Ed. 205–209. San Francisco Press. San Francisco, CA.
13. EGERTON, R. F. 1984. J. Electron Microsc. Tech. **1:** 37–52.
14. EGERTON, R. F. 1975. Philos. Mag. **31:** 199–215.
15. LEAPMAN, R. D. & C. R. SWYT. 1983. *In* Analytical Electron Microscopy. R. Gooley, Ed. 163–167. San Francisco Press. San Francisco, CA.
16. OTTENSMEYER, F. P. & A. L. ARSENAULT. 1983. Scanning Electron Microsc. **4:** 1867–1875.
17. OTTENSMEYER, F. P. 1984. Proc. 42nd Annu. Mtg. Electron Microscopy Soc. Amer. 340–343.
18. EGERTON, R. F., C. J. ROSSOUW & M. J. WHELAN. 1976. *In* Developments in Electron Microscopy and Analysis. 129-132. Academic Press. London.
19. LEAPMAN, R. D. & C. R. SWYT. 1985. Proc. 43rd Annu. Mtg. Electron Microscopy Soc. Amer. 404–405.
20. EGERTON, R. F. 1979. Ultramicroscopy **4:** 169–179.
21. SOMLYO, A. V., H. SHUMAN & A. P. SOMLYO. 1977. J. Cell Biol. **74:** 828–857.
22. SOMLYO, A. P. & H. SHUMAN. 1982. Ultramicroscopy **8:** 219–234.
23. LEAPMAN, R. D. & R. L. ORNBERG. 1983. Proc. 41st Annu. Mtg. Electron Microscopy Soc. Amer. 590–591.
24. ORNBERG, R. L., to be published.
25. DORSET, D. L., A. ENGEL, M. HANER, A. MASSALSKI & J. P. ROSENBUSCH. 1983. J. Mol. Biol. **165:** 701–710.
26. ENGEL, A., R. REICHELT & R. D. LEAPMAN, to be published.
27. SHUMAN, H., P. KRUIT & A. P. SOMLYO. 1983. *In* Microbeam Analysis – 1983. R. Gooley, Ed. 247–251. San Francisco Press. San Francisco, CA.

DISCUSSION OF THE PAPER

E. ZEITLER (*Fritz-Haber-Institut der Max-Planck-Gesellschaft, Berlin, Federal Republic of Germany*): At the beginning you showed this comparison between STEM and fixed-beam filters and there was one point where you said in the filtering case that, if you take pictures at various delays, it might be in register. Could you clarify what you meant by that?

LEAPMAN: If you take two images, there could simply be a drift between one image and the other. That may or may not be a problem.

ZEITLER: I thought you meant the optics changes, because you change your energy levels.

LEAPMAN: That's also possible, but I haven't had enough experience with that up to now.

A. P. SOMLYO (*University of Pennsylvania School of Medicine, Philadelphia, Pennsylvania*): Dr. Zeitler, were you going to suggest that the optics won't change if you change the high voltage?

ZEITLER: Yes, that's exactly the way you would do it, you would change the high voltage.

SOMLYO: On the other hand, Dr. Shuman and I also had trouble with what Dr. Leapman suggested; shift due to microscope drift. This is another commercial to manufacturers to please improve the stages not only for short term stability (vibration), but also for long term drift. Of course, one can also use cross-correlation methods to realign between the consecutive images.

F. P. OTTENSMEYER (*University of Toronto, Toronto, Canada*): I can tell you that there is a shift due to two causes even if the instrument is stable. You have to put plates into the camera and the plate holders aren't always in register. So there's a shift there but that's obviously correctable. There's a fundamental problem that as you irradiate the specimen it shrinks, it stretches, it does things so that you have difficulty in alignment from one image to the other, no matter how you record them, which are the order of 0.2% or one part in 500. You'll notice it on a 512×512 display, because you can get one corner in register and the other one is out by one pixel.

I notice that you were using the established formula AE^{-r} to fit those curves. Now I've looked at a good number of pieces of data, including mine, yours, and everyone else's, and it seems that you yourself and various other people are using different curves to fit. I think Dr. Shuman uses a 4th order polynomial or something like this and I find that that then introduces a systematic error to force the AE^{-r} onto the curve, which, if you have a pre-edge, doesn't quite fit. Do you have a solution to this? Should one just use a power law of any sort?

LEAPMAN: I'm not saying that the AE^{-r} fit is perfect, because I don't think it is. It does provide at least two parameters for the fit and it does take into account some variation in shape. It's almost arbitrary what type of fit you use.

OTTENSMEYER: I agree with the arbitrariness in a sense, because using AE^{-r} one gets a worse fit than if one uses a three-parameter fitting function. The error that's introduced is sometimes bigger than the one that's introduced if you don't use a fit at all or if you use the ratio.

LEAPMAN: We only use it when we don't see systematic errors; if we see systematic errors, then we don't try to image in that way. We can always test that.

SOMLYO: Isn't it true that whether one can use something like an AE^{-r} really depends on what region of the spectrum one is working in, what other edges are near by? Certainly neither you, nor we, nor anyone else is going to use the same AE^{-r} in the presence of a triple plasmon scattering and, let's say, a potassium background, whereas when you are looking at the phosphorus $L_{2,3}$ edge, it may be reasonable. I would think that, unless you had silicon there, that region is reasonably well fitted.

LEAPMAN: That's right.

Elemental Mapping by Energy Filtration: Advantages, Limitations, and Compromises[a]

F. P. OTTENSMEYER

The Ontario Cancer Institute
Toronto, Ontario M4X 1K9, Canada

INTRODUCTION

An ideal electron-probe microanalysis system follows the consequences of each individual electron impinging on the sample, both in terms of the events that occur to that electron, and in terms of the secondary events that are set in motion by its passage. In the broader sense of electron-probe microanalysis whole fields in science have been created by the discovery that such electrons could be induced to form electron micrographs, by transmission through thin specimens or back-scatter from the surfaces of thick specimens. Secondary electrons yielded a plasticity of the surfaces of microscopic specimens akin to our three-dimensional view of the macroscopic world that immediately made the images appear real and familiar. In materials science the effect of charge distribution in semiconductor devices on the electron permitted not only analysis of their structure but also their function, both by absorption measurements of the incoming current and by the effect on the emission of secondary electrons. Energy transferred to the specimen and emitted as photons was found to be exceedingly useful as well in the form both of cathodoluminescence, which was characteristic of molecular relaxation events, and of X ray emission, which formed a signature of the atomic composition of the specimen.

With such a wealth of information available from the action of the electron beam, there seems hardly to be a need for additional techniques, unless one can extend the limits of the available approaches in terms of such parameters as resolution, sensitivity of detection, ease of operation, or even price.

The analysis of the energy lost by the incoming electron in traversing a specimen is a technique that is complementary to all the other analytical approaches in the sense that this deficit in energy, along with the change in momentum, represents the sum of the secondary events that have been set in motion. In general, the measurement of such a cumulative process should make the extraction of analytical information more difficult than the direct measurement of any individual contributing component. In specific instances, however, fundamental, or merely practical, advantages may favor the use of one approach over another.

In what way does electron energy-loss analysis and its extension, imaging with a selected energy loss, offer advantages over existing analytical techniques? And what limits are set by the approach? In brief, the primary advantage is the potential for very high spatial resolution, about 0.5 nm in analysis and elemental mapping. As a corollary of this resolution potential, a very good detection and identification limit is obtained for

[a]Supported by the Ontario Cancer Treatment and Research Foundation, the Medical Research Council of Canada, and the National Cancer Institute of Canada.

small numbers of highly localized atoms. Moreover, this can be obtained with an electron exposure to the specimen of the order of a few C/cm^2, a high number compared to normal electron microscopy, but low by a factor near 100 compared to X ray microanalysis.

The primary drawback is the fact that the electrons must be transmitted easily through the specimen, preferably with no more than one scattering event. This means that specimens should be no thicker than about 0.3 mean free paths for total electron scatter if easy analysis is desired, with the absence or at least the diminution of errors due to multiple scattering effects. This limit may be relaxed in future as acceptable ways are found to correct for plural scattering.

The energy-loss spectrum itself, in which the sought chemical and quantitative information is contained, deserves close attention before accurate qualitative and quantitative information derived from it can be accepted. Overlap of ionization edges occurs for some electronic shells of different elements (e.g. P_L at 132 eV and $Sr_{M,4,5}$ at 133 to 135 eV), and edges can have an abrupt rise at the threshold energy (C_K, Ca_L) or a slow rise over many eV ($S_{L2,3}$, $Ge_{L2,3}$, $Ag_{M4,5}$.) The ionization edge can shift and change in shape even for different physical forms of the same element (diamond, graphite, amorphous carbon) and for different chemical compounds (Si, SiO_2). Equally disturbing for both quantitative and qualitative analysis is the extended fine structure of one element that can mimic a signal from an elemental edge of an atom just higher in atomic number (Si_L and P_L, P_L and S_L, etc).

Finally, a problem arises from the fact that quantification and elemental maps rely on the extrapolation of the pre-edge spectrum underneath the rise of a particular edge. While this is simple in principle, the shape of the spectrum does not always follow the simple inverse power law with energy that is generally assumed.

Nevertheless, a technique with the potential of providing elemental maps with the spatial extent and resolution of the best electron micrograph, and a sensitivity that can be as good as a few tens of atoms, deserves our careful examination of its theoretical and practical limits.

SPATIAL RESOLUTION

A number of spatial resolution values have been measured at different energy losses. With energy-selected imaging at 110 eV and 150 eV, measurements of 0.75 nm[1] and 0.3 to 0.5 nm[2] were the best resolution values obtained for the phosphorus signal in a phospholipid bilayer of a membranous virus in thin sections.

This measurement was questioned in light of early theoretical estimates of the behavior of resolution with energy loss based on the Heisenberg Uncertainty Principle. The first approaches[3] yielded a spatial resolution d at an energy loss ΔE according to

$$d \simeq 2\lambda E/\Delta E \qquad (1)$$

where λ is the wavelength of the electron at an energy E, corrected for relativistic effects. For 80-keV electrons and an energy loss of 150 eV this equation yields a resolution of 4.8 nm (FIGURE 1), obviously in disagreement with the experimental results. A later treatment of the same approach,[31] incorporating a factor of 2π in Planck's constant h, and assuming an impact parameter of 2d instead of d, yielded the relation

$$d \simeq hv/4\pi\Delta E \qquad (2)$$

where v is the relativistic velocity of the electron. At 150 eV the resolution now becomes 0.33 nm (FIGURE 1). A recent more rigorous quantum mechanical treatment by Kohl and Rose[4] produced a different dependence of d with ΔE, and again indicated the potential for lower values of spatial resolution for small ΔE (FIGURE 1). Measurements by Isaacson and Scheinfein[5] over a larger energy range produced values of between 0.4 and 0.8 nm at both 100 eV and 360 eV. Except where limited by instrumentation and techniques, the experimental values are in good agreement with Kohl and Rose using a resolution criterion of 50% included intensity. The full width at half maximum of the theoretical atom intensity distribution follows a different curve, since at low energy losses this distribution takes the shape of a diffuse ring.[4]

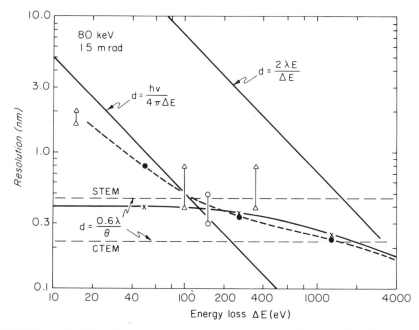

FIGURE 1. Spatial resolution with energy loss. Both *straight lines* are derived from the Heisenberg Uncertainty Principle under different conditions (see text). The *curves* are from a quantum mechanical treatment[4] using as the measure of resolution the full width at half maximum (x), or the inclusion of 50% of the intensity of a point atom (●). The experimental values are from Adamson-Sharpe and Ottensmeyer[2] (O), and from Isaacson and Scheinfein[5] (△). Curves calculated and redrawn after Egerton.[32]

ELASTIC SCATTERING CONTRIBUTION

While the experimental results of Isaacson and Scheinfein[5] and the extrapolation of the 50% intensity curve of FIGURE 1 suggest a potential resolution of 1.5–2.0 nm at a ΔE of 15 eV, lattice images of graphitized carbon and dysprosium oxide with spacings of 0.34 nm and 0.75 nm, respectively, were observed as early as 1977 by Craven and Colliex.[6] These images were not considered a true measure of resolution at this low an energy loss. Instead they were interpreted by Howie[3] to be due to a multiple scattering component which, in addition to the inelastic event, includes at least one high-resolution elastic scatter event. FIGURE 2 indicates that this elastic contribution to inelastic scatter at any given energy loss is fairly small at a thickness corresponding to a mean free path Λ of 0.3 or less, but becomes increasingly important as specimen thickness approaches and surpasses one mean free path for electron scattering. If inelastic events are highly nonlocalized then the final event in a cascade of elastic and inelastic events must be elastic to provide a high-resolution image (FIGURE 2, lower curve). The results of Craven and Colliex[6] were for an estimated thickness of 50 nm, an atomic number of 66 for dysprosium, and a density of 7.8 for dysprosium oxide. Under these conditions a sizeable elastic contribution to the inelastic image is easily obtained.

This interpretation cannot be carried forward to every instance in which high-resolution detail has been observed in an inelastic image. At any given specimen thickness this elastic contribution follows the shape of the single scatter energy loss spectrum in proportion given by the curves of FIGURE 2. For a biological section of

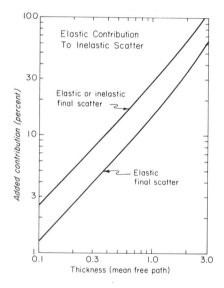

FIGURE 2. Relationship of plural inelastic scatter (up to 5 interactions) to single inelastic scatter with respect to thickness. The plural scatter in this case contains at least one elastic interaction, at any position in the cascade *(top curve)*, or last in the cascade *(bottom curve)*. The data refer to different thicknesses of carbon.[21]

about 30 nm ($\Lambda = 0.3$) this proportion is about 4%. For 30-nm sections of osmium-fixed membranes of murine leukemia virus unstained[1,2] it could be argued that the osmium atoms and the phosphorus atoms in the phospholipid bilayers should contribute a sizeable elastic component due to multiple scattering. This is not observed in the inelastic image at 110 eV,[1] but could form a small percentage of the increase in signal observed at 150 eV, above the phosphorus L edge. In the total membrane, phosphorus contributes about 1.5 at.%. In a well aligned section the phosphorus-containing head groups may make up as much as 7.5% of the atoms in a 0.5-nm thick sheet. The mean free path of this sheet should therefore be commensurately shorter than in the carbonaceous biological structure and embedding medium surrounding it. Nevertheless, even a thickness equivalent to 0.4 mean free paths produces a multiple scatter component containing elastic events that is only 5–6% of the single scatter signal. The observed highly localized changes in intensity of 15 to 20% between images of viral membranes at 110 eV and 150 eV[1,2] must then be due largely to single inelastic scatter below and above the L ionization edge of phosphorus at 132 eV. To assume that this high-resolution signal is due solely to the elastic component of multiple scattering leads to the conclusion that the jump ratio at the phosphorus edge in the presence of 90% carbonlike atoms is the order of 3 over the 0.5-nm sheet, or 0.6 over the entire membrane. The observed spectrum of myelin instead shows the merest inflection at the L edge and a jump ratio of about 0.15 at the K edge.[1]

Thus the 0.3- to 0.5-nm resolution observed at an energy loss of 150 eV on the phosphorus signal of membranes can be considered a fair measure of the spatial resolution of delocalization at this energy. Moreover, this experimental measurement includes not only the effect of nonlocal interactions and instrumental limitations, but also that of radiation damage and the potentially less than perfect alignment of the phosphorus head groups in the phospholipid membrane leaflets. Nevertheless, for practical purposes, any elastic contributions to the localization of a heavier element in a light-atom matrix can be considered qualitatively beneficial. However, for exact quantification this effect must be taken into account.

DETECTION LIMIT

As indicated above, energy-loss analysis both in scanning transmission electron microscopy (STEM) and conventional transmission electron microscopy (CTEM) is characterized by a spatial resolution that is near atomic. Consequently it is obvious that a small number of atoms within a resolution element of the technique suffice to produce a characteristic spectrum. If the stability of the specimen were not limiting, then a single atom should be the lower detection limit. In practice the limitation of beam current, detectors or photographic emulsion, and instrument and specimen stability combine to produce currently practical limits. The lowest numbers of atoms detected and identified in elemental maps in biological specimens to date were about 300 P atoms in nucleosomes with signal-to-noise ratios (S/N) of 30 to 50,[7,8] about 80 P atoms in tRNA with S/N = 9 (Andrews and Ottensmeyer, unpublished), and 40 P atoms in a phospholipid membrane with S/N = 4 ± 1.[2] These all point to a detection limit of about 30–50 P atoms in such specimens. For clusters of uranium about 10–15 atoms have been detected by their energy-loss spectrum.[9] A number of lighter atoms—oxygen, for instance—appear to offer greater difficulties in detection and quantification, becoming volatile due to irradiation under the electron beam.[10] For these, no limits have as yet been established.

In the case of phosphorus and uranium above, the atoms have all been concentrated in very small volumes. If lower concentrations are used, the signal from the desired atoms becomes a smaller proportion to that from atoms of the surrounding matrix, and larger numbers of atoms are required to offset the concomitant lowering of the signal-to-noise ratio. Various limiting concentration values have been observed depending on different circumstances. Generally, values of a few tenths of 1% are considered to be the limit of detection using spectra, although the best reported value is an extrapolated detection limit at about 99% confidence of 0.006%.[11]

SPECTRAL OVERLAP

The energy resolution of electron spectrometers for small acceptance angles (about 1 mrad) can easily be 1 eV or below. Larger acceptance angles are frequently desirable not only to capture information scattered at larger angles from the specimen, but, particularly in the case of imaging filters, to be able to pass a large image through the magnetic field of the spectrometer. For existing designs of imaging spectrometers[12-17] the energy resolution for a full image at an acceptance angle of 15 mrad is approximately 15 eV. However, even if a small acceptance angle is used, an energy band of 5 to 15 eV or more is frequently chosen to increase the signal that forms the image. In most instances this is not a serious degradation, since selected characteristic features in the energy-loss spectrum, such as plasmon peaks or ionization edges, in general are broader than a few eV or are separated by a few tens of eV.

Even this spectral resolution capability is superior to the 150-eV limit of energy dispersive X ray spectrometers (EDXS), and can therefore resolve many signals from elements that overlap too strongly in that technique. But even with a 15-eV resolution

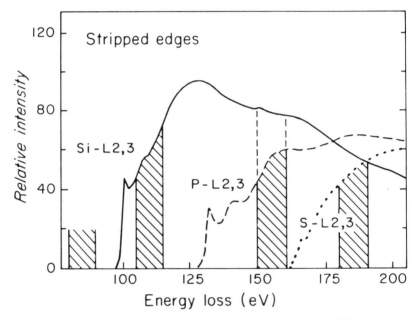

FIGURE 3. Background-subtracted ionization signals from the L edges of silicon, phosphorus, and sulphur. The *hashed areas* indicate regions of measurement for elemental mapping.

there are a number of instances in which overlap occurs. For biological preparations one of the most disappointing facts is the closeness of the potassium L edge at 294 eV to the ubiquitous carbon K edge at 284 eV. Other combinations such as $P_{L2,3}$ at 132 eV, $Ge_{M2,3}$ at 121 and 128 eV, $Sr_{M4,5}$ at about 134 eV, and $As_{M2,3}$ at 141 and 146 eV are more exotic but must be considered when a combination of such elements is possible. Arsenic in cacodylate buffers is a potential complication easily overlooked in the preparation of biological specimens for analysis of phosphorus.

While these examples of direct overlap of characteristic signals are straightforward, a more insidious complication is the spectral structure a few tens eV above the ionization threshold. This is illustrated in FIGURE 3, which depicts the background-subtracted ionization signals of the L edges of Si (Z = 14), P (Z = 15), and S (Z = 16). In all cases of elemental mapping by energy-loss analysis the characteristic signal is measured at an energy above the ionization edge in reference to the signal below the ionization edge. If the elements were present by themselves there would be no difficulty: the Si signal at 115 eV would be measured with respect to the "stripped" background at 80 eV. For phosphorus, 150 and 115 eV would form a corresponding pair, and for sulphur, 180 and 150 eV. However, if only Si were present and the 150-eV–115-eV pair (for phosphorus) were measured, then a clear signal would nevertheless be obtained, which could be mistaken for phosphorus. The presence of Si would be ascertained as above and should be a forewarning against complications in the phosphorus region. But if phosphorus and silicon are both present, it would be very difficult to extract the phosphorus information. Knowledge of the exact spectral shape of the Si signal would be necessary in order to subtract its signal from the total signal above the P L-edge. However, since a given element can have various different spectral features above the ionization edge depending on its chemical environment (e.g. Si versus SiO_2[18]), even this approach is difficult in complex specimens.

This does not mean that a specimen containing two elements with adjacent atomic numbers cannot be analysed or mapped. The element of lower atomic number will have a clear signal. Moreover, the two elements may not be located in identical sites. The high spatial resolution potential of elemental maps using energy-loss electrons means that depositions of different elements separated in space by at least 0.5 nm can be resolved as separate entities even if they have consecutive atomic numbers. Thus, sulphur and phosphorus were easily and unambiguously mapped in sections of cartilage early during mineralization, because spatial overlap was minimal.[19] On the other hand, this clear an interpretation of the presence of sulphur could not be made at a slightly later stage in intramembranous ossification in which the S and P distributions were spatially overlapping.[20] Under such conditions, confirmation from local electron energy-loss spectra must be obtained, either by careful analysis of the L edge structures, or by observation of well separated K edges of the elements (e.g. P_K at 2,146 eV and S_K at 2,472 eV),[1] or one must resort to other techniques, such as X-ray microanalysis, if the size and elemental content of the object of interest permit.

EXTRAPOLATION ERRORS

It is generally assumed that the intensity I of the energy-loss spectrum follows a relationship with energy loss E, that is given by:[21]

$$I = A E^{-R} \tag{3}$$

where R has values between 2 and 5. In a number of specific cases this relationship has been found not to be sufficiently accurate. Grunes and Leapman[22] used a cubic polynomial, and Shuman and Kruit[11] a fourth order polynomial to obtain a better fit; our own data indicated an acceptable fit over 1,000 eV using an exponential decay (FIGURE 4), while Bentley et al.[23] found an excellent fit to their data using a linear fit to a power series of the logarithm of E. Steele et al.[24] modified Equation 3 by adding a cross-section related component fitted at an energy well above the ionization threshold.

While a clear statement of the best relationship may eventually be made for spectra of elements or simple compounds, the proximity of individual ionization edges and the often complex structure above the edges will make the determination of the best approach to analysis of complex specimens quite difficult. The problem is not academic, since it impinges on the accuracy of quantification of any element, and, as a consequence, on the ultimate sensitivity of detection. The experimental determination

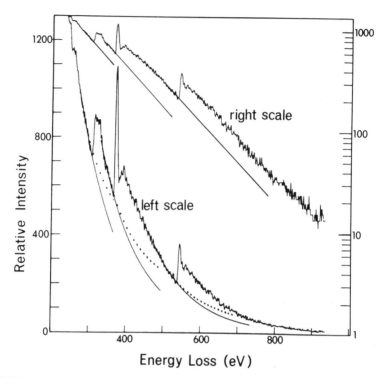

Energy Loss (eV)

FIGURE 4. Experimental electron energy-loss spectrum from a thin section of osteoblasts[1] fitted with an inverse power law *(lower curve, dotted extrapolations)* and a negative exponential *(upper curve and solid extrapolations)* at the ionization edges of C, Ca, and O.

of ionization cross sections is affected by exactly the same difficulty, in addition to the requirement of thin specimens of known composition. So, while atlases of a good number of spectra exist,[18] cross sections are still taken primarily from theoretical approaches.[25,26]

As a demonstration of the size of the errors involved under working conditions, FIGURE 5 shows preliminary experimental measurements of intensity from 300 eV to 360 eV made in the imaging mode on a 30-nm section containing a concentration of 200 mM Ca, characterized by chemical analysis, by neutron activation, and by EDXS. While complete spectra can easily be obtained, the experiment served to mimic the situation of obtaining elemental maps using one image on a characteristic ionization edge and a small number of reference images at energies below the edge.

Equation 3 requires a minimum of two reference images. Using the point near 350 eV (just below the Ca edge) and the measurement at any other single energy from 300 eV to 340 eV, R was found to increase monotonically from 4.33 to 5.17, with a best fit through all points of R = 4.28. This fit is shown in FIGURE 5. It indicates a systematic deviation from the experimental points, suggesting a dependence different from Equation 3. To obtain a measure of the variation in the background extrapolation value at 360 eV under different assumptions, the following two-parameter and three-parameter equations were fitted to the same data.

$$I = a\ e^{bE} \qquad (4)$$

$$I = a + bE + cE^2 \qquad (5)$$

$$I = a + b\log E + c\log^2 E \qquad (6)$$

The results are summarized in TABLE 1. The extrapolation values at 360 eV varied from 17.0 to 17.9 $\mu\mu$A. Since the net Ca signal then varied accordingly from 6.0 to 5.1 $\mu\mu$A, the potential exists for a systematic error of close to 18%. Equation 3 is at one extreme, Equation 5 at the other. The best fit to the data was Equation 5, one of the three-parameter fits. A single-parameter fit of the form

$$I = a\ f(E) \qquad (7)$$

was analysed as well using the lower curve of FIGURE 5 from a neighboring area on the same section, which did not contain Ca. A fit using Equation 7 assumes no form of f(E) for the spectral shape, merely that the two curves in FIGURE 5 are proportional to each other in the region of measurement. For the fit to Equation 7 the average extrapolation value for the Ca background at 360 eV fell between the extremes mentioned above.

If Equation 3, the inverse power law, is assumed to be correct, then the fractional error in the extrapolation using Equation 7 can be calculated according to:

$$\text{fractional error in background} = (1 + \Delta E/E)^{\Delta R} - 1 \qquad (8)$$

where ΔE is the difference in energy between the image at an energy above the selected ionization edge and the energy E below the edge. ΔR is the difference in R between the lower and upper curve in FIGURE 5, both fitted to Equation 3. For the lower curve R = 4.63, making ΔR = 0.35. With typical values for ΔE = 40 eV and E = 320 eV, the fractional error is found to be 4.4%. This is small compared to the potential systematic error of 18% above. But it is a typical error observed. A similar analysis of the results of Chang et al.[27] shows that an error of only 1.5% would occur with a one-parameter fit due to variations in R (2.71 to 2.77) over their images of uranyl-stained catalase.

TABLE 1. Fitting of the Spectral Shape for 200 mM Ca Embedded in a 30-nm Epoxy Section[a]

	Extrapolation at 360 eV[b] $\mu\mu$A	Fitting Error $\left[\dfrac{\Sigma(y - \langle y \rangle)^2}{\text{d.f.}^c} \right]^{1/2}$
One parameter		
a f(E)	17.2	0.42
Two parameters		
A E^{-R}	17.9	0.45
A E^{-R} (3 points only)	17.2	0.12
a e^{-bE}	17.6	0.32
Three parameters		
a + bE + cE^2	17.0	0.13
a + b log E + c log^{2}E	17.2	0.15

[a]See FIGURE 5. Energy window 10 eV; fitting range 300 to 350 eV; extrapolation to 360 eV.
[b]Total signal in Ca spectrum at 360 eV was 23.0 $\mu\mu$A.
[c]d.f. = degrees of freedom.

If one assumes that a three-parameter fit is a more accurate representation of the data, then Equation 3, the most generally used relationship, is the least accurate.

A consequence introduced by the uncertainty of these systematic variations is a limit placed on the sensitivity of the technique. If at 200 mM Ca the signal can have an 18% systematic error, depending on the choice of the equation used for the shape of the spectrum, then under these conditions a 40-mM Ca signal can no longer be observed

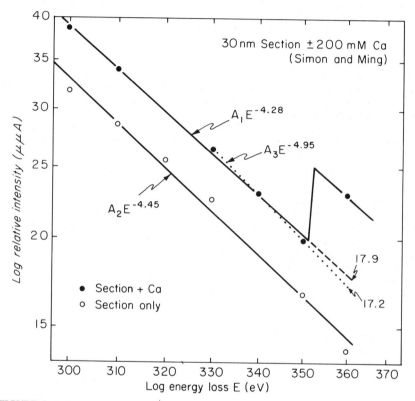

FIGURE 5. Experimental measurements from a 30-nm section of a 200-nM calcium standard (●), and an immediately adjacent portion of the same section without calcium (O). In this log-log plot both sets of data are fitted to Equation 3 *(straight lines).* The *broken line* indicates a fit from 330 to 350 eV only (see TABLE 1).

reliably. It is therefore imperative that a more definitive treatment of the dependence and analysis of energy-loss spectra of complex specimens be formulated.

MULTIPLE SCATTERING

Simple quantitative analysis of energy-loss spectra assumes the absence of plural scatter events. Single scatter is generally associated with a thickness of 1 mean free path, or one scatter event on the average. This is a serious fallacy, since at this thickness no scatter, single scatter, and multiple scatter occur in about equal proportions. For energy-loss analysis lack of scatter is of no consequence. However, multiple scatter events seriously degrade the image and the spectrum.

For spectra, deconvolution algorithms have been used successfully (e.g. REFERENCE 28). For images this approach is not tractable. If "thick" specimens are used, the result is a general decrease in contrast and a noticeable blurring of the specimen image. At the same time a nonlinearity with thickness or density is introduced both in nonspecific regions of the specimen and in the characteristic signal at ionization edges.

For small changes in linearity in signal strength it may be possible to correct the signal by measuring the effective thickness of the specimen at each point using the low-energy portion of the spectrum.[29,30] However, the blurring cannot be counteracted, since it is caused by the convolution of the low-resolution information from the predominant low-energy electrons with the higher-resolution events in the plural or multiple scatter cascade.

Avoidance of multiple scatter is therefore the only recourse in energy-loss imaging and elemental mapping. We have found that a reasonably workable compromise is a thickness of 30 nm in biological sections, corresponding to about 0.3 of a mean free path in carbon. Even at this thickness multiple scatter events still are about 17% as frequent as single scatter events.

CONCLUSION

Elemental mapping by energy filtration, using parallel processing in CTEM fitted with imaging spectrometers, or sequential point by point analysis in STEM, offers a potential spatial resolution of the order of 0.5 nm for most elements of interest that have characteristic ionization edges around 100 eV or higher. Elastic scatter as part of an inelastic multiple scatter component contributes only a small proportion to high-resolution detail for specimens of the order of 0.3 mean free paths in thickness. At greater thicknesses, however, this component can become important. For high concentrations of atoms, as few as 10 uranium atoms or 30–50 phosphorus atoms have been detected. For low concentrations, signals in spectra have been observed above background generally at 0.1 at.% or higher, with the lowest being a value of 0.006%.

Problems in elemental mapping due to multiple scattering are qualitatively seen as a lack of contrast in the energy-loss image, and a blurring due to delocalized scatter of low energy-loss electrons. Quantitatively, the signals, both in characteristic and noncharacteristic regions of the spectrum, become nonlinear with thickness.

Quantification of elements at present suffers most seriously from a lack of experimentally measured ionization cross sections. However, in addition, different formulations of the energy dependence of the electron energy-loss spectrum introduce unknown systematic errors, whose variability makes the results of analyses of low concentrations of elements rather uncertain.

Thus, while for small numbers of concentrated atoms in thin specimens the results of elemental mapping by electron energy-loss analysis can be quite spectacular, low concentrations of elements, and particularly thicker specimens, require a very careful analysis.

ACKNOWLEDGMENT

I thank G.T. Simon for the use of his calcium standard data.

REFERENCES

1. OTTENSMEYER, F. P. & J. W. ANDREW. 1980. J. Ultrastruct. Res. **72:** 336.
2. ADAMSON-SHARPE, K. M. & F. P. OTTENSMEYER. 1980. J. Microsc. **122:** 309.
3. HOWIE, A. 1979. J. Microsc. **117:** 11.
4. KOHL, H. & H. ROSE. 1985. Adv. Electron. Electron Phys. **65:** 173.
5. SCHEINFEIN, M. & M. ISAACSON. 1986. J. Vac. Sci. Technol. **B4:** 326.
6. CRAVEN, A. J. & C. COLLIEX. 1977. J. Microsc. Spectrosc. Electron **2:** 511.
7. BAZETT-JONES, D. P. & F. P. OTTENSMEYER. 1981. Science **211:** 169.
8. HARAUZ, G. & F. P. OTTENSMEYER. 1984. Science **226:** 936.
9. COLLIEX, C. 1982. Proc. 10th Intl. Congr. Electr. Microsc., Hamburg **1:** 159.
10. EGERTON, R. F. 1981. J. Microsc. **123:** 333.
11. SHUMAN, H. & P. KRUIT. 1985. Rev. Sci. Instrum. **56:** 231.
12. CASTAING, R. & L. HENRY. 1962. C. R. Acad. Sci. (Paris) **255:** 76.
13. HENKELMAN, R. M. & F. P. OTTENSMEYER. 1973. J. Microsc. **102:** 79.
14. EGERTON, R. F., J. G. PHILIP, P. S. TURNER & M. J. WHELAN. 1975. J. Phys. E. **8:** 1033.
15. KIHN, Y., G. ZANCHI, J. SEVELY & B. JOUFFREY. 1976. J. Microsc. Spectrosc. Electron **1:** 363.
16. KRAHL, D., K.-H. HERRMANN & W. KUNATH. 1978. Proc. 9th Intl. Congr. Electr. Microsc., Toronto, **1:** 42.
17. EGLE, W., A. RILK, J. BIHR & M. MENZEL. 1985. Proc. 42nd Annu. Mtg. Electron Microscopy Soc. Amer. 566.
18. AHN, C. C. & O. L. KRIVANEK. 1983. EELS Atlas. Gatan, Inc. Warrendale, PA.
19. ARSENAULT, A. L. & F. P. OTTENSMEYER. 1983. Proc. Natl. Acad. Sci. USA **80:** 1322.
20. ARSENAULT, A. L. & F. P. OTTENSMEYER. 1984. J. Cell Biol. **98:** 911.
21. EGERTON, R. F. 1975. Philos. Mag. **31:** 199.
22. GRUNES, L. A. & R. D. LEAPMAN. 1980. Phys. Rev. B **22:** 3778.
23. BENTLEY, J., G. L. LEHMAN & P. S. SKLAD. 1982. Proc. 10th Intl. Congr. Electr. Microsc., Hamburg, **1:** 585.
24. STELLE, J. D., J. M. TITCHMARSH, J. N. CHAPMAN & J. H. PATERSON. 1985. Ultramicroscopy **17:** 273.
25. EGERTON, R. F. 1981. J. Microsc. **123:** 333.
26. EGERTON, R. F. 1981. Proc. 39th Annu. Mtg. Electron Microscopy Soc. Amer. 198.
27. CHANG, C. F., H. SHUMAN & A. P. SOMLYO. 1985. Proc. 43rd Annu. Mtg. Electron Microscopy Soc. Amer. 400.
28. JOHNSON, D. W. & J. C. H. SPENCE. 1974. J. Phys. A **7:** 771.
29. EGERTON, R. F. & S. C. CHENG. 1985. Proc. 43rd Annu. Mtg. Electron Microscopy Soc. Amer. 398.
30. LEAPMAN, R. D. & C. R. SWYT. 1985. Proc. 43rd Annu. Mtg. Electron Microscopy Soc. Amer. 404.
31. SELF, P. G. & P. R. BUSECK. 1983. Phil. Mag. A **48:** 21.
32. EGERTON, R. F. 1986. *In* Electron Energy Loss Spectroscopy in the Electron Microscope. Ch. 5. Plenum Press. New York, NY.

DISCUSSION OF THE PAPER

D. F. PARSONS (*New York State Department of Health, Albany, New York*): How were the viruses embedded and processed? Normally the phospholipid is almost completely extracted in the normal processing.

OTTENSMEYER: The murine leukemia virus was fixed with osmium. And so one of the real worries from a physics point of view was not the lipid content but the mimicking of the phosphorus signal by an elastic osmium signal. This does not seem to be the case, since in myelin sheaths, which were also fixed with osmium, you virtually don't see the myelin sheath arrangement at all at an energy loss of 110 eV, whereas as soon as you go to the phosphorus edge, the myelin figure just shines with the additional phosphorus signal.

PARSONS: But still that sort of processing must have caused a loss of lipids. Presumably if the neutral lipids went completely, the phospholipids would be partly stabilized by the osmium. You have to use special water-soluble embedding methods to retain the lipids in the membrane, and then you could get a very much stronger phosphorus signal.

OTTENSMEYER: That's quite possible. From that you can say that we're not able to give an absolute value for the minimum mass of what we can detect from signal-to-noise; but the experiment provides us with a test specimen for spatial resolution.

PARSONS: Is it expected that at liquid helium temperatures the collective plasmon losses would be qualitatively different? Has that been looked at, and is there any hope for improvement?

J. SILCOX (*Cornell University, Ithaca, New York*): Not much. Dr. Somlyo?

A. P. SOMLYO (*University of Pennsylvania School of Medicine, Philadelphia, Pennsylvania*): Dr. Ottensmeyer, since you questioned the validity of E^{-r} subtraction, I have some questions about the subtraction that you did on the terminal cysternae of the sarcoplasmic reticulum in rather thick cryosections of muscle, where the calcium $L_{2,3}$ edge is very, very small, even though these are very high concentrations from a biological point of view. In this case the Ca $L_{2,3}$ edge is sitting not even on a E^{-r} background, but on a carbon edge fine structure with at least two scattered plasmons, maybe a third. The Ca edge is between the second and third scattered plasmon. Therefore, there is grave suspicion that some of those images you showed may be due to mass effects, rather than representing calcium. I've convinced myself and a few other people of this by using the spectrometer on the microscope in the spot analysis mode. The calcium concentration in the terminal cysternae, 100 mmoles per kilogram is not detectable in a line spectrum mode without serious, relatively sophisticated background subtraction. Therefore, it's unlikely that a subtracted image, using a single pre-edge background subtraction could give reliable compositional information about this concentration of calcium. By the same token, you are showing images that represent compositional information. Dr. Shuman and I and Dr. Chang had worried about the pixel-to-pixel R variation that Dr. Leapman has also shown. I don't know how many pre-edges he collects to determine how to process the STEM images.

We find that even in a simpler structure, such as catalase, two pre-edge images taken at different energies show different crystalline structures. Therefore, unless we do very careful processing to determine pixel-to-pixel R variations, we don't get compositional information about where carbon is, where uranium is, or let alone where phosphorus is, which is a much more slowly rising, smaller edge. What we get is essentially mass density distribution.

OTTENSMEYER: Since you and I took no thickness measurements of your cryosections of muscle, it is difficult to estimate the extrapolation error retrospectively. However, the problem can be serious. In specimens for which we've taken five images in front of the calcium edge, I tried the fit to E^{-r} and found out that it was actually the worst fit among six different trial functions used. The second best was our assumption that the section without calcium had a background similar to the section with calcium. However, that's not always true, because the R value does change. For instance, in your study published in EMSA last year, the R map changed from an R value of 2.71 to 2.77. Now the error in extrapolation due to that is 1.5%. It depends on whether you assume E^{-r} is the correct function or not.

Dr. Leapman published a formula to estimate the extrapolation error if R changes. In most instances, as in your case, the error is only of the order of a few percent. If you consider carbon and silver, the change in R is about 2. Then you are in real difficulty. All of this has to be pointed out; and I agree with you, you have to be very careful. You can't even just plug in E^{-r} and expect to be right.

S. P. LAYNE (*Los Alamos National Laboratory, Los Alamos, New Mexico*): The nitrogen edge will differ slightly, according to how it's bonded. In nucleic acids, as opposed to proteins, there may be a slight difference both in shape and where the edge begins to take off. With the resolution that you have, about an electron volt, have you looked at all to detecting differences in the nitrogen edges between proteins and nucleic acids. Is that a feasible experiment?

OTTENSMEYER: I certainly haven't done it. That's an experiment that should be done in an instrument that gives you the spectrum very much more easily than my instrument does. I don't have a probe-forming lens on my instrument to do that, so the spectra are a little hard to get at. I would refer you to the STEM. Dr. Leapman or Dr. Colliex can probably answer that, give you your chemical analysis of the surroundings, tell you whether there is a difference in near-edge structure or extended fine structure.

C. COLLIEX (*Université Paris-Sud, Orsay, France*): This is a field of growing interest, not only to verify that you have nitrogen, but to know how it is bonded. But to get the signal you need, you have to go to better signal-to-noise ratio, for which you need a higher dose. For instance, a detailed near-edge fine-structure analysis of any oxygen or nitrogen may require a dose about 100 times bigger than the dose which we use to simply detect the edge signal. That is electron consuming. If we can now come to the general discussion, we can note several big problems that are being addressed. There is the problem of limits of detection, for instance. What is very good progress in proving limits of detection is parallel recording; the data of Dr. Shuman are one step ahead of the other data, such as Dr. Leapman's or mine or yours, which were generally at the 1% level of detection. To be able to record in parallel seems to me to be progress by one order of magnitude.

OTTENSMEYER: You are saying we should record spectra in parallel.

COLLIEX: In parallel. That is a great advantage of the STEM geometry, because this will be the method by which to do it.

G. C. RUBEN (*Dartmouth College, Hanover, New Hampshire*): I wonder, Dr. Ottensmeyer, about the size of the bilayer image that you showed us. Do you recall, what was the edge-to-edge thickness of the lipid bilayers?

OTTENSMEYER: The phosphorus atoms in the two lipid bilayer sheets were about 50 Å apart.

RUBEN: That's center to center?

OTTENSMEYER: That's center to center, from the net signal.

RUBEN: What about the outer edge? We measured the outer edge at Dartmouth after just osmication four or five years ago, and we got something like 58 Å. We estimated the size of the membrane without water and without osmium to be about 51 Å. However, we didn't measure to the center of the phosporus line, but to the outside.

OTTENSMEYER: I can't tell you where in that particular slow rise of the measured cross-sectional profile the outer edge is. One would have to define where the edge is, because these images—the dark-field images or these energy-loss images—seem to give you detail outside, which is not seen in stained specimens. Therefore, at the moment it's fairly nebulous where you say the outer edge begins.

PARSONS: I'd like to ask another question about your system. Can you give information about polysaccharide coats of cells? I've always been impressed by the lack of data about this. You look at an electron microscope image, and it appears to have a sharp edge. You do light microscopy with Alcian blue and it has a 1,000-Å rough border round the edge, and it's clear that classical electron microscopy has a way of chopping off the interesting extracellular coat of cells. If you focus in on sulphur, let's say, can you show the polysaccharide coat?

OTTENSMEYER: That depends on the content of sulphur or another specific atom in the polysaccharide coat. Certainly the presence of sulphur or some other atom has revealed structure that was not observed in the pre-edge image.

PARSONS: Sometimes it's a matter of contrast too, because Wilska used normal preparations and some of his 20-kV images showed very nice polysaccharide layers, in material prepared normally, that could not be seen at 50 or 60 kV.

OTTENSMEYER: I haven't specifically looked for polysaccharide layers yet, because most of my collaborators who are interested in sections seem to be interested in diffusible ions. In my own work, I stay away even from those, because they always seem to be gone under the usual preparative conditions by the time you put them in the microscope. So at present I stick with phosphorus that's covalently bound.

Electron Energy Analysis in Emission Microscopy

D.W. TURNER, I.R. PLUMMER, AND H.Q. PORTER

Physical Chemistry Laboratory
University of Oxford
Oxford OX1 3QZ, England

The electron microscopies based on the electron emission processes of field emission, secondary electron emission or photoelectron emission all provide images in which the kinetic energy distributions of the constituent electrons contain information about the chemical nature of the surface of the material under examination. To make use of this information, the energy spectrum of the electrons from a region of the surface small enough to be reasonably chemically homogeneous must be recorded. In this paper we discuss our approach to this problem in the context of the magnetically-collimated photoelectron spectromicroscope and compare these results with the related energy-analysis techniques in the other electron emission microscopies.

Electron images resulting from different electron emission processes show contrast differences which depend on the nature and energy of the source employed. In FIGURE 1 we show two images of the same object (an aluminium surface on which copper has been evaporated through a shadow mask consisting of two grids). In one of these images the excitation source was 21.2 eV photons (He(I) light) and in the other fast argon atoms (3 keV kinetic energy). The shading at the shadow edges illustrates the importance of the penetrating power of different sources in relation to layer thickness. The attenuation depth of UV photons is about 10 nm, and electrons may originate from such depth, whereas the fast atoms interact only with the surface layer of the object. However, in many cases the contrast mechanisms are more subtle than this, particularly for photoelectron imaging. FIGURE 2 is a He(I) photoelectron micrograph of a thin section of human skin mounted on a gold-coated glass cover slip, the sample being fixed and embedded in epoxy before sectioning. The main contrast mechanism in this example is the different probability of capture of photoexcited electrons in chemically different regions. An example of another type of contrast mechanism, a surface-vacuum potential barrier to electron escape, is shown in FIGURE 3. This is the photoelectron image produced by a HeCd laser (442 nm, 2.8 eV) incident on a silicon-based integrated circuit which has first been etched with hydrogen fluoride to remove silicon oxide and metallization and then coated with approximately a mono-layer of Cs. The importance of dopant in determining this contrast is clearly seen by comparing the photoelectron micrograph with the optical micrograph of the same region before the Cs coating.

THE RELATION BETWEEN CONTRAST AND ELECTRON ENERGY

In any electron emission microscopy much information about the contrast mechanisms is contained in the electron energy distribution from the different regions of the surface.[1,2] All sources generate excited or 'hot' electrons, in some cases at the surface, in others many atomic layers deep into the material. The extent to which different sources penetrate the material depends on the nature of the source particle (for

354

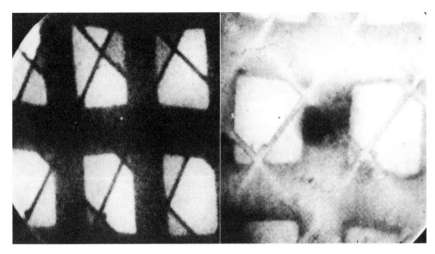

FIGURE 1. Electron images of a copper-on-aluminium pattern; copper evaporated through a shadowing mask of two grids, the finer of 25 μ wire. *Left:* using He(I) photoelectrons $E_e > 11.5$ eV selected. *Right:* using electrons released by fast Ar atom bombardment. Note the magnification and orientation of the specimen to the observer differs in the two cases. A common area is indicated (x).

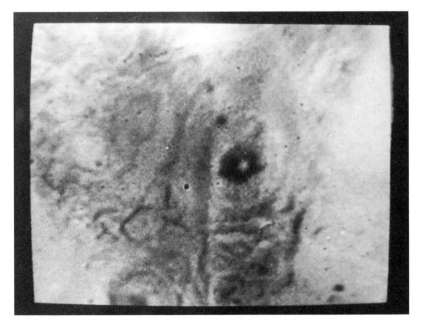

FIGURE 2. He(I) photoelectron image of a human skin section fixed and unstained, mounted in epoxy and attached to gold-coated glass. Specimen temperature $\simeq 150°$ C.

example atoms, ions or photons) and its energy. The efficiency with which 'hot' electrons are generated, on the other hand, depends upon the properties of the material (for example, reflectivity, attenuation length and excitation cross section). Once generated the 'hot' electrons are free to move and the probability of any one of them reaching the surface will depend upon the energy-loss processes, characteristic of the material, it encounters. There are three broad types of energy-loss process—electron-electron collisions, (*i.e.* plasmon scattering with typical losses of a few eV), electron-lattice collisions (*i.e.* quasi-elastic phonon scattering) and electron recapture processes (*e.g.* excitons, radiative deexcitations or cooperative shake-up processes). The relative

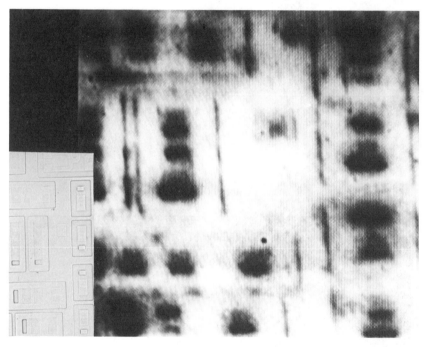

FIGURE 3. HeCd laser (λ = 442 nm, 2.8 eV) photoelectron image of an integrated circuit after stripping with HF and deposition of a Cs monolayer. The plane of the specimen is inclined at 45° to the observer. *Inset:* optical micrograph using normal incidence reflected light of the same region.

proportions of these different processes are very dependent upon the nature of the material. As a general rule we can say that plasmon processes dominate in metallic materials, phonon processes in semiconductors and crystalline organic materials and both phonon and recapture processes are significant in noncrystalline organic or biological material. Thus as a 'hot' electron moves towards the surface it redistributes its excess energy in a manner characteristic of the material. On reaching the surface the electron will only escape if it has sufficient energy remaining to surmount the surface-vacuum potential barrier. Again, as a general rule, we note that, except in the field emission process, barrier tunneling is not significant.

It follows from this discussion, therefore, that if the electron energy spectrum from a region of the surface small enough to be considered homogeneous in its properties is recorded, then detailed information about the local structure, chemistry and electrical properties of the material can be deduced.

PHOTOELECTRON SPECTROMICROSCOPY

The combination of electron spectroscopy with microscopy is well established in both scanning and transmission electron microscopes. In both of these techniques, however, electron energy analysis yields information about the energy losses characteristic of the different atoms in the material under investigation. The contribution of the basic process involved (electron scattering in the scanning microscope and diffraction in the transmission microscope) to the electron energy spectrum is not easily interpreted. However, in the emission microscopies, in particular photoelectron microscopy, the details of the electron energy spectrum potentially give important information about the basic emission processes involved.

In conventional photoelectron microscopy (PEM),[3] photoelectrons emitted from a surface are accelerated by an electric field into an objective electrostatic electron lens which focusses them to form an intermediate image. This image is then magnified by intermediate and projector lenses to form the final image. Magnifications of 3,000X are achieved with lateral resolutions of 10 nm. The sample surface must be as flat as possible to ensure that photoelectrons from any point on the surface experience the same accelerating force. In the photoelectron spectromicroscope (PESM),[4] on the other hand, photoelectrons emitted from a surface are guided by the field lines of an axially-symmetric divergent magnetic field to a phosphor screen to produce a magnified image of the surface. As the image in both cases is formed with all emitted electrons with a range of kinetic energies the resulting photoelectron micrograph is very similar to that obtained in conventional PEM. Magnifications of up to 200X and a lateral resolution of about 1 μm are typical in PESM. As no lenses are involved in PESM there is almost no depth-of-field restriction on the topography of the sample surface.

There are two approaches to energy analysis in emission microscopy.[5] In the first of these, which we may call *image filtering,* the image is formed by electrons with limited range of kinetic energies. This requires an analyzer which can preserve the spatial distribution of the image electrons but only transmit electrons with energies in the selected range. The second approach is to energy analyze all the electrons from a small region of the surface, the *selected area.* The spatial relationships between these electrons is not preserved. It is thus important that the selected area be chemically homogeneous. In this case the analyzer must have an entrance aperture of variable size (and preferably variable shape) and this aperture must be capable of being positioned in either the object or image plane. As, in many cases, the entrance aperture will require to be very small (<40 μm) there are considerable practical advantages in placing this aperture in the image plane.

In PEM the possibilities for energy analysis are restricted by the fact that only one component of the electron velocity is accelerated by the electrostatic field. An attempt at whole-image filtering by rejecting the high energy electrons (with a low-pass energy analyzer)[6] has been reported, but to our knowledge no work on band-pass whole-image filtering or selected-area analysis in conventional PEM has been reported.

In PESM the magnetic field configuration employed has two overwhelming advantages for energy analysis.[7] The divergent axially-symmetric field has the property of aligning the electron velocity vectors with the magnetic field lines. Thus in the low magnetic field region of the PESM, where the field lines are parallel, the electrons are collimated. A magnetic field, unlike an electrostatic field, is conservative, so the energy of the electrons is unchanged during this collimation process. This is particularly advantageous for electron energy analyzers employing planar retarding fields, which, in this particular apparatus, act on the total electron energy and not merely the component normal to the retarding plane (FIGURE 4a).

The second advantage of the magnetic field is that it permits energy analysis by crossed electric-magnetic field techniques (trochoidal analysis, see FIGURE 4b) which in some circumstances (notably in signal-to-noise performance) is preferable to the retardation technique. A third approach to analysis is with pulsed time-of-flight techniques. The discussion of these different approaches has been developed elsewhere.

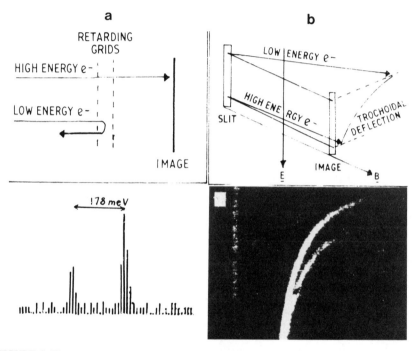

FIGURE 4. Electron energy analysis in a strong magnetic field using (**a**) retardation by an axial electrostatic field and (**b**) transverse (trochoidal) deflection using a crossed electric field.

In the Oxford PESM we have developed whole-image, low-pass filtering (using a retarding electric field) and whole-image, band-pass filtering (using crossed electric and magnetic fields). Selected-area photoelectron analysis has been developed using both retardation and crossed-field techniques.

The present state of the PESM is shown in FIGURE 5 and is more fully discussed elsewhere.[8]

ELECTRON ENERGY ANALYSIS IN THE PESM[2]

The energy analysis of electrons in the low magnetic field region of the PESM is more straightforward in several ways than for more conventional electron emission

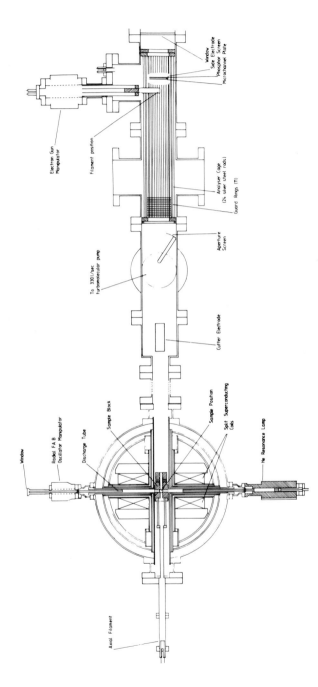

FIGURE 5. The Oxford electron spectromicroscope with facilities to allow clean vacuum operation and selected area energy analysis. In the area labelled "sample block" are a metastable atom source and X ray source anodes. The filaments (not shown) for the latter are outside the magnet off the axis at the left of this figure.

systems, since nearly all of the electron momentum is directed along a common axis. Standard methods of focussing deflection are, however, excluded by the presence of the residual magnetic field. There are then three principal methods available. The first, *retarding field energy analysis,* measures the axial energy of electrons by applying a retarding electric field. The other two measure the electron axial velocity by *electrostatic deflection* with a crossed electric field or by *time-of-flight.* There are a number of possible variations or combinations of these three basic methods.

Retarding Field Energy Analysis

In the presence of an axially-symmetric magnetic field the passage of electrons is determined by the potential distribution, the effect of the transverse components of its local gradients being suppressed by the constraint imposed by the magnetic field, provided the fields change slowly compared to the cyclotron orbit. Only the axial component of the electron velocity is changed by the axial part of the local field. In the PESM, therefore, retarding field analysis can be particularly effective because of this axial collimation of electron velocities. Energy analysis can then be carried out over the whole of an electron image beam, comparable in width with the size of the electrode apertures, without distorting the edges of the image and with an energy scale uniform across the image, provided certain limiting conditions are met:

(1) There are two regions of uniform potential (across the width of the beam) to define the retarding potential difference. Tubes whose length is large in proportion to their diameter suffice.

(2) The magnetic field must not change significantly over distances comparable to the helix pitch of the electron motion in the magnetic field.

The application of this mode is illustrated in FIGURE 6. In the context of the PESM experiment, where an intermediate magnetic field strength of about 0.025 T is found between the high-field region of the superconducting solenoid (about 8 T) and the image recording region (about 0.08 mT), the magnitudes of these terms determine the physical scale of any possible analyser structure. Consequently all of the fields encountered in the apparatus must vary slowly over these magnitudes of distance and in consequence, electrode structures extending over tens of centimeters will be required.

Image Rotations in the Retarding Field

With unlimited space a retarding field could, in principle, be constructed with a large number of field electrodes defining a set of planar equipotential surfaces, normal at all points to the direction of the uniform magnetic field, or curved surfaces for nonuniform magnetic fields. In practice, truncation of these structures produces nonplanar fields due to electrostatic field penetration at their entrance and exit apertures. These equipotentials have components which are perpendicular to the axis of the magnetic field, and so introduce an image rotation which may be radially varying within the aperture. The magnitude of this rotation can be calculated by a method which is described elsewhere.[11]

The electron stream which is passed by such a retarding field can, in principle, be processed to yield the electron flux at a selected energy for each point of its cross-sectional area, giving an energy-filtered image.[9] Alternatively by collecting data from a defined area (say, n pixels) as a function of the retarding field voltage a selected-area energy spectrum can be obtained. The displayed image can be recorded

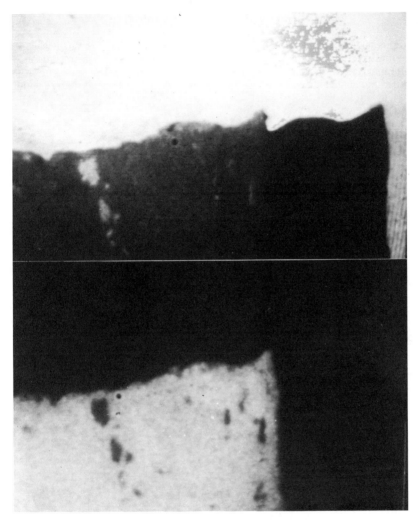

FIGURE 6. He(I) photoelectron images of a silicon wafer/platinum foil "sandwich" specimen. *Above:* with no electron energy selection. *Below:* using only electrons > 14 eV.

by a video camera, digitized and stored in a computer. It is then possible to subtract two such stored images with different cut-off energies, say E_1 and E_2, and thus form a selected energy map of the surface corresponding to electrons with kinetic energies within the narrow energy range $\Delta E = E_2 - E_1$. This forms the basis of XPS elemental distribution maps and chemical-shift maps of surfaces. Alternatively the photoelectron spectrum of a selected area on the surface can be recorded by selecting a region of the image displayed on the phosphor screen and recording the brightness integrated over that region as a function of the retarding voltage. The first derivative with respect to voltage of this function is the electron energy spectrum (see FIGURE 7).

Electrostatic Deflection Energy Analysis

This is achieved by defining with an entrance slit or hole, through which the electrons pass, a region on the object surface to be analyzed. An electrostatic field is maintained perpendicular to the magnetic axis. This produces a region of crossed electric and magnetic fields in which an electron is deflected perpendicular to both the

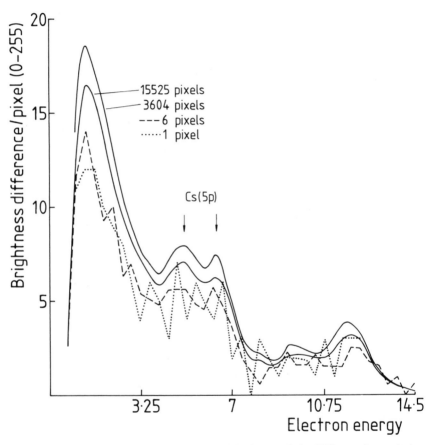

FIGURE 7. He(I) photoelectron spectra obtained in the retardation/differentation mode (see text) for areas of a GaAs surface coated with a Cs monolayer. The areas measured were defined by computer-generated masks on the computer-recorded images. One pixel corresponds to ca. 0.5 μm × 0.5 μm at the object surface.

magnetic and electric field directions. The magnitude of the deflection is a function of the two field strengths (E, B) and the electron energy. In general, a net transverse motion from the undeflected position is given by

$$D = v_d \cdot t$$

where t is the time spent in the crossed-field region and v_d is a constant velocity determined by the fields,

$$v_d = E/B.$$

Thus the energy selection arises from the time-of-flight of the electrons through the deflection region. Slower electrons are in the deflection region for longer than fast ones and are consequently deflected more. The transit time across a crossed-field region of length L is

$$t = L/v_z,$$

thus

$$D = v_d L/v_z,$$

where v_z is the velocity in the direction of the magnetic axis. The dispersion is inversely proportional to the velocity, and thus inversely proportional to the square root of the electron energy. Consequently low-energy electrons will be widely dispersed, whereas high-energy electrons will be grouped closely together.

The trochoidal analyser may be operated in a number of configurations. The obvious mode is to record the dispersed image of the aperture and measure the displacements. This method is well suited to the emission spectromicroscope as the phosphor screen is positioned to display the dispersed image. The resolution is again affected by nonaxial components of the electron momenta when they enter the trochoidal analyser. These lead to increased cyclotron-orbit radii, which spread the image of the slit. Similarly to the retarding-field analyser, the addition of a divergent magnetic region before the analyser has the beneficial effect of orienting the electron motion along the magnetic-field direction and reducing the transverse momentum.

The trochoidal analyser can be used in a scanning mode with the detector at a fixed point in space. Either the electrostatic field can be scanned, causing the displacement to change and the different electron energies to pass over the detector (the side electrode in FIGURE 5), or the electrons approaching the analyser can be preretarded. A voltage scan at the analyser entrance will reduce the electron energies to a point where they are deflected onto the detector.

Both these methods have the advantage of acting as analysers of constant bandwidth, and the scales are linear with the retarding or deflecting voltage. Trochoidal energy analysis is particularly suitable for recording the photoelectron spectrum of a small region or line segment of the surface. Note that since this selection takes place in the low-field region, on the magnified image of the object surface, the aperture can be considerably larger than the selected area on the surface. For example, a 1-mm diameter aperture could be used to analyse a region of 10 μm diameter.

This mode of operation is demonstrated in FIGURE 8, which shows the HeI-generated photoelectron spectra from two 16-μm diameter regions of a GaAs crystal surface on which Cs has been evaporated. FIGURE 9 shows an X ray generated photoelectron spectrum from a 16-μm diameter region of a copper film, a few hundred atoms thick, evaporated onto 6-μm thick aluminium foil. The rear face of the foil receives 20 μA of 3 keV electrons from an electron gun. The spectrum demonstrates an analyser energy-resolution of ca. 0.1 eV.

Time-of-Flight Method

Energy analysis by time-of-flight measurements was earlier proposed[4] for the PESM and has been demonstrated in a low magnetic field gas-phase photoelectron experiment.[10] Timing can be effected by gating the electron stream at two of five points

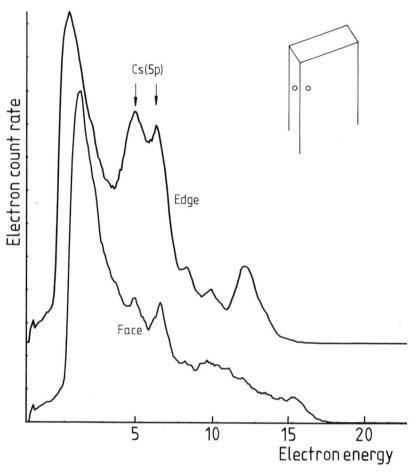

FIGURE 8. He(I) photoelectron spectra obtained in the trochoidal deflection mode (see text) of adjacent areas of the same specimen used for FIGURE 7. In one case the mask (about 16 μm dia) selects a face region, in the other a region of an edge exposed by fracture.

by pulsing:

(a) the specimen potential
(b) an electrode close to it (in the high magnetic field)
(c) an electrode in the intermediate field
(d) the analyser
(e) the detector, channel plate or phosphor screen.

When (a) or (b) are gated the time-of-flight distribution is degraded by the time spread produced by the range of helix angles which exist in the high-field region. This limits the method to low resolving power (E/ΔE) applications. We have shown, however, by computer simulation, that this limitation can be removed by successive applications of axial accelerating and retarding electric fields: accelerating in the high magnetic field region and subsequent retardation in the low magnetic field region. The effectiveness of this approach is illustrated in FIGURE 10 which shows the computed time distribution for a pulsed electron beam (gated at the source) in a 7 T PESM having a total flight path of 1 m. First we show this time distribution under electric field-free conditions and second the effect of preacceleration and subsequent retardation. An energy resolution sufficient for chemical shift studies in the X ray photoelectron spectrum is clearly attainable.

All such timing experiments are, however, inherently inefficient with D.C. (continuous) sources (<0.1% duty ratio in the present case). It seems probable that this new method would become especially attractive as pulsed X ray sources are improved.

THE BAND-PASS IMAGE FILTER

An image can in principle be built up using a sequence of measurements of electrons, selected in energy by the trochoidal analyser, as the electron image is scanned over its entrance aperture. Image scanning is readily effected by using small transverse magnetic fields. However, this method suffers from the disadvantage of all scanning methods in the inefficient use of data. The image subtraction method whilst

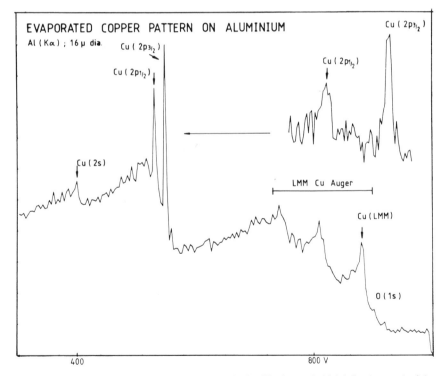

FIGURE 9. The Al(K_α) photoelectron spectrum obtained in the trochoidal deflection mask of the specimen whose image is shown in FIGURE 1. The aperture mask (about 16 μm dia) selects a copper-coated region. Al (K_α) X rays were generated in the rear face of the supporting Al foil by a 3 kV, 10 μA e$^-$ beam incident from the axial filament (see FIGURE 5).

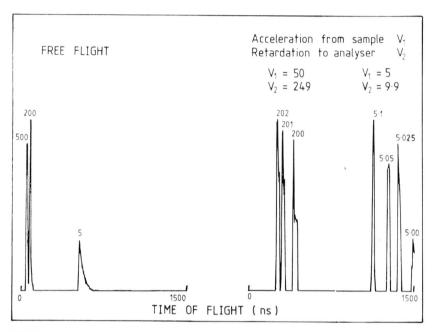

FIGURE 10. Computed time-of-flight distribution for electrons in the PESM divergent magnetic field. *Left:* with no preacceleration from the specimen and no subsequent retardation. *Right:* with an initial acceleration at the specimen (V_1), and subsequent retardation in the low B field region (V_2) (see text).

overcoming this restriction exhibits poor signal-to-noise ratio performance from the digitization limitation.

An ideal image band-pass filter would be one which allowed continuous (in time) image acquisition of a selected electron energy range. In the PESM this could be used to provide, for example, the equivalent of atomic mapping with Auger electrons and ultimately allow chemical shift mapping with X ray photoelectrons.

We recently developed[11] a method for achieving this which employs both trochoidal deflection and retardation.

The simplest approach would be to employ two (or more) stages of trochoidal deflection on the whole image stream with wide delimiting apertures to select a range of beam energies (FIGURE 11, top). For various reasons, however, this is impracticable and a more refined form is now described.

This is a combination of high- and low-pass energy filters with a folded beam geometry (FIGURE 11, bottom). The difference in the high- and low-pass energies defines a fixed band-pass, which is independent of electron energy. The design criteria and performance of this new analyser have been discussed elsewhere,[11] but we summarise the main considerations here.

One problem in designing fixed band-pass energy analysers which operate in a magnetic field, is that there are no elementary transmission low-pass energy filters which will preserve an image. Consequently a retarding-field electron mirror is used which can, in principle, maintain the image geometry.

A band-pass energy filter can, therefore, be constructed by directing the image electrons into an electron mirror then, using crossed-field deflection, to a high-pass filter. The low-pass electron mirror necessitates the use of crossed electrostatic and magnetic fields which serve to transport the image electrons from one stage of the analyser to the next. Unfortunately, in uniform crossed fields, this arrangement suffers from a severe defect. The drift imposed by the crossed fields is a function of the time that the electrons spend in the fields, consequently different energy electrons having different forward velocities are deflected by different amounts. This combined field- and energy-dependent drift has two consequences for the image: a nonlinear shear of the image and an energy-dependent displacement of the image.

When the image is passed through the crossed-field region the electrostatic field gradient across the entrance aperture, due to the electrostatic deflection field, causes

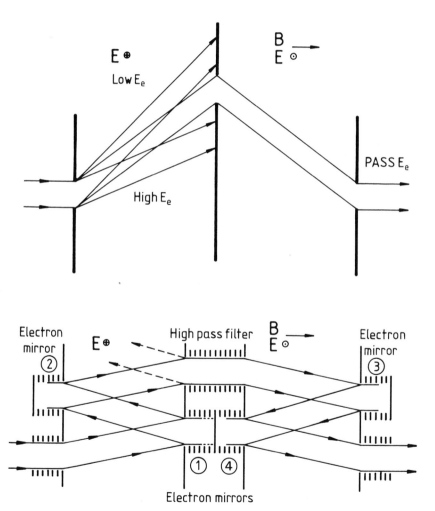

FIGURE 11. Proposed arrangements for an image band-pass analyser (see text). The simplest form (*above*) uses two stages of deflection and an energy selecting aperture. In a more refined form (*below*) crossed-field deflection is combined with high- and low-pass electron energy filters.

the image electrons to be accelerated differently according to their position in the aperture. This causes a nonlinear shear of the image (FIGURE 4b), the magnitude of which depends on the voltage drop across the aperture. Also, image electrons of different energy, even if they enter the crossed-field region at the same initial position, will be displaced by different amounts. This results in a smeared image, unless the band-pass width is very small. The correction of these two imaging defects arising from the use of crossed fields has been described in detail elsewhere.[11] Here we outline the principle features of a successful solution to the problem.

The form of the filter is shown in FIGURE 12 (top). It comprises two stages. The first uses crossed-field deflection to lead the image electrons to a low-pass filter configured as an electron mirror. The mirror is a graded retardation potential. High-energy electrons enter the retardation potential, are slowed and, if their initial energy exceeds the potential of the end plate, finally collide with it and are lost to the system. Electrons with an energy less than the potential applied to the end plate are slowed by the field and finally reverse their direction. These electrons are subsequently guided, by the crossed field, to a second electron mirror then to a high-pass retarding-field filter. The high-pass filter rejects the lower-energy electrons which reenter the crossed-field deflection region and are further deflected and lost by collision with the walls. The higher-energy electrons, lying in the pass-band, proceed into the second stage of the analyzer.

The purpose of the second stage is to cancel the energy-dependent dispersion, return the beam to its original axis and compensate for the image shear. This stage of the filter is identical in construction to the first stage but operates in the opposite direction and acts to bring different electron energy images back into registration and approximately correct the image shear from the first stage. The shear correction depends on the design of the nonuniform electrostatic fields in the crossed-field deflection regions. This is illustrated in FIGURE 13(1) which shows the computation of the final transverse displacement of a vertical bar of monoenergetic electrons. This represents the image resulting from these electrons passing through a narrow slit, aligned along the y-direction, and then passing through a single stage of the filter (i.e. a single crossed-field region). The effect of the second crossed field is approximated by applying the calculated displacement function to its image reflected about the x-direction (see FIGURE 13(2)). The displacement caused by a potential distribution which is symmetric about the potential of the end plates is shown in FIGURE 13(a). FIGURE 13(b) shows the displacement calculated for a field arrangement which has been optimised for minimum dispersion variation. The nonlinear image shear for the fields used in this model can be seen to have been corrected over most of the image.

Initial results have shown that nonuniform electrostatic fields can be found which reduce the nonlinear shear. In some cases, with a narrow pass-band, both the nonuniformity of the shear and the shear itself are eliminated.

The prototype filter illustrated in FIGURE 12 has been tested in a low magnetic field simulation of the normal high PESM field by imaging a thermionic filament to show that by energy filtering an image of just one turn of the helical filament could be selected (FIGURE 12, bottom a–c). This shows the improvement in image fidelity as the deflection field is changed from a homogeneous transverse field to a simple inhomogeneous one, suggested by computer modelling, and provided by adjustment of potentials to a set of plates replacing the single negative plate of the plane condenser structure.

We believe that after further refinement the incorporation of this new form of energy filter in our PESM will allow the recording of images with an energy bandwidth of ca. 0.1 eV at any primary kinetic energy up to ca. 1,000 eV and should make possible for the first time the direct recording of chemical shift and hence chemical structure images through both detailed valence shell and core electron mapping. The spatial

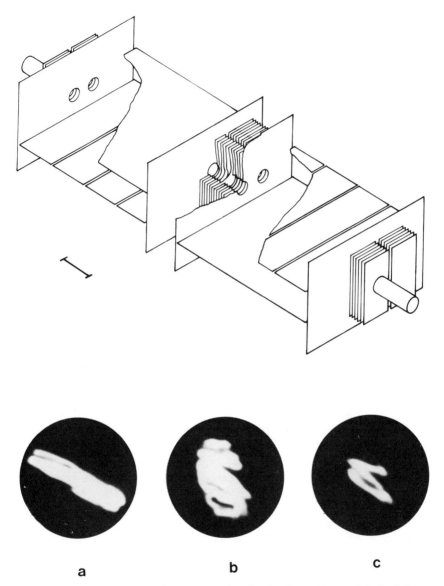

FIGURE 12. Isometric views of a prototype imaging band-pass filter and (*below*) images recorded from a test thermionic filament (**a**) using all electrons and with uncorrected (see text) crossed fields, (**b**) with dispersion-corrected crossed fields and (**c**) with the pass-band set to 0.1 eV.

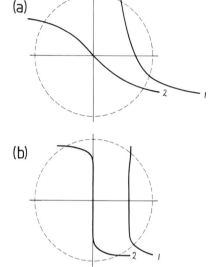

FIGURE 13. Computer simulation of a bar of monoenergetic electrons oriented parallel to an electric field perpendicular to an axial magnetic field after *one (1)* and *two (2)* passes through the crossed-field regions. The electrostatic field configuration is (**a**) that for a parallel plate condenser and (**b**) one optimised for minimum dispersion variation (see text).

resolution possible will, as outlined elsewhere,[2,4] range from <1 μm to about 10 μm depending to some extent on structural form but most strongly on the kinetic energy and magnetic field.

REFERENCES

1. PLUMMER, I. R., H. Q. PORTER & D. W. TURNER. 1982. J. Mol. Struct. **79:** 145–162.
2. TURNER, D. W., I. R. PLUMMER & H. Q. PORTER. 1984. J. Microsc. **136:** 259–277.
3. GRIFFITH, O. H. 1984. Analysis of Organic and Biological Surfaces. P. Echlin, Ed. Chapter 18, 455–476. Wiley. New York, NY.
4. BEAMSON, G., H. Q. PORTER & D. W. TURNER. 1981. Nature **290:** 556–561.
5. TURNER, D. W., I. R. PLUMMER & H. Q. PORTER. 1985. Philos. Trans. R. Soc. London Ser. A **318:** 219–241.
6. CAZAUX, J. 1973. Rev. Phys. Appl. **8:** 371–381.
7. BEAMSON, G., H. Q. PORTER & D. W. TURNER. 1980. J. Phys. E **13:** 64–66.
8. TURNER, D. W., I. R. PLUMMER & H. Q. PORTER. 1986. Philos. Trans. R. Soc. London, accepted for publication.
9. PLUMMER, I. R., H. Q. PORTER, D. W. TURNER, A. J. DIXON, K. A. GEHRING & M. KEENLYSIDE. 1983. Nature **303:** 599–601.
10. KRUIT, P. & F. H. READ. 1983. J. Phys. E **16:** 313–324.
11. TURNER, D. W., I. R. PLUMMER & H. Q. PORTER. 1986. Rev. Sci. Instrum. **57:** 1494–1500. United Kingdom and International Patent Nos. 8513821 and 8530011 pending.

DISCUSSION OF THE PAPER

G. J. BRAKENHOFF (*University of Amsterdam, Amsterdam, The Netherlands*): What is the magnetic field at the specimen?

TURNER: The apparatus that we've used throughout has a maximum value of 8 Tesla. We can work with about 7 and even more commonly at 5, because we're using much helium. Helium usage in our old magnet goes up very rapidly with the field. The magnetic field in the image zone is normally not less than about 10 Gauss. I'm sorry to change my units, but you'll realize that I'm talking about field ratios that are not much greater than about 10^4. The magnification goes up as the square root of the field ratio, so we're talking about magnifications of a hundred or so. The maximum that I can see that's practical is a thousand, but we don't achieve anything like that in our rather old equipment.

J. SILCOX (*Cornell University, Ithaca, New York*): The Tesla is about 10,000 Gauss.

TURNER: Thank you, sir.

BRAKENHOFF: It's exactly 10,000. Will there be any profit in working with super conducting material?

TURNER: What I perhaps ought to have said is that we're looking at what the possible limits may be in the future. It is known how to make a superconductor which will go up to 16 Tesla and this would double the spatial resolution. One could push it just a little bit beyond that with soft magnetic material inserts just to concentrate the fields locally.

M. PETRAN (*Charles University, Plzen, Czechoslovakia*): How can you manage with such very weak fields? How can you manage the interference?

TURNER: Is this the electric field you are referring to or the magnetic field?

PETRAN: Both.

TURNER: And are you worried about the interference caused by the de Broglie wavelength for the electrons or the Larmor motion? There is in fact a kind of interference that is associated with the helical motion of the electrons as they move, particularly when we get to the low-field region, where for an X ray photoelectron that is moving pretty fast, the helix pitch length is about a centimeter. The whole operating principle of the machine depends upon a continuous adjustment of the electric vector as the field lines diverge.

In other words, we are making the assumption that nothing changes rapidly—neither the local electric field nor the local magnetic field vector. They don't change rapidly or significantly within the distance corresponding to the helix pitch. That's a rather crude way of expressing the adiabaticity invariance requirement. Pitch must be very short compared with the speeds at which the fields change. When that doesn't happen, you do see effects that you might call interference-like; the image starts to shift and shimmer, or twist. Provided you control the fields so they remain slowly changing, that is avoided.

D. F. PARSONS (*New York State Department of Health, Albany, New York*): Are there any possibilities of taking resolution further, or are you at a fundamental limit?

TURNER: We know well enough what the fundamental limits are and we are no where near those at the moment. One reason is that with the present magnification the limit arises not at the specimen but at the channel plate which is our ultimate output device and also the phosphor screen. These together produce a spatial resolution of the order of 20–30 μn and if you map that back to the specimen by a magnification of only say 50–100 you are not really probing the fundamental limitation at the specimen. So we do really need to have post-analysis magnification before we finally reach the fundamental magnetic field and electron velocity limitation at the specimen.

Immunophotoelectron Microscopy of the Cell Surface and Cytoskeleton[a]

KAREN K. HEDBERG AND O. HAYES GRIFFITH

Institute of Molecular Biology
Department of Chemistry
University of Oregon
Eugene, Oregon 97403

INTRODUCTION

A major goal of cell biology is an understanding of the structural and biochemical organization of the cell and how this organization relates to function. Electron microscopy provides one fruitful approach to this problem and has generated a wealth of structural information. A major advance took place in this field when immunolabeling technology (originally developed for immunofluorescence microscopy) was adapted to electron microscopy.[1] The resulting technique of "immunoelectron" microscopy allows high-resolution identification of specific cellular components and their positions relative to other components. The familiar electron microscope techniques that make use of immunolabeling are transmission electron microscopy (TEM) and scanning electron microscopy (SEM). Immunolabeling is also applied in an electron microscope technique that is a relative newcomer to cell biology, photoelectron microscopy (PEM).[2] PEM is a uniquely surface-sensitive electron microscopy that provides a different kind of view of the cell than TEM or SEM.[3,4] The essential feature of PEM is that the image of the specimen results from UV light stimulated electron emission (photoemission) rather than from a transmitted, secondary, or backscattered electron signal resulting from exposure to a high-energy electron beam. Typical PEM biological samples are fixed, dehydrated whole mounts observed without staining, shadowing, or a conductive coating. Both the surfaces of cultured cells and cytoskeletal structures exposed after removal of the surface membrane are especially amenable to observation by this technique.[5] In either case cellular organization is revealed primarily by topographical details, which provide a major source of contrast in PEM images.[6] The application of immunolabeling in PEM makes use of a second source of contrast, photoemission contrast. Photoelectron labeling of biological structures depends on the use of markers that are more photoemissive than cellular components, rather than markers that are more electron-dense (as for TEM) or of a recognizable shape and size (as for SEM). The purpose of this paper is to illustrate the use of antibody-mediated photoelectron labeling (immunophotoelectron microscopy) using colloidal gold and silver-enhanced colloidal gold for the localization of specific cell surface and cytoskeletal proteins.

[a]Supported by National Cancer Institute Public Health Service Grant CA 11695 and a grant from the M. J. Murdock Charitable Trust.

MATERIALS AND METHODS

Cell lines MCF-7 (human breast carcinoma) and PtK2 (marsupial kidney epithelium) and rat embryo fibroblasts were provided courtesy of Dr. Lan Bo Chen (Dana-Farber Cancer Institute, Boston, MA). Human foreskin fibroblasts were derived from newborn foreskins by the collagenase method[7] and used at passages 7–12. Cells for PEM were cultured on sterile serum-washed tin-oxide-coated 5-mm glass coverslips as described.[8]

For observation of the intact cell surface (*e.g.*, FIGURE 1A) coverslip cultures were washed with warm isotonic buffer and fixed in 2% glutaraldehyde in cacodylate buffer (100 mM Na cacodylate, 100 mM sucrose, pH 7.4) or HBS (100 mM HEPES, 150 mM NaCl, 1 mM EGTA, pH 7.4). The samples were subsequently dehydrated through a graded series of aqueous ethanol solutions to 50:50 amyl acetate:ethanol and then dried in a stream of warm air from 100% amyl acetate.

Cytoskeletal samples were prepared in a variety of ways. For FIGURE 1B, a culture was exposed to 1% Triton X-100 in 10 mM PIPES, 100 mM KCl, 3 mM MgCl$_2$, 300 mM sucrose, 1 mM EGTA, pH 6.8, for 10 minutes at 4° C followed by washing in detergent-free buffer, and fixation and dehydration as for the sample of FIGURE 1A. For cytoskeletal preparations to be labeled for both immunofluorescence and immunophotoelectron microtubule visualization (see FIGURES 2, 3, and 4), samples were prepared in a detergent-containing microtubule stabilization buffer and labeled by a three-step procedure using a rhodamine fluorescence marker and 20-nm colloidal gold photoelectron markers, as described.[9] All incubations with antibodies or antibody conjugates were for 30–60 minutes at 37° C. For immunofluorescence and immunophotoelectron visualization of actin-containing cytoskeletal elements (*e.g.*, FIGURE 5), cultures were exposed to 0.5% Triton X-100 in PHEM buffer (60 mM PIPES, 25 mM HEPES, 10 mM EGTA, 2 mM MgCl$_2$, pH 6.9) for 10 minutes at 4° C, washed in detergent-free buffer, then fixed in −20° C methanol for 5 minutes. After rehydration in HBS, conditioned medium from hybridoma JLA20 (producing a monoclonal antibody recognizing actin)[10] was applied as the first antibody and the procedure continued as above for microtubule labeling. Cultures in which intermediate filaments were to be labeled for immunophotoelectron microscopy (FIGURE 6) were prepared by detergent extraction as for actin, followed by light prefixation in 0.02% fresh glutaraldehyde in the same buffer and extensive buffer washes, and in this instance exposed to a two-step labeling procedure. After incubation in conditioned medium from hybridoma 163 (producing a monoclonal antibody recognizing keratin-containing intermediate filaments; unpublished results, this laboratory), a second antibody (recognizing the first antibody) bound to colloidal gold markers 6–9 nm in diameter was applied. The final fixation was 2% glutaraldehyde in HBS. This was followed by silver enhancement of the gold markers as previously described.[11] All labeled cytoskeletal samples shown here were dehydrated through graded aqueous ethanol solutions followed by either CO$_2$ critical-point drying or vacuum drying from Freon 113.

Immunophotoelectron visualization of fibronectin on human foreskin fibroblast cell surfaces (*e.g.*, FIGURE 7) was accomplished as described.[11] A three-step labeling procedure was used: conditioned medium from hybridoma HB91 (producing a mouse monoclonal antibody recognizing fibronectin; American Type Culture Collection, Rockville, MD) was applied as the first antibody, followed by rhodamine-conjugated goat anti-mouse Ig's (Cappel-Worthington, Malvern, PA), and finally rabbit anti-goat IgG (E-Y Laboratories, San Mateo, CA) adsorbed to 6–9-nm diameter colloidal gold markers. Samples were fixed in 2% glutaraldehyde in HBS after extensive washes. Silver enhancement of fixed samples and subsequent vacuum dehydration from Freon 113 was as described.[11]

Samples to be observed by both fluorescence and photoelectron microscopy were first examined in the wet state, after fixation but prior to silver enhancement (if used), with a Zeiss fluorescence microscope equipped for rhodamine fluorescence. Photoelectron microscopy was carried out on dehydrated samples with the instrument located at the University of Oregon, Eugene, Oregon.[12] This PEM is an ion-pumped, oil-free, ultra-high vacuum microscope designed to avoid sample contamination. The sample is illuminated by two OSRAM HBO 100 W/2 short arc lamps, and the resulting emitted photoelectrons are accelerated to 30 kV and focused by the three electrostatic lenses[13] of the photoelectron microscope. A 50-micron objective aperture was used. Images were recorded on Kodak electron image film SO-163, and the exposure times used varied from 1 to 45 seconds.

RESULTS AND DISCUSSION

The photoelectron image is formed from electrons released from the outer 5 nm (about the thickness of a phospholipid bilayer membrane) or less of an organic surface.[14] The implication of this for studies on cells is that PEM images reveal specifically those structures that are exposed at the surface of the cell preparation. This is illustrated in FIGURE 1A and 1B. FIGURE 1A is a PEM micrograph of a cultured human carcinoma cell (cell line MCF-7) that was removed from culture, briefly washed with isotonic buffer, and then fixed in glutaraldehyde and dehydrated. The outer cell lamina is relatively intact, and the photoemission from this surface is sufficiently greater than that from the substrate to cause the cell boundaries to be easily recognized. Internal cellular structures are not seen, except where they induce topography in the cell surface. The most obvious example of this is near the center of the cell where the thick nucleus (N) and nucleoli (n) appear as rounded bumps in the cell surface.

In contrast to FIGURE 1A, FIGURE 1B is a PEM micrograph of an MCF-7 cell prepared in a similar manner except that it was exposed to a neutral detergent prior to fixation. This procedure removes most of the outer membrane and releases about half of the total cell protein to leave a "detergent insoluble cytoskeleton."[15] The photoelectron micrograph shows that the resulting cytoskeletal preparation lacks a smooth cell surface and is characterized by a complex of intertwined fibers approximating the shape of the intact cell. The nucleus is still evident and appears to be enclosed in a filamentous network. The visibility of even very fine fibers in these uncoated, unstained cytoskeletons is due to the high sensitivity of PEM to topographic detail. This sensitivity occurs because the photoelectrons are ejected with very low kinetic energies (approximately 1 eV) and their trajectories are readily influenced by microfields at the surface of the specimen.[6]

Although both of the micrographs in FIGURE 1 provide information as to the overall external and internal structural organization of these cells, the additional step of selective identification of specific elements of the cell surface or cytoskeleton would clearly be advantageous for probing the details of this organization. The general strategy for the selective labeling of antigenic sites of interest for PEM is similar to that used for TEM, SEM, and fluorescence microscopy. In the commonly used indirect labeling method, antibodies that recognize particular antigens are allowed to bind to the cell, and these antibodies in turn are bound by antibodies bearing markers. Markers for PEM are more like those used in fluorescence than those used for other electron microscope techniques in that both are based on light-stimulated emission. This is shown schematically in FIGURE 2 for fluorescence (FIGURE 2A) and for

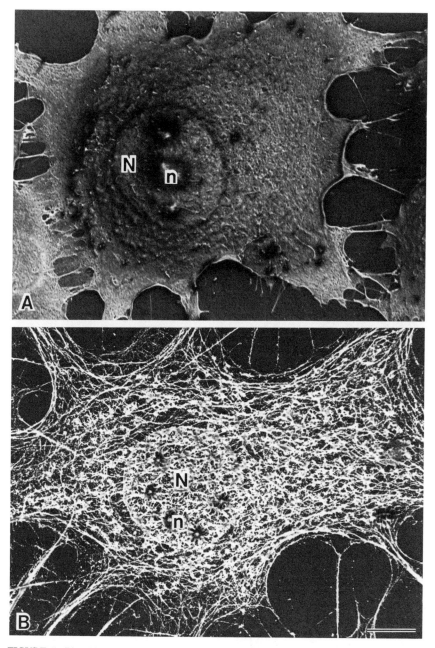

FIGURE 1. Photoelectron micrographs of fixed, dehydrated human breast carcinoma cells showing the intact cell surface (**A**), and after detergent treatment to remove the plasma membrane and expose cytoskeletal structures (**B**). N, nucleus; **n**, nucleolus. *Bar* = 5 μm.

Immunofluorescence Microscopy

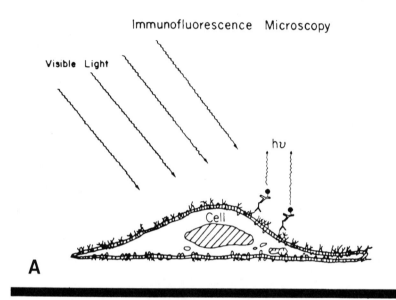

A

Immunophotoelectron Microscopy

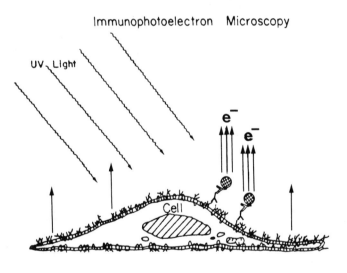

B

FIGURE 2. Schematic cross section comparing immunofluorescence (**A**) and immunophotoelectron (**B**) imaging of labeled cell surface proteins. Both techniques employ light-stimulated emission, but the emitted fluorescence signal is also light, whereas the emitted photoelectron signal is electrons. The low-energy photoelectrons are subsequently accelerated to 30 kV and focused by an electron lens system to form an image.

photoelectron (FIGURE 2B) microscopy, for the example of a cell surface protein. However, this emitted signal is *light* emission in fluorescence (wavy arrows, FIGURE 2A) and *electron* emission in PEM (solid arrows, FIGURE 2B).

Photoelectron markers need to be sufficiently photoemissive to be recognized against the background of photoemission from unlabeled cellular structures. This is shown in FIGURE 2B by the increased density of arrows in the region of the markers compared with the cell surface. Colloidal gold particles, which are used in TEM (because of their electron-dense nature) and in SEM (because the larger particles can be detected by virtue of their spherical shape and electron signal in the back scatter mode) also serve as highly successful markers for PEM because of their photoemissivity.[16] Colloidal gold has been used in several preliminary photoelectron labeling studies of specific sites on biological specimens including red cell membranes,[16] cytoskeletal preparations,[9] and cell surfaces.[11]

Examples of the use of colloidal gold particles for selective labeling of cytoskeletal elements in PEM images are shown in FIGURES 3-5. The cytoskeletons of cultured cells contain three classes of fibers; microtubules (25-nm single filaments), intermediate filaments (10-nm filaments, often in larger bundles), and microfilaments (6-nm individual filaments or bundles). Colloidal gold markers can be successfully used to identify all three classes of cytoskeletal elements in photoelectron images. Microtubule labeling is shown in FIGURES 3 and 4. Detergent-opened marsupial kidney epithelial cells (cell line PtK2) were exposed to a three-step procedure: a first antibody specific for microtubules, a second antibody (recognizing the first) carrying a rhodamine fluorochrome, and a third antibody (recognizing the second) bound to 20-nm diameter colloidal gold particles. FIGURE 3 shows a comparison of the immunofluorescence (3A) and photoelectron (3B) images of the same cell after this double labeling procedure. For orientation, arrows indicate corresponding regions. The same distinctive pattern of microtubules is unambiguously present in both types of images. An obvious difference, however, is that the fluorescent image consists solely of the pattern of the label, whereas the PEM image shows the pattern of both labeled and unlabeled structures. This is shown at low magnification in FIGURES 3B and 4A, in which the microtubules appear as bright fibers against the dimmer, unlabeled cellular background. In addition, in the PEM images both the structures and the label can be further resolved, as shown in FIGURES 3C and 4B. At these higher magnifications the 20-nm colloidal gold markers on the microtubules appear as a highly specific decoration of bright individual dots in the photoelectron micrographs, and the unlabeled structures between microtubules are clearly evident.

FIGURE 5 is a comparison of immunofluorescence (A) and photoelectron (B) labeling for actin near the leading edge of the same rat embryo fibroblast cell. Actin is a major component of the microfilament (6-nm filament) cytoskeletal system, and is also found in a fine network near the cell periphery, especially at the leading edge of motile cells. Unlike the individual, gently curving fibers typical of microtubules, microfilaments tend to occur in fairly straight, roughly parallel bundles. In FIGURE 5A these bundles appear as bright filamentous streaks running both on the long axis of the cell (close to horizontal in these micrographs) and also in an array nearly perpendicular to the long axis. The PEM micrograph (FIGURE 5B) also shows these microfilament bundles, although at higher resolution (arrows). In addition, the PEM image clearly shows the presence of a delicate network of actin labeling between the microfilament bundles, which is seen in the fluorescence micrograph only as a diffuse staining.

The label contrast in FIGURES 3, 4, and 5 is due entirely to the 20-nm colloidal gold markers and is present in spite of the layer of protein (the third antibody), which undoubtedly reduces the amount of electron emission available from gold itself. Recent work from this laboratory has shown that the photoemission from colloidal gold can be

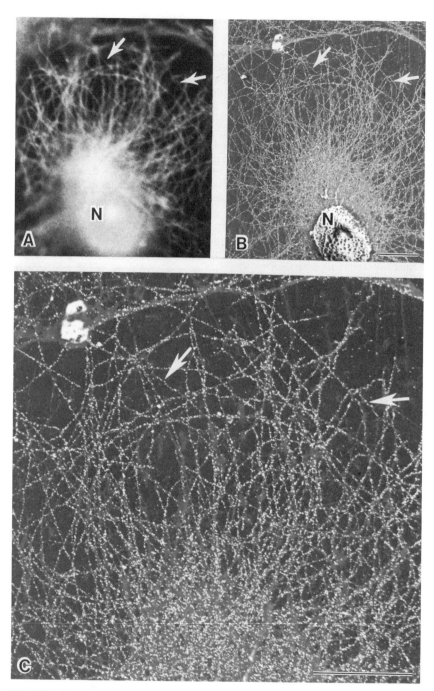

FIGURE 3. Detergent-opened marsupial kidney epithelial cells after labeling for both immuno-fluorescence (**A**) and immunophotoelectron (**B, C**) imaging of microtubules by a three-step labeling procedure. The first antibody recognized microtubules, the second carried a rhodamine fluorochrome as the fluorescence marker, and the third was bound to 20 nm colloidal gold as the photoelectron marker. Samples were first examined in the wet state by fluorescence microscopy and again after dehydration by PEM. Neither marker contributes to the label contrast seen by the other technique. N, nucleus. *Bars* = 5 μm.

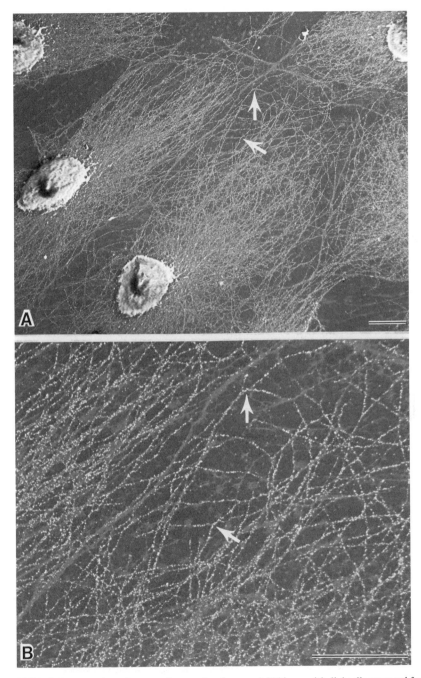

FIGURE 4. Immunophotoelectron micrographs of marsupial kidney epithelial cells prepared for microtubule visualization as in FIGURE 3. (**B**) is a higher magnification micrograph of the cells shown in (**A**). *Bars* = 5 μm.

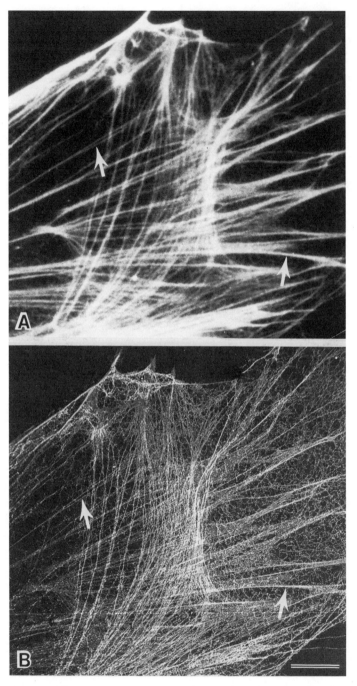

FIGURE 5. Immunofluorescence (**A**) and immunophotoelectron (**B**) micrographs of a portion of the same rat embryo fibroblast labeled as in FIGURE 3 but for actin visualization. *Arrows* indicate corresponding parts of the two micrographs. *Bar* = 10 μm.

further increased by silver enhancement of the labeled sample, providing a means of controlling the label contrast as desired.[11] In brief, silver enhancement is used on labeled, fixed samples to generate a thin coating of metallic silver selectively on the surfaces of the colloidal gold particles. Although there is an increase in the physical size of the marker, the increased label contrast is primarily due to the increased photoemission from the metallic silver coating compared with protein-coated gold.[11] Silver enhancement also allows the use of much smaller colloidal gold particles than would otherwise have been readily detectable by PEM, which has the advantage of reducing possible steric hindrance problems during labeling with the potential result of much greater labeling densities.

Examples of photoelectron labeling with silver-enhanced small (6–9-nm) colloidal gold markers are shown in FIGURES 6 and 7. FIGURE 6A shows a portion of marsupial kidney epithelial cell labeled for intermediate filament visualization. A micrograph at higher magnification is shown in FIGURE 6B. The organization of intermediate filaments in these cells is a lacy tangle of curves and loops, which is strikingly different from that of the microtubule network or the actin-containing microfilament system. Bundles of fibers of different thicknesses (arrows) are encrusted with the silver-enhanced small gold particles and stand out in stark contrast to the unlabeled cellular background (FIGURE 6B). FIGURE 7 demonstrates the use of silver-enhanced 6–9-nm colloidal gold as a photoelectron label for cell surface fibronectin on human fibroblasts. Colloidal gold alone is a less visible marker against the background of the cell surface than it is against the cytoskeleton, but silver enhancement of even small colloidal gold markers provides substantial label contrast on cell surfaces.[11] Fibronectin on normal fibroblasts in culture appears on the cell surface in patches and in fibrillar streaks (FIGURE 7A). These streaks can often be resolved to be as little as one marker in thickness (FIGURE 7B, arrowhead at left) and frequently lie over topography corresponding to intracellular microfilament bundles. This cell surface glycoprotein is also released from cells and incorporated into the extracellular matrix, which is found beneath, between, and around cells. In photoelectron micrographs of cells labeled for fibronectin this matrix is seen wherever it is exposed at the sample surface. Fibronectin-containing extracellular matrix can be seen in the right-hand portion of FIGURE 7B as strings and patches on the substrate and cell. The thin arrows in FIGURE 7B indicate fibronectin-containing matrix fibrils, whereas the thick arrows point to unlabeled fibrils. The specificity of the labeling technique is demonstrated by the clear distinction between labeled and unlabeled structures and regions in these micrographs.

The photoelectron micrographs in FIGURES 1 and 3–7 illustrate the type of information available from direct visualization of surface electron emission from biological specimens, both without and with immunolabeling technology. Immunophotoelectron microscopy can be applied to both cytoskeletal preparations and cell surfaces to identify specific components within the context of unlabeled structures, making it a powerful technique for the study of alterations in cellular organization. This approach is currently being used in our laboratory to study the effects of tumor promoters on cell surface and cytoskeletal components of cultured cells.

ACKNOWLEDGMENTS

We are pleased to acknowledge the invaluable contributions of Dr. G. Bruce Birrell in the development of colloidal gold and silver-enhanced colloidal gold labeling procedures for PEM; and of Mr. Douglas L. Habliston for sample preparation and fluorescence and photoelectron microscopy. We also wish to thank Ms. Loreen Evans

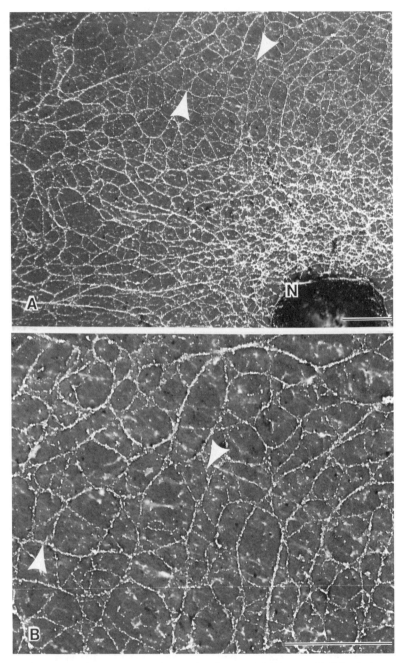

FIGURE 6. Immunophotoelectron micrographs of marsupial kidney epithelial cell detergent-opened and labeled for intermediate filament visualization by a two-step method using a second antibody bound to 6–9-nm colloidal gold markers as the photoelectron label. Subsequent silver enhancement was used to increase the photoemission from the small gold particles. (**B**) is a higher magnification micrograph of the cell shown in (**A**). N, nucleus. *Bars* = 5 μm.

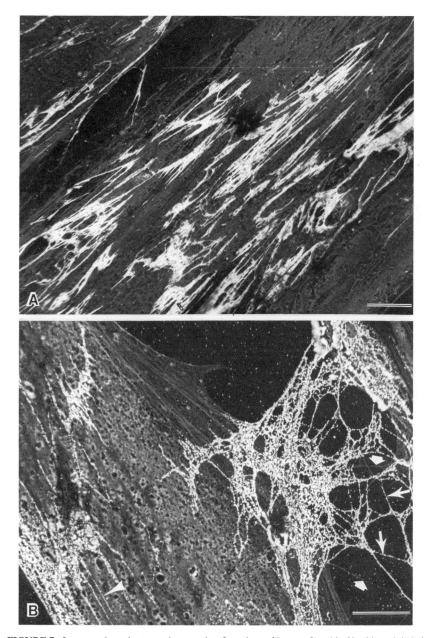

FIGURE 7. Immunophotoelectron micrographs of portions of human foreskin fibroblasts labeled by a three-step method (as for FIGURES 3–5) for cell surface fibronectin visualization using 6–9-nm colloidal gold markers on the third antibody. Subsequent silver enhancement was used to increase the photoemission from the small gold particles. *Bars* = 5 μm.

for expert photographic assistance, and Dr. Gertrude Rempfer and Mr. Walter Skoczylas for continued development of the Oregon photoelectron microscope.

REFERENCES

1. ROMANO, E. L. & M. ROMANO. 1984. Historical perspectives. *In* Immunolabeling for Electron Microscopy. J. M. Polak & I. M. Varndell, Eds. 1–15. Elsevier Scientific. The Netherlands.
2. GRIFFITH, O. H. & G. B. BIRRELL. 1985. Photoelectron microscopy. Trends in Biochem. Sci. **10:** 336–339.
3. GRIFFITH, O. H. & G. F. REMPFER. 1985. Photoelectron imaging in cell biology. Annu. Rev. Biophys. Biophys. Chem. **14:** 113–130.
4. GRIFFITH, O. H. & G. F. REMPFER. 1986. Photoelectron imaging: photoelectron microscopy and related techniques. *In* Advances in Optical and Electron Microscopy. R. Barer & V. E. Cosslett, Eds. Vol. 10. Academic Press. In press.
5. NADAKAVUKAREN, K. K. & O. H. GRIFFITH. 1985. Photoelectron imaging of cytoskeletal elements. Ultramicroscopy **17:** 31–42.
6. REMPFER, G. F., K. K. NADAKAVUKAREN & O. H. GRIFFITH. 1980. Topographical effects in emission microscopy. Ultramicroscopy **5:** 437–448.
7. HILFER, S. R. 1973. Collagenase treatment of chick heart and thyroid. *In* Tissue Culture, Methods and Applications. P. F. Krause & M. K. Patterson, Eds. 16–20. Academic Press. New York, N. Y.
8. NADAKAVUKAREN, K. K., L. B. CHEN, D. L. HABLISTON & O. H. GRIFFITH. 1983. Photoelectron microscopy and immunofluorescence microscopy of cytoskeletal elements in the same cells. Proc. Natl. Acad. Sci. USA **80:** 4012–4016.
9. BIRRELL, G. B., D. L. HABLISTON, K. K. NADAKAVUKAREN & O. H. GRIFFITH. 1985. Immunophotoelectron microscopy: the electron optical analogue of immunofluorescence microscopy. Proc. Natl. Acad. Sci. USA **82:** 109–113.
10. LIN, J. C. 1981. Monoclonal antibodies against myofibrillar components of rat skeletal muscle decorate the intermediate filaments of cultured cells. Proc. Natl. Acad. Sci. USA **78:** 2335–2339.
11. BIRRELL, G. B., D. L. HABLISTON, K. K. HEDBERG & O. H. GRIFFITH. 1986. Silver-enhanced colloidal gold as a cell surface marker for photoelectron microscopy. J. Histochem. Cytochem. **34:** 339–345.
12. GRIFFITH, O. H., G. F. REMPFER & G. H. LESCH. 1981. A high vacuum photoelectron microscope for the study of biological specimens. *In* Scanning Electron Microscopy. O. Johari, Ed. Vol. 2: 123–130. SEM, Inc. AMF O'Hare, IL.
13. REMPFER, G. F. 1985. Unipotential electrostatic lenses: Paraxial properties and aberrations of focal length and focal point. J. Appl. Phys. **57:** 2385–2401.
14. HOULE, W. A., W. ENGEL, F. WILLIG, G. F. REMPFER & O. H. GRIFFITH. 1982. Depth of information in photoelectron microscopy. Ultramicroscopy **7:** 371–380.
15. BELL, P., Jr. 1981. The application of scanning electron microscopy to the study of the cytoskeleton of cells in culture. *In* Scanning Electron Microscopy. O. Johari, Ed. Vol. 2: 139–157. SEM, Inc. AMF O'Hare, IL.
16. BIRRELL, G. B., S. M. ROSE & O. H. GRIFFITH. 1983. Photoelectron microscopy of erythrocyte ghosts; imaging of lectin binding sites with colloidal gold markers. Ultramicroscopy **12:** 213–218.

DISCUSSION OF THE PAPER

D. F. PARSONS (*New York State Department of Health, Albany, New York*): You said that the photoemission quantum yields are as unpredictable as immunofluorescence. I don't understand that. I thought the immunofluorescence unpredictability arose from solvent-quenching effects and various energy exchanges like that. When you are dealing with pure compounds why wouldn't it be predictable?

GRIFFITH: It should be predictable, but it hasn't turned out to be that way. The reason is the intense UV light we use in this experiment.

H. O. JAUREGUI (*Rhode Island Hospital, Providence, Rhode Island*): Colloidal gold particles of less than 10 nm are very sticky. They attach nonspecifically to everything. How do you prevent that? Also, it's very difficult for antibodies or lectins to enter between the cell and the plastic. What do you do to permeate the cell or to enter in between?

GRIFFITH: They're both questions that relate to all forms of microscopy and not just our form. The problem of nonspecificity in small gold particles—there must be 20 labs working on that. It's practically black magic. Getting antibodies underneath without disturbing the membranes is just plain luck, because there are adhesion plaques. If you get the cell early before too many plaques go down, the antibodies can get in. Sometimes as late as you reach confluence, then nothing gets in. So it's timing and luck.

W. W. WEBB (*Cornell University, Ithaca, New York*): I'm interested in your pictures of the cytoskeleton and wonder if you would compare the kinds of information you could get with what comes from transmission electron microscopy with similar gold particle labels? Do you see any discrepancies?

GRIFFITH: The question is how does this compare with TEM. To be honest, we haven't done a thorough comparison where we had the same system under identical conditions with gold or bonded to gold.

J. B. PAWLEY (*University of Wisconsin, Madison, Wisconsin*): Martin Mueller's lab in Zurich has done some very nice work with very small gold labels with the SEM, which also are easily localizable on the surface. The first time I saw a picture of your microscope it was a good thing I was sitting in the front row, because it was a very impressive edifice and probably rather expensive. As the information that comes out from Mueller's lab, for instance, and yours is very similar, it's not obvious why one should choose an emission microscope. There is some advantage to the SEM. For instance, one can make stereo pairs, which I don't believe you can make, is that right? You really can't tilt the sample and get two different angular views.

GRIFFITH: I don't know of a way to do stereo pairs. If there is a way I haven't come across it. We have to look for that. There are advantages, but this isn't going to replace SEM. All these techniques together are barely a match for Mother Nature. But the advantages when you do use it are lower radiation damage, and much higher sensitivity on topography. Also you achieve much higher resolution if you are dealing right at the surface, because it doesn't average down in the cell, rather, it forms a Gaussian image. The SEM forms an image of a lot of information coming from different depths. Photoelectron microscopy has higher surface resolution.

PAWLEY: That's true only compared with SEM at 30 kV. If you do SEM at low voltage or if you make ratios of images between what is collected as secondary and what's collected at back scatter, you can contain your information depth through a quite narrow range.

D. W. TURNER (*University of Oxford, Oxford, England*): The number of comments that have been made about radiation damage made me quickly think what sort of radiation damage one might expect for the kind of imaging that I do. They are very large orders of magnitude difference, at least millions of times less radiation damage to be expected in photoelectron microscopy. I'm not sure whether you'd like to comment upon this, but we can get an image in anything between a tenth of a second and a few seconds for the more difficult ones at our low magnifications, putting in 10^{10} to 10^{12} photons per second, with only 21 eV maximum energy over an area of one square mm.

I did some very quick mental calculations and it suggests to me that the difference in radiation damage, in at least the region where you are in fact collecting every electron that's liberated, is enormously better than in conventional electron microscopy, particularly when analysis is being done. First, because each photon produces one electron and no other damage, it isn't disposing of many different bond dissociations before it's finally lost and second, because of the great efficiency of usage of the resultant electrons. I thought it worth pointing out that there's a very big factor.

GRIFFITH: Ours is not quite so efficient, and we don't know what the radiation damage is. That's why I'm being conservative. So it's our feeling the radiation damage is a lot less, but we do not have as high a yield. We're putting in 6 or 7 electron volt photons. We work right at the threshold, so our yields are not one for one. Your yields may be, but ours are not. But my guess is that the damage is a lot less. The damage is still there. When you go up in magnification, for every factor of ten, you require a factor of 100 in signal-to-noise. It's absolutely a killer for everyone at higher magnification, and we're no exception. We have to face that too.

H. SHUMAN (*University of Pennsylvania School of Medicine, Philadelphia, Pennsylvania*): That's a very good point, but the numbers are very difficult to establish. For a high-energy incident electron the radiation damage is due to the same phenomenon. Obviously it ionizes whatever is there. The point is that it also can produce a high-energy secondary electron, producing cascades and more and more ionization. That's correct, except that for thin specimens the average energy loss of the high-energy electron is roughly 50 volts. So perhaps it's only double the amount of radiation damage. But that sort of calculation requires a great deal of detailed information about the cascade process itself.

J. SILCOX (*Cornell University, Ithaca, New York*): The other problem I sensed coming up in earlier papers was the question of background subtraction and understanding it at various pixels in the image. Is there anybody who wants to tackle that one? Have there been any other additional thoughts that come up in that area that people want to try?

A. P. SOMLYO (*University of Pennsylvania School of Medicine, Philadelphia, Pennsylvania*): In addition to the effects of impact parameter, in delocalizing the information from inelastic images obtained with a transmission electron microscope that has a chromatic abberation coefficient of 2 to 3 mm, one also has to add, perhaps simple minded in the quadrature, the degradation of resolution due to chromatic abberations with a ten-volt window.

F. P. OTTENSMEYER (*University of Toronto, Toronto, Canada*): I had a discussion with Dr. Colliex about that. That calculation has to be handled very carefully as well, because one has to integrate properly over the distribution of electrons at that energy loss over the apperture. One must not in that instance just take the simple chromatic abberation formula.

SOMLYO: What are your results, Dr. Ottensmeyer, for that calculation?

OTTENSMEYER: I haven't done it, but I've been discussing it with Dr. Colliex. I can furnish the data on the distribution of electrons at a specific energy, so we can actually go and measure it properly.

Light Microscopy—A Modern Renaissance[a]

WATT W. WEBB

School of Applied and Engineering Physics
Cornell University
Ithaca, New York 14853

A renaissance in light microscopy is manifest in the present series of papers on light imaging in biology and medicine. Naturally, I am honored by this opportunity to herald the opening of the renaissance. I plan to identify some of the innovative accomplishments of recent years and to point out the technological advances that have enabled them. The papers in this part report important progress in several of the most promising new directions.

I want to complete my introduction by illustrating briefly several *other* new developments in measurement capability by light microscopy from my own laboratory that are being published elsewhere.

Important recent achievements in the biological application of optical imaging have attracted the attention that motivated this conference. They form an impressive list. TABLE 1 is at least representative of this progress but is surely incomplete. These examples are listed in the approximate order that they will be encountered in this part, the first five briefly in my introduction and the remainder in the papers to follow.

What has aroused this resurgence of progress in light microscopy? I believe that it is empowered by new technology derived from laser optics, electro-optical devices, and digital computers. Sensitive, low-noise light detectors of high-quantum efficiency, digital image processing software and hardware, and mechanical manipulators of subnanometer precision and stability make possible the new developments. In addition, biochemistry and immunology have provided highly selective and brilliantly labeled fluorescent markers based on monoclonal antibodies, toxins, and ingenious chemistry. TABLE 2 lists some of the technological developments underlying the new light microscopies.

Many of the reports that follow deal with three-dimensional image reconstruction from optical two-dimensional sections. Many of the biological applications add time as a fourth dimension when the dynamics of living cells are studied. The quantity of information required to represent an image explodes as its dimensionality increases. Consider the information in one pixel, that is, one resolution volume, say, 1 μm cube. Suppose the required intensity resolution is $1/256$, which requires 8 bits or one byte. A single data stream with 16-bit resolution and a data rate at the video frame rate of 30 per second, or 60 bytes per second, is sufficient for a cell surface diffusion experiment. This experiment then generates only about 3,600 bytes of information per minute.

Now suppose we require a two-dimensional image with typical video resolutions of, say, 250,000 pixels per frame. At one-byte resolution only two frames fill a floppy disk! In a dynamic experiment at video rates, a 200-Megabyte Winchester disk operating at

[a]Supported by National Science Foundation Grant DMR-83-03404, National Institutes of Health Grant GM 33028, Office of Naval Research Grant N00014-K-84-0390, and the Biotechnology Center at Cornell University.

TABLE 1. Biological Applications of Modern Light Microscopy

1. *Dynamics of individual fluorescence-marked molecules,* detected, mapped, tracked, and number-resolved with microscopic spatial resolution in real time on living cell surfaces.
2. *Cytoplasmic calcium ion activity measurement* with fluorescent probes with near-microscopic spatial resolution and near-physiological time resolution.
3. *Mapping of cell membrane electrical potentials* with fluorescent probes, with microscope image resolution at slow speeds and electrophysiological time resolution for discrete point sets.
4. *Dynamics of fluorescence-labeled receptors and lipid mobility* on living, motile cells measured by fluorescence fluctuations (FCS) or recovery after photobleaching (FPR) by single point confocal or pattern recording.
5. *Picometer resolution, measurement, and spectral analysis of submicroscopic vibrations* and thermal motion of transparent microscopic organelles (auditory hair bundles) on living cells.
6. *Enhanced contrast imaging* of low-contrast biological structures and submicroscopic particles with sufficient speed for real-time recording of dynamics in living cells at video rates.
7. *Three-dimensional high-resolution image* reconstruction of fluorescence-stained organelles, notably intact chromosomes, from sets of two-dimensional fluorescence microscopy images.
8. *Three-dimensional image reconstruction* of thick biological structures.
9. *Submicroscopic image resolution in near-field images.*

sufficiently high speed is filled in less than a minute. In three dimensions it presently seems unreasonable to cope with dynamic experiments. For a single three-dimensional image reconstruction, computational time on dedicated VAX systems reinforced with array processors still seems to involve hours. Hopefully, more efficient strategies will be reported later.

For many experiments using fluorescent biochemical indicators, two-dimensional images are needed with 16-bit resolution. The important parameter is often derived from the ratios and/or differences of successive images requiring very rapid data storage for subsequent processing and very fast computation for real-time displays.

The demands of problems in cellular biophysics motivate the efforts to extend the capability of light microscopy. Important biological experiments push the limits of the existing technology. As each new experiment pushes back the frontier, it always seems to raise questions that call for faster or more sensitive techniques or for greater resolution. The clear message is that many important cooperative phenomena in cell biology remain hidden by the limitation of our experiments. Each stage of experimental improvement uncovers a deeper hierarchy of facts and questions.

TABLE 2. Technologies Underlying the Renaissance in Light Microscopy

- Lasers and laser technology.
- Electro-optical image intensifiers of high-quantum efficiency.
- Video cameras of low-noise, high-quantum efficiency and stability.
- Charge-coupled imaging devices.
- Precision electro-optical and mechanical manipulators.
- Nanometer microelectronic film technology.
- Low-cost, fast, digital computers and memory.
- Computational software for digital image analysis.
- Specific fluorescent probes, monoclonal antibodies, toxins and brilliant ligands.
- Fluorescence indicators of $[Ca^{2+}]$, V_m, and pH.

There is every indication that the advances in technology will continue to reinforce the capability of light microscopy. Time resolution, optical sensitivity, and accuracy of image-generated information are surely due for orders-of-magnitude improvements within the decade. The beginnings of these improvements will be obvious in the following papers.

The elegant techniques for enhanced image contrast and sensitivity that are described in the following papers are based on video enhancement and image processing combined with exceptional innovation in optical technique. Advances in confocal light microscopy and scanning microscopy, including the enhanced resolution of near-field scanning techniques are reported in detail. These techniques offer the potential for solution of the most difficult imaging problems in cell biology.

Our own research in cellular biophysics has led us to develop capabilities of light microscopy that are not discussed below. I would like to give a few illustrations of some of my favorite results to stimulate interest in these techniques.

Molecular mobility of fluorescence-labeled macromolecules on and in living cells has been susceptible to measurement by observation of the time course of variations of fluorescence intensity at a single point selected by confocal geometry. The techniques of fluorescence correlation spectroscopy (FCS) and fluorescence photobleaching

TABLE 3. Contributors to Reported Research in Applied Physics at Cornell University

Low-density lipoprotein receptor individual-molecule recording, tracking, and counting:
  Larry Barak, David Tank, David Gross, Richik Ghosh

Fluorescent membrane potential and ion activity indicators:
  David Gross, Leslie Loew, Tim Ryan, for V_m
  David Gross, Paul Millard, Clare Fewtrell, for $[Ca^{2+}]$

Hair cell vibration detector:
  Winfried Denk, Jim Hudspeth (collaborator from the University of California at San Francisco)

recovery (FPR) have generated a vast library of data.[1-4] This work has presented an important general puzzle. It is still not known what limits the process of protein motion on cell surfaces. To look more closely at this problem, we have developed a fluorescent analog of one ligand, low-density lipoprotein (LDL) that is so bright that we can see and record the positions and distribution of the *individual* LDL receptor molecules on living cells, determine the distribution of cluster sizes, and track individual receptors as they move on the cell surface.[5-10] High-quantum efficiency allows tracking through many images without photobleaching, but photography still provides excellent images.

Fluorescent probes provide sensitive measures of essential biochemical parameters in cell physiology. Most successful are measures of cytoplasmic calcium ion activity $[Ca^{2+}]$, membrane potential, and pH. Recent interesting results include video pseudocolor difference images recording the spatial variation of cell membrane potentials with microscope resolution.[11-12] Spatial variations are induced by application of electric fields to cells. Cytoplasmic calcium can be measured with the help of fluorophores that change their spectra on binding calcium. Some new unpublished data on the response of rat basophilic leukemia cells to stimulation by cross-linking of the cell surface IgE antibodies, as in allergic stimulation, induces a stochastically delayed but sudden increase in cytoplasmic calcium activity. These effects can be seen in a time series of spatially-resolved images of stimulated cells recorded by David Gross with Paul Millard and Clare Fewtrell in our laboratories.[13]

Finally, I would like to report what I think is a new record in spatial-displacement resolution by light microscopy. To measure the vibration of hair bundles on auditory hair cells, Winfried Denk has developed in my laboratory a device capable of detecting relative displacements of transparent hair bundles of only a few micrometers diameter with a spatial resolution of a few picometers.[14] The hair bundles on hair cells, in the frog sacculus, for example, fluctuate in thermal motion with an r.m.s. amplitude of a few nanometers and a power spectrum that is nearly Lorentzian with a roll off frequency around 200 Hz. This spatial "resolution" of about one ten thousandth of the light wavelength has previously been obtained by interferometry; however, the extraordinary sensitivity for analysis of microscopic objects provides great potential in cellular biophysics. The technique will be reported elsewhere.

ACKNOWLEDGMENTS

I am pleased to acknowledge participation of my colleagues in the reports from our laboratory (TABLE 3).

REFERENCES

1. MAGDE, D., E. L. ELSON & W. W. WEBB. 1972. Thermodynamic fluctuations in a reacting system—measurements by fluorescence correlation spectroscopy. Phys. Rev. Lett. **29:** 705.
2. MAGDE, D., E. L. ELSON & W. W. WEBB. 1974. Fluorescence correlation spectroscopy. II. An experimental realization. Biopolymers **13:** 29.
3. AXELROD, D., D. E. KOPPEL, J. SCHLESSINGER, E. ELSON & W. W. WEBB. 1976. Mobility measurements by analysis of fluorescence photobleaching recovery kinetics. Biophys. J. **16:** 1055.
4. KOPPEL, D. E., D. AXELROD, J. SCHLESSINGER, E. L. ELSON & W. W. WEBB. 1976. Dynamics of fluorescence marker concentration as a probe of mobility. Biophys. J. **16:** 1315.
5. BARAK, L. S. & W. W. WEBB. 1981. Fluorescent low density lipoprotein for observation of dynamics of individual receptor complexes on cultured human fibroblasts. J. Cell Biol. **90:** 595.
6. BARAK, L. S. & W. W. WEBB. 1982. Diffusion of low density lipoprotein-receptor complex on human fibroblasts. J. Cell Biol. **95:** 846.
7. TANK, D. W., W. J. FREDERICKS, L. S. BARAK & W. W. WEBB. 1985. Electric field-induced redistribution and postfield relaxation of low density lipoprotein receptors on cultured human fibroblasts. J. Cell Biol. **101:** 148.
8. GROSS, D. & W. W. WEBB. 1986. Cell surface clustering and mobility of the liganded LDL receptor measured by digital video fluorescence microscopy. *In* Spectroscopic Membrane Probes. L. M. Loew, Ed. CRC Press. Cleveland, OH, in press.
9. GROSS D. & W. W. WEBB. 1986. Molecular counting of low-density lipoprotein particles as individuals and small clusters on cell surfaces. Biophys. J. **49:** 901.
10. WEBB, W. W. & D. GROSS. 1986. Patterns of individual molecular motions deduced from fluorescent image analysis. *In* Applications of Fluorescence in the Biomedical Sciences. D. Lansing Taylor *et al.*, Eds. 405–422. A. R. Liss. New York, NY.
11. GROSS, D., L. M. LOEW & W. W. WEBB. 1986. Optical imaging of cell membrane potential: changes induced by applied electric fields. Biophys. J. **50:** 339.
12. GROSS, D., L. M. LOEW, T. A. RYAN & W. W. WEBB. 1985. Spatially-resolved optical

imaging of membrane potentials induced by applied electric fields. *In* Ionic Currents in Development. R. Nuccitelli, Ed. 263–270. A. R. Liss. New York, NY.

13. MILLARD, P., D. GROSS, W. WEBB & C. FEWTRELL. 1985. Imaging of subcellular Ca^{2+} distribution in secreting basophils using quantitative digital video fluorescence microscopy. *In* Amer. Soc. Cell Biol. Mtg., Atlanta, GA, November 1985 (abstract).

14. DENK, W., W. W. WEBB & A. J. HUDSPETH. 1986. Optical measurement of the Brownian motion spectrum of hair bundles in the transducing hair cells of the frog auditory system. Biophys. J. **49**: 21a.

Computer-Aided Stereoscopic Video Reconstruction and Serial Display from High-Resolution Light-Microscope Optical Sections[a]

SHINYA INOUÉ[b] AND THEODORE D. INOUÉ[c]

[b]*Marine Biological Laboratory*
Woods Hole, Massachusetts 02543
and
[c]*Cornell Engineering School*
Ithaca, New York 14850

INTRODUCTION

The light microscope, despite its limited lateral resolution and long history of use, retains several attributes not readily replaced by other, newer imaging or diffracting devices that provide greater resolution. With the light microscope, we can: readily examine a wide field of view; follow dynamic changes in the specimen nondestructively and in real time; generate contrast in several modes reflecting optical and fine-stuctural characteristics of the specimen; pinpoint composition and distribution of antigenic and other reactive molecules with high specificity; and directly image thin optical sections.

Coupled with video, these attributes of the light microscope are enhanced even further. We can: boost contrast dramatically by electronically increasing the gain and suppressing the signal from stray background illumination; intensify and convert scenes with extremely low luminance into clearly visible pictures; suppress random and fixed pattern noise; speed up or slow down dynamic scenes for closer examination; and electronically extract and digitally quantify selected image features. Aided by these advances we can now: use microscope lenses with excellent correction and NA to yield better resolved images in polarized light, DIC, and anaxial illumination;[1-5] detect, with much improved sensitivity, submicroscopic structures in these modes as well as fluorescence and dark-field microscopy;[6-10] and record fluorescent and other images more rapidly and with less photon damage to the specimen (for an overview of the principle, practices, and promises of video microscopy, see REFERENCE 11). It is no longer even uncommon to see actively gliding, gyrating, transporting, collapsing, or polymerizing nm-diameter molecular filaments captured with a video light microscope projected onto large auditorium screens.

Since with video microscopy we can gain images with good contrast using high-NA, well corrected objective lenses, coupled with matched or nearly matched NA condensers, we can not only achieve maximum lateral resolution but also gain excellent axial resolution, or setting accuracy,[12] that provides very shallow depth of field (see Section 5.5 in REFERENCE 11). Indeed, by using a novel light scrambler,[13] which

[a]Supported by National Institutes of Health Grant 2 R37 GM-31617-05 and National Science Foundation Grants PCM-8216301 and DCB-8518672.

provides homogeneous, full-aperture illumination of high-NA condensers with little loss of illuminance from a concentrated (Hg) arc source, we have been able to attain optical sections in polarized light and DIC images estimated to be only 0.25 to 0.3 μm thick. We have studied such optical sections by first storing them serially in a video laser disk recorder (Optical Memory Disk Recorder; OMDR) and then playing them back in different modes to analyze the three-dimensional distribution of the fine-specimen detail seen at the maximum lateral and axial resolution attained with the light microscope.

This paper describes a method for recording the optical sections and, through a simple digital processing method, rapidly displaying the three-dimensional details of the specimen stereoscopically and in through-focal series with or without stereoscopy. The methods were first demonstrated at the meetings of the American Society of Cell Biology in Atlanta in November, 1985.[14] Preliminary accounts of the surprisingly complex and unanticipated distribution of 5-nm gold-decorated microtubules uncovered by this method in whole-mount dividing plant cells were given briefly at the same meeting and elsewhere.[15,16]

EQUIPMENT

Microscope

We carried out the experiment on a custom-built, inverted polarizing microscope built on a sturdy optical bench. (Photograph and schematic of the microscope are shown in Figs. III-21 and III-22 in REFERENCE 11.) While the principles described below should be applicable to most research-grade light microscopes, they could be tested most readily on our custom-built microscope.

The microscope incorporates: a 4-foot, mehanite optical bench straight to 2.5 μm and vertically supported near its center of gravity; a stable precision (revolving) stage supported directly on the dovetail of the optical bench; a precision coarse and fine adjustment block carrying the objective lenses and analyzer independently of the rest of the body tube; a high-NA (1.35, Nikon) Achromatic Aplanatic condenser rectified for use with low birefringence Plan Apo objectives in polarized light and DIC microscopy; a high-intensity mercury (or xenon) arc illuminator whose homogeneous illumination fills the condenser aperture (using an optical scambler); and a straight, single optical axis between the effective light source and the video camera.

For most of the experiments, we used Plan Apo objectives (Nikon 40/0.95 dry with correction collar; Nikon 100/1.35 and Zeiss 63/1.40 oil immersion). We oil-contacted the slide to the condenser and set its NA to 90 to 95% of the objective NA as observed at the objective back aperture under Koehler illumination. The 100-watt concentrated arc mercury burner coupled with a heat cut and multilayer interference filter provided monochromatic green light of 546 nm. We project an infinity-focused image of the microscope field by a 1.6–6.3X motorized zoom ocular (Leitz Variorthomat) and focused the image by a 300-mm telephoto camera lens (Tokina) onto the face plate of the video camera (see Fig. 5–27 in REFERENCE 11).

Optical Scrambler

To obtain the high illuminance needed for high-extinction polarized light and related forms of microscopy and to simultaneously achieve the homogeneous field *and* aperture illumination desired, we incorporated a novel light scrambler[13] (a single optical fiber; see p. 127 and Figs. III-21 and III-22 in REFERENCE 11). Instead of the highly heterogeneous image of the concentrated arc lamp, the scrambler provides a homogeneous circular patch of light at its exit with little loss of mean luminous density. The exit of the scrambler serves as the effective light source, which is focused with a (8 to 64-mm focal length f/1.9 Angeneux) zoom lens onto the condenser iris. The zoom is adjusted to fill the condenser aperture (thus yielding a small illuminated field at high NA or a larger field at lower NA), satisfying both the field and aperture conditions needed in Koehler illumination. The homogeneous high-intensity illumination at the aperture of the high-NA condenser provides a clean, thin optical section free from the image anomalies associated with complex anaxial illumination (a condition that exists when the concentrated arc image is focused without a scrambler onto the condenser aperture).

Video Camera and Monitor

For the experiments described, we used a monochrome instrumentation video camera (Dage-MTI series 65) equipped with a moderately high-sensitivity, 1-inch

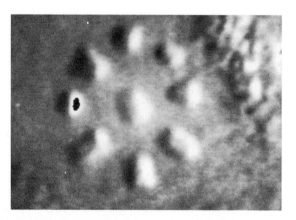

FIGURE 1. Video micrograph of *Chaetopterus* meiotic chromosomes arranged in a characteristic 8 + 1 pattern on the metaphase plate. The live, centrifugally clarified oocyte was viewed along the spindle axis with DIC optics enhanced with video. The spindle diameter is ca. 10 μm.

vidicon tube (Newvicon). Depending on the test, we used the camera with the automatic gain and automatic black (pedestal) controls switched either on or off. The pictures were displayed on 9- or 13-inch monochrome monitors (Panasonic WV-5310 or WV-5410). Pictures for publication were photographed off these monitors (with the V-hold control critically adjusted to gain good 2:1 interlace) on Kodak Plus-X film with a Nikon FM camera equipped with a 55-mm f/2.8 macro lens.

Image Processing

Some of the serial optical sections were recorded on the OMDR (see below) directly, while the contrast of other images was digitally enhanced on line, during the recording, with the Image-I digital image processor[17] (essentially identical to the Image-I system distributed by Interactive Video Systems, Concord, MA).

We developed special programs for the digital image processor to rapidly convert sequential optical sections into stereo pairs as described below.

Recorder

The video image was recorded on a Panasonic Model TQ-2021FBC Optical Memory Disk Recorder. This concentric groove, high-resolution (450 TV lines), monochrome laser disk recorder can hold 10,000 single full-frame video images, which are each recorded in 1/30 sec. In playback, the images can be displayed at rates of 10 frames per second to 2 seconds per frame sequentially, or any of the 10,000 frames can be selected and displayed in 0.5 sec. The OMDR was used to record the serial optical sections as well as the stereo-paired images reconstructed from the serial sections.

DESCRIPTION OF THE NEW METHODS

Model Experiments

Before testing the scheme proposed for stereo-pair reconstruction from serial optical sections taken through the microscope, we tried the following model experiment to test the basic assumption used in the new approach. The assumption was that stereoscopic pairs could be generated from a stack of closely spaced cross sections (of the contour or high-contrast image) of any three-dimensional object, by producing two plane projections of the sections superimposed after appropriate shearing for left- and right-eyed viewing.

As a model we used the intensity profiles of a microscope scene (*e.g.*, of the chromosomes in a living cell shown in FIGURE 1). The intensity profiles were generated by the Image-I processor for every other (horizontal) scan line of the original video image, with the upward deflection of the profile made proportional to the gray value (intensity) of the scanned pixel. In the example shown here, the profile was generated as a series of filled bars (as in a bar graph), each arising from the pixel location along the line scan in the stereogram. Each line scan thus produces a plane section contouring an intensity profile. To draw one member of the stereo pair, the program generated a series of intensity profiles (planes) with the origins of the successive line scans shifted by predetermined increments along the vertical and horizontal axes. Lines hidden by more proximal profiles were removed. The second member of the pair was generated similarly, next to the first image, but with a different value for incrementing the horizontal shift.

FIGURES 2 and 3 show stereo pairs generated in this fashion for cross-eyed and wall-eyed (or with a stereo viewer) viewing. In these stereo pairs each bar, whose height reflects the pixel intensity, is drawn with the same brightness as that of the pixel. Thus, a brighter pixel is represented by a brighter and taller vertical bar, a darker pixel by a darker and shorter bar, each with its base at the pixel location in the stereogram. With this mode of display, the stereogram and original image can also be represented in pseudocolor (FIGURE 4), simply by switching the output look-up tables in the image processor. The same color scale then reflects the pixel gray value in the intensity profile and the original image. Whether displayed in color or in monochrome, the stereo pair of the intensity contour can be generated in about 20 seconds with the software we developed for the Image-I processor (sampling 256 points for each of the 240 H-scans).

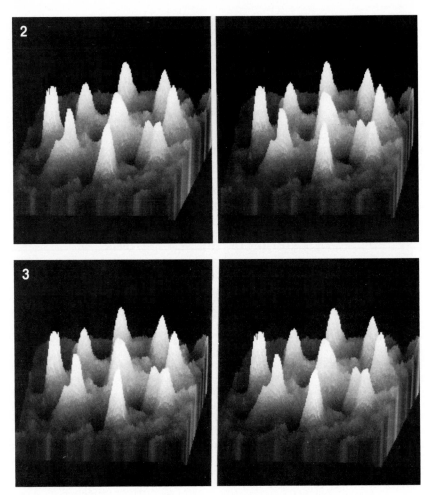

FIGURES 2, 3. Model stereo pairs. The pairs were generated from the scene in FIGURE 1 (in ca. 20 seconds each), with a custom program on the Image-I processor as described in the text. The stereo pair in FIGURE 2 is arranged for cross-eyed viewing, and in FIGURE 3 for wall-eyed viewing or for observation through a stereoscope.

As seen in FIGURES 2 through 4, our scheme, though lacking perspective shrinking of distant objects (but with hidden line removed in these examples) gives striking depth sensation in stereoscopic view.

Stereo Pair Generation from Serial Optical Sections

FIGURE 5 shows the scheme we used for rapidly converting a stack of high-contrast optical sections, stored in the OMDR, into a stereo pair. The rationale is to sum two stacks of sheared images, each with the successive images shifted sequentially to the left or the right. In the summed, sequentially-shifted image, we obtain what amounts to a top view, or plane projection, of a "transparent deck of cards" which, in lateral view, has been sheared into a parallelogram. When (the top of) the stack of cards is sheared to the left and viewed from above, we gain what corresponds to a right-eyed view of the unsheared stack. Shearing to the right gives a left-eyed view of the unsheared stack.

To accomplish this process in practice, we first recall (*e.g.,* the Nth section) from the OMDR, 8-frame-average (to reduce OMDR noise) and store it into the primary frame buffer of the image processor. We then shear the image by panning the image by N x M and N x (− M) pixels as shown in FIGURE 5. The panned left and right images of the Nth plane are summed, pixel by pixel, to the images of planes 1 through (N - 1), already stored into two other frame buffers. If necessary, the contrast of the series of images is exponentially weighted before summing to reduce contrast of more distant planes. Once each member of the full stack is sheared and superimposed, the central half of the resultant images are copied back into the right and left halves of the original frame buffer, which then displays the stereoscopic pair.

Using a custom program with our Image-I processor (using three 512 x 512 x 8-bit frame buffers, a 10-MHz pipe-lined arithmetic logic unit, and video-rate A/D, D/A converter), each left- and right-shifted weighted image is generated from an optical section and stored into its respective frame buffer in a fraction of a second, essentially at a rate determined by the number of frames averaged into the primary frame buffer. Once the full stack of sheared (and weighted) images is stored in the second and third frame buffers, the stereo pair is transferred to the first frame buffer and displayed on the monitor within 0.1 second. Thus, the whole process of converting a stack of 5 to 6 optical sections to a stereo pair is completed in a few seconds.

FIGURES 6 and 7 show samples of high-resolution stereo pairs, constructed by this method from 4 sequential optical sections. Since the optical sections can be obtained at full objective and condenser apertures using Plan Apochromatic objectives, no resolution is lost with this mode of stereo pair generation.

Through-Focal Stereo Pairs

In addition to rapidly converting selected stacks of optical sections into stereo pairs, our aim is to generate stereoscopic views that allow focusing through the specimen or viewing changes in time lapse. We wish to recover such *dynamic* stereo views from optical sections that had previously been stored in real time. In other words, we wish to recover through-focal (Z-scan) and time-lapsed (T-scan) stereo images as shown in FIGURE 8.

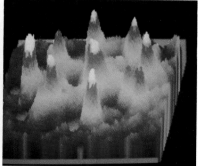

FIGURE 4. Stereo pair of intensity contour displayed in pseudocolor. The color scheme is selected to reflect the brightness of the corresponding pixel. Arranged for cross-eyed viewing.

In our preliminary tests, we have successfully obtained Z-scan stereo pairs of diatom frustules, and 5-nm gold-decorated microtubules in whole-mount *Haemanthus* cells. We first recorded optical sections, serially focused manually with the fine adjustment of the microscope objective lens carrier. The fine adjustment was focused in discrete increments and the image recorded on successive frames (tracks) of the OMDR (sometimes after on-line digital processing with Image-I). Each recording, which in our current setup was triggered manually, is accomplished in 1/30 sec, a single video frame time. In order to use the data also in the mode described in the next section, we started the focal series (at each X-Y specimen location) with an optical section at the closest specimen surface, proceeded to the distal surface, and then backed off again, each in 0.5-μm or 1.0-μm increments until we reached the closest surface again. For a large specimen, we then shifted the stage-micrometer's X or Y coordinate to obtain partially overlapping images. In this way, any volume element of the stored images could later be recalled for montaging or other reconstruction. We generated stereo pairs from the images stored in the OMDR by playing back a series of

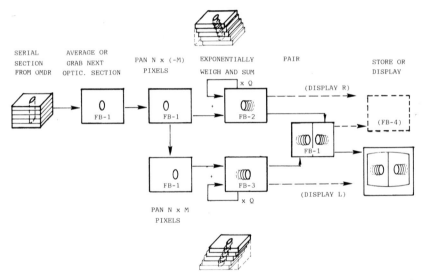

FIGURE 5. Schematic for rapidly converting a stack of serial optical sections into a stereo pair.

sequential optical sections into the Image-I processor as described before. The generated stereo pairs were then stored back on successive tracks of the OMDR.

For through-focal stereo viewing of a particular stack ($S_{x,y,t}$), we first generated a stereo pair from the first two (or three) optical sections, Z_1 and Z_2 (or Z_1 through Z_3), of the stack. The pair was stored on a track of the OMDR. Then on the next tracks we stored the stereo pair generated from Z_2 and Z_3, then Z_3 and Z_4, and so on. In other words, we now have stored, on successive tracks of the OMDR, a series of stereo pairs of overlapping volume elements scanned along the Z-axis of a particular stack of optical sections. When the OMDR is played back onto a video monitor, it displays the desired series of through-focal stereo images. The monitor can be viewed cross-eyed for observers accustomed to this method of stereoscopic fusion, or with the left and right images switched, one can view an underscanned 7 to 9-inch (diagonal dimension) monitor with a stereo viewer. Appropriate lenses are attached to the stereo viewer (*e.g.,* for a stereopticon ca. -3 diopter) so that the monitor picture is not overmagnified.

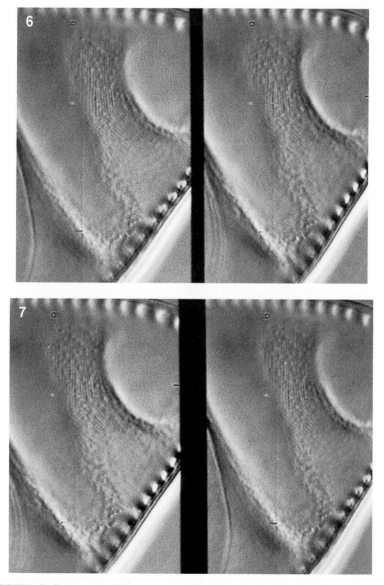

FIGURES 6, 7. Stereo pairs of diatom frustule taken with a 40/0.95 Plan Apo objective lens with matched condenser aperture. The pairs were generated from a stack of 4 optical sections with the Image-1 processor as described in the text. FIGURE 6 arranged for cross-eyed, FIGURE 7 for wall-eyed viewing.

As demonstrated at the meetings mentioned above, one obtains rather striking three-dimensional impressions of the diatom frustule and of the gold-decorated microtubule, chromosome, and vacuole arrays in whole-mount plant cells, by viewing the monitor displaying the stereo pairs focusing through the specimen. Unfortunately these dynamic results cannot easily be shown in journal reproduction. They can, however, be reproduced as video tape (or disk) copies of the OMDR records.

Nonstereoscopic Through-Focal and Montaged Displays

The images stored on successive tracks of the OMDR can be played back at selected rates controlled by an internal microprocessor. With the Panasonic model TQ-2021FBC OMDR we used, the playback rate can be selected between 10 frames per second and 2 seconds per frame. Successive full-frame images appear at the selected rate without intervening glitches in forward or reverse sequence, so that the optical sections recorded sequentially on the OMDR can be played back at varying Z-stepping (focusing) speeds.

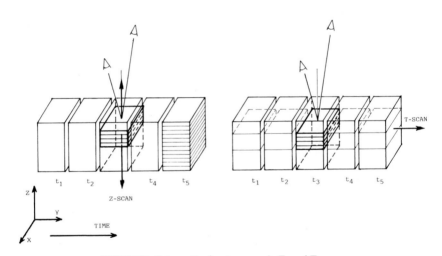

FIGURE 8. Schematics for stereoscopic Z- and T-scans.

The resulting display on the monitor or video projector is a striking series of optical sections that are rapidly through-focused or slowly stepped through at discrete intervals. If desired, the contrast or image features of the display can be further enhanced on-line through the image processor in playback. These displays provide a vivid mental view of the three-dimensional specimen architecture and complement the through-focus stereo view also obtained at full microscope resolution.

Alternatively, once the X, Y, Z, T coordinates of each optical section are recorded together with, or on, the numbered OMDR tracks as planned, we should be able to obtain X-, Y-, or T-scans at any focal level, displayed successively at 0.5-sec or greater intervals. For faster scans (*e.g.,* to speed up lapsed-time recording), a selected sequence could first be stored back, with or without transforming to stereo pairs, on a series of adjoining tracks of the OMDR. With software control for the OMDR and image processor operation, the transfer should take no longer than 1 sec per frame. These tracks could then be viewed at rates of speed chosen during final playback.

Another mode of display that can be recovered from the OMDR is a montage of large image fields obtained at selected focal levels. In fact, the contrast and resolution of the optical sections played back from the high-resolution monochrome OMDR were so superior that the photo montages obtained by patching together photographs of the monitor image (of gold-decorated microtubules) looked as though they were large-area electron micrographs. Details of these results and examples of the serial focal sections (which were shown at the American Society of Cell Biology meetings in Atlanta) will be published elsewhere.

DISCUSSION

For many years three-dimensional wax, cardboard, and plastic models of developing embryos, organs, tissues, and cells have been reconstructed from light and electron micrographs of serial sections to display and analyze their three-dimensional architecture (see Gaunt and Gaunt[18] for extensive discussions of three-dimensional reconstruction and stereo imaging of biological specimens). Several methods for obtaining stereoscopic pairs (with and without video) directly on a microscope equipped with high-NA objectives are discussed in Section 12.7 of Inoué.[11] With these methods it is generally not possible to use the full objective or condenser NA. In addition, methods have been developed to project a series of through-focal video images in "objective three-dimensional space," using an oscillating screen[19] or vibrating mirror[20,21] (summarized in Section 11.5 of REFERENCE 11). The vibrating mirror methods work impressively well on binary images, but further development is needed for application on high-resolution gray-scale images. I have not had a chance to observe de Montebello's very interesting-looking system[19] in action.

Agard and co-workers[22] and Fay et al.[23] examined the contribution of out-of-focus material in fluorescence microscopy and, with digital computers, calculated the intensity distribution that should be observed in "true" optical sections freed from out-of-focus contributions. From these corrected optical sections, they used the computers to reconstruct three-dimensional images of polytene chromosomes in salivary gland cells and calcium-containing structures in smooth muscle cells. The reconstructed images were presented as rotating figures or stereo pairs to display the three-dimensional organization of the sample. In contrast to the time-consuming computation used by these authors, we have resorted to a simple and very rapid approximation for generating stereo pairs.

In serial optical sections, taken in polarized light or in DIC, with the aid of the optical-scrambling illuminator, the thickness of the field in focus (especially of specimens with high spatial frequency) could be made extremely shallow. For example, the dichroic image of a stretch of gold-labeled microtubules was completely invisible in the video-enhanced image when the (60/1.40 or 100/1.35 Plan Apo) objective lens was refocused by 0.5 μm; instead, another stretch of the same, slightly tilted microtubule would come into focus. A focal adjustment of no more than about half of 0.5 μm was required for the ends of the microtubule in adjacent optical sections to be contiguous. In other words, with the optical system used, we estimate the depth of field to be no more than 0.25 to 0.3 μm. This corresponds to the theoretically expected axial resolution, or "setting accuracy," calculated from the equation posed by Francon[12] (see p. 118 in REFERENCE 11), but seldom attained in practice.

Because of the shallow depth of field attained by using the light scrambler, and the desire to reconstruct the stereo images in near-real time, we chose the simple approximation method described. In fact, the approximation seems to work rather well and allowed us, in addition, to observe through-focal series of stereo pairs. Through-focusing the stereo images of shallow volumes is particularly helpful when the specimen structure is complex and the image is not binary (black or white) but consists of several gray levels.

The through-focusing of serial optical sections presented at various rates from the OMDR, but without stereo display, is a useful adjunct to the stereo through-focal display. In principle that is no different than through-focusing and observing the successive optical sections directly through the microscope. But replay from the OMDR provides several advantages in practice. We can enhance or alter image contrast and change the rate of incremental through-focusing after the fact, repeat the observation as desired, and on a display observable by several individuals.

The data stored in the OMDR could in principle also be reconstructed as sections cut in any other plane or rotated as needed, but we have not yet attempted these approaches, which might require larger numbers of frame buffers and a large array processing capability.

Whether with our simplified method or with the more complex reconstructions of Agard, Fay, and their collaborators, "distant" objects in the microscope are not smaller as in macroscopic scenes. In other words, they lack the perspective shrinking of distant objects (unless such a perspective is artificially introduced). Likewise in images produced with high-NA objectives, distant objects are seldom masked by more proximal objects in the field. Thus, two common cues that work together with stereoscopic parallax to produce a sensation of the third dimension in everyday scenes are generally absent in the reconstructed microscope image. In some cases we have digitally decreased the contrast of the more "distant" optical sections used to reconstruct the stereoscopic pairs. The treatment does add to the sensation of depth when up to around 5 optical sections are used to produce the stereo pairs. Further increase in the number of layers added too much detail and confused the image for the type of multiple gray-level images we tested.

For stereo pairs used for through-focusing, 3 or even only 2 sheared optical sections per stereo pair provided images that could be better deciphered than with more sections per scanning volume. In this case, we did not reduce the contrast of the more "distant" sections as we did with the static stack of optical sections encompassing a larger specimen volume.

When stereo pairs are displayed on a conventional video monitor, substantial distortion between the left and right images can be introduced by the monitor itself (especially by nonlinearity of the H-scan; e.g., see Fig. 7–36 A–C REFERENCE 11). However, the curvature of space in the stereogram introduced by such distortion can be reduced by using the central two thirds of the monitor, and the curvature is not particulary noticeable when the stereo pairs are observed dynamically in the Z-scan mode. To prepare photographs of the stereo pair with minimum differential distortions between the two members of the pair (e.g., FIGURES 2–4), we generally zoomed the paired image by 2X, then photographed the left and right images separately centered on the monitor.

The methods we describe here are simple to execute and can be operated in near-real time. From stored sets of serial optical sections taken at lapsed-time intervals, we should also be able to reconstruct stereo time-lapse video images of slowly changing objects (such as dividing cells and developing embryos). The new method should then permit four-dimensional imaging of volume units in depth or in time, selected after recording the serial optical sections of the objects viewed at very high microscope resolution.

In addition to stereoscopic and other displays from serial optical sections, we described here a method for rapidly generating stereoscopically paired, image intensity contours. The stereo pair can be generated in about 20 seconds and displayed in monochrome or pseudocolor, with its gray scale or color matching those of the video scenes. As illustrated, the intensity contours give rise to striking stereoscopic images.

ACKNOWLEDGMENT

We thank Dr. Gordon W. Ellis of the Department of Biology, University of Pennsylvania, for the helpful discussions and suggestions he made during the course of this project.

REFERENCES

1. ALLEN, R. D., J. L. TRAVIS, N. S. ALLEN & H. YILMAZ. 1981A. Video-enhanced contrast polarization (AVEC-POL) microscopy: a new method applied to the detection of birefringence in the motile reticulopodial network of *Allogromia laticollaris*. Cell Motil. **1**: 275–289.
2. ALLEN, R. D., N. S. ALLEN & J. L. TRAVIS.1981b. Video-enhanced contrast, differential interference contrast (AVEC-DIC) microscopy: a new method capable of analyzing microtubule-related motility in the reticulopodial network of *Allogromia laticollaris*. Cell Motil. **1**: 291–302.
3. DVORAK, J. A., L. H. MILLER, W. C. WHITEHOUSE & T. SHIROISHI. 1975. Invasion of erythrocytes by malaria merozoites. Science **187**: 748–750.
4. INOUÉ, S. 1981. Video image processing greatly enhances contrast, quality, and speed in polarization-based microscopy. J. Cell Biol. **89**: 346–356.
5. KACHAR, B. 1985. Asymmetric illumination contrast: a new method of image formation for video light microscopy. Science **227**: 766–768.
6. ALLEN, R. D. 1985. New observations on cell architecture and dynamics by video-enhanced contrast optical microscopy. Annu. Rev. Biophys. Biophy. Chem. **14**: 265–290.
7. KAMIYA, R., H. HOTANI & S. ASAKURA. *In* Prokaryotic and Eukaryotic Flagella. W. B. Amos & J. G. Duckett, Eds. 53–76. Cambridge University Press. London.
8. TILNEY, L. G. & S. INOUÉ. 1982. Acrosomal reaction of *Thyone* sperm. II. The kinetics and possible mechanism of acrosomal process elongation. J. Cell Biol. **93**: 820–827.
9. SCHNAPP, B. J., R. D. VALE, M. P. SHEETZ & T. S. REESE. Single microtubules from squid axoplasm support bidirectional movement of organelles. Cell **40**: 455–462.
10. YANAGIDA, T., M. NAKASE, K. NISHIYAMA & F. OOSAWA. 1984. Direct observation of motion of single F-actin filaments in the presence of myosin. Nature **307**: 58–60.
11. INOUÉ, S. 1986. Video Microscopy. Plenum Press. New York, NY.
12. FRANÇON, M. 1961. Progress in Microscopy. Row, Peterson. Evanston, IL.
13. ELLIS, G. W. 1985. Microscope illuminator with fiber optic source integrator. J. Cell Biol. **101**: 83a.
14. INOUÉ, S., T. D. INOUÉ & G. W. ELLIS. 1985a. Rapid, stereoscopic display of microtubule distribution by a video-processed optical sectioning system. J. Cell Biol. **101(2)**: 146a.
15. BAJER, A. S., J. MOLÈ-BAJER & S. INOUÉ. 1985. Three-dimensional distribution of microtubules in *Haemanthus* endosperm cells. J. Cell Biol. **101(2)**: 146a.
16. INOUÉ, S., A. S. BAJER, J. MOLÈ-BAJER, M. DEBRABANDER, J. DEMEY, R. NUYDENS, G. W. ELLIS, E. HORN & T. D. INOUÉ. 1985b. Microtubules decorated with 5 nm gold visualised by video-enhanced light microscopy. J. Cell Biol. **101(2)**: 146a.
17. ELLIS, G. W., S. INOUÉ & T. D. INOUÉ. 1986. Computer aided light microscopy. *In* Optical Methods in Cell Physiology. P. De Weer & B. M. Salzberg, Eds. Wiley. New York, NY.
18. GAUNT, W. A. & P. N. GAUNT. 1978. Three-Dimensional Reconstruction in Biology. University Park Press. Baltimore, MD.

19. DE MONTEBELLO, R. L. 1969. The RLM synthalyzer technique and instrumentation for optical reconstruction and dissection of structures in three dimensions. Ann. N. Y. Acad. Sci. **157**: 487–496.
20. FUCHS, H., S. M. PIZER, L. -C. TSAI, S. H. BLOOMBERG & E. R. HEINZ. 1982. Adding a true 3-D display to a raster graphics system. IEEE Comput. Graphics App. **2**: 73–78.
21. SHER, L. D. & C. D. BARRY. 1985. The use of an oscillating mirror for three-dimensional displays. *In* New Methodologies in Studies of Protein Configuration. T. T. Wu, Ed. Van Nostrand-Reinhold. Princeton, NJ.
22. AGARD, D. A. 1984. Optical sectioning microscopy: cellular architecture in three dimensions. Annu. Rev. Biophys. Bioeng. **13**: 191–219.
23. FAY, F. S., K. E. FOGARTY & J. M. COGGINS. 1986. Analysis of molecular distribution in single cells using a digital imaging microscope. *In* Optical Methods in Cell Physiology. P. De Weer & B. M. Salzberg, Eds. 51–64. Wiley. New York, NY.

Three-Dimensional Imaging by Confocal Scanning Fluorescence Microscopy[a]

G. J. BRAKENHOFF, H. T. M. VAN DER VOORT,

E. A. VAN SPRONSEN, AND N. NANNINGA

Department of Electron Microscopy and Molecular Cytology
University of Amsterdam
Plantage Muidergracht 14
1018 TV Amsterdam, The Netherlands

INTRODUCTION

The decision to pursue confocal microscopy was born from dissatisfaction with the possibility of electron microscopy to acquire reliable data about the live morphology of biological specimens in the submicron range. This possibility is strongly restricted by the way various specimen-preparation methods affect the apparent structure in the material. The required steps in the preparation (chemical fixation, dehydration and thin sectioning) each introduce their own type of effect. For instance, a volume shrinkage of bacteria has been observed with values up to 50%.[1] The resulting interest in imaging techniques with a better resolution than standard light microscopy, but where the specimen still could be observed live in its natural environment, led us to the application in light microscopy of the confocal principle. With this approach, which has been used before in acoustic microscopy,[2] the resolution limitations of standard light microscopy could be expected to be surpassed if optics of high numerical aperture (N.A. = 1.3–1.4) were used. The actual demonstration of this fact in transmission confocal microscopy by our group[3] resulted in observed point responses of 196 nm at a wavelength of 633 nm and 130–140 nm at wavelengths of 442 and 325 nm.

The confocal principle is also applicable in fluorescence microscopy where, actually due to the incoherence of the fluorescence light, an even higher resolution may result.[4] But even under conditions where practical considerations exclude these resolutions (see below), there still remains the so-called sectioning effect in this mode by which fluorescence contributions from off-focus layers in the specimen are prevented from contributing to the image formation. Such contributions lead in normal fluorescence to a strong reduction of the available contrast.

The serial way in which the data are produced in this type of microscope, together with the sectioning effect, makes the confocal scanning laser microscope (CSLM) particularly suitable for coupling to a computer system. Thus an apparatus results which permits three-dimensional studies of biological specimens at high resolution with relatively simple instrumentation.[5–7] After a description of the scanning microscope (optical part, instrument control, computer system, processing algorithms used) we present a measurement of the spatial point response and a number of applications of three-dimensional imaging in biology. Finally, we will discuss some of the merits and limitations of the various forms of scanning microscopy presently under development.

[a]Supported by the Stichting Technische Wetenschappen (STW) and the Foundation for Fundamental Biological Research (BION/ZWO).

THE CONFOCAL SCANNING LASER MICROSCOPE

Principle

Characteristic for confocal microscopy is that one and the same point in the specimen is both optimally illuminated from a point light source as well as optimally imaged on a point detector. If the optics used are diffraction-limited, it can be shown that at a given numerical aperture the point response of a confocal system will be narrower by a factor of 1.4 in comparison to the conventional one. The demonstration that this property, expected from a theoretical treatment in the paraxial approximation,[8] can be realized in practice with immersion optics of N.A. = 1.3[3] is important, because only at those N.A. can an absolute gain with respect to conventional microscopy be realized. The narrowing of the response transverse to the optical axis is accompanied by a reduction of the depth of field along the axis by the same factor. So, finally, in confocal microscopy the volume element in the specimen that contributes to the image formation is smaller by a factor of $(1.4)^3 \approx 3$ as compared to ordinary microscopy.

The confocal principle also holds in fluorescence, though with two modifications: the longer wavelength of the fluorescence radiation will lead to some reduction of the expected resolution, while the incoherence of the fluorescence radiation may improve the imaging.[4] When a not very small or point detector is used, as is often necessary in fluorescence for reasons of detection efficiency, the effective instrument resolution is determined by the dimensions of the diffraction-limited illumination spot. The illumination-determined resolution in a fluorescence CSLM with a "large" detection pinhole is in any case better than normal fluorescence microscopy because of the wavelength-dependence of the resolution. In addition, the CSLM in fluoresence with such a "large" (but not too large) detection pinhole still possesses the sectioning property. In the normal fluorescence microscope all the radiation generated at different levels reaches the image plane, disturbing and reducing the contrast of the image of the in-focus level in the specimen. In scanning fluorescence microscopy the pinhole used, even if somewhat larger, will suppress quite effectively the out-of-focus contributions, resulting in an image which is only related to the in-focus image plane—a property which we call optical sectioning.

Instrument

In our CSLM we scan the specimen mechanically through the optically defined confocal point. Some aspects of this choice will be discussed later. In FIGURE 1 we show the microscope in the fluorescence mode and give a description of the image formation. The instrument is also equipped with a spectrograph, so that we can select a specific wavelength band for fluorescence imaging. The system contains two computer systems; one small, relatively slow, 8-bits system which takes care of the various instrument-control functions and a faster 16-bits system with a 68000 processor dedicated to image collection, processing and display functions. This second system can perform data operations like stretching, inverting, filtering (gaussian, median) and local contrast enhancement.[5] In addition, we generate from this system a continuous 60 Hz repetition rate viewing image independent of the mechanical scan.

The most useful of the various programs presently available is the one for the generation of three-dimensional images. This program consists of two units; the first automatically acquires a set of images at various heights in the specimen according to the parameters set by the operator. The second unit takes care of the actual three-dimensional stereoscopic image operation by a procedure as described in FIGURE 2. The stereoscopic viewing angles, as determined by the factors s_1 and s_2, can be chosen in such a way that the desired three-dimensional structure is optimally presented.

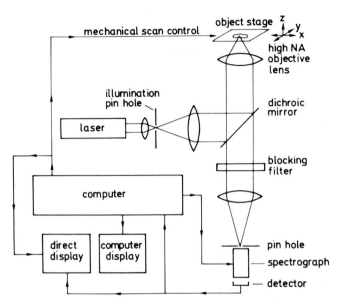

FIGURE 1. Optical layout of the confocal scanning microscope for the fluorescence or reflection modes of operation. Light, originating from the laser-illuminated pinhole is focussed on a certain point of the specimen. This point is subsequently imaged on the detector pinhole. The technique is called confocal since the image of the illumination pinhole and the backprojection of the detector pinhole have a common focus in the specimen. We operate the system at high numerical apertures using for the confocal objective lens a 100X oil immersion lens with N.A. = 1.3. The specimen is scanned mechanically through the confocal point, and the image is generated by the detected intensity, on a display running in synchrony with the mechanical scan. Independent of the mechanical scan a continuous viewing image is generated from the detected fluorescence intensity data stored in a $256 \times 256 \times 8$-bit memory array as a function of position (frame store). In between the detection pinhole and the detector a computer-controlled spectrographic element is present, which enables us to specify the wavelength (and the wavelength band around this wavelength) of the fluorescence radiation to be used for imaging.

Instrument Response

In FIGURE 3 a measurement of the spatial response characteristics of the instrument in fluorescence (under typical operating conditions) is presented. For this measurement we have used the above-described routine to obtain the set of through-focus images through the bead. As can be seen from FIGURE 3, the transverse resolution (width at half intensity) can be estimated to be 220 nm, while the resolution along the optical axis is about 730 nm. The size of the detection pinhole used during this measurement was, backprojected in the specimen plane, 400 nm. To put these values in perspective, we estimate the dimensions of the illumination spot using relations which are strictly speaking only true in the paraxial approximation; the actual dimensions at high N.A. may be expected to be somewhat higher. The width of the illumination distribution transverse to the optical axis is about equal to the value of the diffraction-limited resolution R of a lens system.

$$R = 0.61 . \lambda_0/N.A. \tag{1}$$

To estimate the width of the illumination distribution along the optical axis we can use

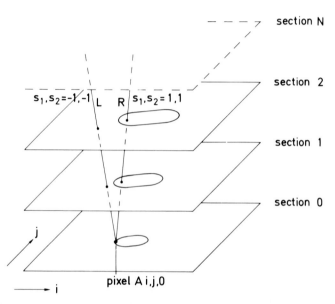

FIGURE 2. In the algorithm for generation of stereoscopic image pairs we assign the maximum value of the pixel values encountered along "viewing" line L and R to the corresponding pixels in the stereoscopic image pair, *i.e.* a pixed value $L_{i,j}$ (i,j: 0..255) will be the maximum of the pixel values $A_{i+s_1k, j+s_2k,k}$ (k: 0..N–1, N number of optical sections) where $A_{i,j,k}$ denotes a pixel of optical section k with coordinates i,j. In cases where s_1 of s_2 are not integers, the compared value is the result of a linear interpolation between the pixels nearest to the coordinate $(i+s_1k, j+s_2k)$.

the relations (adapted from REFERENCE 9):

$$I (u) = ((\sin u)/u)^2/I (0) \text{ with } u = \frac{\pi (N.A.)^2}{2 n \lambda_0} z \tag{2}$$

(z: coordinate along optical axis, n: refractive index immersion medium, λ_0: vacuum wave length light used). We can calculate from Equations 1 and 2 for $\lambda_0 = 483$ nm and N.A. = 1.3 a transverse width of the illumination distribution of 227 nm and along the optical axis a width of 763 nm. We may therefore conclude that the observed

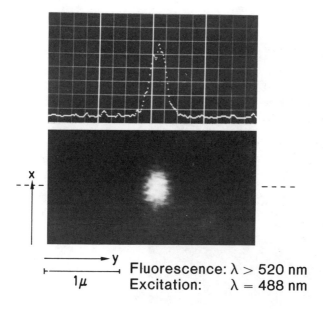

Fluorescence: $\lambda > 520$ nm
Excitation: $\lambda = 488$ nm

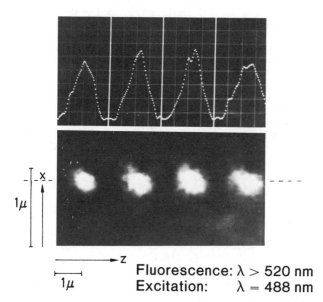

Fluorescence: $\lambda > 520$ nm
Excitation: $\lambda = 488$ nm

FIGURE 3. CSLM instrument resolution in fluorescence as determined by a set of computer-stored images of a 90-nm fluorescent bead taken at mutual distances of 100 nm. Excitation was at $\lambda = 488$ nm and fluorescence light detection at wave lengths above 520 nm. *Top:* Transverse or X-Y resolution (width at half intensity) can be estimated to be 220 nm. *Bottom:* From the set of through-focus images in the frame store 4 X-Z images are generated at Y positions close together within the 90-nm bead. These 4 images were set side by side by the computer in the image presented, and from the line plots through these images a Z-resolution of about 730 nm can be inferred.

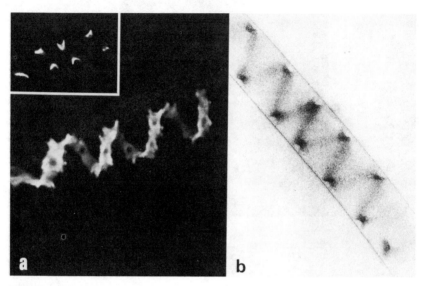

FIGURE 4. Chloroplast of a living green alga *Spirogyra* in water. (a) CSLM autofluorescence image. Excitation illumination λ = 483 nm. Image generated by the algorithm described in FIGURE 3. *Inset* shows one section of the image set during this reconstruction. Image is taken with a 100X immersion objective of N.A. = 1.3. Note the large image field made possible by the mechanical scanning approach. (b) Comparison image as obtained by normal transmission microscopy (N.A. = 0.65). *Bar* = 20 μm.

fluorescent point resolution in our instrument is determined by the illuminating *excitation* radiation and not by the emerging fluorescent radiation. From the fluorescent imaging we would have expected a transverse resolution of about 400 nm, as determined by the size of the backprojected pinhole.

APPLICATION TO BIOLOGICAL SPECIMENS

We present some of the results acquired in our microscope which are chosen to illustrate its capabilities. FIGURE 4(a) shows an image of a living green alga as produced by the above-described three-dimensional procedure. We present just *one* image (with $s_1 = s_2 = 0$) to show that by addition of the images acquired at the different height positions, which have a very short depth of field (see inset), we can in fact produce a high-resolution image with a *very large depth of field*. An image from normal transmission microscopy is also given for comparison (FIGURE 4b). Also note the large field of view in this image acquired in our microscope operating in this and all subsequent images with a N.A. = 1.3 100X oil immersion objective.

The cytoskeleton of a rat hepatoma cell stained with antibodies directed specifically against the keratin is depicted in FIGURE 5. The stereoscopic images clearly show the cytoskeleton with the location of the unstained nucleus visible as a black hole inside this structure, illustrating the point that specific biochemical substances can be localized and visualized in three-dimensions. As a third example we show in FIGURE 6 the mitochondrial network inside a budding yeast cell. This high-resolution image, with image-structural details going down to about 200 nm, presents information which otherwise could only be obtained through normal sectioning techniques (with subsequent microscopy, either light or electron). The specimen imaged is still alive and we expect to be able to follow such a specimen during its subsequent development in time. Finally, we show in FIGURE 7 in the CSLM image of a spore how, by simply tuning the spectrographic element, one can select either a reflection or two different fluorescence images, representing cell components fluorescing at different wavelengths.

COMMENTS AND CONCLUSIONS

We have presented above the mode of operation and some of the results of our scanning microscope based on *mechanical* scanning of the specimen through the focal point. This type of scanning, which we may call *on-axis,* is to be distinguished from other types, which we call *off-axis* and which are characterized by the fact that the information-collecting light beam is moved over the stationary specimen. Image points are addressed which are located away from the optical axis of the instrument, hence the name. The latter type of microscope occurs in two forms: in the first one a fine spot, derived from a laser light source, is scanned rapidly over the specimen in the transverse directions by a mirror (or a rapidly rotating polygon mirror system) placed in the illumination path. Confocal conditions can in principle be realized if the emerging light signal traverses the same light path. The second form is the tandem reflection scanning light microscope (TRSLM) where actually an array of mercury arc lamp-illuminated

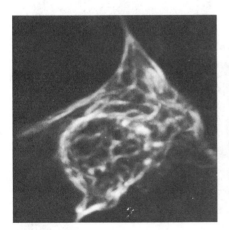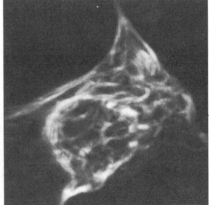

FIGURE 5. The cytoskeleton of a rat hepatoma (cancer) cell stained with fluorescently labeled antibodies directed against one of the constituent proteins (keratin). *Bar* = 10 μm. Excitation λ = 483 nm. Fluorescence detected λ > 510 nm.

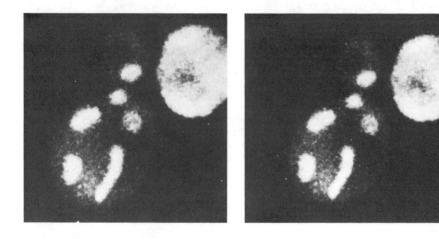

FIGURE 6. Living cell of the budding yeast *Saccharomyces cerevisiae* stained with the fluorescent dye DASPMI in order to reveal the structure of the mitochondrial network. DASPMI is a vital dye, which can be taken up in a living cell without causing this cell to stop functioning. *Bar* = 3 μm. Excitation λ = 476 nm. Fluorescence detected λ > 510 nm.

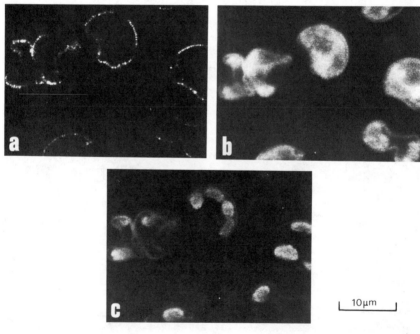

FIGURE 7. Images of one optical section through a field of *Polytrichum commune* spores taken as a function of wavelength with the computer-controlled spectrograph. Excitation wavelength λ = 483 nm. (**a**) Reflection image detection at λ = 483 nm. (**b**) Fluorescence image detection at λ = 550 nm. (**c**) Idem detection at λ = 620 nm. All images bandwidth λ = 30 nm. (a) shows the outer shell of the spore in reflection, (b) the green-yellow fluorescing cytoplasm and (c) the red fluorescent chloroplast organelles inside the spores.

holes in a rotating plate (Nipkov disc) is imaged on the specimen and the emerging signal is imaged on the image plane after passing a corresponding set of holes.[10] The resolution to be achieved in both types of off-axis scanning microscopes depends basically on two factors; the size of pinholes and the quality of off-axis imaging of the lenses used.

In laser-based systems (both on-axis and off-axis) sufficient light is available to use very small pinholes, so that the illumination distribution is set by the diffraction limitations of the lenses used. The detection pinhole can in principle be made sufficiently small so that the available confocal resolution gain can be realized.[3] However, for reasons of light efficiency, often a somewhat larger pinhole is used. In this paper we show in fact that in our instrument of the two possible basic imaging improvements in confocal fluorescence microscopy—(a) resolution determined by shorter wavelength excitation radiation and (b) the full confocal resolution available when very small pinholes are used—at least the first one has been realized. In the TRSLM, however, due to the size of the holes used in Nipkov disc, the illumination distribution is not diffraction-limited. The limitation on the hole size is connected with possibilities of manufacturing precision and the requirement that sufficient light be available. The result is that in this scanning approach it will be very difficult to realize conditions in practice where, apart from the sectioning effect, a basic imaging improvement with respect to normal microscopy is present.

A fundamental advantage of the on-axis technique is that, due to the mechanical scan, for each point in the image the imaging conditions are identical. This is an aspect which is very important for image processing, because one is able to use for the various operations algorithms which are not dependent on pixel position. The mechanical scan may prove limiting for the acquisition of images from large scan fields at high image repetition rates. For small image frames with a reduced number of scan lines spaced at distances of the system resolution, the image repetition rate can, however, be made rather high. For instance, 5 to 10 full resolution images per second of a field of 10×10 μm seem to be well realizable. In on-axis scanning microscopy the aspect of aberrations introduced by the imaging of object points away from the optical axis does not play a role. For the same reason the optics used can be considerably simpler (fewer lens elements, higher transparency). In addition, in on-axis microscopy the image fields are not limited by the field of view of the optics used, but can be much higher, being determined by the amplitude of the mechanical scan. All the laser systems, either on-axis or off-axis, have in common that the laser spot, if of sufficient power, can be used after positioning on a specific point in a live specimen to interact actively with this specimen.

In conclusion we may say that we have realized in the microscope of the on-axis type presented here the capability of generating two- and three-dimensional images at high resolution. We believe that this capability will be invaluable for the study of live biological specimens, especially during their development, not only because of the images generated, but also because the processes observed can be described quantitatively.

ACKNOWLEDGMENTS

We would like to thank F.H.A. Aalst, J.A.W. Kalwij, O. Peters and W. Takkenberg for their technical assistance and V. Sarafis for providing the *Polytrichum commune* specimen.

REFERENCES

1. WOLDRINGH, C. L., M. DE JONG, W. VAN DEN BERG & L. KOPPES. 1976. Morphological analysis of the division cycle of two *Escherichia coli* substrains during slow growth. J. Bacteriol. **131:** 270–279.
2. LEMONS, R. A. & C. F. QUATE. 1970. Acoustic microscopy-scanning version. Appl. Phys. Lett. **24:** 163–165.
3. BRAKENHOFF, G. J., P. BLOM & P. BARENDS. 1979. Confocal scanning light microscopy with high aperture immersion lenses. J. Microsc. **117:** 219–232.
4. COX, I. J., C. J. R. SHEPPARD & T. WILSON. 1982. Super resolution by confocal fluorescent microscopy. Optik **66:** 391–396.
5. VAN DER VOORT, H. T. M., G. J. BRAKENHOFF, J. A. C. VALKENBURG & N. NANNINGA. 1985. Design and use of a computer-controlled confocal microscope. Scanning **7:** 66–78.
6. WIJNAENDTS VAN RESANDT, R. W., H. J. B. MARSMAN, R. KAPLAN, J. DAVOURT, E. H. K. STELZER & R. STRICKER. 1985. Optical fluorescence microscopy in three dimensions: microtomoscopy. J. Microsc. **138:** 29–34.
7. BRAKENHOFF, G. J., H. T. M. VAN DER VOORT, E. A. VAN SPRONSEN, W. A. M. LINNEMANS & N. NANNINGA. 1985. Three-dimensional chromatin distribution in neuroblastoma nuclei shown by confocal scanning laser microscopy. Nature **317:** 748–749.
8. SHEPPARD, C. J. R. & A. CHOUDHURY. 1977. Image formation in the scanning microscope. Opt. Acta **24:** 1051–1073.
9. BORNE, M. & E. WOLF. 1975. Principles of Optics. 441. Pergamon Press. Oxford.
10. PETRAN, M., N. HADRAWSKY & A. BOYDE. 1985. The tandem scanning reflecting light microscope. Scanning **7:** 97–108.

DISCUSSION OF THE PAPER

F. SACHS (*State University of New York, Buffalo, New York*): How do you do the scanning?

BRAKENHOFF: Very simply. We just have a scan table which is driven from this side and the other side. The scan table just rests on a ruby, which scans over a hard metal plate. There is a small spring that keeps it to the hard metal plate.

SACHS: What's the pusher?

BRAKENHOFF: This is just something derived from the voice coil of a loud speaker. There's a sensor, of course, on the movement, and you have a feedback system.

D. A. AGARD (*University of California Medical Center, San Francisco, California*): How many points per second?

BRAKENHOFF: The computer images were acquired at 256 points per scan line and it takes 0.01 of a second to complete a scan line, so that gives an indication. The scan table is scanned at 50 Hz but you only use the movement in one direction, hence the 0.01 second. When you photograph the image directly during the actual scan, the resolution is determined by the CRT (cathode ray tube) in which it is displayed. Incidentally, by varying the scan frequencies and the number of scan lines per image, you can push the microscope to 5 or 10 images per second with small image fields like $10 \times 10 \ \mu M$, and have within these images still the full resolution of the confocal microscope.

A. BOYDE (*University College London, London, England*): The whole family of confocal microscopes has allowed us to overcome a serious problem with light microscopy, that you can't get a high-resolution stereo view with a conventional optical microscope, but it is possible with any of the confocal microscopes that have been discussed.

R. LOMBARDI (*University of Pittsburgh, Pittsburgh, Pennsylvania*): How far into the mouse cerebellum can you go?

BOYDE: That was a 90-μm fixed slide, so you could go all the way through it. And I think in that kind of preparation we could go to 250 μm easily. But one of the limitations which we've forgotten to point out is that one is limited by the free working distance of the objective that may allow you only a few μm after a 170-μm cover slip. So you have to use specially fabricated objectives with a longer free working distance in order to exploit this technique.

A. P. SOMLYO (*University of Pennsylvania School of Medicine, Philadelphia, Pennsylvania*): What is the best resolution that you have been able to demonstrate in your microscope or one of these microscopes?

BOYDE: Apropos of resolution we haven't used any rigid test, but we are sure that we have a resolution which is marginally better than when we use the same objective on a conventional microscope. However, one has to address the fundamental philosophical question of what is resolution. Resolution actually involves being able to see something. With our microscope you can see what you can't see when you look with the same objective on another microscope. And that's true for all the confocal microscopes. Resolution is inextricably bound up with contrast and brightness, both of which are very high in the narrow-focused plane of all the confocal microscopes; therefore, resolution is improved.

E. KELLENBERGER (*University of Basel, Basel, Switzerland*): I actually obtained in Zurich resolution which is definitely better than in the usual conventional microscope, a factor of two or something like that.

BRAKENHOFF: The terms "improved imaging" and "better" are used in conventional microscopy. We should be a bit more objective about things. We should come to an agreement to demonstrate imaging, for instance, on just the point object. One of those small fluorescent beads is quite useful for demonstrating the quality of imaging. You can measure them objectively. You can also describe the imaging of point objects very well theoretically and relate your results to theoretical expectations. With respect to what Dr. Petran said, I greatly believe that you see more in his microscope due to the fact that the point sources from his light are a bit larger than is ideal for confocal imaging. I have the feeling that in your case, the transverse resolution will be about the same as in a conventional microscope, only you see much more because you have your sectioning effect. So what it finally comes down to is volume resolution. And that is terrifically improved, so we should actually be considerably more exact about our definition of what type of resolution we are talking about. When we describe it, we should have test objects which we agree upon.

BOYDE: Yes, I agree with you partly. What we are actually observing in our test objects is not the resolution, as it is defined by the founders of theoretical optics. Higher contrast improves the resolution impression, but actually not the resolution as defined in theoretical optics. So it's actually very difficult to choose the quality of your microscope only from the theoretical optics consideration of the so called resolution defined by Abbe.

Confocal Light Microscopy

T. WILSON

Department of Engineering Science
University of Oxford
Oxford OX1 3PJ, England

INTRODUCTION

The conventional optical microscope is still widely used in many branches of science and technology despite the large number of electron microscopes that are presently available. In many cases it is the absence of a vacuum system and its nondestructiveness which makes optical microscopy attractive. This allows the optical instrument to be used with a wide variety of specimens and materials. The principal limitations of conventional optical microscopy are probably resolution, depth of focus and contrast. These may all be improved by using scanning techniques and especially by using the confocal optical system.[1]

A confocal scanning optical microscope is shown in FIGURE 1. The essential elements of the system are mechanical object scanning, *point* detection and electronic image processing. Of these it is the point detector which gives rise to the confocal imaging. FIGURE 2 shows a typical medium-resolution confocal image which also indicates that mechanical scanning results in images with good linearity and lack of distortion.

IMAGE FORMATION

If we now take a look at the way in which the image is formed in a conventional microscope, FIGURE 3(a), we see that the object is illuminated by a patch of light from an extended source through a condenser lens, imaged by the objective lens and finally viewed through an eyepiece. In this case, of course, the resolution is due primarily to the objective lens. A scanning microscope could then be built using this arrangement, by scanning a point detector through the image plane so that it detects light from one small region of the image at a time, thus building up a picture of the object point by point, FIGURE 3(b). We can take advantage of the principle of reciprocity and build the equivalent microscope of FIGURE 3(c). The point is now provided by using a laser as a light source and a photodetector as the large area detector. This is the form of what we might call a conventional scanning optical microscope. The point source of light illuminates one very small region of the object, while the detector essentially measures the power collected by the second lens. The arrangement shown in FIGURE 3(d) is a combination of those shown in FIGURES 3(b) and 3(c). Here, the point source illuminates one very small region of the object, and the point detector detects light only from the same area. An image is built up by scanning the source and detector in synchronism.

In this configuration we see that both lenses play an equal part in the imaging. We might expect that as two *imaging* lenses are employed simultaneously to image the object, the resolution will be improved; fortunately this prediction is borne out both by calculation and by experiment. This, then, is the arrangement of the confocal scanning microscope. In practical arrangements, of course, it is often more convenient to scan the object rather than the source and detector together.

This resolution improvement may, at first, seem to contravene the basic limits of optical resolution. It may be explained using Fourier imaging theory[1] or by a principle due to Lukosz[2] which states that resolution may be improved at the expense of field of view. The optical system of the confocal microscope takes this principle to the limit; only one portion of the object is imaged by the optics and hence the system has high resolution, but a very small field of view. The field of view in the confocal microscope is increased by scanning. This is the real power of the method; the optical system trades field of view for resolution, but we get the field of view back by scanning.

The confocal system results in purely coherent imaging with an effective point spread function given by the product of those for the objective and collector lenses. Thus the image of an object of amplitude transmittance t is given by

$$I = |h_1 h_2 \otimes t|^2 \qquad (1)$$

where $\otimes$ denotes the convolution operation and $h_{1,2}$ are the point spread functions of the two lenses. The effective point spread function is thus sharper than that of a conventional microscope and with very weak outer rings. The coherent transfer function is therefore twice as wide as for conventional coherent systems. FIGURE 4 shows a comparison between conventional and confocal imaging.

PRACTICAL CONSIDERATIONS

In the designing of a scanning microscope one must decide both how to scan and how small the confocal pinhole should be. We can get a feel for the parameters involved by considering the image of a point object. Equation 1, based on an infinitely small pinhole, gives the image as

$$I = |h_1 h_2|^2 \qquad (2)$$

whereas an analysis with a finite, arbitrary detector, $D(x,y)$, gives the image as

$$I = |h_1|^2 \{D \otimes |h_2|^2\} \qquad (3)$$

Clearly we must therefore choose D to be as small as possible and for true confocal operation it must be small compared with the width of the central portion of $|h_2|^2$.

We are now left with the choice of how to scan. If speed considerations are important then beam scanning is an attractive alternative.[3] However, this does necessitate the use of a highly corrected objective lens. A price one pays for the confocal improvement in resolution is indicated in Equation 1 where we can see that both lenses (and hence both aberrations) contribute equally to the imaging. If we scan off-axis with a scanning beam then we may well have spatial variation in resolution and imaging across an image.

Alternatively we can elect to mechanically scan the object relative to a fixed, on-axis, imaging system.[1,4] This results in space invariant imaging which is ideal for subsequent computer processing. For large, bulky objects, objective scanning may also prove useful.[5]

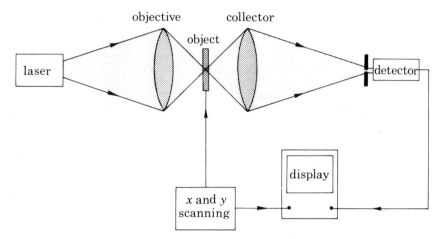

FIGURE 1. Schematic diagram of a mechanically scanned confocal microscopic.

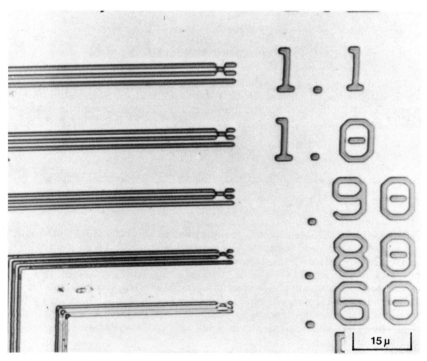

FIGURE 2. Confocal reflection image using a 0.85 numerical aperture objective of a test pattern consisting of narrow lines written in photoresist. A helium neon laser (0.6328 micron wavelength) was used as the light source.

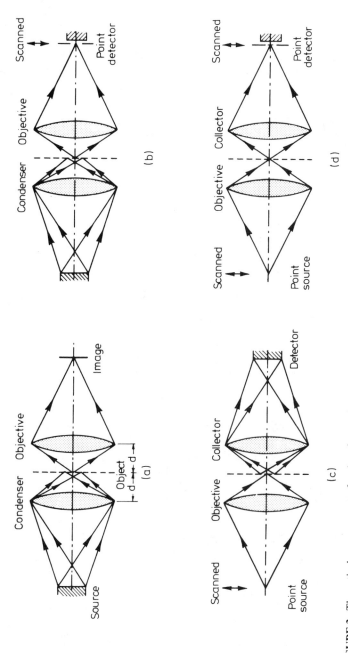

FIGURE 3. The optical arrangements of various forms of scanning optical microscope. (a) Conventional microscope, (b) type 1(a) microscope, (c) type 1(b) microscope, (d) confocal scanning microscope.

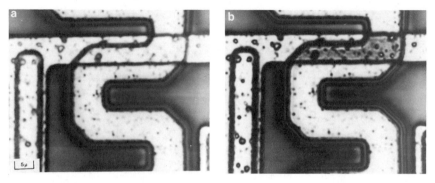

FIGURE 4. A comparison of (**a**) conventional and (**b**) confocal images of the same portion of a microcircuit.

OPTICAL SECTIONING

There are more advantages to be obtained from the use of a point detector in the confocal microscope in addition to the resolution improvement. It is an inevitable consequence, for example, that the microscope images detail only from parts of the objects in the focal plane. This depth discrimination property is not a feature of conventional microscopes and may easily be understood by considering the light from parts of the object outside the focal plane, FIGURE 5. Light from portions of the object outside the focal plane forms a defocussed spot in the detector plane. In a conventional microscope employing a large area photodetector all this light is detected whereas the use of a central point detector in the confocal scanning microscope results in a much weaker detected signal and so provides discrimination against detail outside the focal plane. The effect is illustrated in FIGURE 6 where we have deliberately mounted a planar microcircuit in the microscope with its normal at an angle to the optic axis. FIGURE 6(a) shows the conventional image: only one portion, running diagonally, is in focus. FIGURE 6(b) is the corresponding confocal image: here the discrimination against detail outside the focal plane is clear. The areas which were out of focus in FIGURE 6(a) have now been rejected. Furthermore the confocal image appears to be in

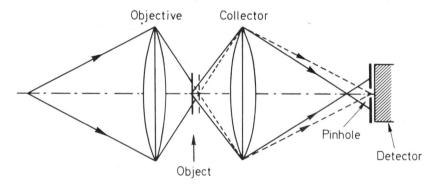

FIGURE 5. Depth discrimination in the confocal microscope. When the object is in the focal plane the reflected light is focussed on the pinhole. When the object is out of the focal plane, a defocussed spot is formed at the pinhole, and the measured intensity is greatly reduced.

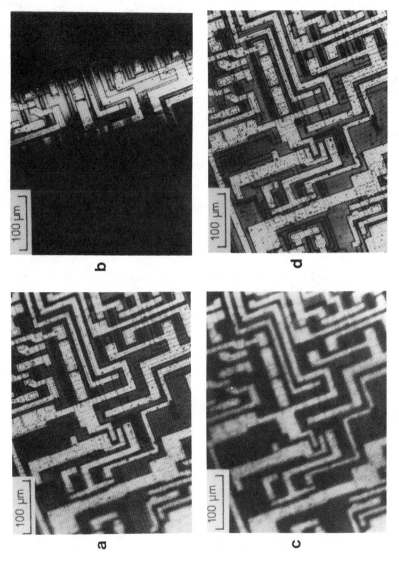

FIGURE 6. (a) Conventional scanning microscope image of a tilted microcircuit: the parts of the object outside the focal plane are blurred. (b) Confocal scanning microscope image of a tilted microcircuit: only the part of the specimen in the focal plane is imaged strongly. (c) Conventional scanning microscope image with axial scan. (d) Confocal scanning microscope image with axial scan.

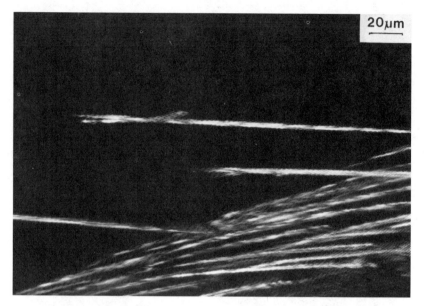

FIGURE 7. The hairs on an ant's leg. Extended focus image with a 0.85 numerical aperture objective and He-Ne laser light.

focus throughout the visible band. This is important as it means that any detail that is imaged efficiently in the confocal microscope will be in focus.

It is now possible to take advantage of this property and combine it with axial scanning to produce a high-resolution imaging instrument with a vastly increased depth of view. For example, if we had moved the specimen of FIGURES 6(a) and 6(b) axially, we would have obtained a corresponding pair of images, but with a different portion of the object brought into focus. Thus we can produce the image of a rough object such that all areas appear in focus by scanning the object in the axial direction

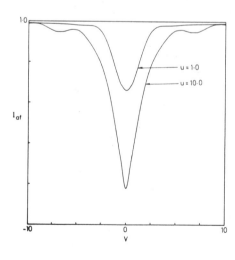

FIGURE 8. The auto-focus image of two phase steps of one and ten optical units each.

with an amplitude sufficiently large for every part of its surface to pass through the focal plane.

We have done this in FIGURES 6(c) and 6(d) for the microcircuit specimen. It is clear that the conventional microscope gives the expected blurred, out-of-focus image, whereas in the confocal mode only the in-focus detail has been accepted by the microscope, thereby producing an image which appears in focus across the entire field of view, FIGURE 6(d). This should be compared particularly with the conventional image of FIGURE 6(a).

The technique is very powerful as it permits the depth of focus to be extended, in principle without limit, while still maintaining high-resolution, diffraction-limited

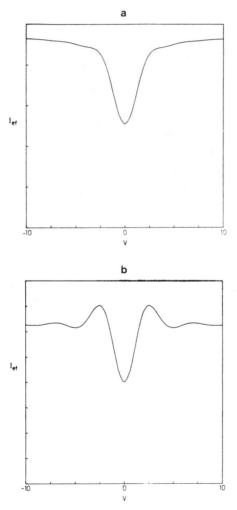

FIGURE 9. The extended focus image of two phase steps of (**a**) one and (**b**) ten optical units each.

imaging. We have already achieved an extension of more than two orders of magnitude in our laboratory. A high-resolution example is shown in FIGURE 7 where the hairs on an ant's leg, with two hairs projecting to the left, have been imaged. The axial distance between the tips of these hairs is 30 microns. The image shows both hairs excellently resolved along their full length, as well as much detail on the leg itself.

As an alternative to the extended focus technique we can form an auto-focus image by scanning the object axially and, instead of integrating, selecting focus at each picture point by recording the maximum in detected signal. The images are somewhat similar, although this method is more sensitive to vibration and noise than the (integrating) extended focus technique. Again substantial increases in depth of focus may be obtained. We present in FIGURES 8 and 9 theoretical extended focus and

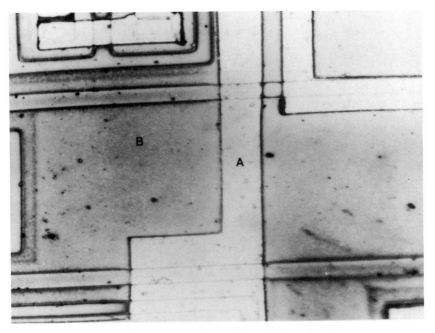

FIGURE 10. Area of microcircuit; (A) metal; (B) semiconductor.

auto-focus images of an object where there is a step in height, but not reflectivity, of $\lambda/2\pi$ (u = 1) and $\lambda/5\pi$ (u = 10) for a lens of unity numerical aperture.[6]

We can go one step further with the auto-focus technique and turn the microscope into a noncontacting surface profilometer. Here we simply measure the axial distances from the auto-focus image. We can demonstrate the technique on a similar microcircuit, FIGURE 10; the profile is presented in FIGURE 11. The shape of the strip is clearly defined and the difference between the surface texture of the metal and the surrounding semiconductor is apparent. The reflectivity of the metal and the semiconductor is also quite different, but it does not affect the height measuring system. This is predicted theoretically as also is the undershoot and overshoot at the edges, FIGURE 12.

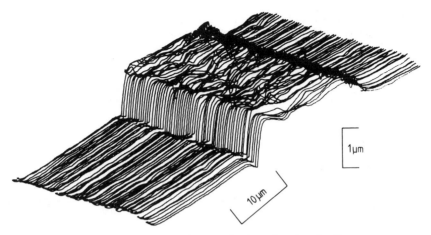

FIGURE 11. Profile of the metal strip on the microcircuit.

The optical sectioning is also the basis of several methods of producing stereo pairs. These are discussed in detail in REFERENCES 7–9.

FLUORESCENCE MICROSCOPY

The scanning microscope is particularly suited to fluorescence microscopy.[10] We can see this easily by considering the conventional scanning microscope where the field just behind the detector may be written $|h_1|^2 f$ where h_1 is the amplitude point spread function of the first lens whose size is inversely proportional to wavelength, (λ_1) and f represents the spatial distribution of fluorescent generation. The resolution in this case

FIGURE 12. The predicted height profile for a unity optical coordinate edge with a 0.5 numerical aperture lens. The reflectivities in case (**a**) were equal on both sides of the step and in (**b**) were in the ratio of one to two.

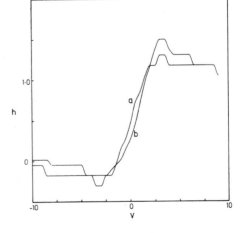

is clearly due primarily to λ_1 whereas in the conventional fluorescence microscope the resolution is due mainly to the fluorescence wavelength (λ_2). This is important because $\lambda_2 > \lambda_1$ means that the scanning microscope always gives better results.

The improvement in the confocal case is naturally greater and we find, for example, that, if we consider two point resolution in terms of the Rayleigh criterion, confocal systems can resolve points a factor of 1.8 closer than a conventional coherent system and a factor of 1.3 closer than a conventional incoherent system.

CONCLUSIONS

We have described the principles of confocal microscopy and several applications. It seems likely that a combination of the depth discrimination properties together with the advantages of scanning fluorescence microscopy will provide a very powerful biological tool.

REFERENCES

1. Wilson, T. & C. J. R. Sheppard. 1984. Theory and Practice of Scanning Optical Microscopy. Academic Press. London.
2. Lukosz, W. 1966. J. Opt. Soc. Am. **57**: 1190.
3. Wilke, V. 1985. Scanning **7**: 88–96.
4. Brakenhoff, G. J., P. Blom & P. Barends. 1979. J. Microsc. **117**: 219.
5. Hamilton, D. K. & T. Wilson. 1986. J. Phys. E **19**: 52.
6. Wilson, T. & A. R. Carlini. 1986. Optik, in press.
7. Brakenhoff, G. J., H. T. M. van der Voort, E. A. Spronsen, W. A. M. Linnemans & N. Nanninga. 1985. Nature **317**: 748.
8. Boyde, A. 1985. Science **230**: 1270.
9. Wijnaedts van Resandt, R. W., H. J. B. Marman, R. Kaplan, J. Davoust, E. H. R. Stelzer & R. Stricker. 1985. J. Microsc. **138**: 29.
10. Cox, I. J., C. J. R. Sheppard & T. Wilson. 1982. Optik **60**: 391.

DISCUSSION OF THE PAPER

W. W. Webb (*Cornell University, Ithaca, New York*): I want to issue a testimonial for confocal microscopy. Years ago when we started studying the diffusion of fluorescence-labeled molecules on cell surfaces, our first method involved counting the number of molecules by measuring the fluorescence intensity in a volume that is defined by confocal microscope geometry. But we didn't scan. So, since about 1970, for these fluorescence photo bleaching experiments and fluorescence correlation spectroscopy experiments we have always looked at a volume defined just as you described it. We found that we could consistently measure intensities in these volumes and see intensity fluctuations to a tenth of a percent, sometimes a hundredth of a percent.

D. Rudolph (*University of Göttingen, Göttingen, Federal Republic of Germany*): Did you ever try to apply two annuli into your microscope to gain resolution? Or is it only a proposal?

WILSON: No, we have not tried that. I think Dr. Brakenhoff has perhaps done more than we have. The property of two annuli is that the point spread function is very narrow. This is good for just one-point objects. When one begins to look at extended objects, one finds that it is not so good after all. However, if you use one annulus and one full lens, that does seem to be good. It turns out that the transfer function is beefed up at the higher frequencies.

F. SACHS (*State University of New York, Buffalo, New York*): The one thing that's been glossed over in these talks on scanning is how you scan well. Would you like to make some comments about how one does that quickly and gracefully?

WILSON: You come up with an idea. If this works, you are happy. Everyone else has an idea, and everyone seems to think his way is best. But basically, you just try until something works.

Applications of Tandem Scanning Reflected Light Microscopy and Three-Dimensional Imaging[a]

ALAN BOYDE

Department of Anatomy and Embryology
University College London
London WC1E 6BT, England

INTRODUCTION

The basic principles of the tandem scanning reflected light microscope (TSRLM) are that the incoming light is provided as a series of mini-beams which scan the field of view and illuminate corresponding patches in the plane of focus more intensely than out-of-focus layers. Reflected light from these patches is imaged on to apertures which are scanning in the focal plane of the eyepiece. The array of apertures which chop the illuminating beam is identical with the array on the observation side: both are scanned in tandem—hence the name Tandem Scanning Reflected LM. The scanning, illuminating apertures are imaged by the objective lens in the focussed-on plane. Reflected light from the patches in this plane is imaged back into corresponding apertures which also lie in the intermediate image plane of the objective. All other reflected light from out-of-focus layers and light scattered from optical surfaces in the microscope is intercepted by the solid portions of the aperture array or by light traps in the microscope head. The TSRLM is used like a conventional LM: one looks through the eyepiece at a steady image. Focussing up and down, the image changes rapidly, and one can construct a mental image of three-dimensional relationships. Contrast and brightness are high only for the in-focus layer. More details of how the microscope functions are given in the previous paper by Petran and Hadravsky *et al.* and in other accounts.[1-5] The TSRLM makes it possible to examine certain features of the structure of semitransparent specimens, up to a certain depth below the surface, by looking into the specimen from outside—and without the need to prepare, by fixation, embedding and sectioning. Preparation can be confined to "peeling" to expose the tissue of interest if it does not lie within the usable, visible range from the external surface, bathing in the appropriate saline or culture medium, and placing under the microscope. Contrast can be obtained from living tissues left in at least a more intact condition than would be possible for optical microscopic methods.

Compared with other methods of vital microscopy, there is no present rival. Petran *et al.*[6] have been able to examine any living tissue which interested them, and to do this with whole animal or whole organ preparations. The TSRLM image of unstained biological objects may appear to be similar to phase contrast and interference contrast images. However, the signal arises from refractive index differences at boundaries between constituents and not from the product of these differences and the optical path length, as in phase and interference contrast. Petran *et al.*[6] indicate that the contrast is in many cases as strong as in silver-impregnated specimens, though having the opposite sense, because this is a reflection microscope.

[a]Supported by grants from MRC and SERC.

428

The most important area of application which we foresaw for the TSRLM was in the examination of both very large (unmicroscopic!) and very dead tissue specimens—namely bones and fossil teeth. Coupled with the proposal to investigate dental fillings at close range, these were the main aims of the grant proposal for which we received the support of the Medical Research Council, which made it possible to build the microscope. To achieve this goal, we had to have established that we could read the internal microstructure of macrospecimens. These pilot studies, made with the Plzen prototype, already clearly showed that our aims were feasible: and these bone, enamel and dentistry projects have taken up most of our instrument time.[7]

FIGURE 1. TSM image of enamel prism cross sections in the superficial enamel of a molar tooth of *Australopithecus boisei*, 2 million years B.P., from East Africa (specimen loaned by Dr. L. Martin). Other than an apparently higher incidence of "prisms within prisms", the structure of this tissue is identical with modern human enamel, even to the inferred dimensions of the ameloblasts which secreted it. 100/1.30 Oil. Fieldwidth 90 µm.

APPLICATIONS

Normal Dental Tissues—Recent and Fossil

Enamel

Tooth enamel is the hardest mammalian tissue, and, as such, in some ways the hardest to study. However, it is also the least likely to change *post mortem,* and may be perfectly preserved in fossils. Details of the structure of enamel within the range of LM resolution allow it to be classified according to 1) the type of prism that it contains—specifically, the cross-sectional shape of the discontinuity in structure resulting from the shape of the secretory pole process of the formative cell: 2) the relative courses of the prisms which relate to the translatory movements of the formative cells: and 3) the circadian incremental growth features usually known as the enamel prism cross-striations or varicosities.[8,9] The cross-sectional shape of the enamel prisms differs in different groups of mammals in ways which make it possible to study taxonomic affinities and differences.[10–13] Such changes are slow to evolve, yet as a living fossil, it is the only tissue in which we could envisage studying evolutionary change at the micromorphological level. In the TSRLM, we have a method which can be used to examine structure in (both recent and) fossil teeth without any risk of damage to valuable specimens. Whole teeth, even in whole skulls and of large mammals can be studied without any form of preparation.

We have studied the enamel structure in members of all the major families of living primates[12] (FIGURE 1). By demonstrating that we can provide useful data by these nondestructive means, we have been able to obtain permission to study many fossil specimens in the custody of the British Museum (Natural History) (B.M.N.H.). It is easily possible, using this approach, to distinguish putative higher primate enamel from that belonging to other orders.

In collaboration with M. Fortelius (Helsinki), we have surveyed recent and fossil rhinoceroses and the enamel of other mammals which evolved some form of vertical prism decussation.[13] With K. Lester (Sydney) we have surveyed enamel structure in recent and fossil marsupials, in which another class of structure—the enamel tubule—is also frequent and can be used to provide a more complete description. P. Bertrand (Paris) has made a survey of recent and fossil elephants. Most of this work has been done on samples which no museum curator would currently dare to commit to destructive analysis, so that we can safely claim that the acquisition of the TSRLM made the study possible.

However, there are examples of its use in the study of conventionally sectioned tissue which are also important. For example, the existence of enamel prism cross-striations is disputed by some, who correctly point out that certain analogous appearances may be artefacts of the sectioning process, or of additive contrast effects

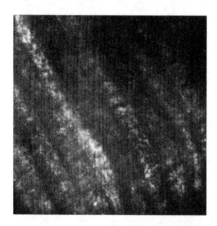

FIGURE 2. Longitudinally intercepted prisms and their cross-striations seen in an oblique transverse section of a human permanent canine tooth. 100/1.32 Oil. Fieldwidth 50 μm.

due to the thickness of a ground section. With the TSRLM, we can focus below any layer in which physical disturbance of the tissue structure might have been caused by sectioning, and that layer is so thin that its image is not disturbed by out-of-focus information. Now it is easily possible to see the real cross-striations (FIGURE 2) where they occur, and to measure the repeat interval to determine the rate of formation of the tissue. It is not necessary to prepare a section—just to cut the tooth once, or to exploit natural, longitudinal breaks. If a cut is permitted, the analysis can be extended by the use of more prominent, longer interval, growth-layer lines to provide an estimate of the age of young mammals.[8] The importance of this information in the analysis of fossil hominid growth patterns has recently been indicated by Bromage and Dean.[14]

Dentine

Dentine structural features are often well preserved in fossils, where the tissue is exposed in worn occlusal facets and in incidental breaks. Comparatively little work has

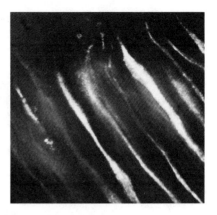

FIGURE 3. Dentine tubules terminating in pointed swellings at the enamel-dentine junction in a fossil equid, *Anchitherium aurelianense.* Cut surface of ectoloph. (Specimen loaned by Dr. M. Fortelius.) 100/1.25 Oil. Fieldwidth 90 μm.

been done to indicate the likely importance of this type of information in studying mammalian evolution, but we have performed enough TSRLM studies to be confident that the limit will not be imposed by the lack of a suitable nondestructive technique (FIGURE 3).

Of particular interest to us, in the TSRLM study of dentine, has been the property of this microscope both to be able to provide a high contrast image of a cut surface of this poorly reflective tissue, while at the same time being able to image the internal structure in layers close to those altered by cutting. Work in our laboratory routinely uses slices of sperm whale dentine as a surrogate bone in studies of the resorptive capabilities of osteoclasts.[15,16] Resorption lacunae in the cut surface—and some details of the smearing caused by cutting—are imaged with high contrast. Images showing the detail of the bottoms of these resorption pits, which are only a few microns deep, show the tubular structure of the bulk tissue in unaffected tissue at the same depth (FIGURE 4).

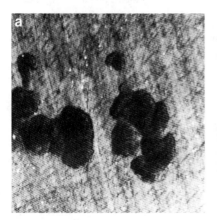
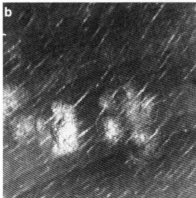

FIGURE 4. Sperm whale dentine, cut surface resorbed by rabbit osteoclasts. (**a**) is focussed at the surface. (**b**) is focussed 6 μm deeper, showing the dentine tubules and reflections from the deeper parts of the resorption lacunae. (Specimen prepared by Dr. Sheila Jones.) 40/1.0 Oil. Fieldwidth 165 μm.

Interfaces between Biomaterials and Bone or Tooth Substance

Materials such as dental fillings and bone cements present real problems when one studies the interface with the biological tissue. In spite of the best attempts to match the physical properties of the man-made materials to the natural tissues, the differences in properties result in the separation of the two in preparing sections for microscopy. The TSRLM enables us to overcome these problems since we only need to be able to look at the interface in an intact excised block of tissue. In certain circumstances the block of tissue does not need to be excised from the living animal.

In the operative removal of enamel caries (FIGURE 5), much importance is now attached to the significance of the smear layer which results from cavity preparation with rotary cutting instruments. Some proponents of some techniques wish to remove this layer by an acid etching procedure. Although only of the order of 1-2 μm thick, we can easily verify the removal of this layer with the TSRLM.[17]

The removal of smear is of importance in bonding the new composite (monomer polymer, plus a filler) restorative materials to dentine. Here we can study the extent of the penetration of the resin monomer into dentine: utilising both the different refractive index and associated reflectance, and fluorescent properties of the monomer, the latter being enhanced by adding suitable fluorophores.[18] The composite resin restorative technique usually involves the intermediary of a bonding agent which, applied as a thin, nonviscous layer before setting, combines chemically with both dentine matrix collagen and copolymerises with the composite resin filler proper. The use of fluorescent labels in one of these agents has allowed us to show how far it mixes into the monomer phase of the composite resin before setting polymerisation occurs. The TSRLM is admirably suited to examining the internal structure of composites, allowing the study of the distribution and packing of the filler particles owing to their difference in refractive index from the resin phase.

In the same way, it is possible to examine the structure and the development of the bond of resin to acid-etched enamel in simulated cavities prepared in freshly extracted teeth. We can study the survival of this interfacial bond under conditions of storage in water, or of applied mechanical stress, to simulate what would happen in the functional use of the tooth. But, even more of direct interest at the present, is that we are able to observe the enamel structure as it is being prepared by the dentist's drill.

My colleague T.F. Watson has elaborated an experimental arrangement whereby a tungsten carbide bur, used at moderately high speeds to cut a cavity in enamel, can be

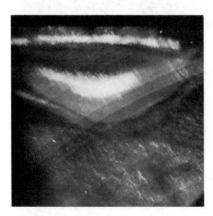

FIGURE 5. Natural carious lesion below the interproximal contact area in a longitudinally cut human premolar. 25/0.65 Oil. Fieldwidth 260 μm.

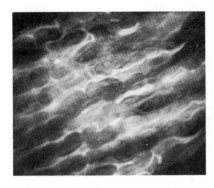

FIGURE 6. Rat femur growth plate cartilge, embedded in polymethylmethacrylate and sectioned longitudinally through the hypertrophic and calcifying zones. (Specimen prepared by Dr. Leroy Klein.) 25/0.8 Glycerine. Fieldwidth 260 μm.

seen end-on, whilst the enamel prism structure being demolished moves towards it. These interactions are videotaped using a conventional television camera, and analysed on playback. These studies have raised the question of how fast we can go in such real-time recording. This answer has not yet been provided or calculated, but we can say, qualitatively, that it must be at least one order of magnitude faster than at present—since we could drive the aperture disc in the TSRLM much faster than is at present the case.

Cementum

Acid etching of root surfaces occurs both accidentally and deliberately in clinical practice. It is possible to determine the depth of the demineralisation of cementum directly using the tandem scanning reflected light microscope.[19] From limited observation so far it would seem that roughly 10 microns of cementum would be demineralised during a 1-minute acid-etch period using 37% phosphoric acid.

The Study of Skeletal Tissues

The study of bone(s) was a third area in which we were sure that the TSRLM would have a major role to play in hard tissue research. Again, we have the choice whether to examine live or dead tissue: we have mainly done the latter.

Cartilage

We have shown that the TSRLM can be used to study both fresh articular cartilage from the surface and cut surfaces of growth cartilage: in joint cartilage, we can show that the native, wet surface is very smooth. Martin Müller (Eidgerössische Technische Hochschule, Zurich) has made more extensive studies of the growth plate cartilage involved in endochondral ossification in freshly removed and longitudinally sliced rat long bones. The method is well suited to visualising the origin and growth of calcospherites—the matrix-vesicle-centred, apatite crystal clusters—which spread through the mineralisable cartilage matrix. The TSRLM can also be used to record through-focus series of images of the chondrocytes at all levels through the growth plate in order to study the dynamics of cell growth in this tissue. (FIGURE 6).

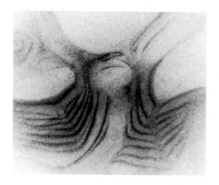

FIGURE 7. Rat tibia cortical bone cut longitudinally. Tetracycline was administered by injection at weekly intervals with one two-week gap.
Print from Kodachrome transparency showing
yellow fluorescent lines as black in this image.
(Specimen prepared by N. Frootko and S.A.
Reid.) 20/0.8 Glycerine. Fieldwidth 310 μm.

Bone and Stereology

We have used the TSRLM to study both the three-dimensional distribution of
osteocytes in bone[20] and to demonstrate details of fine structure in the walls of these
features which were previously only recognised in scanning electron microscopic
(SEM) images of cut, or fractured-open, specimens. In the former, we have explicitly
recognised the merits of this microscope in stereology—defining this as the science and
art of reconstructing three-dimensional information from two-dimensional sections.
Obviously, the TSRLM gives the important advantage that serial images are already
exactly stacked if obtained by through-focussing. We can obtain not only global
three-dimensional distribution, for example, the number of osteocytes per unit volume
of bone, but much more specific information, such as nearest-neighbour distances and
orientation in three-dimensional space.

Bone Growth and Turnover Studies

Knowledge of the remodelling process in human bones is acquired through the
administration of double or multiple tetracycline labels prior to the taking of a biopsy.
In the normal procedure the biopsy is embedded in polymethylmethacrylate (PMMA)
and thin sections are taken for fluorescence light microscopy. Considerable problems
arise in the reconstruction of the three-dimensional details owing to the fact that the
bone has to be cut undecalcified and such sections are usually seriously distorted; the
bone trabeculae are cut at unknown angles, and the sections are separated from one
another, thus making three-dimensional reconstruction time-consuming and unreli-
able.

In studies to date[21] we have shown that tetracycline lines (yellow fluorescence
under UV illumination) can be imaged at considerable depths from the cut polished
block face, allowing us to make three-dimensional reconstructions of growth and
remodelling surfaces using the TSRLM[7] (FIGURE 7): this relates to parallel work in
which we are studying the three-dimensional distribution of bone as a function of its
degree of mineralisation, which relates to age, and to the remodelling process.

Mechanical Aspects of Bone and Articular Cartilage

The TSRLM is unique in that cell spaces (*e.g.* osteocyte lacunae—FIGURE 7) or
stained cells within semitransparent solids such as bone and cartilage may be
visualized whilst the tissue is maintained wet and, if necessary, alive.[22] In the same way

that we experimentally deform teeth, bone and cartilage could be examined under changing mechanical (or osmotic) deformation and the alteration in the cell space and cell shape and volume discovered both within the normal functional range and at failure.

Stereoscopic Tandem Scanning Reflected Light Microscopy

The TSRLM allows the formation of the image of a very thin focal plane deep to the surface of an intact block of translucent material under investigation. No other reflected light microscope giving a real-time image can perform this function. In order to obtain three-dimensional information it is necessary to examine sets of three-dimensional images corresponding to different focal planes. The operator looking down the microscope can have a strong psychological impression of three-dimensional structure, because the focussing up and down can be done rapidly. However, many three-dimensional structures are too complex even to appreciate in this way, and it is not possible to convey the information to a third party.

In our first studies of three-dimensional organisation using this microscope, we recorded images at regular-focus intervals, reproduced these on to photographic plates and made a stack of the plates to reconstruct the object: stereo-pair images can be generated from such a stack.[7] This procedure is costly, time-consuming and not always satisfactory. The information contained in such a stack can be contained in two original images, which correspond to the two directions of view through a real three-dimensional stack.[23] Such images can be obtained directly by focussing whilst photographing. The basis of the procedure is simply as follows: the operator selects the specimen volume to be reconstructed by direct view down the microscope. An image corresponding to one direction of view through this selected volume is recorded whilst changing the mechanical focus of the microscope. The image is therefore an extended-focus view in that direction. The same slice of tissue is then photographed in the same way through the same vertical interval, but along an inclined vertical axis, corresponding to a different view through the volume under study. The two photographs are viewed as a stereoscopic pair to give the three-dimensional effect (FIGURES 8 and 9).

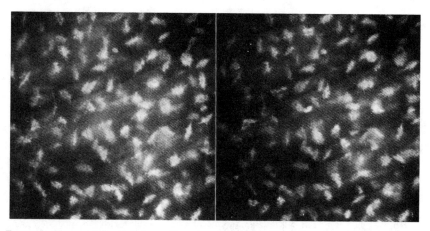

FIGURE 8. Stereo pair, angle between focussing directions = 8°, of rhesus monkey parietal bone, external surface of whole skull. 25/0.65 Oil. Fieldwidth 260 μm.

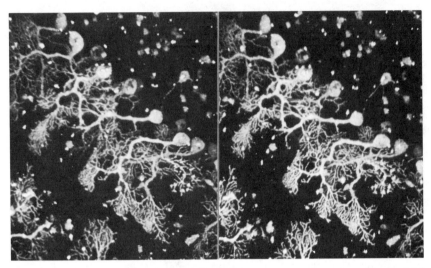

FIGURE 9. Stereo pair, angle 14°, Golgi preparation of mouse cerebellum (loaned by Prof. M. Berry), showing Purkinje cells. 25/0.65 Oil. Fieldwidth 260 μm.

Images recorded in this way have parallel projection geometry, with which we are familiar from SEM usage. We can therefore make exact reconstructions using equipment which we have previously developed.[24] A special stereocomparator transfers X, Y, Z positional information to a microcomputer which is used to calculate the desired parameters. The original purpose of this system was to measure the volume of bone resorbed by individual osteoclasts, using stereo-pair SEM images.[15]

This new stereo-imaging method depends upon and will only work with a TSRLM, because in this microscope only those features which lie in the instantaneous plane of focus contribute any information to the recorded image. Features which are out of focus are not registered, unlike the case in the normal optical microscope. Other confocal scanning optical microscopes[25-29] have the same property of depth discrimination but do not give a real-time image. In the scanned specimen, single beam type of confocal microscope, image reconstruction is necessarily done by computer, and it is not possible to view the specimen to gain an impression as to whether an area of interest is significant before making the three-dimensional reconstruction. In such confocal scanning light microscopes (CSLMs) with a unitary beam, it takes several seconds to form one frame of the image and successive focus levels can only be examined over a long period. In spite of the advantages in image analysis of doing everything by computer, therefore, we feel that the practical advantages of having a real-time image and making the reconstruction by optical techniques are overwhelming. A further significant advantage of the TSRLM is that it works with white light, so that colour effects are retained.

The work done to date with development of the real-time through-focussing stereo method has shown the necessity to avoid the vibrations introduced by an operator turning a mechanical focus control whilst recording a photograph. The through-focussing needs to be done under high-precision, remote control. The most precise systems are those operated by piezoelectric devices, which are also relatively simple to control by computer. The XZ movements can be applied either to the specimen or to the objective lens: the latter would make it simpler to use the microscope to examine bulky specimens, or which may move as a consequence of an experimental manipulation, *e.g.* in a loading test.

Silver-Stained and Fluorescent-Stained Preparations of the Nervous System

To test this new stereo-imaging method we have used Golgi stained prepartions of the central nervous system (CNS) (FIGURE 8), and other silver-stained preparations of the peripheral nervous system to produce high-resolution (by stereo light microscope standards) images of selected three-dimensional volumes within thick sections. The advantage of employing the new method in CNS research will be that analysis and reconstruction of neuronal ramifications will be greatly speeded up.

DISCUSSION

Our two years' experience with the TSRLM allows us to predict that it will be of interest to anyone who already uses light microscopy (LM) to examine surfaces, or who would like to use LM to examine surfaces at high resolution but cannot because the specimen is too translucent and the surface does not reflect enough light at normal incidence. It must interest those who need to look at the internal structure of a translucent material from the outside of a bulk specimen which may (or can) not be sectioned or damaged for any reason, or would like to image a thin layer in a thick sample (bulk, or thick section) and get rid of interfering information from out-of-focus planes (TABLE 1). It will be important in providing the means of making stereoscopic

TABLE 1. Research Areas Where TSRLM Has Already Proven Successful in Trials at University College London

Vital microscopy—Cytology
Zoology—Botany
Anatomy—Histology
Pathology—Histopathology
Physiology—Biophysics
Dermatology—Ophthamology
Neurology—microstructure of the central nervous system
Dentistry—dental materials—caries research
Orthopaedics—bone research—implant materials
Geology—micropalaeontology—fossils—Taxonomy—evolution
Anthropology—Archaeology
Museum curation and conservation
Materials Science
Semiconductor devices
Ceramics Technology
Chemistry—growth of crystals
Construction Materials
History of Art—Paintings—Pottery—Glass
Museum Studies—What is it made of?
Forgeries—Is it genuine?—Is it made of the same stuff?
Forensic Science—identification and typing of surface markings and bulk materials structure
Pulp and paper technology
Food processing industry
Cosmetics industry

images at high magnification, and of extracting three-dimensional information from rough surfaces.

Like the inventors—Petran and Hadravsky—and their original purpose, we have been impressed with the TSRLM as a vital microscope. However, we have also seen that it is a powerful tool in the realm of exploiting the classical, prepared and contrived LM specimen. Although the TSRLM makes it possible to examine structure in thick or uncontrasted specimens, one can see more if the specimen is optimally prepared—and so it is also with the TSRLM. It has a particular advantage in coping with thick sections, removing the need to destroy the three-dimensional continuity and contiguity of the sample. The preparative procedures that most need to be explored are those using reflective or fluorescent stains.

It is probable that the TSRLM will in future stand alongside the other established modes of LM even in the examination of small, prepared samples. The initial feeling of embarrassment of perhaps wasting the time of such a rare microscope in studying specimens which can be studied by the classical LM methods has now gone.

An important aspect of the TSRLM is that it opens up new areas for investigation, by making it possible to see microstructure in macrospecimens. Which of these two facets will be the more important in determining the areas of the expansion of its use will only be determined by practical experience.

It is difficult to show recorded images which really document the usefulness of the TSRLM, because these cannot fully demonstrate the importance of the third dimension in image interpretation. One needs to see the continuously changing image as one focusses up and down through a tissue to appreciate this point. In the case of live tissues—they move, and it is difficult to make the photographic exposure short enough. It is much better to see the live image of the live object. Video image enhancement may prove to be as important in TSRLM as in conventional LM, both allowing much lower contrast features to be seen and for them to be recorded on videotape.

As regards the stereo-imaging potential of the TSRLM, this is matched by the other confocal scanning optical microscopes.[25-29] However, the TSRLM has an advantage in that we see an image in real time and real colour, and the recording progress is direct: *i.e.*, as direct as the usual photomicrographic procedure which we use in everyday research. Furthermore, there are many interesting biological samples which could not be used in the scanned specimen microscopes[25,26,28,29] because they are too floppy or too massive.

ACKNOWLEDGEMENTS

I am grateful for the skilled technical assistance of Roy Radcliffe and Elaine Maconnachie and for help received from Tim Watson, Stephen Reid, Lawrence Martin, and Sheila Jones. All too numerous visitors have, for once, to be thanked for bringing us so many new and interesting challenges.

REFERENCES

1. PETRAN, M., M. HADRAVSKY, J, BENES & A. BOYDE. 1986. *In vivo* microscopy using the tandem scanning microscope. This volume.
2. EGGER, M. D. & M. PETRAN. 1967. New reflected-light microscope for viewing unstained brain and ganglion cells. Science **157:** 305–307.
3. PETRAN, M., M. HADRAVSKY, M. D. EGGER & R. GALAMBOS. 1968. Tandem scanning reflected light microscope. J. Opt. Soc. Am. **58:** 661–664.
4. PETRAN, M., M. HADRAVSKY & A. BOYDE. 1985. The tandem scanning reflected light microscope. Scanning **7:** 97–108.

5. PETRAN, M., M. HADRAVSKY, J. BENES, R. KUCERA & A. BOYDE. 1985. The tandem scanning reflected light microscope. Part 1—The principle, and its design. Proc. R. Microsc. Soc. **20**: 125–129.
6. PETRAN, M., M. HADRAVSKY, A. BOYDE & M. MÜLLER. 1986. Tandem scanning reflected light microscopy. *In* Science of Biological Specimen Preparation. 85–94. SEM, Inc. AMF O'Hare, IL.
7. BOYDE, A. 1985. The tandem scanning reflected light microscope. Part 2—Pre-MICRO 84 applications at UCL. Proc. R. Microsc. Soc. **20**: 130–139.
8. BOYDE, A. 1964. The structure and development of mammalian enamel. PhD. thesis. University of London.
9. BOYDE, A. Amelogenesis and the structure of enamel. *In* the Scientific Foundations of Dentistry. B. Cohen and I.R.H. Kramer, Eds. 335–352. Heinemann. London.
10. KORVENKONTIO, V. A. 1934–35. Mikroskopische Untersuchungen an Nagerincisiven inter Hinweis auf die Schmelzstruktur der Backenzahne. Histologisch-Phyletische Studie. Ann. Zool. Soc. Zool. Bot. Vanamo (Helsinki) **2**: 1–274.
11. BOYDE, A. & L. B. MARTIN. 1982. Enamel microstructure determination in hominoid and cercopithecoid primates. Anatomy and Embryology **165**: 193–212.
12. BOYDE, A. & L. MARTIN. 1984. A non-destructive survey of prism packing patterns in primate enamels. *In* Tooth Enamel IV. R. W. Fearnhead & S. Suga, Eds. 417–421. Elsevier. Amsterdam.
13. FORTELIUS, M. 1985. Ungulate cheek teeth: developmental, functional and evolutionary interrelations. Acta Zool. Fenn. **180**: 1–76.
14. BROMAGE, T. A. & C. M. DEAN. 1985. Re-evaluation of the age of death of immature fossil hominids. Nature **317**: 525–527.
15. JONES, S. J., A. BOYDE, N. N. ALI & E. MACONNACHIE. 1985. A review of bone cell and substratum interactions. Scanning **7**: 5–24.
16. BOYDE, A., N. N. ALI & S. J. JONES. 1985. Optical and scanning electron microscopy in the single osteoclast resorption assay. Scanning Electron Microsc. **3**: 1259–1271.
17. WATSON, T. F. & A. BOYDE. 1984. The tandem scanning reflected light microscope (TSRLM) in conservative dentistry. J. Dent. Res. **63**: 512.
18. BRANNSTROM, H., B. TORSTENSON & K. J. NORDENVALL. 1984. The initial gap around large composite restorations in vitro: the effect of etching enamel walls. J. Dent. Res. **63**: 681–684.
19. BOYDE, A. 1985. Anatomical considerations relating to tooth preparations. *In* Posterior Composite Resin Dental Restorative Materials. G. Van Herle & D. C. Smith, Eds. 377–403. 3M Co. St. Paul, MN.
20. HOWARD, V., S. A. REID, A. BADDELEY & A. BOYDE. 1985. Unbiased estimation of particle density in the tandem scanning reflected light microscope. J. Microsc. **138**: 203–212.
21. BOYDE, A. & S. A. REID. 1986. 3-D analysis of tetracycline fluorescence in bone by tandem scanning reflected light microscopy. Bone **7**: 148–149.
22. BOYDE, A., M. PETRAN & M. HADRAVSKY. 1983. Tandem scanning reflected light microscopy of internal features in whole bone and tooth samples. J. Microsc. **132**: 1–7.
23. BOYDE, A. 1985. Stereoscopic images in confocal (tandem scanning) microscopy. Science **230**: 1270–1272.
24. HOWELL, P.G.T. & A. BOYDE. Three-dimensional analysis of surfaces. *In* The Analysis of Organic and Biological Surfaces. P. Echlin, Ed. 325–349. Wiley. New York, NY.
25. COX, I. J. & C. J. R. SHEPPARD. 1983. Digital image processing of confocal images. Image and Vision Comput. **1**: 52–56.
26. WIJNAENDTS VAN RESANDT, R. W., H. J. B. MARSMAN, R. KAPLAN, J. DAVOUST, E. H. K. STELZER & R. STRICKER. 1985. Optical fluorescence microscopy in three dimensions: microtomoscopy. J. Microsc. **138**: 29–34.
27. CARLSSON, K., P. E. DANIELSSON, R. LENZ, A. LILJEBORG, L. MAJLOF & N. ASLUND. 1985. Three-dimensional microscopy using a confocal laser scanning microscope. Opt. Lett. **10**: 53–55.
28. WILSON, T. with C. SHEPPARD. 1984. Theory and Practice of Scanning Optical Microscopy. Academic Press. London.
29. VAN DER VOORT, H. T. M., G. F. BRAKENHOFF, J. A. C. VALKENBURG & N. NANNINGA. 1985. Design and use of a computer controlled confocal microscope for biological applications. Scanning **7**: 66–78.

In Vivo Microscopy Using the Tandem Scanning Microscope[a]

M. PETRAN,[b] M. HADRAVSKY,[b] J. BENES,[b] AND A. BOYDE[c]

[b]Department of Biophysics, Faculty of Medicine
Charles University
30166 Plzen, Czechoslovakia
and
[c]Department of Anatomy and Embryology
University College London
London WC1E 6BT, England

INTRODUCTION

The tandem scanning reflected light microscope (TSRLM) was invented to permit the examination of internal structure in live bulk tissue (*e.g.* brain, muscle, eye). Our TSRLM was the first of the confocal scanning optical microscopes to function. It is very simple. It needed to be simple, because we had to fabricate all the parts of the microscope ourselves.

The main problem in the microscopy of living specimens is to separate the light which carries information from that which does not carry any, because once an image has been formed one cannot separate these two components. Thus it must be a first principle in such a device that the light carrying information and that carrying disinformation (if we may coin such a word) must be separated at the very beginning. It is another elementary principle if one wants to use a microscope in physiology, that because the object will have a definite thickness, one must be able to observe the object from its surface using superficial illumination. Under these circumstances, one must allow the light to enter into much deeper portions of the object or specimen than those from which one might wish to form images. One therefore has a discrepancy between the volume of tissue which is illuminated and the volume from which one can form a sharp image, which corresponds to a thin layer close to the level at which the objective lens is focussed (FIGURE 1b). Light scattered in the illuminated volume above and below the focus plane will also enter the image in a conventional reflection microscope: this fraction of the light serves to overwhelm and destroy the image from the focussed-on plane. The illumination of the layers above and below the desired plane of focus can be limited by restricting the diameter of the field of illumination. It can be very easily demonstrated experimentally that contrast improves as the field of illumination is narrowed. However, the advantage of the improvement in contrast on restricting the field is countered by the reduction in the width of the field. Eventually, therefore, one gets a very small field of view with good contrast, but it has no connection to the neighbouring features in neighbouring fields.

This difficulty might be overcome by scanning. However, by illuminating only a

[a]The University College London tandem scanning reflected light microscope project was supported by the Medical Research Council. AB gratefully acknowledges travel funds from the Royal Society, and the British Council in the early stages of this work.

440

small field and then scanning it over the entire, broader field of view, one runs into another problem. In this case (which is the one considered in FIGURE 1a) the very small, very well lit, and very well focussed-on field in the object plane is surrounded with a halo deriving from the parts of the cones of light which are lying remote from the level of interest. Reconstructing the image by scanning results in the addition of all the halo light to that carrying information from the plane of focus and, therefore, in a reconstruction of the original condition: in which nothing can be seen in the plane of focus. This problem can be overcome by scanning a second time in the field of observation in the field of the image. We call this tandem scanning.

THE TANDEM SCANNING REFLECTED LIGHT MICROSCOPE

The kind of scanning which we need to use can in fact be achieved in several different ways.[1-9] We chose the simplest kind based on the development and miniaturisation of the so-called Nipkow disc or wheel invented one hundred years ago.[1-6] This device consists of a series of openings arranged as spirals in the peripheral portion

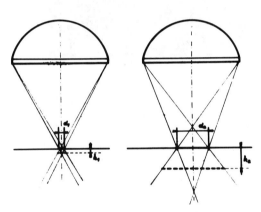

FIGURE 1. Diagrams showing the reduction in the volume illuminated due to reducing the diameter of the field of illumination.

of a thin disc. These openings constitute the apertures which limit the illumination of the sharply defined spots in the plane of focus in the specimen. In our case the design of the aperture disc has been considerably modified, so that openings in the opposite side of the disc also collect the light from exactly designated places in the object plane.

FIGURE 2 shows an area of our aperture disc, a circle in the middle of which would approximate the field of view in the intermediate focus plane of the objective lens of our reflected light microscope. The spots are actually beams of light coming through openings in the periphery of the aperture disc. The objective lens forms images of these spots in the focussed-on plane within the specimen.

The construction of the rest of the microscope is such that light from the illuminated spots in the object plane is directed to be focussed on corresponding, conjugate aperture holes on the opposite side, and equidistant from the centre, of the aperture disc. Under these circumstances a very high portion of the light which passed the second, observation side of the disc was scattered or reflected in the focussed-on plane. All other light scattered from planes above or below the plane of focus is intercepted by solid portions of the aperture disc.

FIGURE 2. Photograph in transmitted light of a portion of the periphery of a recent aperture disc.

The arrangement of the holes in the aperture disc is such that when the disc is rotated, the spots of light cover in succession the whole field of illumination and the whole field of view. Each single spot, as the disc turns, scans a single line. Since each spot has to be matched by a conjugate observation point on the image-forming side of the microscope, each has an exact conjugate partner on the same diameter and at the same radial distance from the center of the disc but on the opposite side of the disc. The

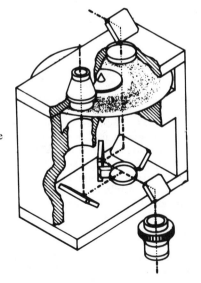

FIGURE 3. Diagram showing the construction of the head of the TSRLM.

main problem in our microscope then, is how to make the corresponding pairs of points—that is to say, all the points in the field of illumination and field of observation—correspond to each other. The disc has central symmetry. The pattern of holes on opposite sides of the disc are brought into congruency with mirrors. The arrangement of mirrors in the microscope head can be seen in FIGURE 3. One of these mirrors is semireflecting.

The schematic view of the microscope head in FIGURE 3 shows it as it actually is. The disc carrying the pattern of holes (shown in FIGURE 2) rotates in very high quality bearings. A final lens in the illuminating system lies immediately above the disc on the illuminating side. The mirror above the objective lens serves the purpose of directing the light downwards, assuming that the microscope will be used with immersion objectives and immersion fluids. Light returning from the specimen is deflected by the beam splitter and the final mirror to arrive at the observation side of the disc. The eyepiece in the microscope is of the Ramsdem type, focussed on the aperture disc in front of the eyepiece itself.

The transmission of the aperture disc is approximately one per cent. All light coming through the aperture holes and reflected in the object plane returns quantitatively through holes on the image side of the disc. All light which is reflected elsewhere, on optical components and outside the focussed-on level in the specimen, is suppressed

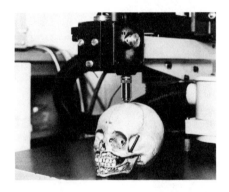

FIGURE 4. Photograph of the TSRLM at University College London being used to examine a skull.

to a very small fraction, which is, in fact, the transmission factor of the disc or approximately one per cent.

PRACTICAL FEATURES

Although the real microscope is contained in a box measuring $110 \times 120 \times 100$ mm, we have built recent models on to the end of an arm which also supports two light sources and which is motor driven up and down a column bolted onto a large, heavy base. The maximum specimen height which can be accommodated beneath the microscope configured in this way is 70 cm, making this a very versatile arrangement.

The microscope is brought roughly into focus on a specimen by using the motor and/or a manual, coarse adjustment which also raises and lowers the whole supporting arm. Two mm of fine focus are then available using a micrometer which alters the height of the objective and which is used to adjust the position of the optical section through the specimen.

The microscope is designed to accept all objective lenses having a standard Royal

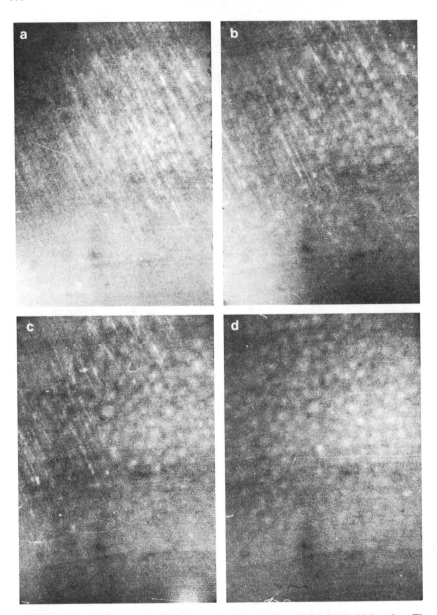

FIGURE 5. A through-focus sequence through the central part of a fresh chick retina. The figures were made in the period 1974–1976 with an aperture disc consisting of 12,000 holes and giving 300 scanning lines. Micrographs were taken using a 40/0.75 water immersion objective, Lomo Leningrad, a 10X eyepiece, and 16-mm format photography. The field width is in each case 152 μm. (**a**) is focussed in the superficial region of the nerve fibre layer. (**b–d**) show successively deeper layers of the retina when the ganglion cells become visible. (b) is 4 μm deeper than (a). (c) is 8 μm deep, (d) 16 μm deeper than (a).

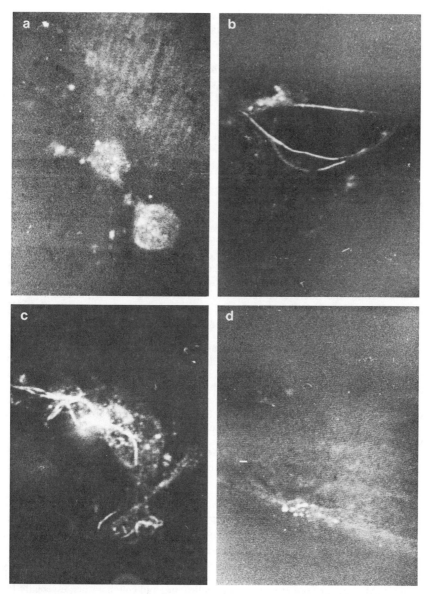

FIGURE 6. Various neural elements in the chick retina. The figures were made during the same period and to the same specifications as those in FIGURE 5. In (**a**) the retina is strongly tilted. In the *top right-hand corner* one can see the superficial nerve fibre layer and in the *lower part of the field* the ganglion cell layer, as seen through the overlying layer of fibres. (**b**) is taken from the periphery of the retina and shows connections of fibres to ganglion cells. (**c**), also from the periphery of the retina, shows a large ramified ganglion cell. In (**d**) the strongly reflecting and refracting features are pigment in the processes of rods and cones as seen through all the overlying layers of the chick retina.

Microscopical Society thread and 160 mm working distance. The particular choice of lens is obviously determined by the investigation being carried out. Dry lenses may be best suited to surface examination. For optical sectioning deep with respect to the specimen surface, immersion lenses are essential in order to reduce the effects of strong reflections at the surface. The choice of immersion medium is dictated by circumstances. Glycerin may be chosen for fluorescence work, and it is imperative to use water immersion lenses for the examination of live biological tissue.

After the light-scattering properties of the specimen, the main determinant of the depth to which optical sectioning can be achieved using the TSRLM is the lens which is used. High aperture oil immersion lenses usually have a working distance which accommodates a cover slip and little more. Thus, in the absence of coverslip, we can usually section down to 180 μm below the specimen surface before the objective and specimen come into physical contact. Lower numerical aperture lenses with longer working distances permit sectioning to a greater depth. The wavelength of the light being used also affects the sectioning depth, so that images can be obtained at greater depths with longer wavelengths: we are able to work into the near infra-red. For image recording, the TSRLM is equipped to accept either 35-mm or television cameras; either may be simply fitted over (or after) the eyepiece. The use of a television camera in conjunction with the TSRLM has some benefits. Firstly, images may be viewed on a monitor by several people at once. Also, they may be relayed to an image-analysing computer for image processing, or to a video tape recorder. In the absence of being able to demonstrate the microscope in operation, videotaped moving pictures showing continuous optical sectioning may be the best way of presenting results. Suitable television cameras can also be used for imaging in the ultra-violet or infra-red.

As will already be realised, the idea of our configuration of the TSRLM is that anything can be looked at, and that the surface layers of translucent objects can be looked into. In the former case, an image is only formed when part of a (reflective) surface is nearly in focus. This image shows much higher contrast than in a conventional reflected light microscope and a sharp reversal near focus.

With translucent specimens, the TSRLM allows us to pick out the image due to weak reflections at one particular depth. By focussing up and down one can trace structures in depth, but the images from each selected focus level are far more discrete than from a conventional light microscope.

FIGURE 4 is a photograph of the TSRLM at University College London. FIGURES 5 and 6 show typical applications of the prototype microscope in Plzen dating from the mid-1970s.

ACKNOWLEDGMENTS

We gratefully acknowledge the help given and interest shown by the Pro-Rector of the Charles University, E. Klika; the Dean of the Faculty of Medicine in Plzen, F. Macku; the Vice-Dean, P. Sobotka; and many other officials of the University and Faculty. Also all our present and previous colleagues and collaborators, especially R. Galambos (La Jolla, CA), M. D. Egger (New Brunswick, NJ), M. Sallam-Sattar (Cairo), R. Kucera (Hodonin, CSSR) and F. Franc and P. Vesely (Prague).

REFERENCES

1. EGGER, M. D. & M. PETRAN. 1967. New reflected-light microscope for viewing unstained brain and ganglion cells. Science **157:** 305–307.

2. PETRAN, M., M. HADRAVSKY, M. D. EGGER & R. GALAMBOS. 1968. Tandem scanning reflected light microscope. J. Opt. Soc. Am. **58:** 661–664.
3. PETRAN, M., M. HADRAVSKY & A. BOYDE. 1985. The tandem scanning reflected light microscope. Scanning **7:** 97–108.
4. PETRAN, M., M. HADRAVSKY, J. BENES, R. KUCERA & A. BOYDE. 1985. The tandem scanning reflected light microscope. Part 1—The principle, and its design. Proc. R. Microsc. Soc. **20:** 125–129.
5. PETRAN, M., M. HADRAVSKY, A. BOYDE & M. MÜLLER. 1986. Tandem scanning reflected light microscopy. *In* Science of Biological Specimen Preparation. 85–94. SEM Inc. AMF O'Hare, IL.
6. BOYDE, A., M. PETRAN & M. HADRAVSKY. 1983. Tandem scanning reflected light microscopy of internal features in whole bone and tooth samples. J. Microsc. **132:** 1–7.
7. CARLSSON, K., P. E. DANIELSSON, R. LENZ, A. LILJEBORG, L. MAJLOF & N. ASLUND. 1985. Three-dimensional microscopy using a confocal laser scanning microscope. Opt. Lett. **10:** 53–55.
8. WILSON, T. with C. SHEPPARD. 1984. Theory and Practice of Scanning Optical Microscopy. Academic Press. London.
9. VAN DER VOORT, H. T. M., G. F. BRAKENHOFF, J. A. C. VALKENBURG, & N. NANNINGA. 1985. Design and use of a computer controlled confocal microscope for biological applications. Scanning **7:** 66–78.

Scanning Optical Microscopy at λ/10 Resolution Using Near-Field Imaging Methods[a]

M. ISAACSON, E. BETZIG, A. HAROOTUNIAN,
AND A. LEWIS

School of Applied and Engineering Physics
Cornell University
Ithaca, New York 14853

INTRODUCTION

Throughout this conference we have been discussing various aspects of the spatial and spectral imaging of components of biological systems. In fact, most of the imaging or microcharacterization methods discussed in this symposium and presently in use are fundamentally limited by the wavelength of the exciting radiation used. Furthermore, as nature would have it, in general, the smaller the wavelength of the radiation probe, the greater the structural damage to the sample under study. For example, optical microscopy (and spectroscopy) can be reasonably nondestructive if visible wavelengths are used, but the spatial resolution achievable is only of the order of one-half micron or so. Even with the new confocal scanning methods recently developed (e.g. REFERENCE 1) the resolution is only about 2,500 Å, and that using near UV radiation. Thus, for this method to achieve higher spatial resolution, more destructive, shorter wavelength UV radiation needs to be used.

Of course, there are proponents at this symposium who circumvent that resolution limit by using high-energy electrons. There, using conventional electron microscopes of 100 kV to 300 kV accelerating voltages, electron wavelengths of .037 to .0197 Å are achievable. The aberrations of the lenses used limit the spatial resolution to about 100 times the wavelength (a few Å or so), but that is still significantly better than that achievable with conventional light microscopy. The problem is that electrons of such high energies are extremely destructive (e.g. REFERENCE 2) and the sample preparation required to view the object in vacuo requires extreme care.

There are other proponents who would use soft X rays rather than electrons for their imaging (see REFERENCE 3), but again radiation damage to the sample can play a deleterious role.

An alternative way around this wavelength limit to resolution is that described by O. H. Griffith (e.g. REFERENCES 4, 5) using photoelectron microscopy. In this case, light is incident on the sample, but one images the resultant photoelectrons. Thus, the resolution achievable is ultimately limited by the photoelectron wavelength and not the incident light. Since the wavelength of an electron is less than that of a photon of the same energy ($\lambda_{electron} = (150/E)^{1/2}$ whereas $\lambda_{photon} = 12,399/E$ where E is the energy of the photon or electron), this method can achieve resolutions almost two orders of

[a]Supported by National Science Foundation (NSF) Contract ECS-8410304 and US Air Force Contract AFOSR 84-0314. Test patterns fabricated at National Research and Resource Facility for Submicron Structures (NSF Grant ECS-8200312). EB supported as IBM Graduate Fellow.

magnitude smaller than the incident light wavelength. Unfortunately, the sample still needs to be put in vacuo and the incident light has to have sufficient energy to break bonds and liberate photoelectrons—so the method is partially destructive.

Thus, it would appear that if we want to view biological systems with minimal damage to their structure (*i.e.* nonionizing radiation and no need to place the sample in a vacuum) we must relinquish high spatial resolution.

There is, however, another way around the fundamental resolution limit imposed by diffraction (or by the uncertainty principle if you wish to look at it that way). This is the method of near-field imaging—and it has general applicability for all wave probes. The method is based upon the fact that unlike standard imaging systems, which rely on far-field optics, the distribution of radiation in the near field of an object can be highly collimated, and that degree of collimation is not dependent on the wavelength of the radiation to any great extent. In fact, more than a decade ago, Ash and Nicholls[6] had demonstrated that resolution of $\lambda/60$ was attainable at microwave wavelengths ($\lambda = 3$ cm) using scanning near-field imaging.

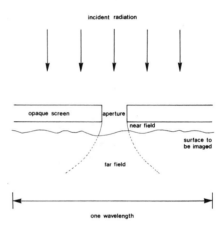

FIGURE 1. A schematic representation of the collimation in the near field of radiation emanating from a subwavelength aperture. The regime in which the radiation has spread out is known as the far-field (or *Frauenhoffer*) regime. All the microscopy reported in the other papers of this symposium are constrained by the limits of far-field diffraction.

BASIC PRINCIPLES OF NEAR-FIELD MICROSCOPY

The details of near-field optical imaging have been discussed before (*e.g.* REFERENCES 7, 8, 9). We present here the salient features of the method. The basic principle is outlined in FIGURE 1. Here, incident radiation is depicted as being normally incident on a screen that is opaque to the incident radiation. This screen contains a small (subwavelength diameter) aperture. Because the screen is opaque, the radiation emanating through the aperture and into the region beyond the screen is first collimated to the aperture size rather than to the wavelength of the radiation employed. This problem is discussed from a theoretical point of view in REFERENCE 8 and in that study it is shown that in this "near-field" regime, a radiation beam exists that has a dimension commensurate with the aperture size and independent of the wavelength. Eventually, as one goes further away from the aperture, there is a marked divergence of the radiation until the angular divergence is independent of distance from the aperture and dependent only upon the ratio of the wavelength to the aperture size. This is the far-field regime, which occurs at distances greater than a wavelength from the aperture. The radiation pattern is no longer a geometrical projection of the aperture.

In order to apply this collimation phenomenon to microscopy, an object (such as membrane) is placed in the near field relative to an aperture. In the case shown in FIGURE 1, the aperture acts as a light source whose size is not limited by standard geometrical (far-field) optics considerations. If this light source (*i.e.* aperture) is scanned relative to the object, the detected light can be used to generate an image. (Note that in the case shown here, we could detect a reflected light or fluorescence if the object were fluorescently labeled.) Since the spatial resolution is dependent upon the aperture size rather than the wavelength, the resolution limit now depends upon how small we can make such apertures and whether a sufficient intensity of light can

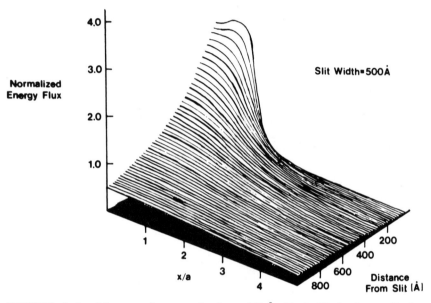

FIGURE 2. A plot of the energy flux emanating from a 500-Å wide slit. The flux is normalized to the energy of the plane wave incident on the slit. This normalized energy flux is plotted as a function of both the distance from the screen and the distance across the slit. $x/a = 0$ is the slit center and $x/a = 1$ is the slit edge. (From Betzig *et al.*[8] Reprinted by permission from *Applied Optics*.)

pass through such apertures. We have demonstrated that one can easily "see" light passing through apertures as small as 300 Å in diameter,[10] and we can fabricate apertures even smaller than that.[11] Thus, the only other parameter we need to know is how close an object has to be to the aperture in order to be in this "near field."

We have performed complete electromagnetic calculations for the case of an infinitely long slit in a thick opaque screen[8] and have found that the collimated region extends to at least half the slit width. This can be seen in FIGURE 2. Although these calculations have been done for a slit case, similar (though more approximate) calculations have been done for the case of a circular aperture,[12] and a qualitatively similar result is obtained. The main difference between a long slit and an aperture is that the slit in a thick screen can support a propagating mode, whereas the aperture can only support evanescent modes if its diameter is much less than the wavelength. Given

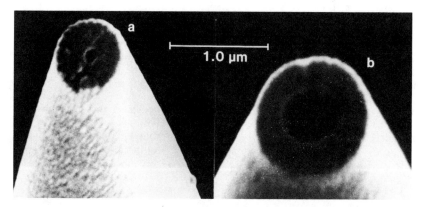

FIGURE 3. Scanning electron micrographs of metallized pipettes with various size apertures.

this caveat, if we use the information from FIGURE 2, we see that in addition to the aperture being placed close to the object, that distance must be closely controlled, since the intensity falls off exponentially from the aperture. Thus, to realize our subwavelength optical microscope we must: 1) be able to position the object within a couple hundred Å from the aperture (of ≈ 500 Å in diameter), 2) control that position to 10 Å or so, and 3) scan the object relative to the aperture in steps at least one half smaller than the aperture diameter.

It was for these reasons that it took a decade from Ash and Nicholls' first demonstration of near-field microwave imaging (1972)[6] for optical near-field imaging to begin. Since our initial publication on optical near-field imaging a few years ago,[13] there have been several groups working on the use of near-field collimation for imaging with optical[14,15] and far infrared wavelengths.[16,17] In this paper, we will concern ourselves only with the methods used at Cornell for implementing the near-field scanning optical microscope (NSOM). The interested reader can refer to the literature for discussions of other possible methods.

FIGURE 4. A schematic of the one-dimensional prototype NSOM instrument.

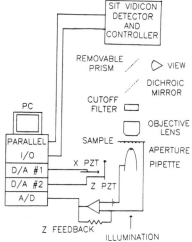

EXPERIMENTAL STUDIES

To construct a near-field scanning optical microscope with 500 Å resolution at visible wavelengths we must pay attention to position movement at the 10 Å scale. In addition, we must eliminate all sources of vibration and acoustical noise that could disturb positional accuracy. By using the scanning method, we do not have too much difficulty in detection since detection is done in the far field. Therefore, conventional optics and standard low-level detection systems can be used.

We have chosen to use piezoelectric scanning and positioning for our fine motions. Such mechanisms are suitable for the <1 Å movement level in the scanning tunneling microscope,[18] and xy stages with better than 10 Å positional accuracy are commercially available.[19] A more detailed description of the mechanics of scanning is given in REFERENCE 20.

Perhaps more crucial than the scanning is the fabrication of the near-field apertures. These act as as either the source or the detector of the near-field radiation. The main problem in near-field imaging is the finite depth of field resulting from the limited extent of the collimated radiation in the near field. Thus, our apertures have to be situated at the tip of a long, narrow probe in order to be able to access recessed

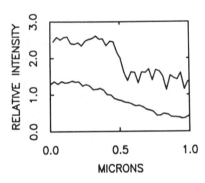

FIGURE 5. NSOM line scans across a 500-Å thick chromium step on glass obtained with white light radiation (λ_{max} = 5,700 Å). The *upper trace* is from a 1,000-Å aperture diameter, the *lower trace* from a 6,000-Å aperture diameter. (From Harootunian *et al.*[23] Reprinted by permission from *Applied Physics Letters.*)

regions in rough surfaces. (One must remember that the finite depth of field means that near-field imaging is essentially a surface probe).

To address this problem, we have developed a simple method of aperture fabrication involving the formation of small apertures in the tips of highly tapered, metallized glass pipettes. This method is a simple extension of the technique used for fabricating patch clamp pipettes.[21] Currently we can fabricate pipettes with diameters less than 1,000 Å using a two-stage method of gravity pulling. Apertures as small as 500 Å have been made in this fashion.[7,20] These pipette apertures have been characterized by light transmission[22] as well as by scanning electron microscopy (SEM). The pipettes are metallized at the tip and the sides by about 500 Å of aluminum to ensure sufficient opacity.[23] A typical pipette shape can be been in FIGURE 3. Our pipettes have a large diameter that rapidly constricts at the tip. Thus, through most of the pipette length, the light exists in a propagating mode above the cut-off frequency. Only in the thin metallized tip region is a region of evanescence reached.

In order to accurately assess the resolution of any NSOM instrument, it is essential to have test patterns that are well characterized. To produce test patterns for our experiments we fabricated masks in Si_3N_4 membranes using electron beam microfabrication methods.[20,24] These membrane masks were then placed on glass slides and

FIGURE 6. Line scans across a 500-Å thick chromium step on glass. *From bottom to top: first,* a densitometer trace of a scanning electron micrograph; *second,* an NSOM scan of the same edge obtained in transmission (λ_{max} = 5,700 Å) plotted along with a theoretical curve; *third,* a theoretical conventional scanning optical microscope (SOM) scan (λ = 5,700 Å, NA = 0.55) plotted with points of an experimental optical microscope scan of the same chromium step taken with the same collection optics used for the NSOM scans; and *fourth,* a theoretical SOM scan across a Cr step using a 1.4-NA objective (with λ = 5,700 Å). All scans were normalized to an intensity change of unity across the step.[23]

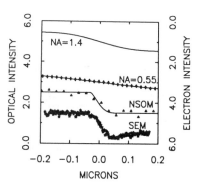

material was evaporated through the mask onto the glass in a contact printing process. Patterns with regular feature sizes as small as 500 Å were fabricated by this method using aluminum, chromium, and perylene (a fluorescent dye). In order to assess the quality of our test patterns, they were all characterized by scanning electron microscopy. (However, for the tests using the perylene patterns, the characterization was performed after the NSOM measurement, since electron-beam irradiation destroys the perylene fluorescence).

In order to test the NSOM concept, we assembled the prototype system shown in FIGURE 4. This system only allows for one-dimensional scanning, but the z position (*i.e.* the distance of the pipette aperture from the sample) has a feedback control based upon measurement of the tunneling current between the pipette and the sample.[23] Light was passed through the pipette illuminating the aperture from below. The collected light (in transmission) was focused onto an SIT vidicon detector.

FIGURE 5 shows a transmission scan over a single 50-nm thick chromium step on glass for two different size apertures. The distance between the 12% and 88% transmission points scale with the aperture size, demonstrating the near-field collimation dependence on aperture diameter. A direct comparison between near-field optical (NSOM) scan across such a step and an SEM scan is shown in FIGURE 6, where we can see that the resolution of the NSOM and SEM scans are comparable.[23] These scans represent the respective beam profiles convoluted with the chromium step edge. For the data obtained in FIGURES 5 and 6, the sample was illuminated through the pipette aperture using a 100 W tungsten lamp with a maximum intensity of 5,700 Å.

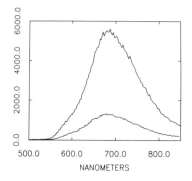

FIGURE 7. Fluorescence from 500-Å thick perylene on glass (excited by green light). The *lower trace* is the transmission fluorescence if the perylene is coated with 50 Å of Cr.

Of more importance in a biological context is the ability of the near-field scanning optical microscope to operate in the fluorescent mode, thus providing *in situ* chemical mapping at resolutions well below the fluorescing radiation wavelength. To assess this capability we fabricated a fluorescent grating of perylene[3,4,9,10] (perylenetetracarboxylic dianhydride). The perylene was illuminated by green light, and the red fluorescence (maximum intensity at 6,900 Å; see FIGURE 7) that passed through a 5,800 Å cut-off filter was transmitted to the detector. The NSOM fluorescence scan of this grating is shown in FIGURE 8(A) along with a densitometer trace of a scanning electron micrograph of the same grating (B). The sharpest maximum-to-minimum change in the fluorescence line scan is thus seen to be comparable to the SEM scan indicating an NSOM resolution of about an order of magnitude smaller than the fluorescence wavelength.[23]

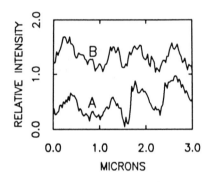

FIGURE 8. (A) A fluorescence NSOM scan (λ_{max} = 6,900 Å) of a perylene grating shown along with (B) a densitometer trace of an SEM scan of the same grating.[23]

Thus, we have demonstrated in both transmission and fluorescence using a prototype one-dimensional scanning system that near-field scanning optical microscopy (NSOM) can attain a lateral resolution of almost $\lambda/10$. This opens up the unique possibilities of optical resolution of biological samples of about 500 Å using a visible light.

CONCLUSION

In summary, near-field imaging offers us the possibility of circumventing the far-field diffraction limit of resolution. 500-Å resolution (in transmission or fluorescence) using visible light appears feasible. A two-dimensional NSOM is currently under construction to take advantage of this unique possibility for biological imaging. This new instrument will have fast feedback by monitoring the local capacitance between the pipette aperture probe and the sample,[25] and the combination of fluorescence and capacitance will allow us to simultaneously acquire a topographical and fluorescence map of the sample surface at 500-Å resolution. Thus, NSOM shows great promise as a high-resolution, nondestructive, *in situ* spectral probe, which will extend the range of optical microscopy into a spatial resolution regime previously inaccessible.

ACKNOWLEDGMENTS

We wish to thank Dr. E. Kratschmer for technical assistance in the grating fabrication.

REFERENCES

1. WILSON, T. & C. SHEPARD. 1984. Theory and Practice of Scanning Optical Microscopy. Academic Press. London.
2. ISAACSON, M. 1977. *In* Principles and Techniques of Biological Electron Microscopy. M. A. Hayat, Ed. Ch. 1. Van Nostrand-Reinhold. New York, NY.
3. RUDOLPH, D., G. SCHMAHL, B. NIEMANN, W. MEYER-ILSE & J. THIEME. 1986. X ray microscopy—state of the art and expected developments. This volume.
4. GRIFFITH, O. H. & G. REMPFER. Advances in Optical and Electron Microscopy. R. Buer & V. E. Cosslett, Eds. Vol. 12. Academic Press. London. In press.
5. HEDBERG, K. K. & O. H. GRIFFITHS. 1986. Immunophotoelectron microscopy of the cell surface and cytoskeleton. This volume.
6. ASH, E. A. & G. NICHOLLS. 1972. Nature **237**: 510.
7. BETZIG, E., A. LEWIS, A. HAROOTUNIAN, M. ISAACSON & E. KRATSCHMER. 1986. Biophys. J. **49**: 269.
8. BETZIG, E., A HAROOTUNIAN, A. LEWIS & M. ISAACSON. 1986. Appl. Opt. **25**: 1890.
9. DURIG, U., D. W. POHL & F. ROHNER. 1986. J. Appl. Phys. **59**: 3318.
10. LEWIS, A., M. ISAACSON, A. HAROOTUNIAN & A. MURAY. 1984. Ultramicroscopy **13**: 227.
11. MURAY, A., M. ISAACSON, I. ADESIDA & B. WHITEHEAD. 1983. J. Vac. Sci. Technol. **B1**: 1091.
12. LEVIATIN, Y. (private communication).
13. LEWIS, A., M. ISAACSON, A. MURAY & A. HAROOTUNIAN. 1983. Biophys. J. **41**: 405A.
14. POHL, D. W., W. DENK & M. LANZ. 1984. App. Phys. Lett. **44**: 652.
15. FISCHER, U. CH. 1985. J. Vac. Sci. Technol. **B3**: 386.
16. MASSEY, G. A. 1984. Appl. Opt. **23**: 658.
17. MASSEY, G. A., J. A. DAVIS, S. M. KATNIK & E. OMON. 1984. J. Opt. Soc. Am. **A1**: 1259.
18. BINNIG, G. & H. ROHRER. 1982. Helv. Phys. Acta **55**: 726.
19. SCIRE, F. E. & E. C. TEAGUE. 1978. Rev. Sci. Instrum. **49**: 1735.
20. LEWIS, A., E. BETZIG, A. HAROOTUNIAN, M. ISAACSON & E. KRATSCHMER. *In* Spectroscopic Membrane Probes. L. Loew, Ed. CRC Press. Cleveland, OH. In press.
21. SAKMANN, B. & E. NEHER, Eds. 1983. Single Channel Recording. Plenum Press. New York, NY.
22. HAROOTUNIAN, A. 1986. Ph.D. dissertation. Cornell University.
23. HAROOTUNIAN, A., E. BETZIG, M. ISAACSON & A. LEWIS. 1986. Appl. Phys. Lett. **49**: 674.
24. ISAACSON, M., A. MURAY, M. SCHEINFEIN, I ADESIDA & E. KRATSCHMER. 1984. Microelectron. Eng. **2**: 58.
25. MATEY, J. R. & J. BLANC. 1985. J. Appl. Phys. **57**: 1437.

DISCUSSION OF THE PAPER

D. RUDOLPH (*University of Göttingen, Göttingen, Federal Republic of Germany*): Does the resolution of your microscope change as a function of the distance of the aperture from the sample?

ISAACSON: Yes, you have to be in the near field, and the near field essentially means some fraction of the aperture diameter. So for using, 5,000-Å irradiation, and 500-Å diameter aperture, we have to be within 100 Å or so from the aperture. But to do that type of positioning is almost trivial compared to the type of positional accuracy that people do with the scanning tunneling microscope, which is 0.1 Å. This distance requirement essentially implies that our method looks at the surface. We're looking at the topography as well as the chemical, and we're getting topographical and chemical information out from the surface.

F. SACHS (*State University of New York, Buffalo, New York*): How do you control the distance from the scanner to the aperture?

ISAACSON: Do you mean from the surface? Either of two ways. One can either use tunneling, in which you have to get the aperture quite close to the surface or you can measure the capacitance (between the sample and the pipette aperture).

SACHS: But we're talking about living systems under water. This may be a little tough.

ISAACSON: Our system is only a surface probe. Things away from the surface are in the far field. The intensity from the far field is greatly reduced by the time it gets to the aperture. And you can prepare samples to have minimal fluid over the surface.

SACHS: I understand. But where the surface is exactly is a temporal problem. If you have a cell, for instance, with a lot of membrane, you have a lot of junk on top. Exactly where is the surface? What signal are you feeding back if this is biological? If this is anticipated for wet specimens, how do you do this?

ISAACSON: You can use the capacity between the probe and a specific surface, or in some cases one can measure the conductivity directly.

A. BOYDE (*University College London, London, England*): There is a method of specimen preparation for that microscope—biological specimens—which would work right off. That would be to have your specimen pre-stained with the fluorescent stain embedded in polymethyl methacrylate and then finished off flat. Then look at the fluorescence from that very transparent optical medium.

X Ray Microscopy—State of the Art and Expected Developments[a]

D. RUDOLPH, G. SCHMAHL, B. NIEMANN,
W. MEYER-ILSE, AND J. THIEME

Forschungsgruppe Röntgenmikroskopie
Universität Göttingen
D-3400 Göttingen, Federal Republic of Germany

INTRODUCTION

X ray microscopy has several advantages compared with light and electron microscopy. For photons of visible light and electrons, biological tissue provides little natural contrast. Hence, for light microscopy, specimens normally are treated with chemical dyes to provide sufficient contrast for image generation. In electron microscopy the high reactivity of electrons prohibits examination of thick biological specimens in their natural state. On the other hand, the interaction cross section for electrons depends very weakly on the atomic number and hence, contrast for image generation has to be provided by staining with heavy metal atoms or by other artificial methods, *e.g.* coating with thin metal layers. All these procedures carry the risk of altering the composition or ultrastructure of biological tissue, and the question of whether results of examination are real or artificially introduced by preparational methods has always to be addressed. So, in many cases additional experiments are necessary to prove the validity of microscopic results.

Contrary to the above-mentioned faults, there is in the soft X ray wavelength region a strong dependence of the absorption of the atomic species and density of matter. In the wavelength region between 2.3 nm and 4.4 nm there is about an order of magnitude difference between the absorption coefficients for water and organic components such as protein, for example. This difference provides a natural contrast mechanism for the investigation of wet biological materials. The situation is comprised in FIGURE 1, which shows the atomic cross sections for X radiation and electrons versus wavelength. These cross sections for soft X rays are about one order of magnitude less than those for electrons at energies used for electron microscopy and hence, offer the possibility of investigating whole cells of several micrometers thickness in the natural state by providing sufficient absorption for image generating contrast in microscopic dimensions. In addition FIGURE 1 shows that for soft X rays the interaction with matter is dominated by photoelectric absorption, whereas the cross sections for elastic and inelastic scattering are about five orders of magnitude less. This makes evident firstly that there are no contrast-reducing scattered photons, *e.g.* from the zone plates used as imaging elements in the wavelength region suitable for X ray microscopy, and secondly that at shorter wavelengths in the medical X ray domain there is not sufficient contrast from submicron structures.

[a]Supported by the Bundesministerium für Forschung und Technologie.

THE X RAY OPTICAL SYSTEM AND THE X RAY MICROSCOPE

FIGURE 2 shows the X ray optical arrangement. The polychromatic synchrotron radiation meets the condenser zone plate at a distance of 15 m from the source, which is the tangent point to the electron beam in the storage ring. The condenser zone plate, with a diameter of 9 mm and a focal length of 304 mm for 4.5-nm radiation, generates a reduced image of the source. The object is placed in an object chamber. Together with the free diameter of the object chamber the condenser acts as a linear

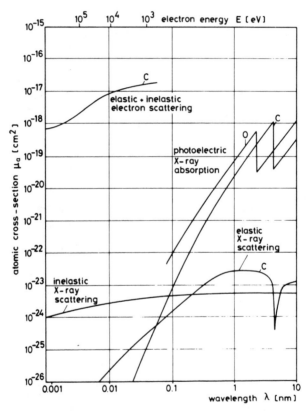

FIGURE 1. Atomic cross sections for X ray scattering, X ray adsorption and electron scattering for oxygen and carbon. The values for photoelectric absorption and elastic scattering for wavelengths larger than 0.6 nm are taken from REFERENCE 1.

monochromator and illuminates the object with quasi-monochromatic X radiation. A micro zone plate (55.6-μm diameter, f = 0.69 mm at 4.5 nm) generates an enlarged image of the object in the image field. A central stop at the condenser zone plate prevents contrast-reducing zero order radiation of the condenser and the micro zone plate from reaching the image field. The enlarged image can be viewed using a microchannel plate to convert the X radiation into the visible or it can be photographed

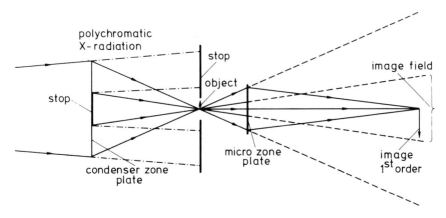

FIGURE 2. The X ray optical arrangement.

directly. A detailed description of the zone plates used as X ray optics is given in REFERENCES 2 and 3.

As in normal air the X radiation is absorbed to $I/I_0 = 1/e$ in a layer of 1.56 mm thickness at 4.5 nm and in a 0.57-mm layer at 2.4 nm, the microscope has to be operated under vacuum. For examination of wet biological specimens in their natural state an environmental object chamber has been developed in which the specimen is separated from the vacuum by thin polyimide foils transparent to X rays. FIGURE 3 shows the schematic mechanical arrangement of the X ray microscope and FIGURE 4 a photograph of the X ray microscope at the Berliner Elektronenspeicherring-Gesellschaft für Synchrotronstrahlung m.b.H. (BESSY storage ring) in Berlin. For detailed description of the microscope and the object chamber see REFERENCE 4.

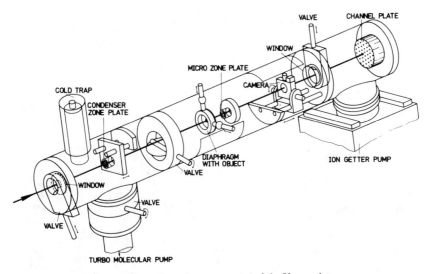

FIGURE 3. The schematic arrangement of the X ray microscope.

RESULTS

FIGURE 5 shows an image of a part of a human fibroblast made with the X ray microscope at 4.5 nm and with an X ray magnification of 330X. The fibroblast was critical-point dried without any staining.

Images of wet cells, *e.g.* a human epithelial cheek cell and spores of the Australian moss *Dawsonia superba* were made.[5] Investigations of the viability of spores of *Dawsonia superba* under X irradiation were started.[6]

ASPECTS AND FUTURE DEVELOPMENTS

Up to now for X ray imaging amplitude zone plates have been used with a theoretical maximum first order diffraction efficiency of 10%; in practice values of about 5% have been achieved. Higher values of diffraction efficiency can be obtained by using phase zone plates (up to about 20% for laminar phase structures). Phase zone plates with laminar phase structures are under development.

Zone plates with higher diffraction efficiency permit shorter exposure times; in addition a larger amount of the X radiation that loads the object is used for image generation, leading to a reduced radiation dose applied to the specimen.

Up to now a coarse focussing of the object has been done by one looking at the X radiation converted to the visible by the microchannel plate and subsequently taking a few fine focussing photos with slightly different focus setting. A further improvement will be done by introducing an optical microscope into the X ray microscope for prefocussing. Prefocussing with visible light will reduce the X radiation load to the object to the amount necessary for taking just one X ray image.

In the existing X ray microscope at BESSY the image is directly viewed by converting the X radiation to the visible using a microchannel plate, or it is directly

FIGURE 4. The X ray microscope.

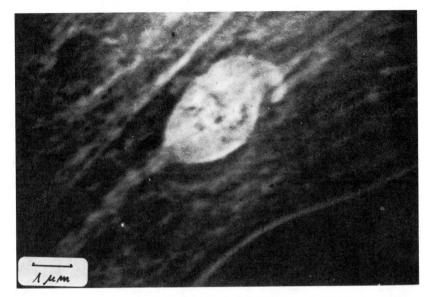

FIGURE 5. Part of a human fibroblast, critical-point dried, imaged with 4.5-nm radiation.

photographed. Using semiconductor imaging devices such as CCD's will provide higher detective quantum efficiency and the possibility of electronic image processing. Such devices are under development; preliminary results are given in REFERENCE 7.

Lastly, it should be mentioned that projects have been started to develop laboratory X ray sources[8] making the X ray microscope for special applications independent of the storage ring facilities.

SUMMARY

The Göttingen X ray microscope at the electron storage ring BESSY in Berlin and X ray microscopy experiments with biological specimens are described. A look ahead to future developments—optics for higher resolution and with better efficiency, detector devices with higher detective quantum efficiency, and the development of a laboratory X ray microscope with a plasma X ray source—is taken.

ACKNOWLEDGMENTS

For a joint effort to compare the results of different microscopic methods Dr. P.C. Cheng, IBM Th.J.Watson Research Center, Yorktown Heights, NY, USA, provided us with critical-point dried human fibroblasts (see FIGURE 5).

REFERENCES

1. HENKE, B. L. 1981. Low energy X ray interactions: photoionization, scattering, specular and Bragg reflection. B. L. HENKE *et al.* The atomic scattering factor, f1 + f2, for 94 elements

and for the 100 to 2000 eV photon energy region. Am. Inst. Phys. Conf. Proc. **75**. Low Energy X-Ray Diagnostics.
2. SCHMAHL, G., D. RUDOLPH, P. GUTTMANN & O. CHRIST. 1984. Zone plates for X-ray microscopy. In X-Ray Microscopy. G. Schmahl & D. Rudolph, Eds. Springer Series in Optical Sciences. Vol. 43: 63–74. Springer-Verlag. Heidelberg.
3. GUTTMANN, P. 1984. Construction of a micro zone plate and evaluation of imaging properties. Ibid. 75–90.
4. RUDOLPH, D., G. SCHMAHL, B. NIEMANN & O. CHRIST. 1984. The Göttingen X-ray microscope and microscopy experiments at the BESSY storage ring. Ibid. 192–202.
5. SCHMAHL, G., B. NIEMANN, D. RUDOLPH, P. GUTTMANN & V. SARAFIS. 1985. X-ray microscopy: experimental results with the Göttingen X-ray microscope at the electron storage ring BESSY in Berlin. J. Microsc. **138**: 279–284.
6. NIEMANN, B., V. SARAFIS, D. RUDOLPH, G. SCHMAHL, W. MEYER-ILSE & P. GUTTMANN. 1986. X-ray microscopy with synchrotron radiation at the electron storage ring BESSY in Berlin. Proc. Int. Conf. X-ray and VUV Synchrotron Radiation Instrum. Stanford, 1985. Nuclear Instruments and Methods, in press.
7. GERMER, R. and W. MEYER-ILSE. 1986. X-ray TV-camera at 4.5 nm. Rev. Sci. Instrum., in press.
8. HERZIGER, G. 1984. X-ray emission from a 1 kJ plasma. In X-Ray Microscopy. G. Schmahl & D. Rudolph, Eds. Springer Series in Optical Sciences. Vol. 43: 19–24. Springer-Verlag. Heidelberg.

DISCUSSION OF THE PAPER

E. MANDELKOW (*Max-Planck-Unit for Structural Molecular Biology, Hamburg, Federal Republic of Germany*): What was the time course of the radiation damage? How long did it take for the cells to become dead?

RUDOLPH: Under this radiation, about two minutes (>5 · 10^6 Photons/μm^2, that is, a dose of about 5 × 104 Gy).

MANDELKOW: So two minutes after irradiation you did the test?

RUDOLPH: No, we did the test within the range of 5 to 10 minutes.

MANDELKOW: And what you applied, of course, is a functional test of the cells' viability. What about the structures inside the cell? Did you see any damage to organelles?

RUDOLPH: No, the results give only indications of viability.

A. P. SOMLYO (*University of Pennsylvania School of Medicine, Philadelphia, Pennsylvania*): What's the best spatial resolution you have obtained on a test object with the new zone plates?

RUDOLPH: Up to now we have a resolution of 500 Å. Zone plates under development should have about 200 Å.

E. ZEITLER (*Fritz-Haber-Institut der Max-Planck-Gesellschaft, Berlin, Federal Republic of Germany*): What wavelength did you use?

RUDOLPH: Up to now we have used 45 Å. That's about 270 eV. But for biological objects it's better to use 24 Å. We are just preparing our zone plates to switch over to this better wavelength.

D. F. PARSONS (*New York State Department of Health, Albany, New York*): Have you adjusted wavelengths to give you differential contrast between carbon and oxygen?

RUDOLPH: Not yet. We know that it is an interesting problem.

Microanalysis with a Soft X Ray Scanning Microprobe[a]

C. JACOBSEN,[b] J. M. KENNEY,[b] J. KIRZ,[b] I. McNULTY,[b]
R. J. ROSSER,[b] F. CINOTTI,[c] H. RARBACK,[c] AND D. SHU[c]

[b]Department of Physics
State University of New York
Stony Brook, New York 11794
and
[c]National Synchrotron Light Source
Brookhaven National Laboratory
Upton, New York 11973

INTRODUCTION

In order to explore the potential of soft X ray microanalysis, we have, in collaboration with scientists from IBM and the Brookhaven National Synchrotron Light Source (NSLS), constructed a scanning soft X ray microprobe, which is operating on the U15 beamline of the 750 MeV VUV storage ring at the NSLS.[1] A schematic drawing of the instrument is shown in FIGURE 1. A toroidal grating monochromator, of resolving power 300, focuses a selected wavelength onto a pinhole. A Fresnel zone plate is used to form a demagnified image of this pinhole in the plane of the specimen. The specimen is scanned through this point in a raster fashion while the transmitted X rays are detected using a flow proportional counter. This data is stored by computer and displayed on a television screen. The ultrahigh vacuum requirements of the synchrotron source are far removed from the ideal conditions for biological cells, so the specimen and source are decoupled in two stages. A 150-nm aluminum contamination barrier separates the ultrahigh vacuum region from a relatively low vacuum area, keeping out hydrocarbon contamination, but providing no mechanical strength. A small 120-nm thick silicon nitride window then separates the vacuum from the atmospheric pressure region. Helium is flowed over the specimen, allowing the soft X rays to be transmitted in this region. Having the sample in a helium environment at atmospheric pressure permits relatively simple handling. The spatial resolution of the instrument is determined by the size of the pinhole image, which depends on the quality of the zone plate. At present, 0.2 μm is the limit, but work underway at IBM and King's College, London, should enable an order of magnitude improvement within the near future.

EXPERIMENTAL RESULTS

The microprobe has been used to demonstrate the potential of elemental mapping by computer subtraction of X ray absorption images. So far we have concentrated our attention on calcium. It is particularly suitable, since it is biologically important and

[a]Supported in part by Department of Energy Contract DEAC0276CH00016 (Brookhaven) and by National Science Foundation Grant DMB-8410587 (Stony Brook).

present in bone in high concentrations. Moreover, it has large atomic resonances, near the L III edge, at 3.58 nm, which is in the region of the spectrum that is readily accessible to our microprobe. To date we have produced elemental maps of calcium with a spatial resolution of 0.2 μm, detecting concentrations of as little as 5 percent by weight in specimens 0.2 μm thick.[2] FIGURE 2 shows some of these results. This capability is currently being used by one of us (F. Cinotti) to investigate the calcium distribution in pathological bone.

DISCUSSION

The potential advantage of this technique over the more sensitive electron-probe methods is the possibility of doing the element mapping on relatively thick specimens of light elements at atmospheric pressure in an hydrated state. This would be of particular relevance to biological material. It is, therefore, a useful exercise to consider what elements are biologically important and which of them could be detected.

Calcium in bone is an important exception to the normal distribution of elements in biological tissue. The primary constituents of living cells, *i.e.*, those occurring with concentrations between 1 and 60 percent, are H, C, N, O, and P. Of these only C, N, and O have absorption features in the 2.3 to 4.4-nm region and are therefore candidates for mapping in the hydrated state in our present microscope. Each, however, has its difficulties in practice. Although many of these difficulties may be particular to our experimental setup, we believe it is useful to discuss them in order that others can learn from our problems, and to focus attention on the technical improvements that could lead to improved microprobes.

Carbon is difficult to observe in the existing apparatus because of carbon contamination on our optical elements and the fact that the efficiency of the toroidal grating monochromator is reduced by about a factor of ten in this region, compared with its performance at 3.2 nm. The contamination, in addition to contributing to the reduced flux, provides a complex spectrum, making the comparison of images taken at even slightly different wavelengths more problematic. The reduced flux means it is not possible to collect enough data in a reasonable time to give good statistical accuracy. Higher flux and a slightly different microchromator design, which will be available on the undulator-based microprobe, should solve these problems.

Nitrogen edge mapping is potentially extremely interesting, as it should differentiate clearly between protein and lipids.[3] A difficulty in our setup is the presence of two

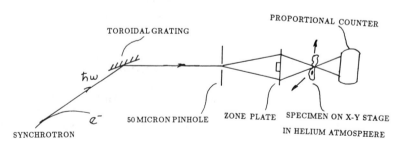

FIGURE 1. A schematic drawing of the scanning X ray microscope as set up at the U15 beamline.

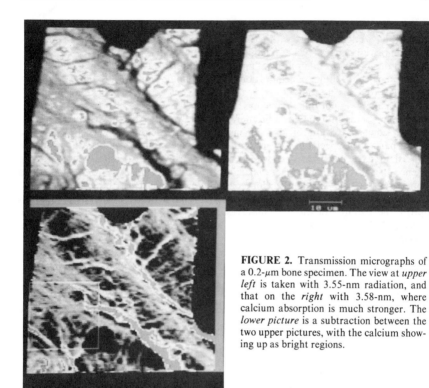

FIGURE 2. Transmission micrographs of a 0.2-μm bone specimen. The view at *upper left* is taken with 3.55-nm radiation, and that on the *right* with 3.58-nm, where calcium absorption is much stronger. The *lower picture* is a subtraction between the two upper pictures, with the calcium showing up as bright regions.

silicon nitride windows, one of which acts as a vacuum barrier, the other being the substrate on which the zone plate is fabricated. In principle, one could measure the absorption of the windows and subtract it out. However, the lower flux on the absorbing side and the extra errors introduced make this a nontrivial task. The vacuum barrier performs an important function by maintaining a one-atmosphere pressure difference between the specimen and the vacuum system, while introducing only 120 nm of absorbing solid material. This leaves the majority of the soft X ray flux available for probing the specimen. Practical elemental mapping of nitrogen in hydrated specimens would be aided by finding a nitrogen-free substitute for the silicon nitride windows. The higher flux from the undulator will reduce the constraint on window thickness and should allow the use of alternative materials.

Oxygen mapping of hydrated specimens would seem unlikely because of the variable thickness of water, which is primarily oxygen, covering the specimen. However, it may prove useful as a monitor of the hydration of the specimen. By taking absorption readings on either side of the oxygen edge and hence determining the oxygen content at various points in the specimen before and after the imaging of the sample, it should be possible to detect any change in the hydration state. It would not even be necessary to refocus the microprobe, as the spatial resolution of this measurement is not critical.

The secondary constituents of cellular material, *i.e.* those occurring in concentrations between 0.05 and 1 percent, are Na, Mg, S, Cl, K, Ca, and Fe. Although these concentrations are well beyond the capability of the current instrument, it is worth considering them, as the undulator version of the microprobe should have an element detection capability of 0.01 percent by weight.[4] Of these secondary constituents, only Ca and K have edge structure within the 2.3 to 4.4-nm range. Both these elements have large atomic resonances—Ca at 3.58 nm and K at 4.24 nm.

With the undulator microprobe it may become possible to do elemental mapping on calcium in normal cellular concentrations. This would be of interest to those concerned with such problems as the activation of muscle contraction, which is mediated by calcium in an as yet not fully understood way.

Mapping potassium may be a bit more difficult. We have tried, without success, to see the K edges in 0.5-μm thick specimens of KI using the present microprobe. It is a difficult part of the spectrum to work with because the flux is down by a factor of ten, so the data have poor statistics when compared with the Ca data. Moreover, the presence of the carbon near-edge structure and the second order oxygen features complicates the analysis. The carbon contamination is on both the grating and the windows. Better vacuum conditions, higher flux, and differential pumping could all improve the visibility of the potassium resonances. The undulator-based microprobe will have all these advantages and should be able to create K maps of hydrated samples at normal cellular concentrations.

CONCLUSIONS

Elemental mapping of hydrated biological specimens using soft X rays is at the moment restricted to calcium. While this is of great interest to those who study bone and teeth, a more generally interesting biological element would be nitrogen, which would allow strong contrast between proteins and lipids. A nitrogen-free substitute for the silicon nitride windows would be helpful for such studies. Carbon and potassium mapping would appear to require either the removal of the carbon contamination or a large increase in the flux. Oxygen mapping may prove to be a useful way of checking on the integrity of environmental cells by monitoring the water content. The arrival of the undulator-based microprobe, which is currently under construction, should, with its three orders of magnitude greater flux, provide much better statistics and so extend the range of elements and the sensitivity of detection. It may also provide the possibility of observing physiological levels of calcium in biological specimens in an hydrated state.

ACKNOWLEDGMENTS

We are grateful to our colleagues at Brookhaven, Stony Brook, and IBM who have made this work possible. In particular we would like to thank D. Sayre, R. Feder, W. Thomlinson, and G. Schidlovsky for their important contributions.

REFERENCES

1. RARBACK, H., J. M. KENNEY, J. KIRZ, M. HOWELLS, P. CHENG, R. COANE, R. FEDER, P. J. HOUZEGO, D. KERN & D. SAYRE. 1984. *In* X-ray Microscopy. G. Schmahl & D. Rudolf, Eds. 203. Springer. Berlin.
2. KENNEY, J. M., C. JACOBSEN, J. KIRZ, H. RARBACK, F. CINOTTI, W. THOMLINSON, R. J. ROSSER & G. SCHIDLOVSKY. 1985. J. Microsc. (Oxford) **138**(3): 321.
3. KIRZ, J. & H. RARBACK. 1985. Rev. Sci. Instrum. **56**(1): 1.
4. KENNEY, J. M., C. JACOBSEN, J. KIRZ, R. J. ROSSER, F. CINOTTI & R. RARBACK. Biol. Trace Elem. Res., to be published.

DISCUSSION OF THE PAPER

D. F. PARSONS (*New York State Department of Health, Albany, New York*): How long does it take to get the complete frame?

ROSSER: At the moment it's about 30 to 40 minutes for a high-resolution picture. We're hoping that with the increased flux from the new undulator devices it will be down to a minute. In fact the computer data collection starts to become the limiting factor, rather than the flux.

PARSONS: How much absorption is there across the nitride window?

ROSSER: We lose about 50% of the flux. The real problem with the silicon nitride windows is that it has nitrogen in it, so you get a large nitrogen edge.

PARSONS: Why don't you use differential pumping across apertures instead?

ROSSER: Because the specimen is very fragile.

J. B. PAWLEY (*University of Wisconsin, Madison, Wisconsin*): Could you tell us more about what aberrations limit you. You are using a zone plate, but zone plates are not perfect and are limited by practical difficulties. You say you feel you can increase the resolution by essentially a factor of ten from where you are. Why do you think that? What is the limitation, how are you going to overcome it, and what might prevent you from overcoming it?

ROSSER: The limitation is the fineness of the outermost zones, so the real limitations are questions of micro fabrication. You are talking about things that are about 90 μm across with a few hundred lines and the line spacing here relates directly to the resolution. To get 100-Å resolution, you have to have lines that are spaced apart by less than 100 Å.

PAWLEY: I'm aware of that, but presumably you will be limited by diffraction.

ROSSER: Using a wavelength of between 23 and 44 Å we are still quite a way away from the diffraction limit of the wavelength used.

PAWLEY: So you get to 100 Å at a 24-Å wavelength with an aberration-free lens of considerable numerical aperture. The other thing I'd like to point out is that the nearest microscope to be compared to an X ray microscope would not be a light microscope, but a UV microscope. Although you don't get elemental analysis with such a microscope, you can in fact see protein, because protein is very strongly absorbent at about 2,800-Å wavelength. I don't see how you are ever going to know more than protein. That is to say, if you got protein you have carbon, hydrogen, nitrogen, oxygen altogether anyway. You won't be able to tell one from the other, so it's not clear to me why it's worth this immense amount of extra effort.

ROSSER: You are talking about an order of magnitude increase in resolution over something like an ultraviolet microscope, if you get down to the 100-Å range. We hope in the next month to be testing zone plates that approach that sort of resolution.

E. MANDELKOW (*Max-Planck-Unit for Structural Molecular Biology, Hamburg, Federal Republic of Germany*): I'll make one comment to Dr. Parsons' question; I think the 99 other uses of the synchrotron would probably not allow one to have a differential pumping system. My question is: can you make use of the polarized, of the nearly pole polarization of the beam to visualize oriented structures in a cell?

ROSSER: It's an interesting question, but we've not thought about it at all in any detail.

A. BOYDE (*University College London, London, England*): A microscope is only as good as the specimen that you put into it, a fact already known to the inventor of the microscope, Antony van Leeuwenhoek 300 years ago, who sold his microscopes with the specimens in them! The relevance of this to this paper is that, unfortunately, the results regarding calcium distribution in bone are inexplicable in terms of what we already know about bone structure and the mineralization process. I suggest that the most likely explanation of that is that you haven't been working with the right specimen.

ROSSER: At the moment we're obviously using specimens to test the microscope. I am no biologist and know nothing about the bone preparation.

W. W. WEBB (*Cornell University, Ithaca, New York*): What determines your elemental resolution? Also, since you quoted a much better elemental resolution with the undulator, I'm wondering whether the 1987 generation of undulators will be too high-powered for your silicon nitride window to tolerate?

ROSSER: It's primarily a statistical thing. It's just a question of collecting enough photons, and at the moment the limitation is just the time to get the 5% resolution. It's a 40-minute picture, and on a synchrotron that's about the limit of what you can do. And the increased flux from the undulators is what should make these elemental analyses easier. At the moment the window is not seeing very much flux at all. There's no evidence of breakage or damage. We have windows that have been used over 3 to 4 years with no problem.

D. RUDOLPH (*University of Göttingen, Göttingen, Federal Republic of Germany*): What is the resolution of your scanning device and the reproducibility? Is it a piezoelectric device?

ROSSER: Yes, a simple piezoelectric device, which seems to be very stable. The limitation seems to be how accurately you can put the voltage on, which in turn is limited by D to A conversion. We've got a 12-bit D to A converter. That defines the voltage you can get on. We hope that we can accurately position it to 300 Å. We check the position. At the moment we have LVDT's to check the position, and we also have a new laser interferometer that's going to go on the X-1-T device to rapidly check exactly where we are.

H. SHUMAN (*University of Pennsylvania School of Medicine, Philadelphia, Pennsylvania*): If instead of using the soft X rays available at Brookhaven you had a hard X ray probe, what kind of sensitivity would you get then?

ROSSER: A hard X ray probe is being planned, but the optics are more difficult to make. It's very much more difficult to get the high-aspect ratios to focus hard X rays with the same degree of accuracy.

One of the reasons for working at these particular wavelengths, between the oxygen and the carbon absorption edges, is the supposed water window. There you can have specimens that are hydrated, and you get a lot of differentiation between the oxygen and the carbon. So for hydrated specimens there is good reason for trying to work at these particular wavelengths, rather than going any shorter.

SHUMAN: You gave some estimates of concentration detectability limit, but you did not say whether these were limited in any way by, shall we say, matrix effects. It's one thing to detect 5% calcium in a uniform solution of plastic and calcium; it's another thing altogether to do it in a cell, where we have all sorts of other things changing the density—the so called R controversies. Do these sorts of limits apply to your microscope? Are there mass thickness effects, as well as elemental effects?

ROSSER: Absolutely. You have exactly the same problem, so these estimates are based on the assumption that these slices have been sliced fairly evenly.

SHUMAN: Is it sensitive to mass per se? To Z per se, as well as the characteristic?

ROSSER: Yes, the mass makes a difference. We had high hopes of detecting potassium, because potassium also has the giant resonance. As you are probably aware, all of the metals beyond sodium have these giant resonances—everything that stands next to the transition element. But we ran into similar problems. In the electron energy loss you also get the carbon near edge structure washing out the potassium, and it makes detection very difficult. This is exactly the same problem.

G. J. BRAKENHOFF (*University of Amsterdam, Amsterdam, The Netherlands*): Before I come to my actual question I would like to ask what that LVDT stands for?

RUDOLPH: Linear voltage difference transducer.

BRAKENHOFF: In microscopy we actually are interested in volume resolution, not so much in transverse resolution and, if you consider this, then the problem takes on a completely different dimension. In biology you want to relate structures with respect to each other in three dimensions. If you have a small numerical aperture, you haven't large depths of field and even if your transverse resolution is completely small, your volume resolution is still not quite so good. Second, you finally want to look at live specimens. If you look at the limitations, you find that it's the signal-to-noise ratio in your image. You have to get a certain amount of count on your detector in order to get your signal out of a certain contrast ratio.

Quite often you are not so much interested in the carbon, or hydrogen, or whatever. What you are interested in is if there is a protein, and if there is a protein what type of protein is there, etc.

RUDOLPH: May I give a partial answer to your question? In fact, it's even more difficult. The irradiation depends not only on the focal depth and on the beam divergence, but also on the local situation in the cell. The nucleus is much more sensitive to irradiation than outside of the nucleus. We hope, on the basis of our viability experiments, that we can look at these spores without killing them at a resolution of 100 Å and signal-to-noise ratio of two.

BRAKENHOFF: I wonder if spores are very good reference.

RUDOLPH: Certainly not. We have to begin with some biological tissue, and spores are very simple to handle. Therefore we choose them for the first experiment. We are aware that spores are not the most interesting biological objects.

BRAKENHOFF: Not the point that they are interesting. You should keep in mind that the sensitivity of eukaryotic cells is at least a factor of a thousand higher, possibly even more.

RUDOLPH: We hope to gain a factor of at least a hundred, but I'm pushing forward the development of optics. If we go from 2% efficiency, what we have now, to about 20% in phase zone plates, and if we go from 5 to 9% efficiency, detective quantum efficiency, in the detector that we have now with the photo plates, to about 60 or maybe 80% in better sensors, then we may gain at least a factor of a hundred. That's directly what we gained in resolution and what we gained in dose reduction. Therefore, our most exciting problem is to increase the sensitivity and the quantum efficiency of the detectors and to increase the efficiency of the zone plates. Before we did not do that. We know that all examination of living biological tissue is very problematic. It will not work.

Index of Contributors

Page numbers in italics refer to Discussions of the Papers.